Be AND SHELL STARS

INTERNATIONAL ASTRONOMICAL UNION
UNION ASTRONOMIQUE INTERNATIONALE

SYMPOSIUM No. 70

(MERRILL-McLAUGHLIN MEMORIAL SYMPOSIUM)

HELD AT BASS RIVER, MASSACHUSETTS, U.S.A.,
15–18 SEPTEMBER, 1975

Be AND SHELL STARS

EDITED BY

ARNE SLETTEBAK

Perkins Observatory
Ohio State and Ohio Wesleyan Universities
Delaware, Ohio, U.S.A.

SPRINGER-SCIENCE+BUSINESS MEDIA, B.V.

1976

Library of Congress Cataloging in Publication Data

Main entry under title:

Be and shell stars.

    (Symposium—International Astronomical Union ; no. 70).
    Includes bibliographies.
      1. B stars—Congresses.   2. Shell stars—Congresses.
I. Slettebak, Arne. II. Series: International Astronomical
Union. Symposium ; no. 70.
QB843.B12B4   523.8   76-19102
ISBN 978-90-277-0700-0   ISBN 978-94-010-1498-4 (eBook)
DOI 10.1007/ 978-94-010-1498-4

Filmset by J. W. Arrowsmith Ltd., Bristol, England

TABLE OF CONTENTS

DEDICATION   IX

PREFACE   XI

SCIENTIFIC ORGANIZING COMMITTEE   XIII

LIST OF PARTICIPANTS   XV

INTRODUCTORY ADDRESS BY MIROSLAV PLAVEC   1

PART I/OBSERVATIONS OF Be STARS

J. B. HUTCHINGS / Spectra and Photometry of Be Stars (*Review Paper*)   13

W. P. BIDELMAN and A. J. WEITENBECK / A Survey of Hα in the Brighter, Northern Be Stars   29

R. SCHILD and W. ROMANISHIN / A Study of Be Stars in Clusters   31

H. HUBERT and M. Th. CHAMBON / Emission Features of Several Be Stars as Related to their Luminosity Class and Spectral Type   33

V. DOAZAN / Rapid Variations in the Spectra of o And, γ Cas, and χ Oph   37

J. D. R. BAHNG / Rapid Variations of Hα in Be Stars   41

H. E. BUTLER / Stellar Photometric Observations at Hα through a Narrow-Band Interference Filter   51

R. S. POLIDAN and G. J. PETERS / Spectroscopic Observations of Be Stars in the Near Infrared   59

P. HARMANEC, P. KOUBSKÝ, J. KRPATA, and F. ŽĎÁRSKÝ / New Observational Data Concerning 4 Her and ζ Tau   67

G. J. PETERS / The Surface Gravities of Be Stars   69

A. M. DELPLACE and M. Th. CHAMBON / A Comparative Study of the Shell of ζ Tau and 48 Lib   79

L. HOUZIAUX and Y. ANDRILLAT / A Model for the Shell of HD 50138   87

R. HERMAN and H. HUBERT / Observations Récentes de HD 200120   95

R. VIOTTI and P. KOUBSKÝ / Singly-Ionized Iron Emission Lines in the Spectra of Early Type Stars   99

A. S. SHAROV and V. M. LYUTY / Pleione as a Variable Star   105

R. E. SCHILD / Energy Distributions of Be Stars (*Review Paper*)   107

PART II/Be STARS AS ROTATING STARS

A. SLETTEBAK / Be Stars as Rotating Stars: Observations (*Review Paper*)   123

S. R. HEAP / Ultraviolet Observations of Rapidly Rotating B-type Stars   137

S. A. FROST and P. S. CONTI / The Relationship of the Oe to the Be Stars   139

A. FEINSTEIN / Hα and Hβ Measures as Related to Be Star Rotation   149

PART III / NEW OBSERVATIONAL TECHNIQUES

C. R. PURTON / Radio Observations of Be Stars (*Review Paper*)   157

S. R. HEAP / Ultraviolet Observations of Be Stars (*Review Paper*)   165

J. M. MARLBOROUGH and T. P. SNOW, JR. / A Survey of Mass Loss from Be and Shell Stars, Using Ultraviolet Data from *Copernicus*   179

K. G. HENIZE, J. D. WRAY, S. B. PARSONS, and G. F. BENEDICT / Ultraviolet Si IV/C IV Ratios for Be Stars   191

A. M. DELPLACE and K. A. VAN DER HUCHT / The Near Ultraviolet Spectrum of ζ Tau   197

W. M. BURTON and R. G. EVANS / Spectroscopic Observations of the Be Stars η Cen, γ Cas, and ϕ Per   199

G. J. PETERS / The Far Ultraviolet Spectra of υ Cyg and μ Cen   209

J. P. SWINGS / Photographic Infrared Spectroscopy and Near Infrared Photometry of Be Stars (*Review Paper*)   219

D. BRIOT / Paschen Decrements in Be Stars   227

D. A. ALLEN / The Classification of Faint Be Stars   229

G. V. COYNE, S. J. / Polarization in Be Stars (*Review Paper*)   233

I. S. MCLEAN and D. CLARKE / Polarization Measurements Across the Balmer Lines of Be and Shell Stars   261

R. POECKERT and J. M. MARLBOROUGH / Intrinsic Linear Polarization of Be Stars as a Function of $\upsilon \sin i$ 277

PART IV / LINE FORMATION IN EXPANDING ATMOSPHERES

D. G. HUMMER / Line Formation in Expanding Atmospheres (*Review Paper*)   281

R. I. KLEIN / Radiative Transfer in Dynamic Stellar Atmospheres   313

S. R. HEAP / Si II Lines in the Shell of ζ Tau   315

J. SURDEJ and J. P. SWINGS / The Complex Structure of the Ca II H and K Lines in the Spectrum of the A0ep Star with Infrared Excess HD 190073   321

S. KŘÍŽ / Theoretical Emission-Line Profiles Computed at Ondřejov   323

N. F. VOJKHANSKAYA / Motions in the Shells and Atmospheres of V923 Aql and EW Lac and their Manifestation in the Spectrum   327

PART V / MODELS

J. M. MARLBOROUGH / Models for the Circumstellar Envelopes of Be Stars (*Review Paper*)   335

D. N. LIMBER / On the Possible Role of Magnetic Fields in the Dynamics of the Be Phenomenon   371

B. M. HAISCH and J. P. CASSINELLI / Theoretical Wavelength Dependence of Polarization in Early-Type Stars   375

R. L. KURUCZ and R. E. SCHILD / The Possible Role of Radiative Acceleration in Supporting Extended Atmospheres in Be Stars 377

PART VI / SINGLE VERSUS BINARY STARS

P. HARMANEC and S. KŘÍŽ / Duplicity of Be Stars as Seen from Ondřejov
(*Review Paper*) 385
R. S. POLIDAN / On the Detection of Binary Be Stars 401
G. J. PETERS / Evidence for the Existence of Mass-Exchange Binary Be Stars
from Periodic Spectral Variations 417
E. M. HENDRY / Toward a Model for the Be Binary System ϕ Per 429
M. PLAVEC / Final Remarks on the Binary Hypothesis for the Be Stars 439

PART VII / GENERAL DISCUSSION

GENERAL DISCUSSION BY ALL PARTICIPANTS 447

CONCLUDING REMARKS BY W. P. BIDELMAN 453

APPENDIX: A List of Early-Type Shell Stars (*compiled by W. P. Bidelman*) 457

This volume and the symposium of which

it is a record are dedicated to the memory of

PAUL W. MERRILL (1887–1961)

and

DEAN B. McLAUGHLIN (1901–1965)

whose observations of Be and shell stars stimulated

much of the later work

PREFACE

The International Astronomical Union Symposium No. 70 on Be and Shell Stars, the Merrill-McLaughlin Memorial Symposium, was held in Bass River (Cap Cod), Massachusetts, U.S.A., from September 15th through 18th, 1975. Fifty-three astronomers from Argentina, Belgium, Canada, Czechoslovakia, France, Israel, the United Kingdom, the United States, and the Vatican attended and participated in the Symposium. This volume, which parallels the actual program closely, contains the papers presented at the Symposium plus most of the discussion following the papers.

New observational techniques and fresh theoretical ideas have resulted over the past few years in a renewed interest in Be and shell stars. At IAU Symposium No. 51 on Extended Atmospheres and Circumstellar Matter in Spectroscopic Binary Systems, the Otto Struve Memorial Symposium, in Parksville, British Columbia, Canada, three years ago, a number of participants expressed the wish to organize a symposium on Be and shell stars. If we wish to identify an official 'Father of IAU Symposium No. 70', it would be Mirek Plavec who, in his capacity as President of IAU Commission 42 (Photometric Double Stars) requested and received the cooperation of Commissions 29 (Stellar Spectra) and 36 (Stellar Atmospheres), suggested an Organizing Committee, and wrote to the IAU General Secretary in 1973 requesting that the IAU Executive Committee approve the proposed conference as an IAU Symposium. Approval was granted at the IAU General Assembly in Sidney, where a number of members of the Organizing Committee first met to make preliminary plans for the Symposium and drew up a tentative program. Subsequently, Anne Cowley, as Secretary of the Scientific Organizing Committee, also served as a kind of 'one-person local organizing committee', and personally handled much of the planning and arrangements for this Symposium. The success of IAU Symposium No. 70 is due largely to her efforts and those of the other members of the Scientific Organizing Committee.

I should like to acknowledge also financial support from the International Astronomical Union and from the National Science Foundation which provided travel assistance for ten of the Symposium participants. The Dominion Astrophysical Observatory and the Ohio State University were also most generous in providing many services necessary for planning and carrying out this Symposium.

The contributions by Paul Merrill and Dean McLaughlin to our understanding of Be and shell stars are treated in Mirek Plavec's Introductory Address, but I would like to acknowledge my personal debt to each. It was my privilege to have known Merrill while I was a guest investigator at the Mount Wilson Observatory in the 1950's, and I met McLaughlin often at meetings of the Ohio Neighborhood Astronomers. It was always a pleasure and a source of stimulation to discuss spectroscopic problems, and specifically emission-line spectra, with each of them.

The framework of IAU Symposium No. 70 is formed by the review papers, chosen to represent the most important aspects of Be and shell star research, with special emphasis on recent observational techniques. Contributed papers were also given following each review paper, but these were necessarily shorter because of limitations of time. The program was planned from the outset to leave ample time for discussion and was generally successful in this respect. A special session was designated for general discussion at the end of the Symposium.

All of the discussion was recorded on tape by George Sonneborn and subsequently transcribed, after editing, by Mrs Delores Chambers. We are grateful to both for their help. We are also indebted to the chairpersons of the various sessions for keeping all of us on the track; these were R. Herman, J. F. Heard, D. N. Limber, A. P. Cowley, L. Goldberg, J. P. Swings, and A. Feinstein. The hard-working Symposium participants were allowed one afternoon off to visit Nantucket Island, where Dorrit Hoffleit very kindly provided a tour of the Maria Mitchell Observatory for us.

I would also like to express my appreciation to my wife, Connie, for her help and advice at various stages of preparing for the Symposium and editing these Proceedings.

The cooperation and help of the management and staff of the Red Jacket Beach Motor Inn, where the Symposium was held, is also gratefully acknowledged.

Finally, I would like to thank all the participants for their friendly cooperation both during the Symposium and in preparing these Proceedings for publication.

ARNE SLETTEBAK

THE SCIENTIFIC ORGANIZING COMMITTEE

A. Slettebak, Chairman, Ohio State University, U.S.A.

A. P. Cowley, Secretary, University of Michigan, U.S.A.

A. A. Boyarchuk, Crimean Astrophysical Observatory, U.S.S.R.

M. Hack, Trieste Astronomical Observatory, Italy.

R. Herman, Observatoire de Paris, France.

D. G. Hummer, JILA, University of Colorado, U.S.A.

J. B. Hutchings, Dominion Astrophysical Observatory, Canada.

J. M. Marlborough, University of Western Ontario, Canada.

M. Plavec, University of California at Los Angeles, U.S.A.

J. Sahade, Institute of Astronomy and Space Physics, Argentina.

LIST OF PARTICIPANTS

J. D. R. Bahng, Northwestern University, Evanston, U.S.A.
P. K. Barker, LASP, University of Colorado, Boulder, U.S.A.
W. P. Bidelman, Case Western Reserve University, Cleveland, U.S.A.
D. Briot, Observatoire de Paris, France.
W. M. Burton, Culham Laboratory, Abingdon, U.K.
H. E. Butler, Royal Observatory, Edinburgh, U.K.
O. Cardona, JILA, University of Colorado, Boulder, U.S.A.
P. S. Conti, JILA, University of Colorado, Boulder, U.S.A.
A. P. Cowley, University of Michigan, Ann Arbor, U.S.A.
G. V. Coyne, S. J., University of Arizona, Tucson, U.S.A., and Vatican Observatory, Vatican City-State.
A. M. Delplace, Observatoire de Paris, France.
V. Doazan, Observatoire de Paris, France.
A. Feinstein, La Plata Observatory, Argentina.
S. A. Frost, JILA, University of Colorado, Boulder, U.S.A.
R. F. Garrison, David Dunlap Observatory, Richmond Hill, Canada.
L. Goldberg, Kitt Peak National Observatory, Tucson, U.S.A.
A. Gulliver, David Dunlap Observatory, Richmond Hill, Canada.
B. M. Haisch, JILA, University of Colorado, Boulder, U.S.A.
P. Harmanec, Ondřejov Observatory, Czechoslovakia.
R. J. Havlen, E.S.O., Hamburg, Germany.
S. R. Heap, Goddard Space Flight Center, Greenbelt, U.S.A.
J. F. Heard, David Dunlap Observatory, Richmond Hill, Canada.
E. M. Hendry, Northwestern University, Evanston, U.S.A.
K. G. Henize, Johnson Space Center, Houston, U.S.A.
R. Herman, Observatoire de Paris, France
H. Hubert, Observatoire de Paris, France.
D. G. Hummer, JILA, University of Colorado, Boulder, U.S.A.
J. B. Hutchings, Dominion Astrophysical Observatory, Victoria, Canada.
W. Kalkofen, Center for Astrophysics, Cambridge, U.S.A.
R. I. Klein, NCAR, Boulder, U.S.A.
D. W. Latham, Center for Astrophysics, Cambridge, U.S.A.
E. Leibowitz, Tel Aviv University, Israel.
H. Leparskas, University of Western Ontario, London, Canada.
D. N. Limber, University of Virginia, Charlottesville, U.S.A.
D. J. MacConnell, Wesleyan University, Middletown, U.S.A.
J. M. Marlborough, University of Western Ontario, London, Canada.
D. Massa, LASP, University of Colorado, Boulder, U.S.A.
I. S. McLean, University of Glasgow, U.K.
D. D. Meisel, New York State University College, Geneseo, U.S.A.

P. D. Noerdlinger, Michigan State University, East Lansing, U.S.A.
G. J. Peters, University of California, Los Angeles, U.S.A.
M. Plavec, University of California, Los Angeles, U.S.A.
R. Poeckert, University of Western Ontario, London, Canada.
R. S. Polidan, University of California, Los Angeles, U.S.A.
C. R. Purton, York University, Downsview, Canada.
R. E. Schild, Center for Astrophysics, Cambridge, U.S.A.
A. Slettebak, Ohio State University, Columbus, U.S.A.
T. P. Snow, Princeton University Observatory, U.S.A.
T. Snijders, Goddard Space Flight Center, Greenbelt, U.S.A.
G. H. Sonneborn, Ohio State University, Columbus, U.S.A.
J. P. Swings, Institute d'Astrophysique, Cointe-Ougree, Belgium.
I. Thompson, University of Western Ontario, London, Canada.
A. Young, San Diego State College, U.S.A.

INTRODUCTORY ADDRESS

MIROSLAV PLAVEC

Dept. of Astronomy, University of California, Los Angeles, Calif., U.S.A.

Abstract. The early history of the studies on Be stars is reviewed. The importance of keeping in mind a basic model of a Be star is emphasized, and the binary star model is suggested as one serious possibility at least for some Be stars.

The name *Be Star* was introduced by Commission 29 of the International Astronomical Union in 1922 at the first General Assembly of the Union in Rome. Walter S. Adams was President of the Commission at that time, but it was Henry N. Russell who actually presided over the meeting. Besides these two, the Commission had seven more members.

The Be stars were of course known before they were given the name. A good many of them were recognized when the HD catalog was compiled, and they were usually classified as Bp. In fact, the first two stars with emission lines were discovered by Angelo Secchi very soon after he began his systematic work on the first classification of stellar spectra. In 1866, Secchi noticed that Hβ was in emission in γ Cas and β Lyr. These two stars can therefore be considered as prototypes or symbols of sorts. At first, they were considered to be rather similar. Otto Struve probably more than anyone else separated them as different objects, while recently at least some of us have felt that there is more similarity than diversity between these two.

The first systematic observing program on emission line stars was, as far as I know, started by Ralph H. Curtiss in 1911 at Ann Arbor in Michigan. The very first publication (Curtiss, 1916) resulting from this project was devoted to γ Cassiopeae. I am sure many of you will be delighted to read Curtiss' argument why this star had been selected:

It was hoped that the analysis of this simple spectrum would assist in the study of more complicated cases.

No doubt Curtiss soon learned the lesson that a spectrum with a few visible lines is not necessarily a simple spectrum. In fact, γ Cas has so far been the only Be star to display truly large changes in the integral light. I remember that when I first looked at the constellations in 1937 or 1938, Cassiopea was even more beautiful than now, having one star of the first magnitude.

In this context, however, it is good to recall the words of Curtiss' successor, McLaughlin (1932):

Unusual developments may conceivably occur at any time in the spectrum of any one of them, furnishing us with an additional example of what a stellar atmosphere can do.

The series of Michigan publications on Be stars, published to 1929 by Curtiss and then by McLaughlin, is very interesting. In their *Surveys of the Brighter Be Stars*, one finds almost all the stars that still intrigue us today: γ Cas, β Lyr, ϕ Per, ψ Per, Pleione, ζ Tau, β Mon, even HR 2142 . . ..

A. Slettebak (ed.), Be and Shell Stars, 1–10. *All Rights Reserved*

Curtiss and McLaughlin soon recognized that many of the Be stars showed rather pronounced spectral variations, and that these variations often were cyclic or almost periodic. The concept of the E/C and V/R variations emerged from their papers, and McLaughlin spent a good deal of his life following these changes and attempting to explain them. In his enthusiasm and devotion to these stars, he had a good peer in Paul W. Merrill. The phrase "... what a stellar atmosphere can do" originally came from Paul Merrill (1929).

Paul Merrill became interested in stellar spectra, and in particular in the peculiar ones, when he participated in the large radial velocity program of the Lick Observatory. Before he started his own program in 1917–1919, however, he was able to acquire laboratory experience with the then new red-sensitive plates. While McLaughlin watched the emission lines, Merrill was much more attracted to the enigmatic spectra of the shell stars. The term *shell star* was apparently introduced by Otto Struve, but to him it was merely a convenient abbreviation for 'a star with an extended atmosphere'. Thanks mostly to the innumerable articles by Paul Merrill, the meaning of the term became narrower and more definite.

Paul Merrill covered dozens of pages of the Astrophysical Journal and of the Publications of the Astronomical Society of the Pacific with detailed descriptions of the various shell spectra, of their cycles and episodes. However, he was very reluctant to write a summarizing and interpretative article. His articles whose titles promise a review are extremely concise. Perhaps the most review-like article is the one entitled 'Stars with Expanding Envelopes', published in Volume II of Beer's *Vistas in Astronomy*. Even here, Merrill is extremely cautious and simply indicates that the phenomenon is very complex. His attitude is best expressed by the following sentence, to which McLaughlin alluded:

The immediate outcome of this investigation is an example of what a stellar atmosphere can do. A complete interpretation of the observations in terms of physical conditions in the star will not be made at once – at least not by the author.

McLaughlin did attempt to interpret his observations in terms of a geometrical model. This model was a modification of the basic explanation of the Be star phenomenon made by Otto Struve in 1931. In that famous article, Struve made two essential observations: "Apparently, rapid rotation sponsors the occurrence of wide emission lines", and again "Rapid rotation seems to be prerequisite to the appearance of bright lines". As to the actual mechanism that leads to the eventual formation of the lines, Struve says:

Sir James Jeans has shown that under certain conditions a rapidly rotating gaseous body may become lens-shaped and throw off matter at its sharp equatorial edge. It is therefore reasonable to expect that Be stars in extremely rapid rotation will eject gaseous matter at the equator. A gaseous ring will be formed and the system will resemble in appearance the planet Saturn. (Struve, 1931).

It was this rather too artificial jump from a rotationally unstable star to a detached circumstellar ring that enabled McLaughlin to suggest a qualitative explanation of the V/R variation. In his summarizing article on 'The Bright-Line Stars of Class B' (McLaughlin, 1961), he examines and rejects several possible models, and eventually (without too much enthusiasm), decides in favor of an elliptical ring. The model was recently revived by Su-Shu Huang (1973) and widely popularized in the June 1975 issue of *Sky and Telescope*.

It is rather easy to speak about the first generation of investigators, although even here we are committing a serious crime of distorting history if we speak only about the four leading astronomers. To summarize the more recent history of our subject is much more difficult. As in all branches of astrophysics, the quiet pleasant creek of scientific progress has turned, since World War II, into an impressive broad river, or at times into a wild torrent. It is better to talk about problems than about people.

I think in the field of Be and shell stars we have three specific problems that must be solved if we want to understand these objects: (1) line formation in an extended atmosphere; (2) relation between the rapid rotation of the central star and its Be character; (3) origin and dynamical support of the extended atmosphere. These problems are far from being solved; nevertheless, considerable progress has been made. Much of what has been done can be found summarized, reviewed, discussed and occasionally questioned and challenged in the proceedings of several recent meetings. Thus, the broad problem of stellar rotation and its influence on stellar spectra, stellar atmospheres, and stellar structure is thoroughly examined in the Proceedings of the IAU Colloquium No. 4, edited (quite appropriately) by Slettebak (1970) under the title *Stellar Rotation*. Another similar Colloquium, No. 2, was devoted to line formation in extended atmospheres, and its proceedings were published by Groth and Wellmann (1970). Although primarily devoted to the circumstellar matter in binary stars, the IAU Symposium No. 51 at Parksville had many contributions on the general problem of circumstellar envelopes. The proceedings, edited by Batten (1973), have the additional advantage of being published three years after the former two books, which means a lot considering the present rate of progress in astrophysics. Nevertheless, it was at Parksville that many of us agreed that another Symposium devoted specifically to Be stars and shell stars would be very valuable.

At times one feels that the Proceedings of various conferences, extremely valuable as they are, lack the unifying spirit and the personal flavor of a monograph. Fortunately, two good recent monographs deal rather extensively with our stars: Underhill (1966) covers them as an important group among *The Early Type Stars*, while Hack and Struve (1970) see them as one kind of the *Peculiar Stars*, which in the second volume of their *Stellar Spectroscopy* are given twice as much space as the *Normal Stars* in the first volume: a fact telling us something either about the nature of the stars, or about the human nature of the astronomers.

Naturally, no book can be written and published fast enough, and for the most recent accomplishments one must look into the current journals. Among the most recent and most outstanding achievements, I would like to mention: infrared photometry, and the discovery of an excess in the infrared radiation from many Be and related stars; the far ultraviolet observations from rockets and satellites, and the resulting discovery of stellar wind in luminous early-type stars; a vast progress in the treatment of the effects of rotation upon the spectral lines and stellar atmospheres; and a gradual but steady expansion into the field of extended atmospheres of the sophisticated methods developed originally for more normal atmospheres. If we add to these the substantial improvement in all kinds of spectroscopic, photometric, and polarimetric observations, as well as a similar progress in all aspects of theoretical astrophysics, we can be sure that new and fundamental discoveries in our field are coming.

I wonder how well I managed to conceal my effort to evade my assigned task to summarize the more recent history of our field. I claim that it cannot be done in a short talk, and, more important, that it is not necessary. This history is still alive, we live in it. The leading investigators are either personally present, or represented by their close collaborators or pupils. We will hear from them.

Thus I think I can abandon the historical approach here. Instead, I would like to invoke the brilliant introductory talk to the Magnetic Star Symposium delivered by Dr Bidelman ten years ago. I hope he will permit me to quote from his address (Bidelman, 1967):

> After all, in some ways, an introduction is a little bit like an overture to an opera. For instance, the overture isn't expected to be taken very seriously, because people are just supposed to be sliding – mentally or physically – into their seats. Furthermore, all the overture does is to give you bits and snatches of the tunes that are to be played, with ample elaboration, later on during the opera.
>
> At first I felt that this meant that I really shouldn't say very much. Then it occurred to me, following the analogy a bit further, that it often happens that the overture is the only part of the opera that survives.

I hope Dr Bidelman will permit me to exploit his allegory. His Introduction then was really a masterpiece which can be studied independently and with great profit. I will not attempt to do the same. There are operas where the orchestra plays just a few bars of introduction, the curtain goes up, and that's all for the overture. Nevertheless, if done properly, these few bars of introduction can bring us just to the middle of action and excitement on the stage. The problem is, how to create the proper atmosphere. I know a playwright who toys with the idea of opening his next play by firing two cannon shots from the stage into the audience. He claims that this is the

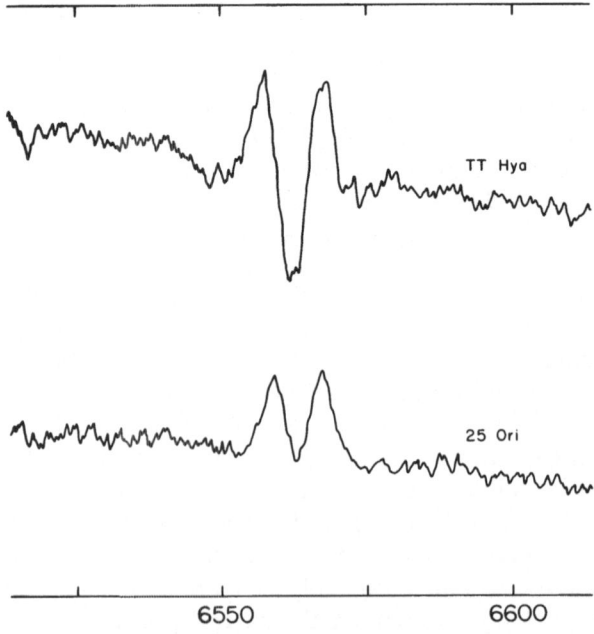

Fig. 1. Density tracings of similarly exposed spectrograms near $H\alpha$. The wavelength scale is in Ångströms. Original scale was 17 Å mm^{-1}.

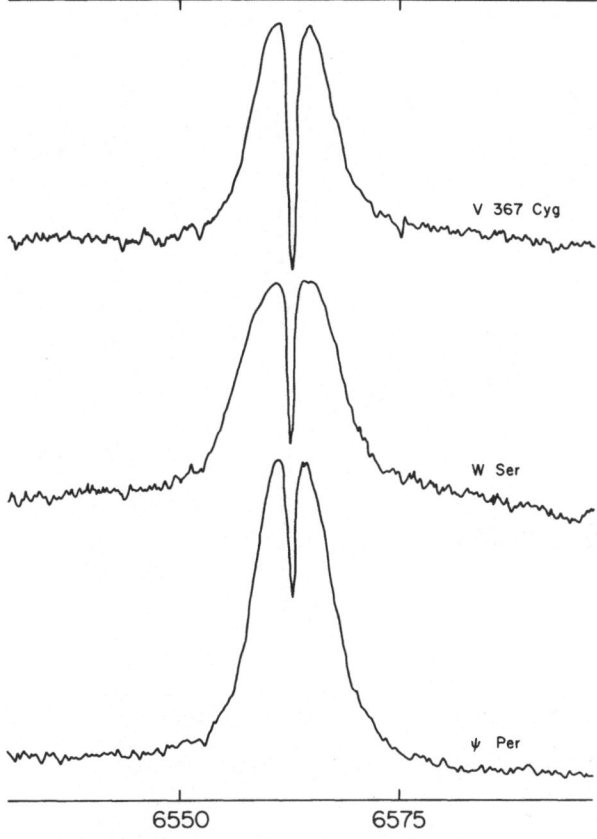

Fig. 2. Density tracings as in Figure 1.

only sure way to make everybody forget his daily problems and concentrate on what is happening on the stage. We do not have the necessary requisites here, so I will show you a few slides instead.

The pictures show the profiles of the Hα line obtained with the Varo image tube of the Lick Observatory by Peters, Polidan, and myself. In Figure 1, you have a classical Be star, 25 Ori, and an eclipsing binary, TT Hydrae. In Figure 2, we compare a rather typical shell star, ψ Persei, with two eclipsing binaries, V 367 Cygni and W Serpentis. Figure 3 shows another shell star, 88 Herculis, and the eclipsing binary RZ Scuti. Figure 4 shows a Be star η Cen sandwiched between two eclipsing stars, RX Gem and AU Mon. And, finally, Figure 5 shows the Be stars κ CMa and HR 3034 and the spectroscopic binary HD 698. This last figure shows two Be stars with single, narrow emission lines; and we can see that a similar case can be found among close binaries, too, but naturally we have to search among spectroscopic binaries, not among eclipsing binaries.

The five figures are meant to convey to you the extreme similarity between certain Be stars, shell stars, and close binaries. Now we understand in principle the mechanism by which the circumstellar emission or absorption lines are formed in

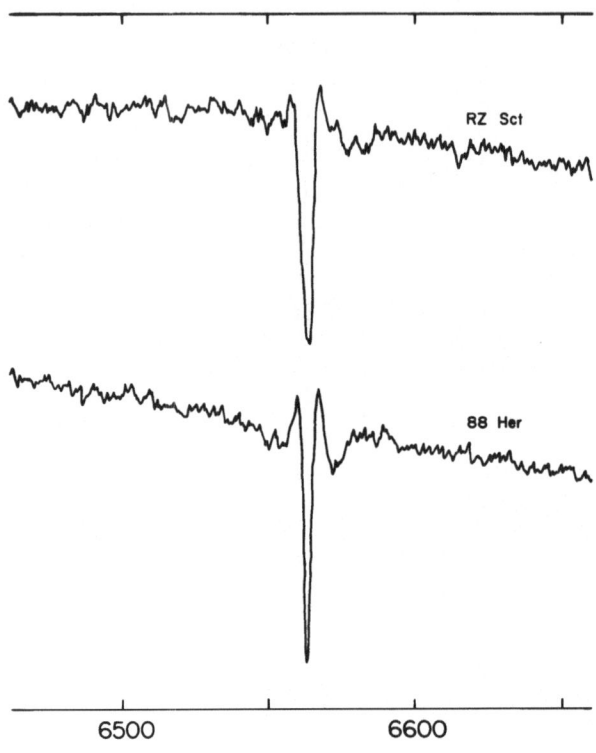

Fig. 3. Density tracings as in Figure 1.

close binaries. The cooler star in such a system is unstable and transfers mass to the hotter and brighter component. The streaming gas carries an excess angular momentum, and therefore forms a disk or ring around the other star. It is in this disk that most of the circumstellar emission or absorption lines originate. There is no problem in this case with the mechanical support for such a disk. The star inside the disk accretes only a fraction of the surrounding gas; however, if the star is relatively large, the impact is quite oblique and accelerates the rotation of the surface layers of the accreting star. Thus we get something similar to the Be star. Here, however, the envelope is not the result of an instability of a rapidly rotating star, as in Struve's picture; rather, the rapid rotation of the star's photospheric layers is a byproduct of the formation of the envelope.

We can go even farther and explain more by the binary model. The eclipsing binary star TT Hydrae displays an apparently periodic V/R variation during one orbital period of 7 days. We believe that this variation can be explained in terms of an uneven distribution of the circumstellar gas, and by the periodically changing geometrical aspect of the system. Much work remains to be done to convert this crude qualitative picture into a quantitative model. Nevertheless, I think that the binary model may very well explain many observed Be stars and shell stars.

It is really the central problem of all our investigations to understand how the extended circumstellar envelope we observe in the Be stars has been formed and how

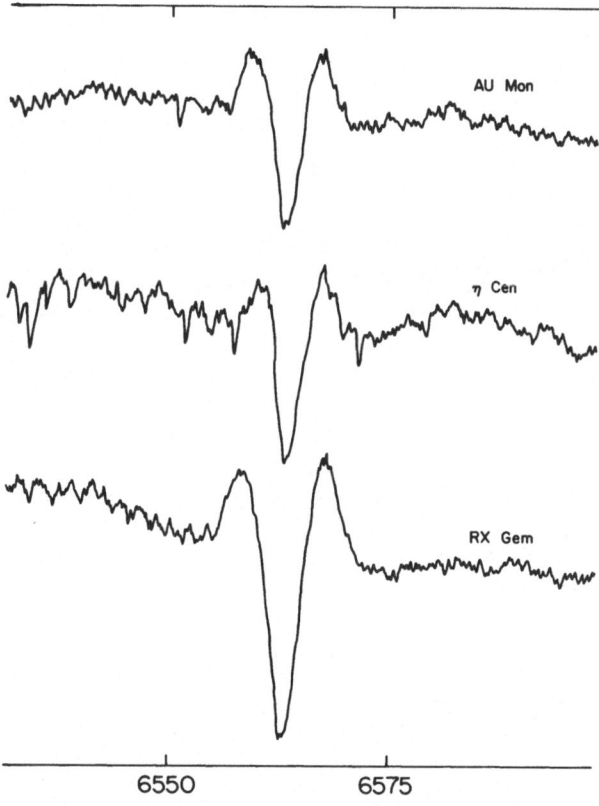

Fig. 4. Density tracings as in Figure 1.

it is maintained and supported. It is rather surprising that sometimes very vague concepts are accepted as a satisfactory explanation. This makes even more commendable the systematic efforts by Limber and Marlborough (1968) to clarify the question. They have developed the canonical picture proposed by Struve of a rotationally unstable star, but they emphasize that an additional force must be postulated. This model is often – rather crudely – presented in a form similar to Figure 6: a rapidly rotating star (which should therefore be flattened!) is surrounded, in the equatorial plane, by a flat disk. Some of you may prefer Huang's model of an elliptical ring, as shown schematically in Figure 7. This model is designed to explain the observed V/R variation. The origin of the ring is thought to be basically the same as in Struve's model. To me, it is rather difficult to imagine how such a ring could be formed and maintained. Vague doubts of course cannot eliminate these models, they can only stimulate further search for alternatives. The model of an interacting binary, shown schematically in Figure 8, is in my mind such an alternative. This model will be presented in more detail later on by Drs Harmanec, Polidan, Peters, and myself; however, our time will not come until the very end of the Symposium, which is the time to be tired and sleepy. This is why I am trying to call attention to it right now in the Introduction. If all the alternatives of the basic picture of a Be star are kept in

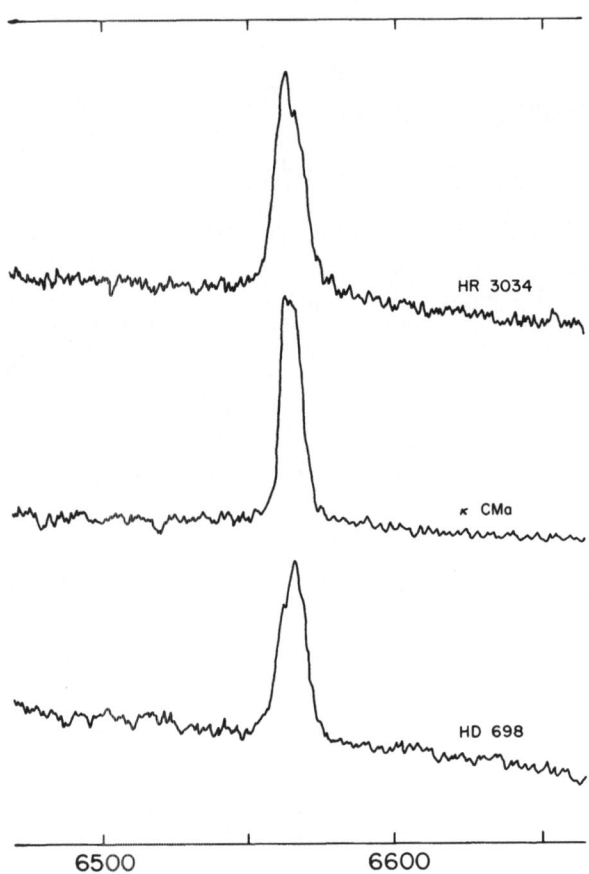

Fig. 5.   Density tracings as in Figure 1.

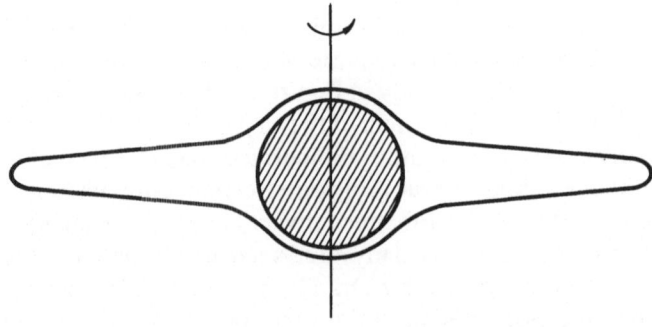

Fig. 6.   The classical model of a Be star is usually given in this form, although the rapidly rotating star should be considerably flattened.

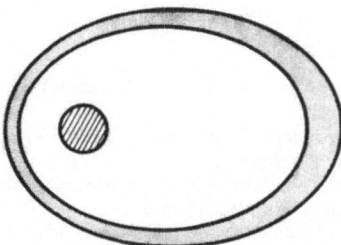

Fig. 7. This very schematic picture of an elliptical ring is often used to represent the Be star model proposed by McLaughlin and by Huang.

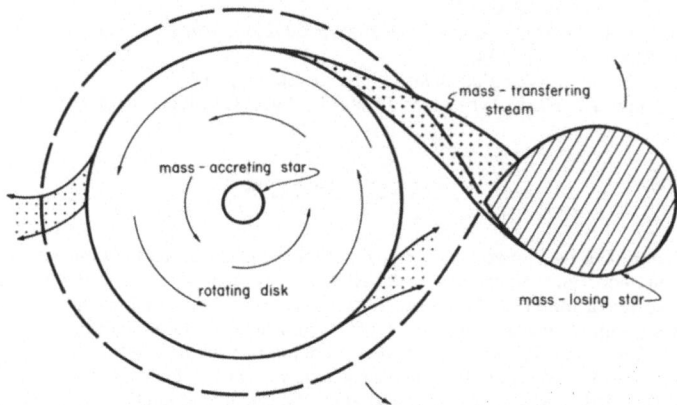

Fig. 8. Be star as an interacting binary: Gas streams from the large and cool giant into the disk surrounding the mass-accreting star. Additional weaker streams may be present as indicated.

mind during all the discussions, we will gather many stimulating thoughts and criticisms.

I heard Otto Struve narrate the story of his famous 1931 article on the Be stars: "The revolving-ring of gas hypothesis, in rapidly rotating stars, was written in the course of a rather long evening." (Struve, 1958). It is a fascinating story, quite fitting for the astronomical giant of the twentieth century, as Struve was recently called by Dr Huang. But it is also a challenging story. Many brilliant scientists are working on the problem of the Be stars nowadays; many combined evenings should in the end be a good match for that one in 1931. It would be very nice to say: "A very substantial progress in our understanding of the Be stars was accomplished in the course of a rather long Symposium." Perhaps that much cannot be achieved at any meeting; truly revolutionary ideas are more likely born in the solitude of long evenings. Nevertheless, this Symposium can certainly stimulate new thoughts and new projects. I hope it will.

Acknowledgements

My thanks are due to Mrs Marietta L. Eaker for typing the manuscript, and to Miss Katherine E. Sedwick for drafting the illustrations.

References

Batten, A. H. (ed.): 1973, 'Extended Atmospheres and Circumstellar Matter in Spectroscopic Binary Systems', *IAU Symp.* **51**.

Bidelman, W. P.: 1967, in R. C. Cameron (ed.), *The Magnetic and Related Stars*, Mono Book Corp., p. 29.

Curtiss, R. H.: 1916, *Publ. Obs. Univ. Michigan* **2**, 1.

Groth, H. G. and Wellmann, P.: 1970, *Spectrum Formation in Stars with Steady-State Extended Atmospheres*, Nat. Bureau Stand, Special Publication 332.

Hack, M. and Struve, O.: 1970, *Stellar Spectroscopy – Peculiar Stars*, Observ. Astronom. Trieste.

Huang, S. S.: 1973, *Astrophys. J.* **183**, 541.

Limber, D. N. and Marlborough, M. J.: 1968, *Astrophys. J.* **152**, 181.

McLaughlin, D. B.: 1932, *Publ. Obs. Univ. Michigan* **4**, 198.

McLaughlin, D. B.: 1961, *J. Roy. Astron. Soc. Canada* **55**, 13.

Merrill, P. W.: 1929, *Astrophys. J.* **69**, 378.

Slettebak, A. (ed.): 1970, *Stellar Rotation*, D. Reidel Publ. Co., Dordrecht-Holland.

Struve, O.: 1931, *Astrophys. J.* **73**, 94.

Struve, O.: 1958, in *Etoiles à raies d'émission*, Univ. de Liège, p. 12.

Underhill, A. B.: 1966, *The Early Type Stars*, D. Reidel Publ. Co., Dordrecht-Holland.

DISCUSSION

Schild: I would like to ask Dr Plavec how many Be stars are mass-exchange binaries.

Plavec: This is an opportunity to give another half-hour talk. It depends very much on how stable is the disc formed by transfer of material. There are two competing schools of brilliant theorists who cannot agree on that. One group claims that those discs, for example, around X-ray binaries are dominated by very large viscosity. They cannot explain what the source of viscosity is: whether it is local magnetic fields or turbulent viscosity. They only claim that the viscosity is very large. Thus when the stream of material is turned off the disc should disappear almost immediately. That would mean that such a disc could exist only as long as the other star supplies the material. In that case, the other star must be fairly large, and would have a fairly large probability of eclipses. And you can calculate that out of one hundred Be stars, fifteen should display eclipses, which has not been observed. In this case you would have to say that the binary hypothesis could account only for a certain fraction, not terribly large, of all the Be stars. We have direct evidence that it does explain some because we have observed the cool components, but it probably does not account for all of them. On the other hand, if the other school is correct, the disc has very low viscosity and the material is being transferred to the surface of the star by shock waves. In that case the ring will dissipate very slowly and when the mass transfer is finished the disc will stay there for a relatively long time. In that case it may happen that the disc is still there when the star has finished its mass transfer and the star is contracting and will become a helium star or a white dwarf star. In that case you cannot expect that there will be eclipses, because the star is already quite small. If this case is correct, it might be that the majority of the Be stars can be explained in terms of the binary star model. At least the objection of eclipses is not there.

Doazan: Are there many eclipsing binaries known which have lost their shell emission line characteristics?

Plavec: Probably. I think the phenomenon is variable, for example, in RW Tau and other stars. An unfortunate factor is that there has been no systematic observation of the emissions in eclipsing binaries. As in Be stars, the emission is mostly present at Hα and Hβ, and those lines are not too frequently observed. The observations are only occasional. Somebody studied the system, obtained the radial velocity curve, and forgot about the system for another 20 years. Therefore, our information is very scattered. Recently we discovered the sudden flaring up of emission in U Cep and it really does seem to be a variable phenomenon.

PART I

OBSERVATIONS OF Be STARS

SPECTRA AND PHOTOMETRY OF Be STARS

(*Review Paper*)

J. B. HUTCHINGS

Dominion Astrophysical Observatory, Victoria, B.C., Canada

1. Introduction

My task is to review the spectroscopy and photometry of Be stars. In the wide range of topics this covers, I will naturally be biased by my own interests and knowledge. I hope however at least to mention all forms of observed behaviour within these limitations, but note that as we have specific sessions on energy distributions, rotation, observations at unusual wavelengths (i.e., not from 3500 to 6700 Å), and polarisation, I shall deal less thoroughly, if at all, with these topics. I shall also attempt to incorporate the results of the speakers who follow in this session, in order to save duplication of individual introductions and allow the later presentations to concentrate more on the relevant discussions. I shall deal with nine aspects of Be stars behaviour, under the headings below.

2. Rotation

The rotation of Be stars has been a subject of some controversy for many years. It is clear that the stars are, generally, rapid rotators and together with the evidence that the emission originates in an equatorial disk, led to the early suggestion that the stars lose mass by rotation at breakup velocity. This simple idea however meets with theoretical difficulties: unknown behaviour of a star under effective surface g of zero; lack of an ejection mechanism, outward transport of angular momentum. Collins has shown that line profiles do not correspond to breakup velocity of rotation and the need for a mass-loss mechanism has led to suggestions that the phenomenon may occur before the rotational velocity reaches breakup.

An investigation of the distribution of $v \sin i$ among Be stars by Stoeckley (1968) showed that current values of $v \sin i$ were clearly wrong and that allowance for gravity darkening effects produced a large correction. He was then able to fit the distribution with values of v corresponding to a half Gaussian centred on breakup, but still with half of the distribution lying below 475 km s^{-1}. However, his arguments do not consider the consequences of the peak lying at lower velocities. More recently, Massa (1975) has deduced a distribution (of about the same width) centred on $v = 350$ km s^{-1}, for all B stars which have shown emission. A significant point of his analysis is that *no* Be stars rotate at breakup velocity. It is clear that selection effects and heterogeneous estimates of $v \sin i$ can easily distort the picture, but these and other investigations point to the conclusion that Be stars probably do not need to rotate at breakup velocity. The distribution of Be stars with spectral type can be explained qualitatively by considering the emission line disk to be ejected by

A. Slettebak (ed.), Be and Shell Stars, 13–27. All Rights Reserved
Copyright © 1976 by the IAU.

radiation pressure in the ultraviolet, in the low gravity equatorial regions, and some theoretical models (Marlborough and Zamir, 1975) have been constructed along these lines.

Further work on the rotational distribution has been done by Hardorp and Strittmatter (1970) and Bernacca (1970). These may be discussed in detail in the session on rotation, but for now it is sufficient to note that all these investigations indicate that while Be stars rotate rapidly, they are not necessarily at breakup velocity.

3. Special Spectral Characteristics

Here, as in most of this discussion, I restrict remarks to the usual photographic region of the spectrum, and leave discussion of other wavelengths to appropriate later sessions.

Because of their high rotation, lines in Be star spectra are often very broad and shallow and difficult to make measurements on. Radial velocity studies of useful accuracy are possible only on very low dispersion spectra with most measuring instruments. Wide scan oscilloscope devices, such as ARCTURUS at the Dominion Astrophysical Observatory, are necessary to make direct measurements on high quality spectra. Photoelectric scanners, TV cameras and other digital devices have also been used with variable success (see Figures 1 and 2).

Study of the lines present in some Be spectra has suggested that the stars correspond to a mixture of spectral types spreading over two or three subtypes. While classification procedures for such objects are not at all standard, and may be misleading, this does suggest that an appreciable temperature (and gravity) gradient may exist across the disk of the star. A quantitative estimate of these effects would be a valuable tool in deducing i and v_{rot}, and perhaps distinguishing between different gravity darkening laws.

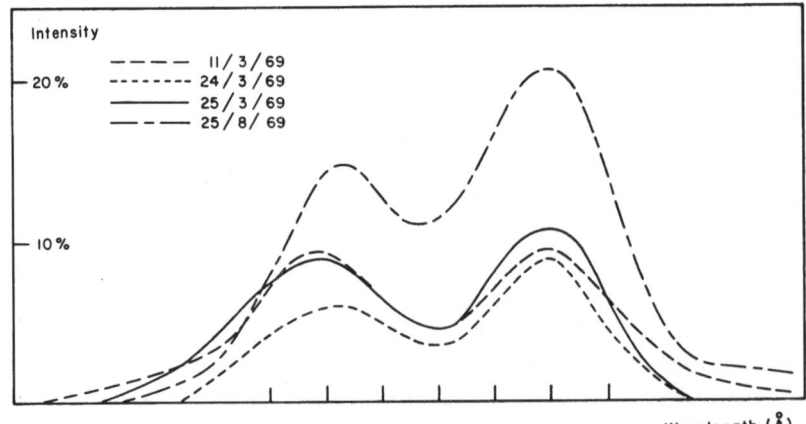

Fig. 1.   Photoelectric scans of Hγ in γ Cas showing night-to-night changes and general increase in intensity over an interval of months.

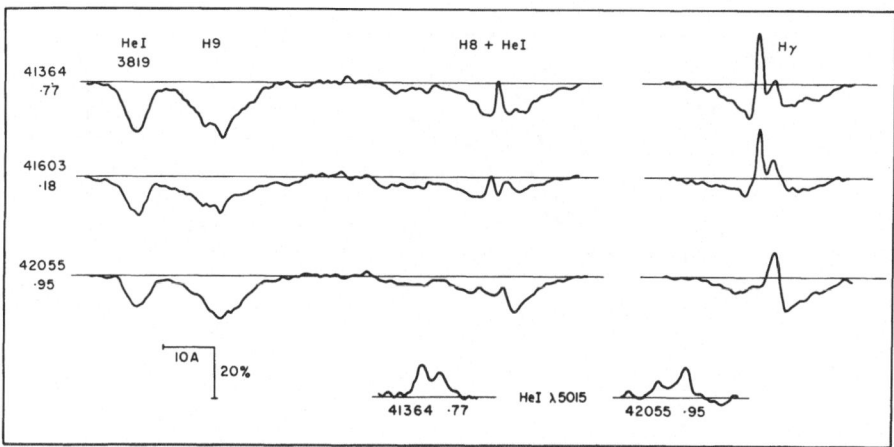

Fig. 2. Rectified spectrographic profiles from X Persei at different times. Julian Days and phases in possible 580 day orbital period are shown.

Another spectral peculiarity which is hard to detect is the appearance of emission in He I, and perhaps other lines. He I emission is seen in $\lambda\lambda$ 5875, 6678 in some Be stars, and occasionally at λ 5015 (Figure 2). It is rarely seen in other lines, although it has been suspected in κ Dra, and very probably is the cause of asymmetry in He I lines in X Per. Mg II λ 4481 emission peaks were also detected marginally in κ Dra (Hutchings *et al.*, 1971).

The peak separation in these cases is always larger than in the Balmer lines and presumably indicates formation in lower regions of the envelope where rotational velocity is higher, if angular momentum is conserved. The order in which the He I lines progressively show signs of emission is similar to that seen in the P Cygni supergiants and presumably indicates similar deviations from LTE level populations and transitions. In general, the lines which go most easily into emission, show lower peak separation.

A similar situation is seen in the Balmer lines. A progression is always found in peak separation, down the series (Figure 3). This is presumably because Hα is formed principally in the outer regions of the disk where it is optically thinnest and where rotational velocity is lowest (provided the rotation is not rigid-body). Down the Balmer series, separations increase as lines are formed closer to the photosphere. It is possible that such progressions can give us information from their slopes and zero points. One may expect these to depend on i, $\omega(H)$, $V(h)$, $\rho(h)$ and ρ_0, and if some of these are known or guessed from other information, careful measurement of the progression may be useful. I shall discuss later some predictions of simple geometrical models. Observationally, there is no clear correlation of the progression with i (as far as it is known), spectral type, or V/R ratio. It is noteworthy that the slope of HR 2142 is around the same for both shell and non-shell phases, although the separations are different. Also that the slope of X Per is much lower than any other I have seen. On the few stars where Balmer emission is seen far down the series, it appears that a maximum separation is reached at about H7 or H8, and this

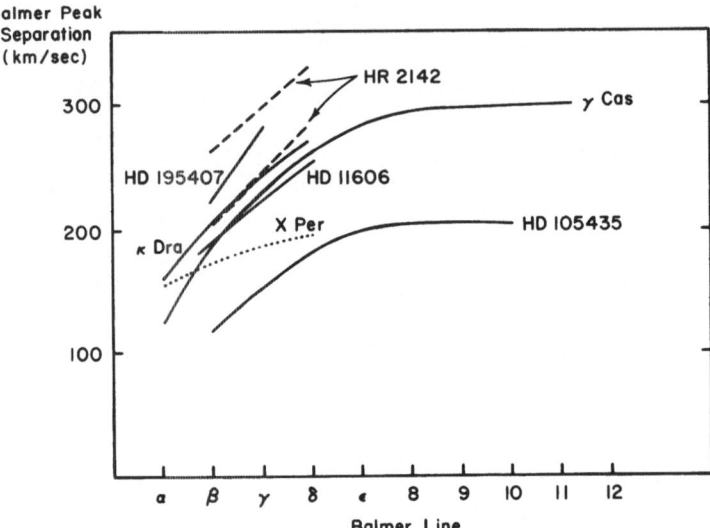

Fig. 3. Progression of Balmer peak separations in Be stars. Slope is similar for all spectral types and values of *i* represented, with exception of X Per.

presumably is a measure of the photospheric value of rotation. The conversion to $v \sin i$ is, of course, model dependent.

A further spectroscopic phenomenon is the width of the emission lines. In particular, the width of Hα is usually considerably greater (in velocity units) than Hβ. An investigation on a small sample of stars by Ringuelet-Kaswalder (1965) suggested that the V/R variables are qualitatively different from other Be and shell stars by (1) having a higher $W(H\alpha)/W(H\beta)$, (2) having no correlation of $W(H\alpha)$ and $W(H\beta)$ and (3) having main sequence luminosities rather than the one magnitude overluminosity of other Be stars. My feeling is that these results should be confirmed by work on a larger sample of stars, before attempting to explain them.

It is my general impression and experience that careful measurement of special spectral characteristics (line strengths, depths and widths, emission profiles and peak separations) can take us a long way towards defining or deciding between different models and parameters for Be stars, which in turn may lead towards a basic understanding of the Be phenomenon.

4. Periodicities in Emission Lines

Emission lines show V/R changes, velocity changes of edges and centre usually with characteristic times of hundreds to thousands of days. However, abrupt changes occur in the behaviour and it is rare to find more than two or three complete cycles before discontinuities occur (Figure 4). Some Be stars on the other hand show no such changes. There are also changes in the emission intensity which are generally unrelated and occur in time scales of tens to hundreds of days.

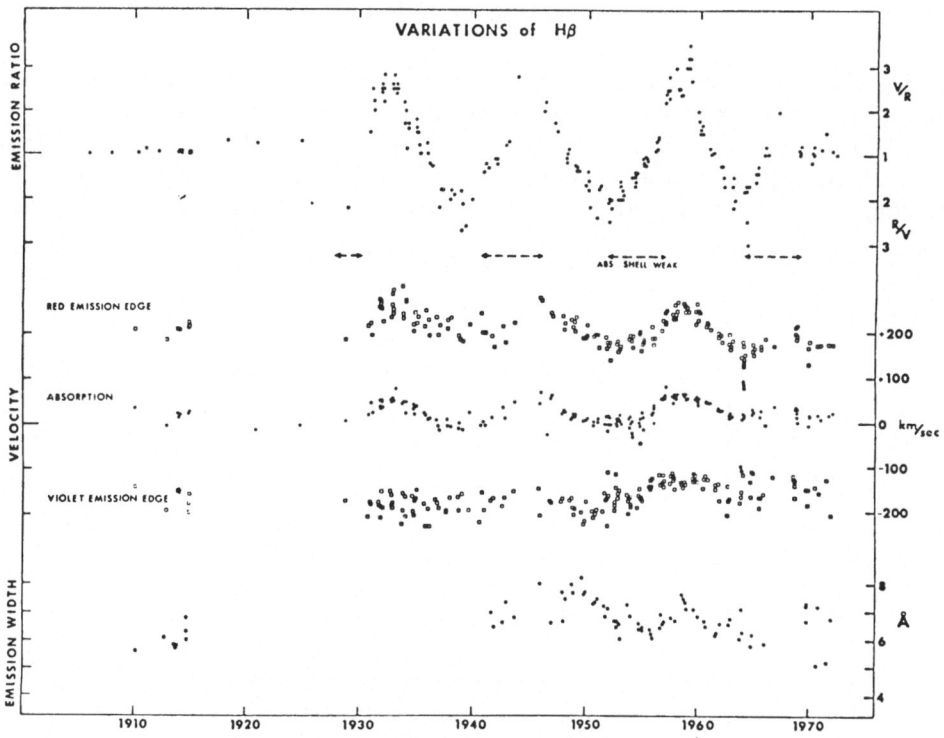

Fig. 4.   Hβ line profile variations in β^1 Mon (by Cowley and Gugula, 1973), showing their discontinuous nature.

These changes may be ascribed to 3 possible causes. (1) Orbital binary motion of the Be star. This explains velocity changes, but only explains V/R changes if there is significant interaction between the stars, either in the form of tidal deformation of the line emitting region, or by emission from a gas stream between the stars, or by a phase dependent mass ejection mechanism. The lack of strict periodicity, discontinuities and periods of no change are serious problems in this explanation. (2) Changes are periods of mass ejection and mass accretion by the central star. These explain V/R changes and velocity changes if they are small. They cannot easily account for large velocity amplitudes as these would lead to regular P Cyg profiles. Also, the long periods involved make the accretion stage difficult to explain by fall-back, as mass will have left the system. Therefore a source of infalling matter, such as a mass-losing companion, is required. (3) Changes are caused by precession of an elliptical emission ring. This can explain all the observed phenomena, at least qualitatively. Discontinuities occur when fresh ejection forms a new ring. However, there are theoretical difficulties with this model which will presumably be discussed later. It is also difficult to explain emission intensity changes and very rapid changes in this model. Again, a possible way out is to involve streaming from a companion. These points are central ones for our discussion, but it seems at this point that we may well have to accept an explanation which embodies all of these effects in some way.

5. Rapid Variations

As detectors have improved, providing increased signal-to-noise and time resolution, it has become possible to study small scale spectral variations in a number of stars. Among the more spectacular results were those on Be star emission lines, where changes in times as short as one minute have been claimed (e.g. Bahng, 1971; Hutchings *et al.*, 1971) (Figure 5). It is often impossible to check such results by older

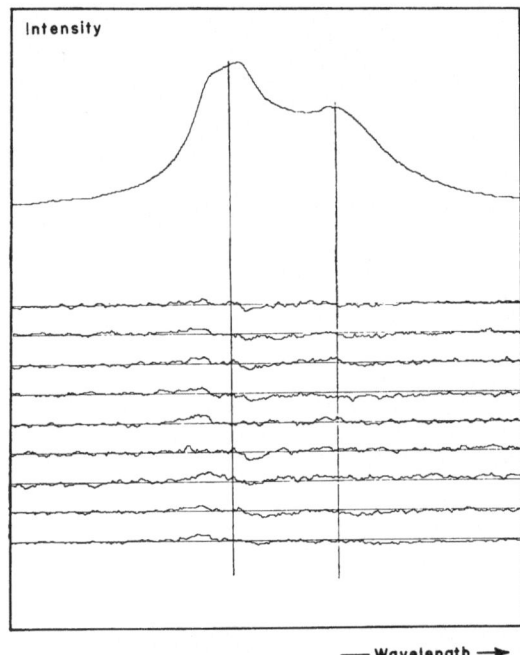

Fig. 5.   Hβ in γ Cas. Isocon scans. Mean profile over 1 h and differences between this and sequential 6 min means.

tested methods so that there is sometimes an unknown instrumental contribution to such changes (monitoring seeing and guiding, with a slow scanner, signal averaging with a fast scanner, electronic instabilities and read-out noise in TV and solid state detectors). The general consensus is that many Be stars at some time show line profile and strength changes, which occur on time scales from ~1 h to 1 min (which is the present limit) and from more than 25% to a lower observational limit of 2–3%. The confidence with which such changes can be measured is a function of the equipment used and the extent of the changes themselves, so that at present little is known beyond the fact that they occur. No pattern of behaviour is apparent and no periodicities have been observed. It is not known what stars or types of star are more likely to show such changes, nor how frequently. Clearly, a full understanding of the processes will involve a very extensive observational program and probably sophisticated data reduction. Finally, it is not clear just how much we will gain from such a study.

The significance of such rapid changes may be clarified by simultaneous observation in different lines. This will show whether changes occur through the whole envelope or in a small part of it or whether different changes occur simultaneously in different parts. Some observations by Doazan suggest that changes occur in all lines at once, in time scales of 15–30 min, but others do not. These and shorter times scales imply disturbances which are very small, or which travel with speeds approaching c. If they occur simultaneously in different lines, they are more likely to be light-time effects. Perhaps a level pumping mechanism, and radiative de-excitation triggered by a flare, or pulse of radiation from the photosphere? Only more extensive, accurate and rapid monitoring will answer such questions. The only other datum at present is that variations seem to be most significant ($\Delta e/e$) in weaker emission lines, suggesting that disturbances occur preferentially in lower strata of the disk or envelope.

Looking at changes lasting longer than 1–2 h, we find that structure (e.g. extra absorptions) in emission sometimes occurs for these times. These may represent dynamic phenomena in a moving envelope: dispersal of density inhomogeneities, or rotation of the envelope. Recurrent, but discontinuous, periodicity in changes in the shape of emission in γ Cas (~0.7 days) has been attributed to the rotation of the envelope with variable, inhomogeneous emission intensity (Hutchings, 1970).

6. Photometry

The photometry of Be stars has been even more patchy and sporadic than spectroscopy. Here again, it has been found that variability exists on time scales from minutes to years and it is difficult to know (a) the true or undisturbed state of the star and (b) whether variations are in any way periodic.

In general, Be stars lie about one magnitude above the main sequence, and have an ultraviolet excess. These properties have been explained by various people as gravity darkening effects which move the star to later type and hence make them apparently bright for this type. However, it is not clear how much of the effect is a true evolutionary one and we will return to this question later. Once again, the study of mass-exchange in binaries may clarify this point considerably.

As far as the study of variability is concerned, most Be stars have been found to vary over some tenths of magnitudes – often more during shell activity. Feinstein (1968) found that 40% of a sample of southern Be stars showed variations of $\geq 0^{m}.06$ in V and 30% showed colour changes in $U-B$ of this size. Later work roughly confirmed these figures. Ferrer and Jaschek (1971) found a higher fraction to be variable: 60% variable in V and $B-V$ and 35% in $U-B$. Their data were taken over a longer time (17 years) and used a larger sample.

The long periods of variation and irregular non-periodic activity make it very hard to obtain 'light curves' for Be stars. An example is the star HD 187399 (which I have referred to before). A peculiar light curve is found which can be explained in its general shape by the distortion of the primary star to its Roche limit, in an elliptical orbit (Figure 6). This again should encourage us to investigate others (e.g., HR 2142), as any repeating photometric variation contains a lot of new information and light curve synthesis programs allow much of this information to be extracted.

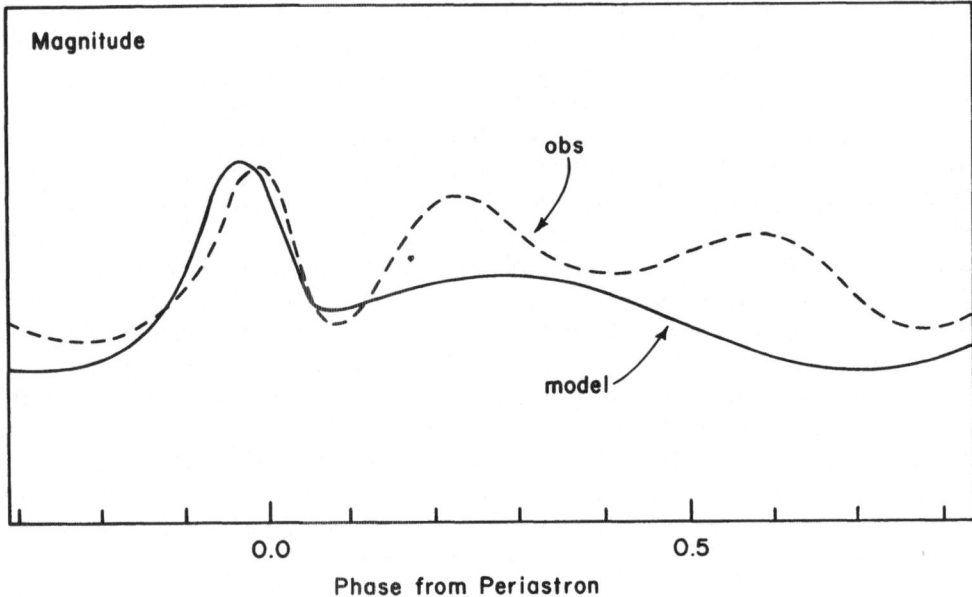

Fig. 6. Light curve of HD 187399 and calculated light variation (amplitude ≃ 0.1 mag.) of primary distorted to Roche limit at periastron passage. (Note that Harmanec (see p. 392) has alternative explanation.)

Looking to shorter time scales, little hås been done. Plageman (private communication) made simultaneous rapid measurements of Hβ and continuum changes and found marginal activity with no definite periodicities. However, much more work of this nature needs to be done to be conclusive. Sanyal (private communication) has reported finding line variations in the Balmer and He I lines in HD 57219, but no activity in the nearby continuum, using special filters.

7. Shell Spectra

The distinction between Be and shell stars has often been a confusing one to me, and I can find no clearly defined criteria in the literature. I suspect that the confusion is not restricted only to myself and that the terms are used somewhat loosely by a number of people. There are shell stars, shell phases, and shell spectra. What I hope to discuss briefly is the spectrum seen when sufficient matter exists around a star to affect its spectrum in a way which is different from the Be phenomena discussed so far. This implies in the context of Be stars that at times or in some stars, either sufficient mass is ejected by a non-continuous force to form a detached envelope, or such an envelope is formed by some unusual accretion event. A shell may however be simply a very extended Be star envelope. Spectroscopically a shell is seen by the existence of sharp and unusually deep absorption lines of hydrogen, He I and of metals such as Si II, Fe II, Ti II. The usual emission lines are often enhanced. The absorptions may be asymmetrical, and show different radial velocities, particularly as a Balmer progression. Velocities may be positive or negative with respect to the stellar velocity and may change over periods of weeks, months, or years. The shell lines show, in fact, most of the behaviour characteristics of Be spectral lines, and the same types of explanations may be put forward to account for them. It is clear from the line depths that densities are higher and radiation more dilute than in normal Be envelopes. The velocity gradients have been interpreted as due to ballistic and collisional slow-down and/or increase in density with height by interstellar material, or of accelerating infall. This distinguishes shells from steady state envelopes in which outward velocities may accelerate under a constant driving force. Thus, the denser regions may lie on the outside of the shell, and evidence for this is found in the velocity progressions, line asymmetries and widths of the shell absorptions, between ions and along the Balmer series.

In particular the shell phase of Pleione, which lasted several years, has been analysed by several workers. Limber (1969) showed that the observed velocity of the higher Balmer lines and strengths may be reproduced by a model with slowly increasing mass ejection and a sharp cutoff, but with an increasing velocity with height. Malborough and Gredley (1972) however showed that this model was unable to reproduce the Balmer progression in the later shell stages. They conclude that it is not clear that the basic model is right, although some way of incorporating emission in the velocity field of the shell may improve the picture. The alternative model they suggest is that of two line forming regions with different Balmer decrements and velocities, but no one has explored this seriously yet.

The steady shell of 1 Del has been modelled by Marlborough and Cowley (1974). A good fit to Hα was obtained by a model again with outward acceleration $(dV/dw > 0)$, and extending to $\geq 30 R^*$. (A different approach by Kitchin (1973) led to similar conclusions about the extent of the Be envelopes in general, even without shell spectra, but his analysis seems to depend on lines formed in the outer envelopes.)

8. Binary Be Stars

As there is a session devoted to the question of whether Be stars are all binaries or not, I do not intend to discuss this hypothesis. However, it is certainly true that several Be stars are known to be binaries, and their behaviour is determined by this circumstance. I shall therefore discuss briefly a few remarkable Be binaries (Table I).

TABLE I

Some Be star binaries

Name	Period	K (km s^{-1})	V_0 (km s^{-1})	e	$f(m)$
187399	28 d	105	−20	0.39	2.6
4 Her	46 d	11	−15	0.38	0.006
173219	58 d	124?	25	0.15	11.4
HR 2142	81 d	20?			
88 Her	87 d	10	−12	0.16	0.008
AX Mon	232 d	52	6	0.02	3.0
X Per	580 d	66	−50	0	18
β^1 Mon	12.5 y	30?		0	13

HR 2142 was discovered by Peters (1972) to have a periodic shell phase, during which positively displaced absorptions were seen in Balmer and He I lines. These repeat every 81 days, and thus suggest a binary origin. Radial velocities of the broad primary absorption suggest an orbital ($e > 0$) motion, which, however, needs to be confirmed. Two explanations are possible here, which only a good orbit can resolve: mass-ejection at periastron by the primary, or mass-accretion seen along the gas-stream. The present velocities make either possible.

This again calls to mind the binary HD 187399 (Hutchings and Redman, 1973a) which shows a strong shell effect at periastron in an $e \sim 0.4$ orbit. There seems to be a good case for roche lobe overflow in this system, so that it may be expected in others. The star is not a Be star in the sense we mean at this meeting so I will not discuss it further than this.

HD 173219 is a Be star with a 55 day orbit (Hutchings and Redman, 1973b), with several peculiarities. Chief of these is that the value of K is very different for different lines, so that the mass function is somewhat uncertain. Also, there is no clear explanation for the effect (either tidal or heating) and the secondary is mysteriously unseen.

A good case exists for the binary nature of 4 Her (Harmanec et al., 1973), with a period of 46 days and $e \sim 0.4$, on the basis of velocity variations alone, and the system will be discussed further by Dr Harmanec at this meeting. He also has evidence for an 87 day period, $e = 0.16$ orbit for 88 Her. All these binaries are fairly well separated and presumably have evolved beyond core hydrogen exhaustion before any mass exchange. This *may* be significant in explaining the origin of the Be phenomenon.

Finally I would like to mention X Persei. This star has been suggested (Hutchings et al., 1975) as the optical counterpart to a weak X-ray source (3U0352 + 30). Radial velocities of the broad absorptions show periodic variations of large amplitude and period 580 days, implying the presence of a very massive unseen secondary. Emission line velocities are antiphased and of much lower velocity, and the Balmer decrement and peak separation progression much lower than normal. It is possible that *some* emission arises from a high density region near the secondary, accounting for these differences, and that this object is a massive collapsed star. If this is so, it may require that the primary is losing mass to and not gaining it from such a companion.

Few other Be stars show strictly periodic spectral changes or light changes. It has been suggested that some of the cyclic changes represent binary motions (e.g., β^1 Mon by Cowley and Gugula, 1973) which are not always seen for some reason. No Be star is known to show eclipses, again presumably because of the high separations in these systems.

Any further points on Be binaries should probably await the discussion reserved for this topic. It is clear that the detection and study of Be binaries is difficult and there may be considerable complications. An attempt has been made at the Dominion Astrophysical Observatory (Gower, private communication) to detect the presence of faint late type companions to Be stars, by obtaining very high signal-to-noise television scans at $\sim \lambda 6400$ Å, where there are many late type spectral lines. A numerical convolution of the spectra with standard late type star scans is very sensitive to the presence of a blended late type spectrum in the data. Preliminary results for upper limits to a late type companion are as follows: γ Cas (0.01), κ Dra, 4 Her, 48 Lib (0.007). 4 Her is a known binary whose probable late type contribution is 0.005, so we need to increase the accuracy of the observations to make it sufficiently sensitive to be useful.

9. Statistics and Evolution

As the Be phenomenon is variable in its strength, estimates of the number of Be stars are likely to be underestimates. The numbers found range from the ~8% found in clusters by Schild to ~20% for all Be stars by Massa. The distribution by spectral type is highest for early B, lower for late B, and falls off beyond B9 and perhaps earlier than B0. It is possible that the Be phenomenon depends on the existence of radiative pressure to support a disk-like envelope. The effect falls with increasing spectral type and at very early types it may be strong enough to blow away such a disk. These arguments favour the single star hypothesis, but can be incorporated into accretion as well.

Further statistics and surveys relevant to this question will be reported by Drs Hubert and Bidelman at this meeting. The evolutionary stage of Be stars is perhaps one of the important points for discussion here. I have mentioned the work of Hardorp and Strittmatter and others who claim that gravity darkening and aspect effects will move Be stars from the ZAMS to their observed positions. Schild (1966) has claimed that even with these effects the stars (and especially the Bex stars) lie on the hook of the evolutionary track. Model atmosphere analysis to be reported by Peters at this meeting has suggested log g values of 3.5–4.0 which also suggest that some lie off the ZAMS. Finally, the question of mass-exchange in binaries is one tied in with evolutionary radius changes. Most models suggested are evolved off the ZAMS to beyond hydrogen-core burning.

Schild (1966) has introduced the sub-class of Be stars known as 'extreme'. These are characterised by being basically similar and stable, showing stronger emission and having weaker absorption lines. They are also slightly redder than 'normal' Be stars. More recently, in a study of Be stars in clusters, Schild and Romanishin (1976) have claimed at this meeting that Hα emission and Bex characteristics are found most strongly at the end of core H burning, supporting the claim that Bex stars are a class of Be stars whose mass loss is caused or enhanced by core contraction. These claims are difficult to make convincingly and the idea is regarded as controversial by some. I hope Dr Schild will present his arguments and that we may have some profitable discussions in the next session.

10. Expanding Models

Finally, I should like to discuss some geometrical models for Be envelopes in the sense that they apply directly to observational quantities. Dr Marlborough will discuss the details and theoretical justification of these and other models in due course. The class of models I want to consider are extended equatorial disks in which angular momentum/unit mass is conserved and in which line emission and absorption are simply proportional to the density. The density falls in a way necessary to preserve continuity. The free parameters to be explored are thus largely geometrical: velocity of outflow (constant with height, as small gradients produce very similar results), radial extent, spread out of orbital plane, and inclination of line of sight. It is of interest in comparing models with observation to know what quantities the observed features are sensitive to. Model profiles were computed in a grid, using three geometries: (1) disk extending 10% from equatorial plane, with maximum thickness 1 R_* and (2) disk 30° from plane, extending 1.5 R_* (3) disk covers whole star, but extends only to 1.3 R_* away from equator. The results are shown in Figures 7 and 8 and may be summarised as follows:

(1) R/V is a function of expansion velocity and is about the same for all models. The increase in R/V with expansion is fastest for i small.

(2) The peak separation *varies* with expansion velocity similarly for all models, but is larger for the thicker disk. The separation increases by ~15% when expansion goes from 0 to 60 km s^{-1}, for all i.

(3) The emission EW is the same for all i, unless electron scattering is present, or absorption in the line of sight.

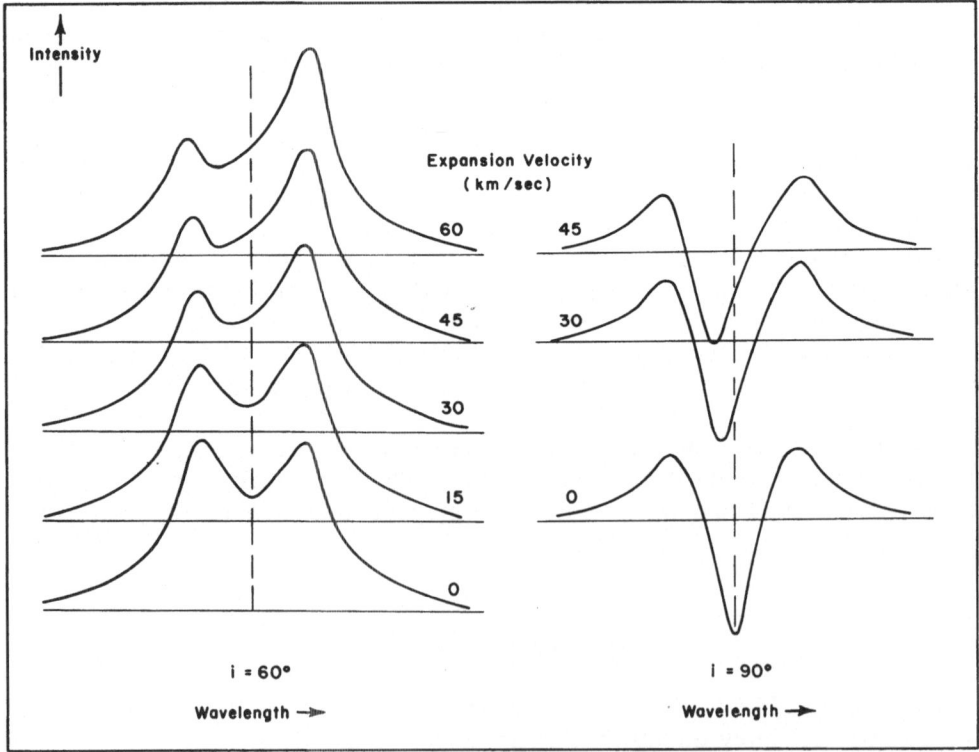

Fig. 7.   Calculated Balmer line profiles for equatorial expanding envelope (±10°, maximum thickness 1 R_*, extending to 5 R_*, conservation of angular momentum).

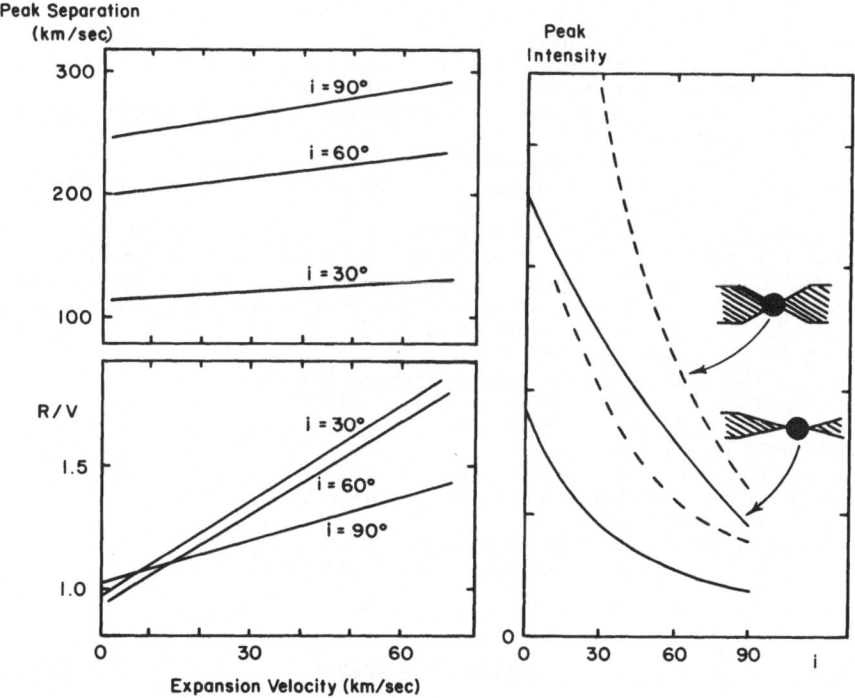

Fig. 8. General behaviour of calculated double peaked emission lines for expanding envelopes.

(4) The peak intensity varies with i similarly for all models. At $i = 0$ it is two times that at 30°, four times that at 90°.

(5) The separation varies with radial extent of the envelope similarly for all i, and less for the thick disks.

(6) The central dip is less marked for thick disks (aside from absorption in the line of sight). Expansion gives a shallower minimum, and so does an increase in i (aside from absorption).

The properties of a model can be deduced from such a grid and compared with a grid of observed stars. Naturally other data and considerations must be used to decide on the validity of such models, unless complete incompatibility with observation eliminates them entirely. It seems that many features of Be star profiles are explainable in terms of such models and we now need a grid of good observations of different stars to test them. In particular, I hope we can help to use the information contained in V/R ratios, peak separations and emission line widths to test such models and eventually lead to a better understanding of the mechanisms of the Be star envelopes.

References

Bahng, J. D. R.: 1971, *Astrophys. J.* **167**, L75.
Bernacca, P. L.: 1970, in A. Slettebak (ed.), *Stellar Rotation*, D. Reidel Publ. Co., Dordrecht-Holland, p. 227.

Cowley, A. P. and Gugula, R.: 1973, *Astron. Astrophys.* **22**, 203.
Feinstein, A.: 1968, *Z. Astrophys.* **68**, 29.
Ferrer, L. and Jaschek, C.: 1971, *Publ. Astron. Soc. Pacific* **83**, 346.
Hardorp, J. and Strittmatter, P. A.: 1968, *Astrophys. J.* **153**, 465.
Hardorp, J. and Strittmatter, P. A.: 1970, in A. Slettebak (ed.), *Stellar Rotation*, D. Reidel Publ. Co., Dordrecht-Holland, p. 48.
Harmanec, P., Koubskey, P., and Krpata, J.: 1973, *Astron. Astrophys.* **22**, 337.
Hutchings, J. B.: 1970, *Monthly Notices Roy. Astron. Soc.* **150**, 55.
Hutchings, J. B.: 1971, *Monthly Notices Roy. Astron. Soc.* **152**, 109.
Hutchings, J. B. and Redman, R. O.: 1973a, *Monthly Notices Roy. Astron. Soc.* **163**, 209.
Hutchings, J. B. and Redman, R. O.: 1973b, *Monthly Notices Roy. Astron. Soc.* **163**, 219.
Hutchings, J. B., Auman, J. R., Gower, A. C., and Walker, G. A. H.: 1971, *Astrophys. J.* **170**, L73.
Hutchings, J. B., Crampton, D., and Redman, R. O.: 1975, *Monthly Notices Roy. Astron. Soc.* **170**, 313.
Kitchin, C. R.: 1973, *Monthly Notices Roy. Astron. Soc.* **161**, 389.
Limber, N. D.: 1969, *Astrophys. J.* **157**, 785.
Marlborough, J. M. and Cowley, A. P.: 1974, *Astrophys. J.* **187**, 99.
Marlborough, J. M. and Gredley, P.: 1972, *Astrophys. J.* **178**, 477.
Marlborough, J. M. and Zamir, M.: 1975, *Astrophys. J.* **195**, 145.
Massa, D.: 1975, preprint.
Peters, G. J.: 1972, *Publ. Astron. Soc. Pacific* **84**, 334.
Ringuelet-Kaswalder, A. E.: 1965, *Publ. Astron. Soc. Pacific* **75**, 323.
Schild, R.: 1966, *Astrophys. J.* **146**, 142.
Schild, R. and Romanishin, W.: 1976, this volume, p. 31.
Stoeckley, T. R.: 1968, *Monthly Notices Roy. Astron. Soc.* **140**, 149.

DISCUSSION

Plavec: Can we stop for a moment to discuss this problem of V/R variation? I would like to point out that first of all, it would be nice to know whether the periodicity of the phenomenon is typically years. It is possible that it is so, but I don't know if the observational evidence is really very convincing, because when you look at the old observations, as made by Curtiss and McLaughlin, what strikes you is that the observations were made regularly each year once or twice in the same season. This type of observation made once a year certainly fails to detect periodicities of the order of decades or months. I wonder whether we are really sure that the typical periodicity is years?

Hutchings: In *some* cases (e.g., X Per or β^1 Mon) it seems clear that the variation with 'periods' of some years is definitely present, as coverage on such a time base is good. In other cases, of course, Plavec's remarks are very true, and there may be shorter periodicities present which have been missed.

Peters: HR 2142 was on McLaughlin's observing list for about 30 years, and he missed the periodic shell activity. He also missed the periodicity in V/R in Hβ because he only obtained one plate per year. Since the period is 81 days and the duration of the shell phase is 7 days, at least 10 plates per year would have been required to find the periodic shell structure.

Snow: I question the reality of the ultraviolet excesses often attributed to Be stars, because I believe that the ultraviolet extinction is often over-estimated. Reddening due to excess infrared emission can lead one to expect far more ultraviolet extinction than is actually present; thus the appearance of an apparent ultraviolet excess can be due to an over-estimate of the extinction, rather than to a true excess of flux in ultraviolet wavelengths. An example of this is χ Oph, whose $B - V$ color implies $E(B - V) \approx 0\overset{m}{.}6$, but which can be shown to have a true interstellar extinction corresponding to a color excess of $E(B - V) \approx 0\overset{m}{.}35$ (Snow, *Astrophys. J.* **198**, 361, 1975). *Copernicus* data on a few bright Be stars show that in most cases the flux in the ultraviolet compared to that in the V bandpass matches the distribution in normal stars of similar MK class, if an intrinsic reddening of the order $0\overset{m}{.}2$ in $B - V$ can be assumed for the Be star.

Plavec: I would like to point out two facts. You just mentioned that the envelope extends rather far from the star: 30 or so stellar radii which might be 100 solar radii from the surface of the star. On the other hand, there have been several articles written, for example, by Boyarchuk where he derives that the emission and absorption features are formed within two or three stellar radii, which is inconsistent with what was found by Marlborough. Boyarchuk's data are derived from the assumption that the angular momentum is conserved in outer layers so he measures the width of the individual lines and derives the distances. It seems to me that the results based on radiation transfer in the lines indicate very clearly that the assumption is wrong. Because of this discrepancy, I think one has to conclude that it is not true that the angular momentum is preserved in the envelope. And another point: we all are used to the idea that shell stars are Be stars at which we look edge-on, so that the flat disc as seen in the equatorial plane is seen at great geometrical depth. But we should remember, for example, the case of 17 Lep, where Anne Cowley found that the inclination is really not near 90°, but something like 20° or 30°. So we are looking at a system very far from the equatorial plane, yet there is plenty of material there for we see strong shell lines. Therefore in some cases, at least, we must accept the fact that the system is surrounded by a cloud which is certainly extended very far away from the central plane.

Doazan: I think that almost all the lines in X Per are in emission. How did you measure the radial velocities of the absorption lines?

Hutchings: We measured the broad underlying absorption in the Balmer lines from about H8 to H12 or so, using our wide scan oscilloscope machine. In these lines the emission is present only weakly in the center and on good (usually IIIa-J) plates we could get good agreement between individual lines velocities.

Henize: If I may comment on distribution with spectral type, the data from my southern Hα survey (now in press) may be of interest. It was something of a surprise to find that peak frequency among 1232 stars with spectral types (nearly all are HD types) occurred at spectral type B8 (207 stars) in contrast to the classical data by Merrill which showed an apparent peak at B3 (my data show only 110 stars of this type). Most of my stars lie between $m_v = 8.0$ and 10.0 and therefore are somewhat affected by the systematic misclassification of B8 and B9 stars in the HD and HDE catalogues. This effect shifts the peak to class B6 or B7 on the MK system but it is very difficult to completely explain the peak away on this basis. Rather, it seems likely that the B8, 9 peak is explained partly by the greater proportion of B8, 9 stars among the fainter stars caused by interstellar absorption and partly to my ability to observe weaker emission lines than could be observed by the early surveys. Since the B8, 9 stars show considerably weaker emission than the B0–B2 stars, it should be expected that my survey will show a greater proportion of B8, 9 stars. That the emission-line stars with HD spectral class B8 are significantly later than those with HD spectral class B0–B2 is borne out both by their distribution in galactic latitude and by the differing distributions of emission-line intensity.

Hutchings: It is important to eliminate supergiants and selection effects in discussing the distribution of rotational Be stars. Schild and Massa have attended to these points carefully.

Cowley: With regard to Henize's comments, if HD spectral types were used, N. Houk finds for the faint stars that many early B's are classified as later B's because of the strength of the interstellar K line, which Miss Cannon did not recognize as non-stellar. It would be better to use the new types of Houk before this statistical result is accepted. (**Note added in proof:** Analysis of the frequency vs spectral type of the Be stars in the *Michigan Catalogue of Revised HD Spectral Types* shows the same frequency distribution for bright and faint stars.)

Heap: I have a comment concerning the spectral type of ζ Tau. I found that the ultraviolet line spectrum of the star suggested the atmospheric parameters, $T_{eff} = 27\,500$ K, log $g = 4.0$. This value of T_{eff} is a good 10 000° higher than estimates based on visual studies, e.g., the spectral type B4 IIIp (Lesh, 1968), and the observed Balmer jump (Schild *et al.*, 1971). After studying some visual data, I found that there is no real discrepancy between the effective temperatures indicated by the visual and ultraviolet regions of the spectrum: the lines which suggested to Lesh a late spectral type (e.g., strong Mg II 4481) are really *shell* lines, and the strong Balmer jump observed by Schild *et al.* is due to *shell* absorption. Hence, both the visual and ultraviolet spectrum of ζ Tau indicate a high effective temperature – around 27 000°, if you use Mihalas' non line-blanketed models.

A SURVEY OF Hα IN THE BRIGHTER
NORTHERN Be STARS

WILLIAM P. BIDELMAN and ANTHONY J. WEITENBECK

Warner and Swasey Observatory, Case Western Reserve University, Cleveland, Ohio, U.S.A.

Abstract. Hα observations have been made of the stars brighter than $m_v = 7.5$ and north of $\delta = -30°$ contained in the three classical Be-star catalogues of Merrill and Burwell. At least one observation of each star was made at each of the two epochs 1958/9 and 1970/2 at the Lick Observatory and at the Warner and Swasey Observatory respectively. The dispersions used were 88 Å mm^{-1} at Lick and 66 Å mm^{-1} and 18 Å mm^{-1} (with image tube) at Warner and Swasey.

Results were as follows: The total number of objects observed was 215. Of these, 32 were supergiants and will be discussed later. Another 16 stars were deemed abnormal in some manner (helium and/or forbidden emission, composite spectrum, markedly violet-displaced Hα emission component: an interesting group of five stars including the nebular variable AB Aurigae). Thirteen of the stars showed no emission at either epoch. Of the 154 presumably normal low-luminosity stars showing emission, 70 displayed shell structure in Hα at one or both epochs, though in over half of these cases the shell absorption was considered 'weak'. Thirty of the 70 stars varied in their shell characteristics between the two epochs. The total number of stars whose Hα line varied in some respect was 43, or 26 percent of the 167 normal low-luminosity stars observed. Quite a number of the stars observed have been reported to show shell characteristics by others but were not so noted by us.

Eight additional Be stars not contained in the Merrill–Burwell catalogues have also been observed. Seven of these are shell stars; the other is the unusually high latitude object HD 127617.

It is a pleasure to acknowledge that many of the Lick observations were made by Dr Jack E. Forbes, a graduate student at the time in the Department of Astronomy at the University of California, Berkeley.

DISCUSSION

Conti: I want to be clear as to your definition of a shell. Do you call a shell star one with an Hα line showing double emission and central absorption?

Bidelman: Yes. But you must remember that my dispersion is not particularly high, so any such structure that I see would no doubt be very conspicuous at coudé dispersion. Since I have not done spectrophotometry, however, I cannot tell you the depths of the Hα reversals observed, though in most cases they must be as low as the continuum and in some well below it.

Peters: We have been surveying Be stars at Hα using the Lick Observatory cooled 40 mm Varo image tube. Our dispersions range from 11 Å mm^{-1} to 33 Å mm^{-1}. One aspect of our project has been to obtain light exposures on baked IIIa-J plates to look for structure in Hα. We have found that the profiles of Hα in Be stars are far more complicated than was previously believed to be the case. Often times we observe fine structure which does not persist from one observation to the next.

Bidelman: Your point about making the exposures light is certainly a good one. In many cases, we had to reobserve the star to get an exposure light enough to see structure at Hα. It is also true that if the emission is narrow the chance of seeing structure in it is less. I might add that practically every shell star I had has a rotational velocity of at least 300 km s^{-1}.

Doazan: Are the variations which you observed based on eye estimates?

Bidelman: Yes.

A. Slettebak (ed.), Be and Shell Stars, 29. *All Rights Reserved*

A STUDY OF Be STARS IN CLUSTERS

R. SCHILD

Center for Astrophysics, Cambridge, Mass., U.S.A.

and

W. ROMANISHIN

Dept. of Astronomy, University of Arizona, Tucson, Ariz., U.S.A.

Abstract. Calibrated spectrograms at Hα of 566 stars in 29 young galactic clusters led to the detection of 41 Be stars in clusters. Using cluster membership we have inferred ages and intrinsic $(B - V)$ colors of Be stars to permit a discussion of their evolutionary states.

Rotating stars can become Be stars in their early hydrogen burning evolution away from the main sequence. Both the fraction of stars showing hydrogen emission and the strength of emission appear to vary little during the first 80% of post main sequence evolution. However, at the onset of gravitational core contraction, both the fraction of stars showing emission and their mean emission strength undergo a fourfold increase. Many stars in the core contraction phase develop an intrinsic $(B - V)$ excess of 0.15 mag. due to the H^- free-bound continuum radiation. Because of the great strength of Hα emission and the short time duration of the effect, the extreme Be stars would be excellent probes for studies of spiral structure and would also serve as probes for studies of ages and distances of extragalactic systems.

Analysis of the corrected colors of the Be stars in clusters suggests that the Lucy and Solomon (1970) mechanism for reduction of effective surface gravity by ultraviolet resonance line scattering is probably important for the hotter Be stars. However, a discrepancy exists for the cooler stars between the predicted and observed colors.

References

Lucy, L. and Solomon, P.: 1970, *Astrophys. J.* **159**, 879.

DISCUSSION

Conti: I was particularly struck by the absence of any strong correlation between Hα emission strength and τ, the 'core-contraction' parameter. There were only a few stars, just at τ = 1, with strong Hα. Could this just be those earliest B stars with wind enhanced envelopes? Or possibly 'mass exchanging' binaries with enhanced emission, or slightly modified M_v and $B - V$ parameters?

Schild: I worried about this, but I can't imagine why the binaries would all have τ between 0.8 and 1.0, although perhaps there is a reason.

Plavec: There may be a reason. Recently, Ulrich and Burger at UCLA studied the accretion of the material on a binary component. When the rate of mass transfer gets higher the accreting star becomes overluminous and is displaced away from the main sequence. What you would expect in these systems, if they are really exchanging material, is that the observed Be star, when it is surrounded by this material, will also be displaced away from the main sequence into the region where the stars normally are when they are leaving the main sequence. So that is one possible explanation.

Young: Does your study suggest that there is a significant difference in the frequency and appearance of Be stars among early B stars that are in galactic clusters, vs those that are field stars?

Schild: That is a difficult question to answer, because relatively few complete surveys of this kind have been made for field stars and you have all the problems of systematic effects such as the intrinsic $(B - V)$ and $(U - B)$ color excesses in the field stars which are basically avoided by going to the clusters. Thus I cannot give any kind of reasonable answer to your question.

Young: In the field of galactic clusters there are a number of workers who have shown that there is a remarkable inverse correlation between high frequency of binaries in clusters and low rotation. Does your sample overlap theirs so that you can say something about this?

A. Slettebak (ed.), Be and Shell Stars, 31–32. All Rights Reserved

Schild: Unfortunately I cannot. Helmut Abt and collaborators have observed primarily A-type stars and I am an observer of B-type stars, so he tended to select clusters by different criteria. I cut off at about the temperature where Abt starts, and there is essentially no overlap.

Cowley: Recent work by Crampton at the Dominion Astrophysical Observatory has shown that the correlation that Abt found between stellar rotation and binary frequency in clusters may not be supported by more recent observations.

Henize: Since my southern Hα survey probably includes the most uniform data available on the field stars, let me comment that, even though my survey data to a fixed magnitude limit show a peak frequency at HD class B8, the emission frequency per hundred stars of the same spectral class still shows a clear peak at B2. I find that one star in five shows emission at B2 while only one star in 30 shows emission at B8,9. This is quite different from the statistics Schild finds in his cluster data.

Slettebak: In your earlier paper in which you define extreme Be stars, you define them also in terms of sharp helium lines, which implies small rotation. Do you have any information from this sample about line widths?

Schild: No. These spectrograms are of relatively low quality. They were Hα spectrograms taken with an image tube and were widened to only 0.2 mm. I did not try to measure line widths. Dr Slettebak is fully aware that I had the idea of two kinds of Be stars back in 1966. There are now a number of criteria which I have used to identify extreme Be stars: I made infrared observations, continuum scans, and spectroscopic observations, and I am still having the greatest difficulty in really showing that there are two separate, different kinds of Be stars (the core-contraction stars and the normal post-main sequence stars). It is very difficult to tie together all these different kinds of observations in one consistent definition which applies to all Be stars over the temperature range 10–30 000 K.

Henize: Do your data or your theoretical concepts lead to a prediction of how emission-line strengths are distributed for a particular spectral class? I ask this because my survey data show that weak emission lines are conspicuously infrequent among the B0–B2 stars, thus suggesting an all-on or all-off mechanism for triggering the emission.

Schild: I have no information.

EMISSION FEATURES OF SEVERAL Be STARS AS RELATED TO THEIR LUMINOSITY CLASS AND SPECTRAL TYPE

H. HUBERT and M. Th. CHAMBON

Observatoire de Meudon, France

1. Introduction

At the Haute Provence Observatory, a survey of the Be stars brighter than the seventh magnitude was initiated by Herman and her collaborators twenty years ago.

With this material, the relationship of emission features of Be stars to their luminosity class and spectral type is investigated. At the present time, about fifty B2 and B8–9 emission stars have been studied. The plates were taken at the 1.20 meter telescope with dispersion 77 Å mm^{-1} at Hγ, 120 Å mm^{-1} at Hβ and 300 Å mm^{-1} at Hα. All these Be stars have been classified by Herman-Rojas (1973–1974). Their classification is based on the hydrogen lines. Our study of the B3 emission stars was published this year (Delplace and Hubert, 1975) and the results will only be summarized. For the B8–9 emission stars the results will be given in detail.

2. The B2 Emission Stars

About twenty five B2 emission line stars were studied. Using the Herman-Rojas (1974) and Rountree-Lesh (1968) classifications, we have been able to distinguish three groups of stars:

(1) *First group*; two sub-groups are found:
   (a) the stars exhibit strong emission of the higher Balmer lines – Metallic emission lines are present.
   The emission is observed to be fairly stable.
   The time scale of the emission features is long: about 30 or 40 years.
   (b) the stars exhibit strong emission in the higher Balmer lines, but metallic emission lines are faint or missing. The emission features are not as stable; a hydrogen shell is often present which is enhanced during the minimum of the emission.
   The time scale of the emission variations is longer than or equal to 16 years.
   In the first group, the stars would generally have luminosity class V or V–IV.

(2) *In the second group*, the stars exhibit emission only in the first Balmer lines. The emission features are not as stable. In these stars the emission lines disappear and then reappear, giving: Be → B → Be.
   The time scale of the emission variation is about 12 years.
   In the Herman-Rojas and Rountree-Lesh classifications these stars would generally have luminosity class IV–III.

A. Slettebak (ed.), Be and Shell Stars, 33–36. All Rights Reserved
Copyright © 1976 by the IAU.

(3) *The third group* includes the typical shell stars; they would have luminosity class III.

3. B8–B9 Emission Stars

Twenty-three B8–B9 emission stars were studied. As for the B2 stars, a relation between the emission features and the luminosity class is investigated.

Four stellar groups are found.

(1) *In the first group* the stars exhibit rather strong emission at Hα. Hα is always in emission and Hβ also. The emission variations would be very faint and large. On our plates they are chiefly visible at Hβ. Though these stars have been observed for thirty to fifty years, no time scale of the emission variation can yet be given.

During the maximum of the emission, some Fe II emission lines in the long wavelength range are observed. A faint hydrogen shell is also present before and during the maximum of the emission and some enhancements of the Mg II and He I lines are also possible.

All these stars have homogeneous emission features. In the Herman-Rojas classification, they would generally have luminosity class V (Table I).

TABLE I

Group 1 of B8–9 emission stars

Star	Herman-Rojas classification	Morguleff classification	Miczaika classification	Rotational velocity (Slettebak)
HD 6343	B8 IV-V	–	–	–
HD 9709	B8 V	–	–	350
HD 18552	B8 V	B9 V	–	320
HD 21641	B9 V	–	B8 V	160
HD 23552	B8 IV ?	–	–	250
HD 47054	–	–	–	270
HD 53416	B9 V	–	–	–
HD 192044	B8 V	B8 V	–	350
HD 196712	B7–8 V	–	–	250
HD 207232	B8 IV ?	–	–	330

(2) *In the second group* the emission features are not stable. The emission in these stars disappears and reappears (giving Be → B → Be) but when the emission is present, Hα is strong (Hubert, 1971, 1973).

The time scale of the emission variations is longer than forty-five years. During the transition phase (B → Be), Na I, Ca II, and Mg II lines are enhanced.

In the Herman-Rojas classification these stars would have luminosity class V–IV or IV (Table II). We recall the case of Pleione (B7IV), for which the time scale of the emission variations is about 60 years (Delplace and Hubert, 1973).

TABLE II

Group 2 of B8–9 emission stars

Star	Herman-Rojas classification	Morguleff classification	Miczaika classification	Rotational velocity (Slettebak)
HD 164447	B8 IV	B7 IV	–	250
HD 142926	B8 IV–V	–	B8 V	350
HD 210129	B8 IV	–	B8 V	200
HD 216057	B8 IV–V	B8	–	370

(3) *In the third group* the stars exhibit very faint emission on a strong absorption at Hα. Unhappily these stars have not been observed for a long time and, on our plates, the dispersion is not sufficient to give a time scale of the emission variations. These variations are very faint, but perhaps shorter than for the first group.

In the Herman-Rojas classification these stars would generally have luminosity class IV–III (Table III).

TABLE III

Group 3 of B8–9 emission stars

Star	Herman-Rojas classification	Morguleff classification	Miczaika classification	Rotational velocity (Slettebak)
HD 6811	B9 III	B8 III	B8 III	70
HD 175511	B9 V?	–	–	–
HD 175869	B8 IV–III	–	–	–
HD 195554	B9 IV	–	–	250
HD 205551	B9 III	–	–	200

(4) *In the fourth group* the stars exhibit a faint emission at Hα (sometimes also at Hβ) or an absorption.

The time scales are certainly very large, because the absorption phase is about 40 years.

In the Herman-Rojas classification these stars would generally have luminosity class III (Table IV).

TABLE IV

Group 4 of B8–9 emission stars

Star	Herman-Rojas classification	Morguleff classification	Miczaika classification	Rountree-Lesh classification	Rotational velocity (Slettebak)
HD 144	B9 III–II	B8 III	B8 III	–	170
HD 23302	B8, 5 IV–III	B7–8 III	B6 III	B6 III	230
HD 23630	B8 III	B8 III	B7 III	B7 III	210
HD 23850	B8 IV–III	B8 III	–	–	–
HD 24479	A1 III	–	B9 III	–	110

4. Conclusions

(1) For a given luminosity class, the emission features are much more important for B2 than B9 stars.

(2) When the stars are going through the phases Be → B → Be, the time scale of the emission variations is longer for B9 stars than B2.

(3) Emission features probably vary with luminosity class.

(4) The Be phenomenon is present

    (a) when the B stars are leaving the main sequence

    (b) when the B stars are at the end of the post main sequence contraction.

References

Delplace, A. M. and Hubert, H.: 1973, *Compt. Rend. Acad. Sci. Paris* **277**, 575.

Delplace, A. M. and Hubert, H.: 1975, *Astron. Astrophys.* **38**, 75.

Herman, R.: 1973, in Ch. Fehrenbach and B. E. Westerlund (eds.): 'Spectral Classification and Multicolour Photometry', *IAU Symp.* **50**, p. 17.

Herman, R.: 1974, private communication.

Hubert, H.: 1971, *Astron. Astrophys.* **11**, 100.

Hubert, H.: 1973, *Astron. Astrophys. Suppl.* **9**, 133.

Rountree-Lesh, J.: 1968, *Astrophys. J. Suppl.* **16**, 371.

RAPID VARIATIONS IN THE SPECTRA OF

o And, *γ* Cas, AND *χ* Oph

V. DOAZAN

Observatoire de Paris, France

o And

o And is a shell star which has lost its shell characteristics several times since 1897. It exhibits short time scale light variations with a period of about one day. Its binary nature is doubtful according to Olsen (1972). We have detected rapid spectroscopic variations in October 1973 when *o* And had no envelope, that is, when it appeared as a 'normal' B star. Spectrograms of 7.2 Å mm^{-1} dispersion were obtained at the Haute Provence Observatory with exposure times ranging from 12 min to 52 min.

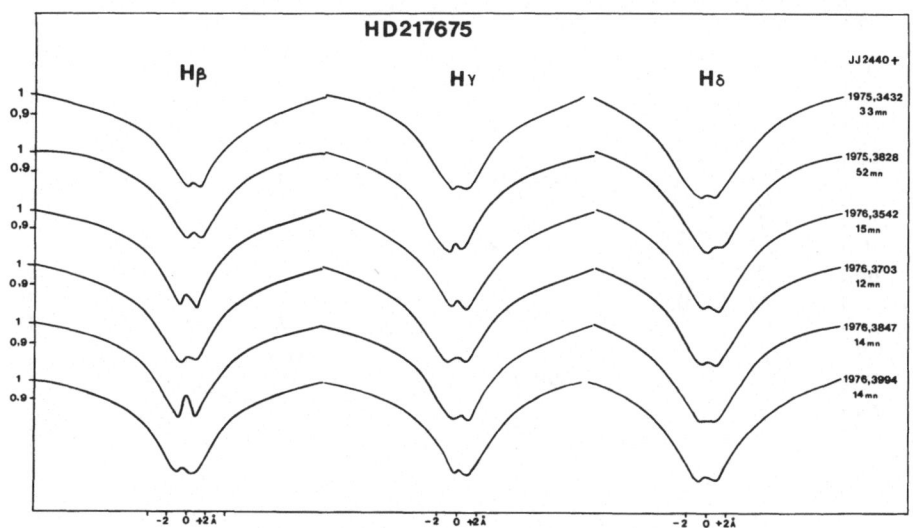

Fig. 1.   Hβ, Hγ and Hδ line profiles for *o* And.

The variations in the hydrogen lines can be seen even by visual inspection of the plates. We notice in Figure 1 that:

(1) The variations of the profiles are irregular and occur in about 15 minutes.

(2) The variations of Hβ, Hγ, and Hδ show some correlation for the first four spectra but this correlation does not persist for the last two spectra.

(3) The observed line variations cannot account for the smallest light variations detected previously in the B band.

(4) On lower dispersion spectra *o* And would have been considered as a 'normal' B star.

A. Slettebak (ed.), Be and Shell Stars, 37–40. All Rights Reserved
Copyright © 1976 by the IAU.

These observations show that even when *o* And has lost its shell characteristics, it does not behave as a normal B-type star: short time scale variations occur in the lines. In consequence rapid activity does not depend on the size of the envelope.

γ Cas

We have obtained a series of spectra of 7.2 Å mm^{-1} dispersion during five consecutive nights in October 1973, with exposure times ranging from 4 to 11 minutes. Our

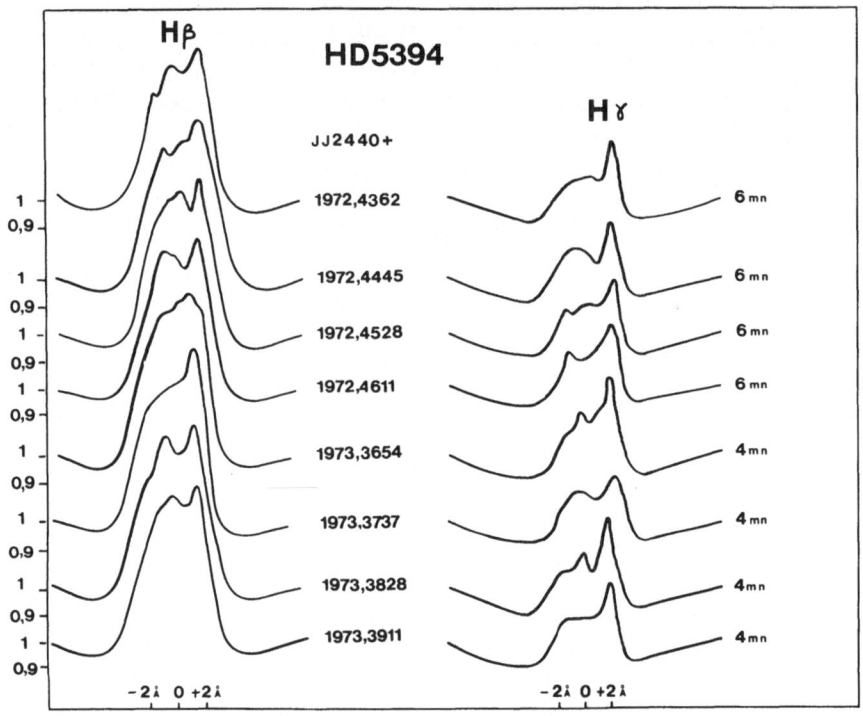

Fig. 2. Hβ and Hγ line profiles for γ Cas.

aim was to detect an eventual correlation between the variations of Hβ and Hγ. We notice in Figure 2 that:

(1) The variations are very similar to those reported by Hutchings in 1969.

(2) The variability affects essentially the emission peaks. Consequently fluctuations of the *V/R* ratio occur.

(3) No correlation could be detected between the variation of Hβ and Hγ.

χ Oph

The intensity of the emission lines and the *V/R* ratio of χ Oph show long term variations. Low dispersion spectra obtained since 1959 show that the most noticeable changes concern the metallic emission lines, which disappear completely in about five years. According to its low value of $v \sin i$ χ Oph is considered as a pole-on star. This fact is particularly interesting since in this case rapid variations cannot be the consequence of rotation. We have obtained a series of spectrograms with a dispersion of 12.2 Å mm^{-1} during four consecutive nights in July 1972, with exposure times ranging from 16 to 25 minutes. The hydrogen emission lines, which are usually double, show a complex structure with three components. This structure was observed in a few cases for Hγ by McLaughlin (1932).

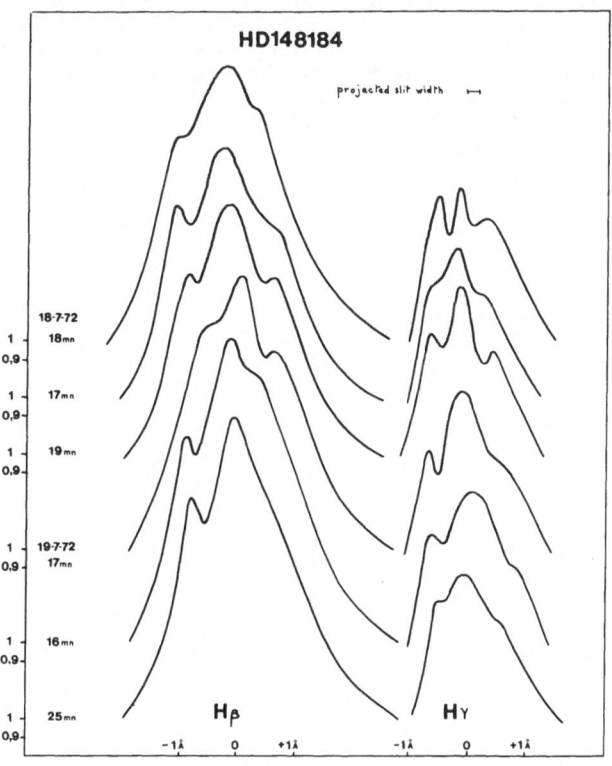

Fig. 3.   Hβ and Hγ line profiles for χ Oph.

We notice in Figure 3 that the triple structure of Hβ and Hγ varies irregularly, being more or less pronounced. This variability could explain the fact that very few observations of the three components have been reported until this time.

In summary, the observations of o And, γ Cas and χ Oph show that rapid spectroscopic variations occur as well in a Be star which has lost its envelope, like o And, as in a pole-on star like χ Oph and as in a rapid rotator like γ Cas. It is thus tempting to conclude that rapid spectroscopic variations (1) do not depend on the size of the envelope (2) are not a consequence of the star's rotation.

V. DOAZAN

References

Hutchings, J. B.: 1969 in L. Detre (ed.), *Non-Periodic Phenomena in Variable Stars*, Academic Press,
    Budapest, p. 191.
McLaughlin, D. B.: 1932, *Publ. Obs. Mich.* **4**, 175.
Olsen, E. H.: 1972, *Astron. Astrophys.* **20**, 167.

DISCUSSION

Swings: In addition to the three classes of Be stars in which rapid variations are observed simultaneously
in several Balmer lines, I wish to add peculiar Be stars. For example, let me mention the B2 IVep star
HD 45677: variations in the complex profile of Hγ and Hδ were observed first from night to night, then in
the course of one night, with a time scale of the order of less than one hour: one of the components in
absorption at $V = -90$ km s^{-1} completely disappears. The observations, which were obtained at the
Haute Provence Observatory and at the European Southern Observatory, are described and illustrated in
a paper published in *Astronomy and Astrophysics* (**26**, 443, 1973).

RAPID VARIATIONS OF Hα IN Be STARS

J. D. R. BAHNG*

Astronomy Dept., Northwestern University, Evanston, Ill., U.S.A.

Abstract. Photoelectric spectrum scans of Be stars were analysed to study the short-term variations of Hα emission strengths. In ζ Tau, α Col, PP Car, and δ Cen, definite variations of a few percent with time scales of 1 to 3 minutes were found. These variations do not exhibit any periodicity.

1. Introduction

Recently, several reports have been published in which rapid spectral changes in Be stars with time scales of a few minutes were observed. Profiles of hydrogen emission lines were found to vary in a time scale as short as one minute (Hutchings *et al.*, 1971). Other studies (Bahng, 1971; McBeath, 1974; Sanyal, 1974) indicate variations in the total emission strengths of hydrogen lines with a time scale of one to ten minutes. The profile changes observed by Hutchings *et al.* appear to be accompanied by some changes in the total line strengths as well.

It is quite clear that these rapid changes cannot be explained in terms of the binary orbital motions (Delplace, 1970), a geometry connected with the asymmetric distribution of gases in a circular ring (Huang, 1972), or an elliptical ring (Huang, 1973). It is equally unlikely that they are related to oscillations or pulsations of the star or shell, since the observed time scale is much too short. As yet, no satisfactory theoretical models exist to explain these rapid phenomena. From the time scales involved, Mihalas (1974) concluded that they imply *hydrodynamic* interactions; a model proposed by Marlborough and Zamir (1975) may be relevant. On the other hand, some unknown *radiative* processes (see e.g. Prendergast and Spiegel, 1973) may be responsible for these changes.

It is extremely important to establish first the reality of these rapid changes. If these variations do indeed occur, then the following questions must be answered. (1) What are the precise time scales? (2) What is the nature of the variations? Are they periodic, recurring, or completely random flickering? (3) What is the amplitude or range of variations? And finally, (4) Are there any common characteristics in these, or any correlations with the known parameters of these stars?

For some time, a study has been underway to investigate these problems with regard to the total emission strengths of hydrogen lines in early-type stars.

2. Observations

Observations were made of four Be stars in December 1973 at Kitt Peak National Observatory, and of five Be stars in February 1974 at Cerro Tololo

* Visiting Astronomer, Kitt Peak National Observatory and Cerro Tololo Inter-American Observatory, which are operated by the Association of Universities for Research in Astronomy, Inc., under contract with the National Science Foundation.

A. Slettebak (ed.), Be and Shell Stars, 41–49. All Rights Reserved
Copyright © 1976 by the IAU.

Inter-American Observatory. In both cases, a two-channel low-resolution photo-electric spectrum scanner was used at the Cassegrain focus of the 91-cm telescope. The entrance apertures were 31″ in diameter separated by 4′.5 for Kitt Peak observations, and 18″ separated by 2′.6 for Cerro Tololo observations. The range of spectrum covered in each scan was λ6530–6600 Å in the first order. The band pass as defined by the width of exit slit was 4 Å for the Kitt Peak observations and 2 Å for the Cerro Tololo observations. The step size, the interval between two successive data points in the spectrum, was 2 Å.

The observing procedures were similar to those employed in an earlier study (Bahng, 1975). That is, each program star was observed continuously for about two hours to obtain a time series of spectrum scans. The Kitt Peak scanner could not be used in the automatic mode, so that data recording on a magnetic tape and the commencement of the next scanning operation had to be initiated manually. For this reason, each scan in the time series observations for the northern Be stars could not be made at a precisely uniform time interval. A number of normal main sequence B stars were also observed as comparison stars.

TABLE I

Variations of Hα equivalent widths – Northern stars

Star	Date (UT) 1973	Number of scans	$\bar{W}$ (Å)	σ_o (Å)	σ_p (Å)	$q = \sigma_o/\sigma_p$	$\bar{q}$
1 Per	Dec. 15	3	4.38	0.21	0.14	1.50	
B2 V	Dec. 16	3	4.48	0.37	0.20	1.85	
	Dec. 17	5	4.33	0.17	0.18	0.94	
							1.43
121 Tau	Dec. 14	3	5.14	0.26	0.20	1.30	
B3 V	Dec. 15	15	5.02	0.24	0.14	1.71	
	Dec. 16	7	4.94	0.34	0.20	1.70	
							1.57
48 Per	Dec. 14	6	−20.11	0.58	0.24	2.42	
B3 Ve	Dec. 16	6	−20.21	0.55	0.20	2.75	
	Dec. 17	13	−20.17	0.52	0.20	2.60	
							2.59
ζ Tau	Dec. 14	97	−19.90	0.67	0.25	2.68	
B2 IVp	Dec. 15	80	−20.44	0.66	0.17	3.88	
	Dec. 16	39	−23.01	1.04	0.30	3.47	
	Dec. 17	105	−23.21	1.01	0.28	3.61	
							3.41
ν Gem	Dec. 16	16	−2.49	0.30	0.19	1.58	
B7 IVe	Dec. 17	14	−2.29	0.21	0.19	1.11	1.35
κ Dra	Dec. 15	35	−13.13	0.28	0.16	1.75	
B7 p	Dec. 16	25	−12.89	0.39	0.17	2.29	
							2.02

TABLE II

Variations of Hα equivalent widths – Southern stars

Star	Date (UT) 1974	Number of scans	$\Delta t^a_{(Sec)}$	$\bar{W}$ (Å)	σ_o (Å)	σ_p (Å)	$q = \sigma_o/\sigma_p$	$\bar{q}$
χ Car B2 IV	b	b	b	3.01	0.18	0.16	1.13	
σ Cen B2 V	b	b	b	3.62	0.14	0.15	0.93	
ε Cen B1 V	Feb. 7	52	67.5	2.95	0.20	0.16	1.25	1.10[c]
α Col B8 Ve	Feb. 7	99	100.5	−8.57	0.47	0.17	2.77	
	Feb. 9	68	100.5	−8.48	0.60	0.18	3.33	3.05
κ CMa B2 Ve	Feb. 8	32	337.3	−17.66	0.39	0.19	2.05	2.05
ω Car B7 IV	Feb. 7	65	202.4	−5.20	0.26	0.16	1.63	
	Feb. 9	66	202.4	−5.17	0.25	0.16	1.56	1.60
PP Car B5 Ve	Feb. 8	75	202.4	−30.24	0.61	0.22	2.77	
	Feb. 11	36	202.4	−29.12	0.74	0.20	3.70	3.24
δ Cen B2?V?pe	Feb. 11	76	67.5	−32.64	1.02	0.25	4.08	
	Feb. 11	45	67.5	−32.75	0.75	0.25	3.00	3.54

[a] Time interval between successive scans.
[b] Standard stars, 2 scans each night.
[c] Mean from 3 standard stars.

3. Results

Since the spectral resolution is too low, the line-profile study cannot be made. Nor is it possible to study the V- and R-component separately. Instead, only the equivalent widths of Hα, W, were obtained from each scan. For each time series, the mean equivalent width, $\bar{W}$, and a quantity q, defined by $q = \sigma_o/\sigma_p$, were computed. Here σ_o is the usual rms value of observed W from $\bar{W}$, and σ_p is the expected rms value computed on the assumption that the scatter is due to the photon-counting statistics alone. Comparison of the q values for Be stars with those for normal B stars will serve as a basis for deciding whether the observed variations in W are significant.

The results are summarized in Tables I and II. The individual time series are shown in Figures 1 through 5. In these figures, the larger error bars give $\pm\sigma_o$, and the smaller error bars $\pm\sigma_p$ from the mean. For the southern stars, power spectrum analyses were performed. The mean power spectra weighted according to the number of scans in each time series, are shown along with the corresponding time series. The solid lines are power spectra computed from the observed data, while the crosses indicate the power spectra when a sinusoidal trace signal with a semi-amplitude of 0.5 Å was superimposed on the observed data.

J. D. R. BAHNG

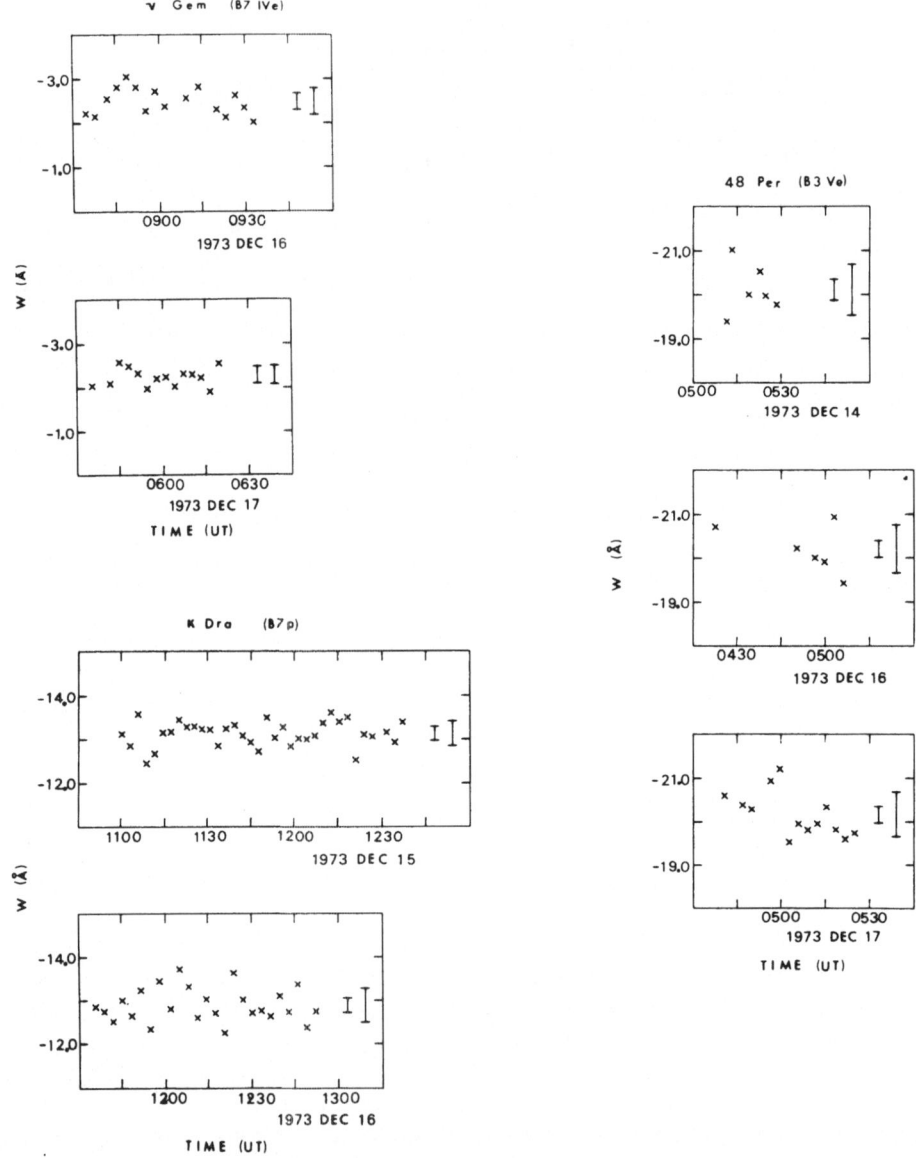

Fig. 1.   Hα equivalent width as a function of time for ν Gem, κ Dra, and 48 Per. The error bars are
observed rms (larger) and the computed photon rms (smaller).

4. Discussion

In order to assess the significance of variations in W, the values of q for each Be star
are compared with the q for the comparison stars. For these stars, the values range
from 0.93 to 1.85 with the mean of 1.5 for the northern, and 1.1 for the southern
stars. We shall adopt an arbitrary criterion that if the value of q is larger than those

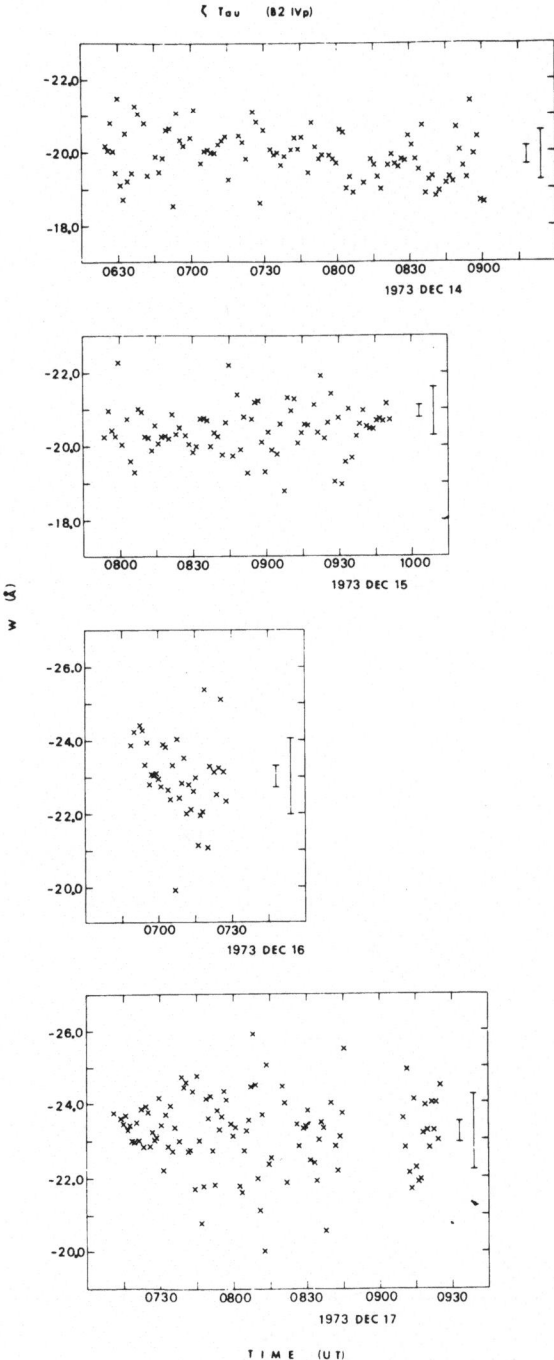

Fig. 2.   Hα equivalent width as a function of time for ζ Tau. The error bars are observed rms (larger) and
the computed photon rms (smaller).

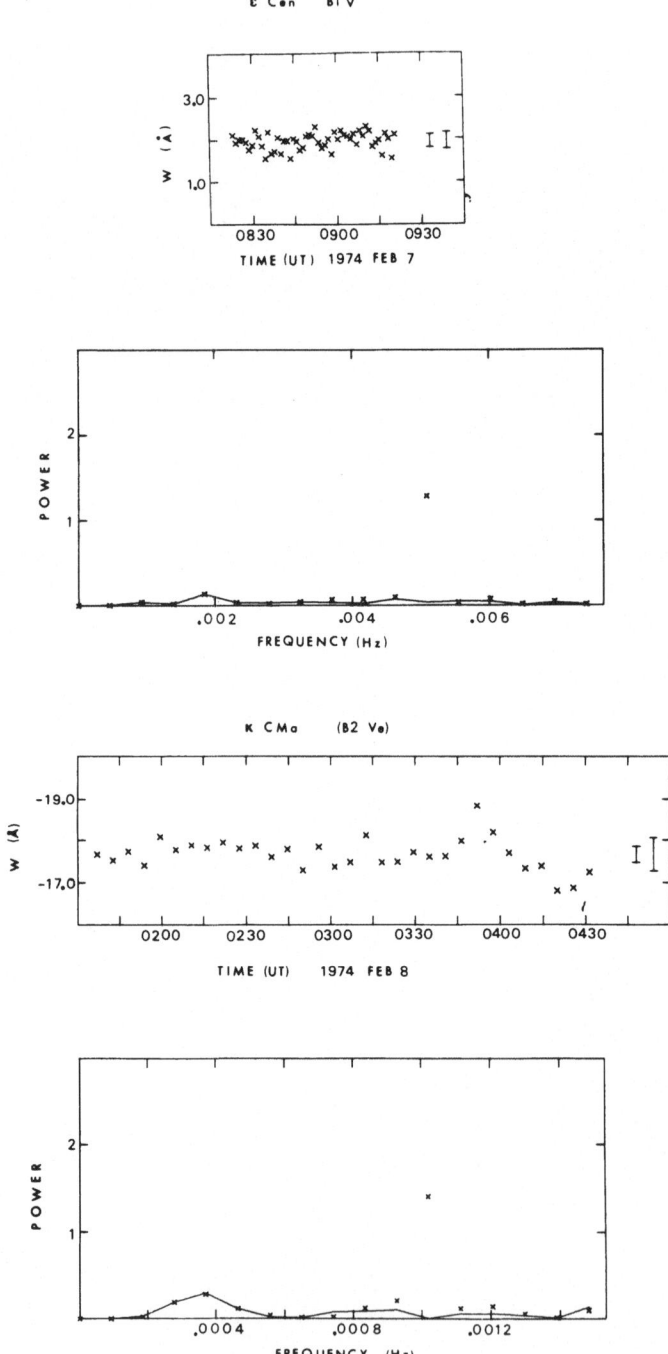

Fig. 3.   Hα equivalent width as a function of time, and the power spectra for ε Cen and κ CMa. The error
bars are observed rms (larger) and the computed photon rms (smaller). In the power spectra, the
crosses indicate power spectra with a pure sinusoidal variation of 0.5 Å semi-amplitude superimposed
on observed data.

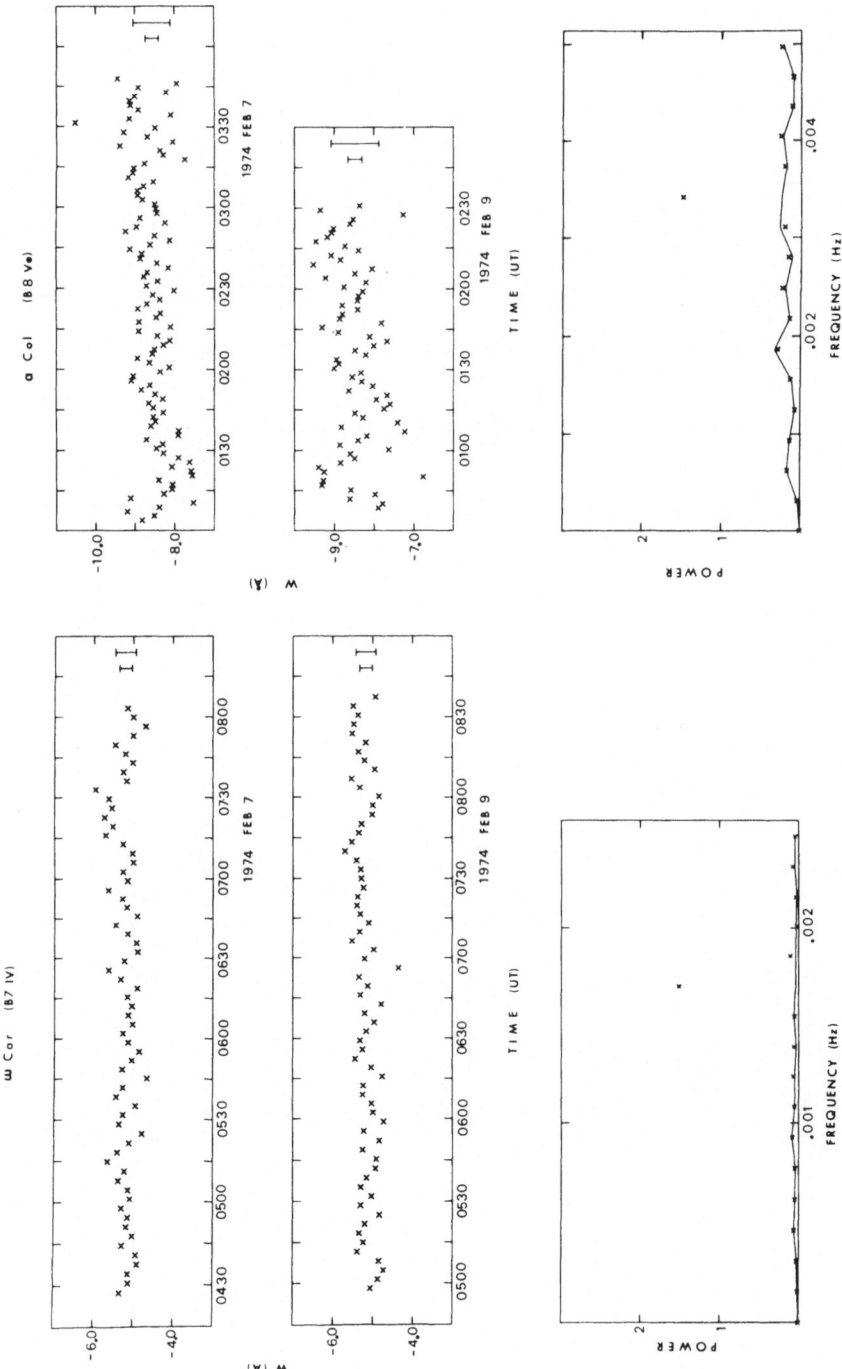

Fig. 4. Hα equivalent width as a function of time, and the power spectra for ω Car and α Col. See the caption for Figure 3 for explanation.

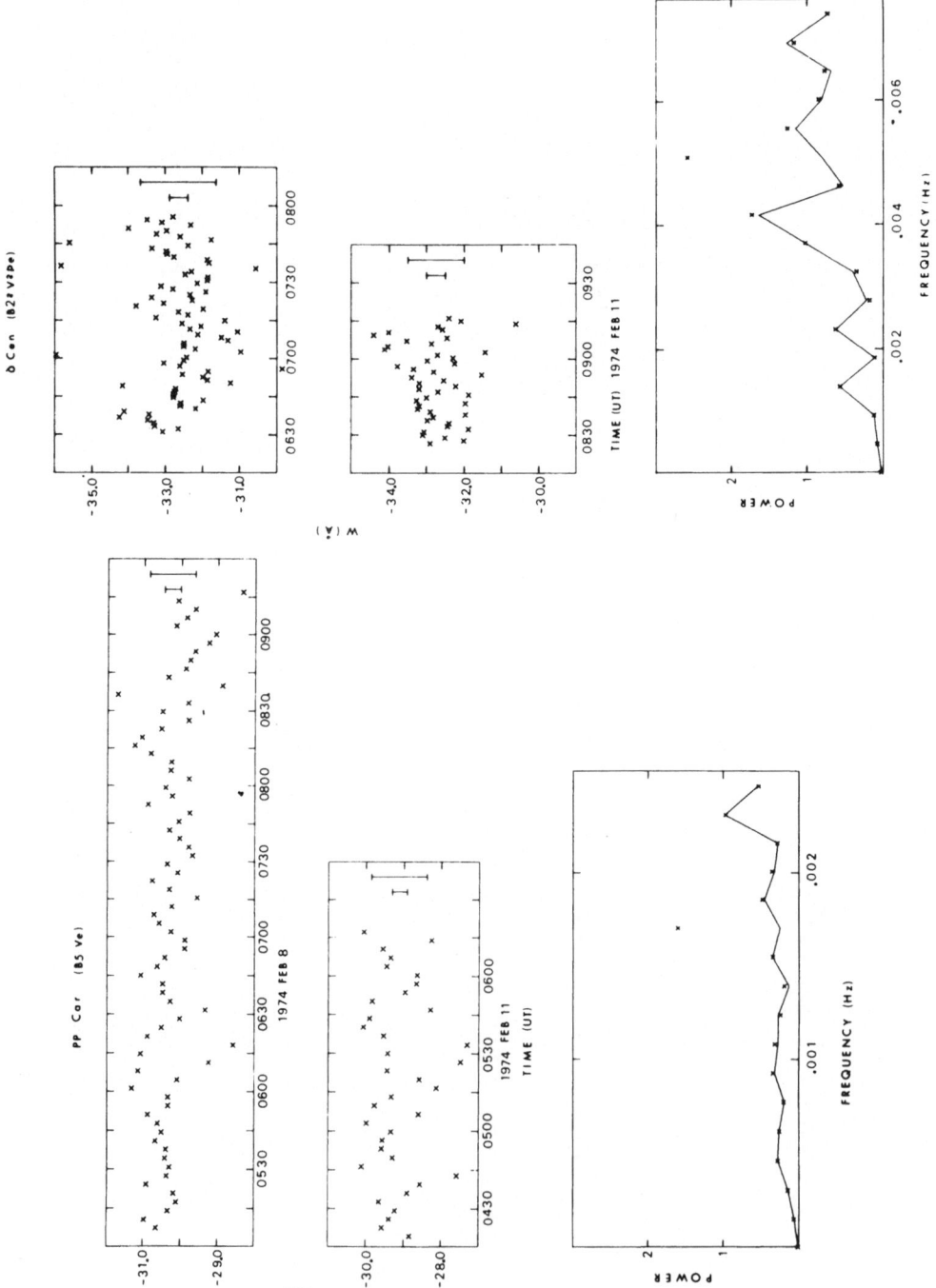

for the standard stars by more than a factor 2, the variations are significant. On this basis, ζ Tau, α Col, PP Car, and δ Cen are judged to show significant variations in the emission strength of Hα. In addition, for ζ Tau, there is a steady increase in the emission strength over four nights, from $W = -19.9$ to -23.2 Å. The amplitudes of variations are a few percent (rms), with the shortest time scale being the time interval between two successive scans, on the order of 1–3 min. The mean power spectra show that there is no periodicity in any of these variations. In the remaining Be stars in these studies, the variations are either absent (ν Gem and ω Car), or marginal (48 Per, κ Dra, and κ CMa).

The amount of variations found here is rather small, which makes it difficult to establish the reality of these rapid changes with absolute certainty. However, accumulated evidences indicate that there are short-term variations in the Balmer line emission strengths in some Be stars. The data are not sufficient to draw even preliminary conclusions regarding any characteristics which may be ascribed to those stars showing the variations. From the time scales involved, and the fact that the amplitudes of variations are small, it is evident that the variations are localized phenomena. Further observations of many more objects are required to answer the questions raised at the beginning of this paper.

References

Bahng, J. D. R.: 1971, *Astrophys. J.* **167**, L75.
Bahng, J. D. R.: 1975, *Monthly Notices Roy. Astron. Soc.* **170**, 611.
Delplace, A. M.: 1970, *Astron. Astrophys.* **7**, 459.
Huang, S.-S.: 1972, *Astrophys. J.* **171**, 549.
Huang, S.-S.: 1973, *Astrophys. J.* **183**, 541.
Hutchings, J. B., Auman, J. R., Gower, A. C., and Walker, G. A. H.: 1971, *Astrophys. J.* **170**, L73.
McBeath, K. B.: 1974, 'Rapid Variations of Balmer Line Strengths in the Spectra of Be Stars', Northwestern University (Ph.D. Thesis).
Marlborough, J. M. and Zamir, M.: 1975, *Astrophys. J.* **195**, 145.
Mihalas, D.: 1974, *Astron. J.* **79**, 1111.
Prendergast, K. H. and Spiegel, E. A.: 1973, *Comments Astrophys. Space Phys. (Part C)* **5**, 43.
Sanyal, A.: 1974, *Bull. Am. Astron. Soc.* **6**, 460.

STELLAR PHOTOMETRIC OBSERVATIONS AT Hα
THROUGH A NARROW-BAND INTERFERENCE FILTER

HUGH E. BUTLER

Royal Observatory, Edinburgh, United Kingdom

Abstract. The paper describes the results of stellar photoelectric observations made with a small telescope and a $3\frac{1}{2}$ Å wide filter centred at Hα.

The method proved to be a sensitive one for detecting change in the spectrum but did not allow differentiation between changes in the Hα profile and wavelength changes due only to the Doppler effect. It was possible, however, to recognise and qualitatively to eliminate the Doppler changes due to the Earth's Annual Motion.

Of ten Be stars examined three, although in strong emission, showed little change over the periods of observation, the longest of which was 9 months. The other seven showed large and random changes.

Warnings are given about the difficulty of tuning such a filter exactly to Hα and the need to monitor accurately the filter for long term changes in passband and throughput.

1. Introduction

In general, this was an exploratory programme designed to study stellar Hα variations using a 3.5 Å wide filter, a photoelectric photometer and a small telescope: observations were made in Edinburgh between August 1972 and September 1973.

In particular, the work covered:

(a) assessing the particular instrumental set-up;

(b) determining whether rotation periods could be detected in late type stars on the assumption that Hα emission from limited areas of a star's surface would modulate the observed intensity of the Hα line in phase with the rotation period;

(c) studying variation characteristics of Be stars at Hα;

(d) studying a small number of regular variables at Hα.

This short paper is mainly concerned with (a) and (c); reference will be made to (b) and (d) but it is intended to describe these in more detail elsewhere.

2. The Observing Equipment

A 12-in. telescope was built especially for this work. It was a short Cassegrain instrument with a final beam of focal ratio of 15. The photometer used a 2' aperture, a filter slide thermostatted to about ±1 °C, a cooled EMI photomultiplier with an S-20 cathode, photon counting electronics and print out. Emphasis was laid on ease of operation for one observer for long periods in complete darkness.

Two filters only were used:

(a) a Baird interference filter, 3.5 Å halfwidth at half height;

(b) a similar filter 60 Å wide, always used in conjunction with a 10% transmitting neutral density filter.

A single parameter was derived from the observations, this being the ratio (R) of starlight through the two filters, due allowance having been made for dark and sky

A. Slettebak (ed.), Be and Shell Stars, 51–58. All Rights Reserved
Copyright © 1976 by the IAU.

counts. The purpose of the neutral density filter was to make the counts through the two filters of the same order of magnitude so as completely to avoid any possible trouble due to non-linearity of counting. In practice the values of R ranged from 0.67 (for strong emission) to 2.70 (for broad and strong absorption).

It will be realised that while accurate centring of the passband of the broad filter is unimportant, the reverse is true for the narrow filter. This latter was set by tilt and temperature to be as close as possible to 6562.8 Å. Guaranteeing close correspondence proved difficult with the equipment to hand, but it was felt at the time that this correspondence was better than 0.5 Å, a value which was considered acceptable compared with a filter half width of 3.5 Å. Once the filter was installed and set in the equipment every effort was made to keep it undisturbed. In fact, except for some experiments involving changes in tilt which the equipment allows, the filter was not moved, cleaned or even examined for 11 months. Very soon after, however, a change became apparent and the series had to be abandoned.

3. The Observations

The observing list included 43 stars of all spectral types and all brighter than magnitude 5.0.

Conventional comparison stars were not used but two circumpolar stars (see Table I) were chosen for intensive observation with a view to monitoring any changes in the equipment and to give long runs for statistical purposes.

TABLE I

Circumpolar comparison stars

Star	Spectral type	Mag.	Dec.	Ecliptic lat.	Nights observed
HR3751	K3 III	4.44	81°	60°	80
HR7750 (κ Cep)	B9 III	4.38	78°	75°	58

Figure 1 shows observed values of R for the 43 stars, the individual ranges of variation being shown as vertical bars. The period over which the observations for any star extended varied considerably depending, as usual, on declination and the relation between right ascension and the sidereal time of the short summer nights. Cases of very few observations are noted in the figure.

Figure 2 plots, for four stars (including the two detailed above), the observed ratios R against date. The smooth curves are drawn by eye.

Data are plotted for ten known or suspected e-type stars in Figures 3 and 4, the curves, again, being drawn by eye.

In addition, various night-long runs were made on stars of known, short period variability. Figure 5 shows the results of an 8 hour session on β Cephei.

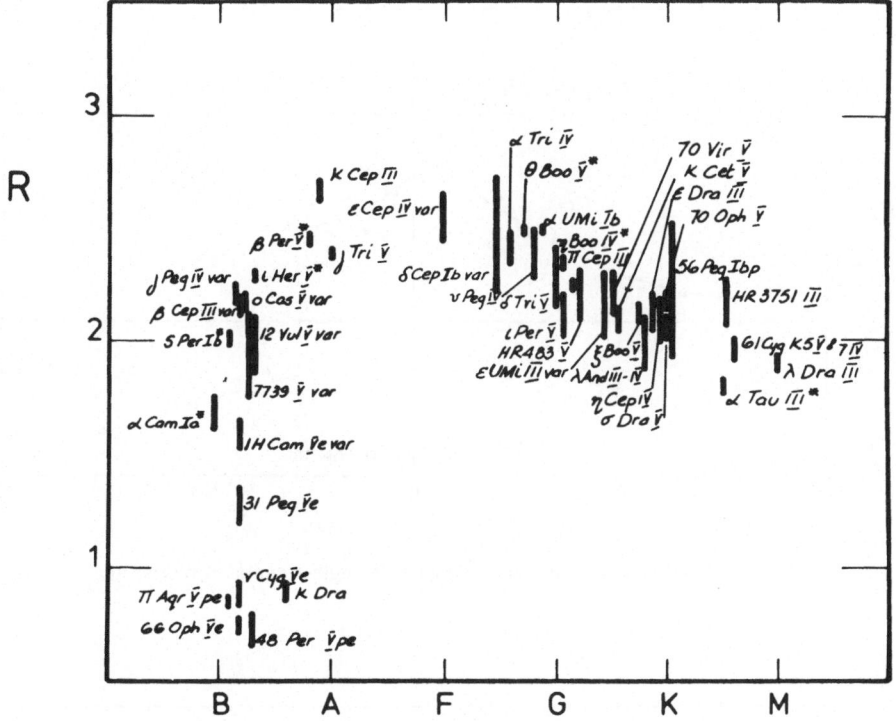

Fig. 1.   Ranges of *R* for all the observed stars plotted against spectral type.

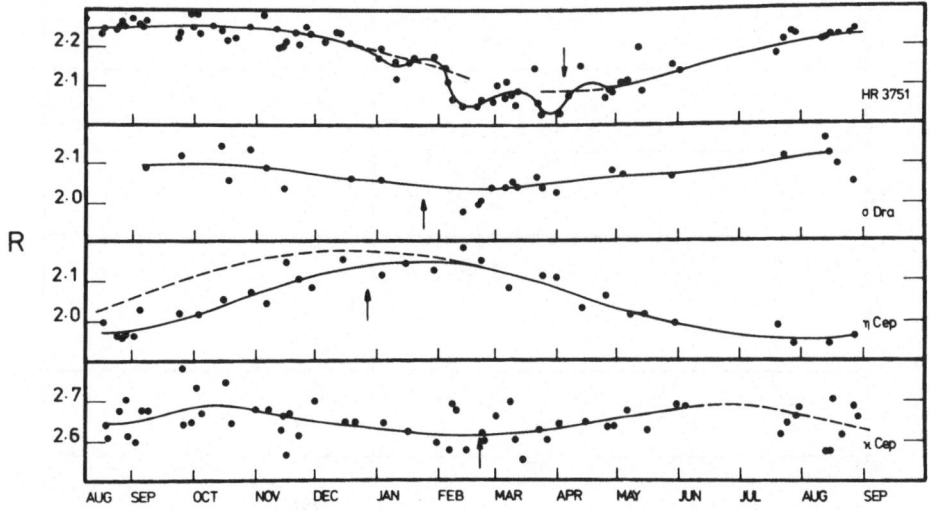

Fig. 2.   Values of *R* for the four stars HR 3751, σ Dra, η Cep, and κ Cep plotted against date.

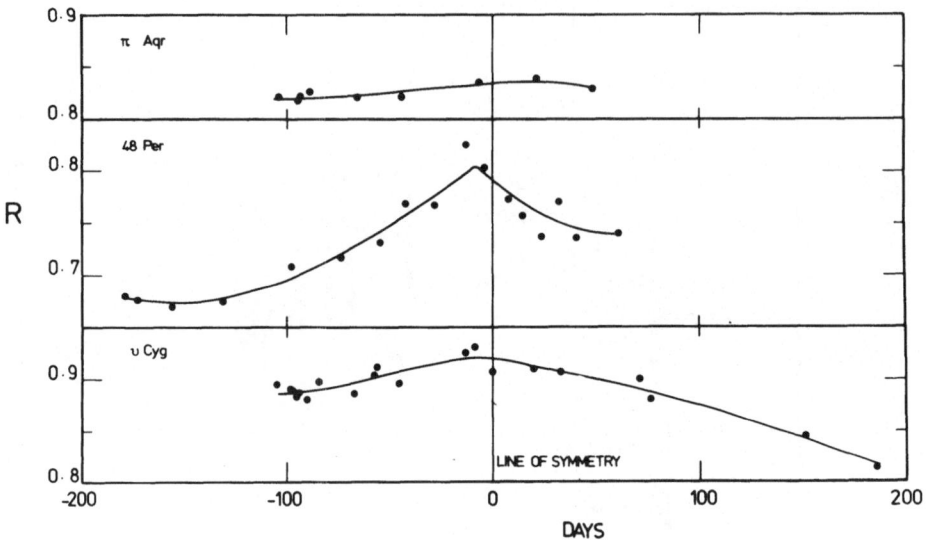

Fig. 3   Values of R for three Be stars having Hα strongly in emission, plotted against time.

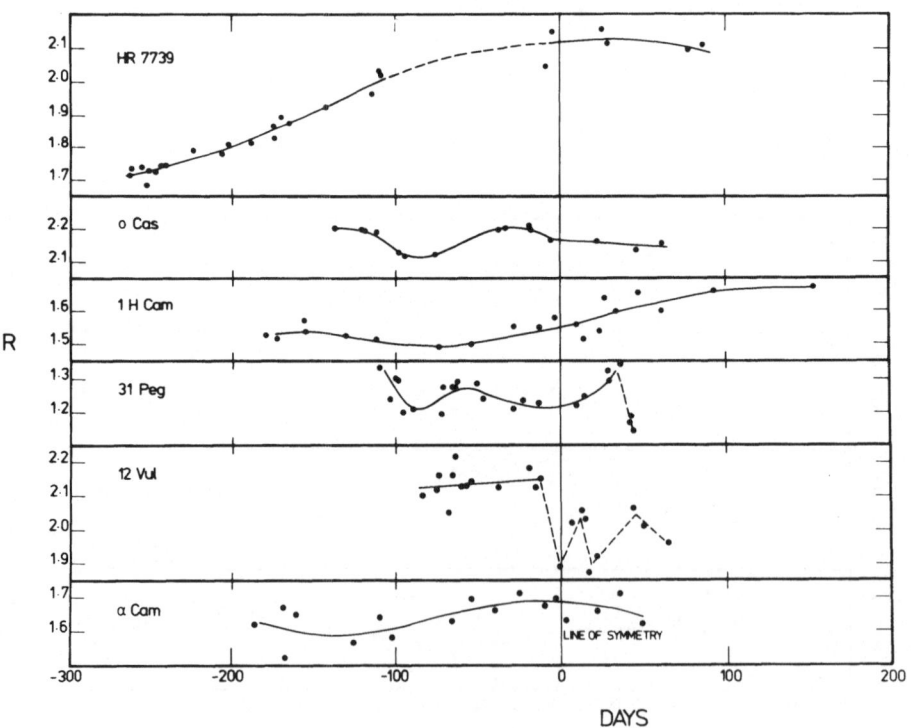

Fig. 4.   Values of R for six Be stars plotted against time.

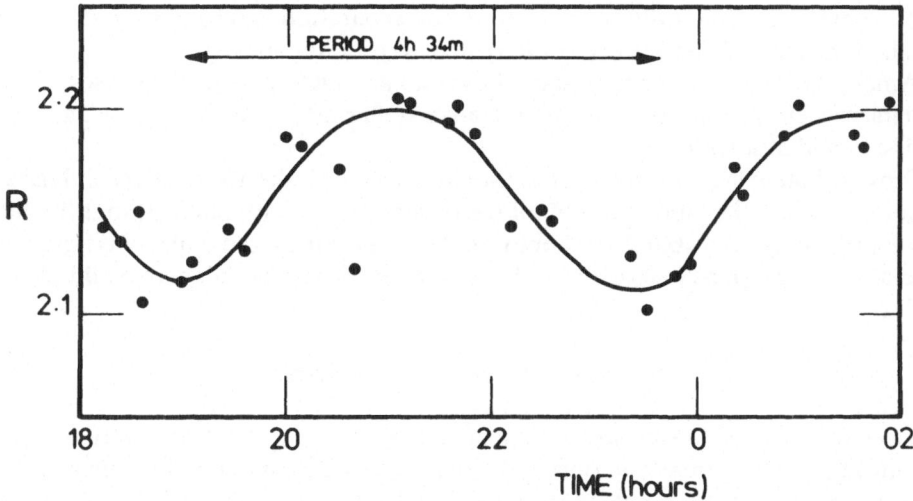

Fig. 5.   Values of R for β Cep plotted during eight hours of the night 13–14 November 1972.

4. Effect of Doppler Shift, Particularly that Due to the Earth's Annual Motion

A narrow-band filter, if set near to the centre of an emission or absorption line, is evidently a sensitive device for the detection of any change in wavelength due, for example, to the Doppler effect.

There is good, but not certain, reason to believe that the Doppler effect consequent upon the pulsation of β Cep causes the curve of Figure 5.

Another such variation will come from the Earth's Annual Motion around the Sun which produces an apparent sinusoidal change in each star's radial velocity of amplitude approximately 30 cos (ecliptic latitude) km s^{-1}, equivalent to 0.66 cos (ecliptic latitude) Å at Hα, with a period of one year. This brings about relative movement of stellar spectrum and filter passband which is perhaps best seen as the filter profile scanning across the Hα profile. Assume that the Hα line is symmetrical and in absorption. Consider first the case when the filter passband is exactly centred on the laboratory value of the wavelength of Hα (6562.8 Å) and the star has zero radial velocity. Starting from the moment of maximum red shift of the spectrum, the filter profile starts from rest, crosses the centre of the Hα line and stops again after 6 months: it then returns across it during the next 6 months. This will lead to a plot of R with time showing two equally spaced humps and two troughs – i.e. to a period of 6 months.

If, on the other hand, the filter passband is appreciably offset from 6562.8 Å, it will not cross the centre of the Hα profile at all during one year, only encroaching into one wing of Hα, and so the observed variation in r will show only one hump and one trough in a year, i.e. a period of 12 months.

If we consider cases with the offset decreasing toward zero, the peak of the 12 monthly curve first becomes double, then the separation of the twin peaks increases to the point that, when the offset is zero, the separation is 6 months and we have reached the case of 6 month periodicity first considered above.

Among the stars that were observed over an adequate period were cases of wide and narrow Hα profile, positive and negative stellar radial velocity, as well as Hα in emission and absorption.

Provided there are no large intrinsic stellar variations, the above effect is clear and rational. A careful examination of all the observations shows, in fact, that the filter passband was offset by 0.6 Å, rather more than was anticipated. Once this figure was available, crude measures of the radial velocities of individual stars could also be deduced.

5. Intrinsic Stellar Variations

The reduced data of 14 stars are to be seen in Figures 2, 3 and 4. Each plotted point is the mean of all data observed during the night – usually 4 independent measures. In Figure 2, the symmetry points of the variation due to the Earth's Annual Motion are marked with arrows at the appropriate dates. In Figures 3 and 4, the curves have been moved sideways (and the actual dates thereby lost) so that symmetry of all will be about the same vertical line marked 'line of symmetry'.

Figure 2 shows the data for four stars with Hα in absorption – three of spectral type K and one of type B. They include HR3751 and κ Cep, the two circumpolar stars chosen for intensive observation. κ Cep shows no indication of intrinsic variation but its curve is of interest because it shows one of the intermediate cases due to the Earth's motion, mentioned earlier. The main minimum is in late February and there are two maxima, one clearly evident in the first week of October and the second, far less so, in June.

HR3751 shows an unexceptionable 12 month variation but, incidentally, does just seem to show 3 periods of about 40 days which could be the hoped for evidence of rotation mentioned in (b) in the Introduction.

σ Dra has a high ecliptic latitude (81°) and little effect of the Earth's motion is to be expected, nor is seen. There is slight evidence for intrinsic changes in Hα. The fourth star, η Cep, appears at first sight to show the correct response to the Earth's motion and to exhibit no stellar variation. However, the maximum does not occur at the correct time (shown by the arrow). To bring the curve to a position such as the dashed line, requires some stellar change and/or considerable asymmetry in the Hα profile.

Figure 3 shows three stars with Hα in emission. The curves have been moved sideways so that one or other of the symmetry points for the Earth's motion coincide in time with the vertical line. It can be seen that the general shape can be explained by Earth's motion. For υ Cyg, only the last two points show any divergence from a perfectly smooth curve, and for those the observations were made when the star was low in a bright sky. The individual points of 48 Per do show rather more variation about the smooth curve than is to be expected. The curve of the third star, π Aqr has the general shape of the other two stars.

Of the stars of Figure 4 (presented in the same way as those of Figure 3), three are classed as e-type, two others have high values of υ sin i, and one is of spectral type O9.5Ia, and all show some degree of emission.

It will be seen that except for HR7739, intrinsic stellar variations swamp any changes due to the Earth's motion. Judged by the degree of internal consistency of the individual readings (during one night, and from night to night and also from star to star) there is real variation with a time scale of the order of one day for 1HCam between +10 and +40 days, 12 Vul between 0 and +50 days, and for α Cam throughout the period of the observations. No deductions can be made about shorter period variability.

6. The Practical Use of a Narrow-Band Interference Filter

Heating a filter of this sort increases the passband wavelength, tilting decreases it and, as a corollary of the latter, decreasing the f-number of the beam through the filter likewise decreases the passband wavelength and somewhat alters its shape.

It is usual to specify a filter having a central value positioned a small amount longwards of the desired wavelength. The amount can, to a certain extent, be calculated, but it is essential to confirm, in practice, that the final setting is correct. This needs some, or all, of the adjuncts of a photometric laboratory.

In the present instance, two methods were used:

(a) the thermostatted filter was placed in front of the slit of a solar spectrograph. The photographed spectra at different filter tilts confirmed that the chosen tilt gave the correct passband;

(b) the complete photometer was fed with light from a low pressure hydrogen lamp, and readings were taken for different tilts.

As stated, it was felt that the tilt setting gave the correct passband to an accuracy better than 0.5 Å whereas the stellar observations showed it to have an error of 0.6 Å. Halfway though the observing season the filter was checked and did not seem to have changed in any way. The work had begun in August 1972. By August 1973, readings on a number of stars which did not seem to show any intrinsic variation, were systematically different from what they had been a year earlier. The difference was small and considered acceptable. However, during September and October this difference rapidly increased and in November it was assumed that one of the optical components was changing and the observations were discontinued.

On repeating the checks there was no convincing evidence that the passband had changed although there did appear to be more variation across the narrow-band filter (corresponding to slight visible patterning) than had previously been noticed. The cause could have been due to a change in either of the filters (in shape of passband or in general loss of transparency) or to a change in the grey filter; it was not possible to determine which.

7. Conclusions

(1) A 12-in. telescope, a 3.5 Å Hα filter and a photon counting photometer are together satisfactory for monitoring variations in stars brighter than the fifth magnitude: such as instrument can be dedicated to one programme for a long period.

(2) It is important to have access to a photometric laboratory for setting the filter

equally important to check both filter passbands and their through-put accurately at regular intervals.

(3) The method is a sensitive one for detecting *change* in the radiation passing through the Hα filter but it does not allow one to differentiate between changes in the Hα profile and wavelength changes in the spectrum due to the Doppler shift.

(4) Of the Be stars observed, three, although in strong emission, showed negligible variation while six others showed appreciable changes with a timescale of the order of one day.

(5) The instrument is particularly suitable for observing short period changes: as no comparison star is used, very efficient use can be made of the observing time. It is, in fact, intended to search with it for rapid changes (1 min to 1 h) in a few Be stars.

DISCUSSION

Bidelman: Wouldn't it be worthwhile to use a somewhat wider filter? In many stars the emission is appreciably wider than 3.5 ångström.

Butler: I chose the width to deal with late types, solar types, where it very roughly corresponds to the pass band of the profile of Hα in the sun. For the emission lines, of course, it is rather nice to have a narrow filter. Also, the wider the filter, the more the noise creeps in.

McLean: At the University of Glasgow, Scotland, we have developed two instruments employing narrow band Hα and Hβ interference filters, viz., a line profile scanning photometer and a line profile scanning polarimeter. Line profile scans are obtained by controlled tilting of the interference filters in collimated light beams. Excellent stability and accuracy in recording line profiles is obtained with the photometer (developed mainly by T. H. A. Wyllie), in which the filters are housed in a thermostatted enclosure. This instrument also employs continuous guidance on the star image and a fixed monitor channel which is ratioed with the scanning channel. Small and rapid changes in intensity and/or wavelength shifts are easily identified. Only for γ Cas has variability been observed at Hβ on a time scale of minutes (Clarke, D., McLean, I. S., and Wyllie, T. H. A.: 1975, *Astron. Astrophys.*, in press).

Harmanec: It would be useful to make Hα and Hβ measurements simultaneously.

Butler: I agree. I had toyed with the idea of using a dichroic filter for this purpose.

SPECTROSCOPIC OBSERVATIONS OF Be STARS
IN THE NEAR INFRARED

R. S. POLIDAN and G. J. PETERS

Dept. of Astronomy, University of California, Los Angeles, Calif., U.S.A.

Abstract. The results of near infrared spectroscopic observations of over 120 Be stars are presented. In particular, the O I lines at λ 7774 and λ 8446 Å and the Ca II triplet (λ 8498 Å, λ 8542 Å, and λ 8662 Å) are discussed. The presence of emission at O I λ 7774 Å is shown to correlate with Fe II emission. Shell absorption at this line is found to be similar in appearance to that found in the photospheric spectra of supergiants. The O I λ 8446 Å line, the result of a fluorescence with Lyman β, is briefly discussed in connection with Be envelope dimensions. The paradox presented by the presence of strong Ca II triplet emission without H and K emission is discussed. The implications of this phenomenon on the structure of Be envelopes and possibly the binary nature of some Be stars is discussed.

Evidence indicating the presence of cool companions to the Be stars HD 218393 and HR 894 is presented.

1. Introduction

In 1972 we, in collaboration with Dr M. Plavec, initiated a program to study the red and near infrared spectra of Be stars, shell stars, and interacting binary stars. Earlier investigations of Be stars in this region by Hiltner (1947), Slettebak (1951), and Andrillat and Houziaux (1967, 1972) revealed that a systematic, quantitative, study of emission line stars in this region could yield important information on the structure of their envelopes and possibly on the relationship between Be stars and interacting binary stars.

2. Observations

The observations were made using the Lick Observatory 40 mm cooled Varo image intensifier in conjunction with the 24-in. (61 cm) Coudé Auxiliary Telescope. A substantial number of observations were also made with the 120-in. (305 cm) telescope through the cooperation of Dr M. Plavec. Over 500 near infrared plates of over 120 Be stars have been obtained during the three years of observation. The spectrograms in the region λλ 7500–8200 Å were taken at dispersions of 34 Å mm^{-1} (85% of the total) and 12 Å mm^{-1}. In the region of the higher order Paschen lines (λλ 8200–8700 Å), ninety percent of the stars were observed at 23 Å mm^{-1} and 10 percent at 34 Å mm^{-1}. All spectra were widened to 0.6 mm at the plate. The characteristics of the Lick Observatory Varo tube have been described by Zappala (1971). The spatial resolution of the tube is comparable with that of a I-N plate. However, the Varo tube's high degree of freedom from small scale irregularities combined with the high contrast and low noise of III-aJ plates produce a spectrum of considerably higher quality than is possible from I-N plates.

A. Slettebak (ed.), Be and Shell Stars, 59–65. All Rights Reserved

3. Data

The near infrared region ($\lambda\lambda$ 7500–8800 Å) of Be stars is dominated by the Paschen lines $P12$ and higher, the neutral oxygen lines at λ 7774 Å and λ 8446 Å, and by the infrared calcium triplet (λ 8498 Å, 8542 Å, and 8662 Å). The Fe II emission line at λ 7712 Å is seen in those stars exhibiting Fe II emission in the blue/violet spectral region. This region is devoid of metallic shell lines except for the N I triplet near λ 8680 Å. For the remainder of this talk we shall deal only with the results for the two oxygen lines and for the calcium triplet.

3.1. O I λ 7774 Å

The neutral oxygen triplet at λ 7774 Å (λ 7772, 4, and 5) arises from the transition between the metastable ^5S level and the ^5P level. As a consequence of the metastability of the lower level this feature has been shown to be a useful luminosity indicator in non-emission line stars (see, for example, Osmer, 1972). Our results for Be stars show that twenty percent display emission at O I λ 7774 Å and twenty percent have enhanced (shell) absorption. The remaining sixty percent show only the absorption lines expected from the underlying star's photosphere. Figure 1 shows

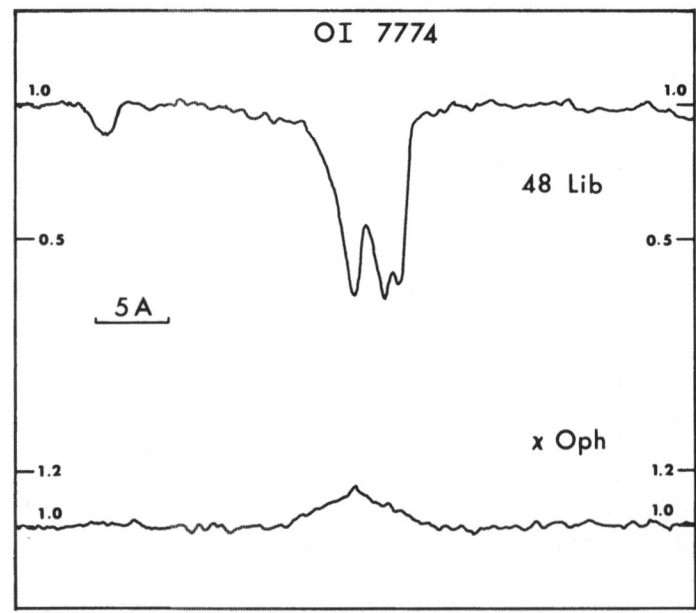

Fig. 1.   Intensity tracings of the O I λ 7774 Å region in the shell star 48 Lib and the Be star χ Oph. Ordinate is in intensity units normalized to the continuum. Original dispersion: 12 Å mm^{-1}.

tracings of the region around O I λ 7774 Å in the objects showing the strongest emission (χ Oph) and the strongest shell absorption (48 Lib).

All stars which show O I λ 7774 Å emission are observed to have Fe II emission. However, the reverse is not necessarily true. Objects do exist for which Fe II emission is seen without accompanying O I λ 7774 Å emission (e.g. β^1 Mon). Most O I emission objects also display strong Hα emission with little or no structure. A correlation was found to exist between the strength of the O I λ 7774 Å emission and the Fe II λ 7712 Å emission line (Figure 2). A correlation also exists between the width of the O I emission and the width of the Fe II feature. Both also correlate with $v \sin i$ of the underlying star.

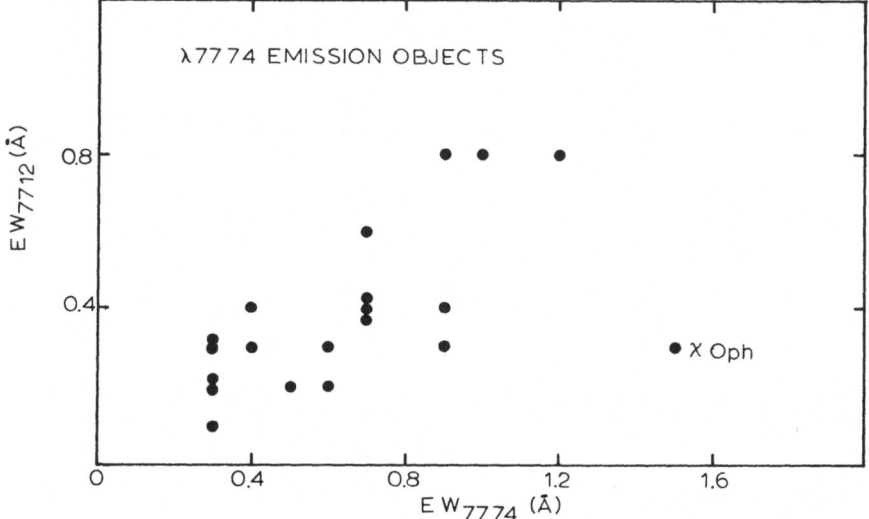

Fig. 2.   Observed correlation between the equivalent widths of Fe II λ 7712 Å and O I λ 7774 Å.

The tentative conclusion is that the region giving rise to the O I emission is the same as that giving rise to the Fe II emission.

Because of the strong metastability of the lower level the O I λ 7774 Å line is a useful indicator of a shell star. Stars exhibiting only a weak metallic shell in the blue show significant enhancement of this line (e.g. o And). In the case of stars with strong metallic shells (e.g. 48 Lib, Figure 1), the O I triplet is seen in absorption stronger than is observed in the most extreme supergiants. In general the appearance of the line is similar to what is seen in normal giants and supergiants. This suggests that in some cases the physical characteristics and structure of a shell star envelope may be quite similar to the atmosphere of a giant or supergiant star.

3.2 O I λ 8446 Å

Emission in this line was found in 75% of the Be stars surveyed. The explanation for the high percentage of emission was pointed out by Bowen (1947): the line is the result of a fluorescence with $L\beta$. Kitchin and Meadows (1970) investigated the expected correlation between the emission strengths of Hα and O I λ 8446 Å. Our results support their conclusion that the correlation does exist. However, we have found a few exceptions to the relation (e.g. 31 Peg).

Kitchin (1970) has used the O I λ 8446 Å line to obtain dimensions for the emitting regions around Be stars. His model is based on the assumption that the line is optically thin. Our observations show that in all but the sharpest lined stars ($v \sin i > 150$ km s^{-1}) a strong reversal is present at the line center. This is inconsistent with the assumption that the line is optically thin. Therefore, envelope dimensions based on an optically thin O I λ 8446 Å should be regarded with caution.

Figure 3 shows tracings of the Paschen region (λλ 8400–8575 Å) in the Be star χ Oph and the shell star 48 Lib.

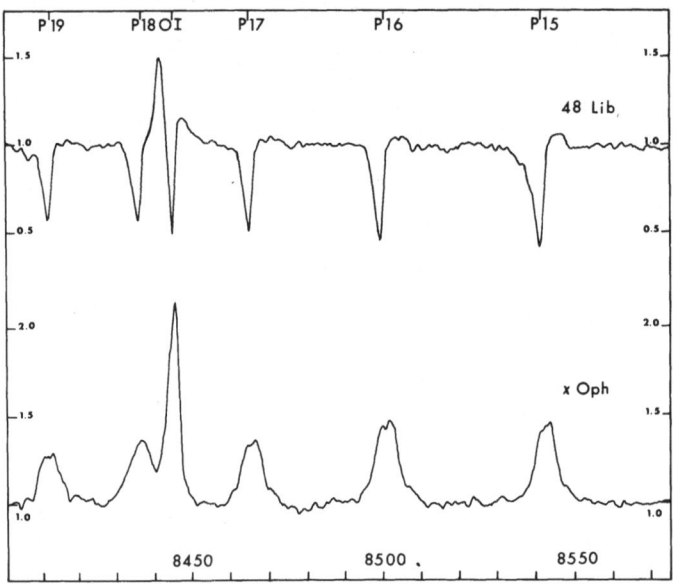

Fig. 3.   Spectra of the shell star 48 Lib and the Be star χ Oph in the region of the higher Paschen lines. The weak absorption distorting the blue wing of P15 in 48 Lib is Ca II λ 8542 Å. Ordinate is in units of intensity normalized to the continuum. Original dispersion: 23 Å mm^{-1}.

3.3. Ca II (λ 8498 Å, 8542 Å, and 8662 Å)

The calcium triplet was found in emission in less than 20% of the objects surveyed. This is a lower percentage than has been reported by the earlier investigators. The probable reason for this is illustrated in Figure 4. All three calcium lines are severely blended with Paschen lines: Ca II λ 8498, 8542, and 8662 Å are respectively 4.5, 3.3, and 3.9 Å blueward from P16, P15, and P13. In stars with strong Paschen emission or a strong Paschen shell, apparent weak Ca II emission is seen on lower dispersion plates (our 34 Å mm^{-1} plates and those of the early investigators). However, higher dispersion (23 Å mm^{-1}) plates taken at the same time do not confirm the emission; they indicate only Paschen emission.

The presence of Ca II triplet emission in Be stars presents an interesting paradox. The lines arise from the ^2P to ^2D transition. The upper level (^2P) is the same as for the H and K lines; the lower (^2D) level is strongly metastable. The triplet lines are very subordinate transitions to the H and K lines, however, in *no* case do we observe

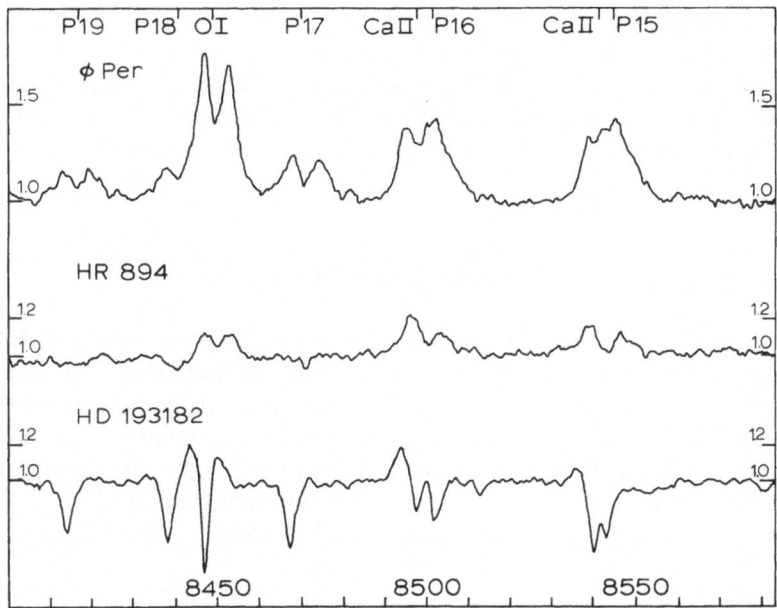

Fig. 4.   Intensity tracings of the region of the higher Paschen lines in the Be stars ϕ Per and HR 984 and the shell star HD 193182. Note the strong emission at the Ca II lines λ 8498 Å and λ 8542 Å. Ordinate and original dispersion same as in Figure 3.

significant H and K emission in a Be star showing the triplet in emission. The triplet lines themselves do not appear normal: all have the same intensities rather than the expected 1 : 9 : 5. This implies that the gas giving rise to the emission is optically thick to the triplet. This, combined with the greater flux at λ 3900 Å than λ 8500 Å, reduces the problem of the lack of H and K emission. It, however, does *not* eliminate it. One still requires a reduction of H and K emission with respect to the triplet of by a factor of at least 2 to 10. Calcium triplet emission also does not correlate with spectral type, emission envelope strength, the presence of dust or cool gas (as inferred from the far infrared flux), nor does it participate in variations seen in the rest of the envelope.

The best explanation for the presence of calcium triplet is to invoke the existence of dense cool ($T \sim 5000$ K) gas not associated with the visible envelope. The source of this gas is as yet unknown. However, the existence of visible cool ($T_e \sim 4$–5000 K) giant companions to five of the objects showing triplet emission plus the presence of binary-like periodic activity in most of the other triplet emission objects suggest some connection with binary nature. A possible explanation is that the Be stars showing Ca II triplet emission are actually semi-detached interacting binary stars. The source of the cool gas is the mass losing contact component. A full discussion of the Ca II emission in Be stars along with its binary implications will be presented later in this symposium.

The lower level of the infrared calcium triplet is a strongly metastable level. In stars with extended envelopes one would expect to see these lines as strong shell lines. This is not seen. Calcium triplet shell lines are seen only in the cooler shell stars (e.g.

14 Com, 1 Del) or the most extreme shell stars (e.g. 48 Lib). It is possible to understand this in terms of photoionization from the metastable level by the far ultraviolet photons ($\lambda \lesssim 1219$ Å). The metastable level has a large photoionization cross section, hence, no build up of atoms in this level is expected except in the stars with insufficient ultraviolet flux for photoionization or stars with very extensive envelopes.

4. Be Binary Stars

One of the primary purposes of this survey was to investigate the possibility that some Be stars are intereacting binary stars with cool companions. Two such systems are already known: 17 Lep and AX Mon. While it turns out that this region, λ 8300 to 8800 Å, is far from the optimum place to look for secondaries (this too will be discussed later in the Symposium), we would like to announce the discovery of two more Be binary stars. The well known periodic Be/shell star HD 218393 (Doazan and Peton, 1970) was found to contain an early K-type giant secondary (Figure 5).

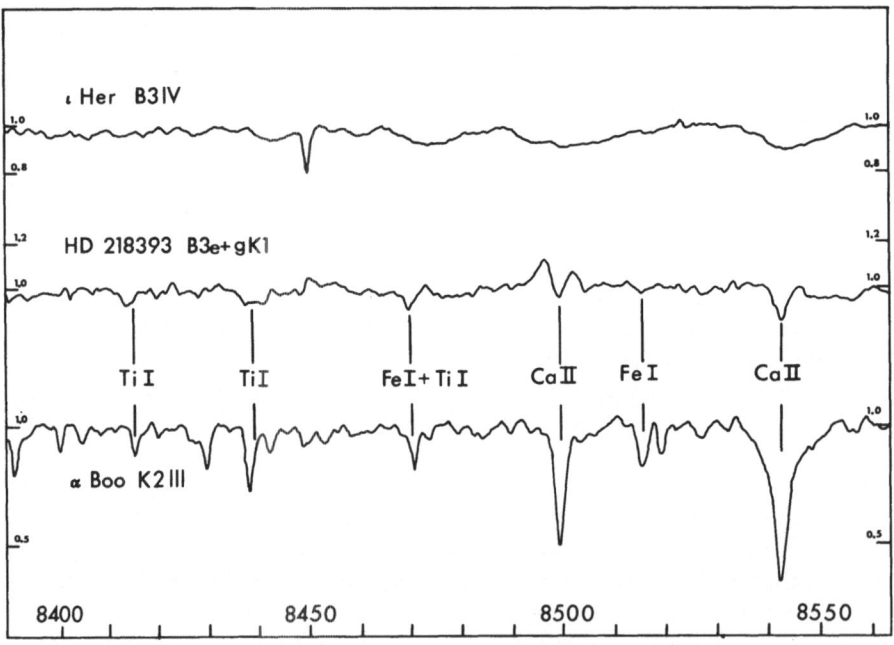

Fig. 5. The infrared spectrum of HD 218393 compared with that of ι Her (B3IV) and α Boo (K2III). The most prominent lines of the companion are marked. Ordinate and original dispersion same as for Figure 3.

The late B-type (B8V) emission star HR 894 (=HD 18552) was also found to contain a cool, in this case approximately G9, companion (Figure 6). Both these systems will be discussed in detail and compared to AX Mon, 17 Lep and other possible Be binary stars later in this Symposium.

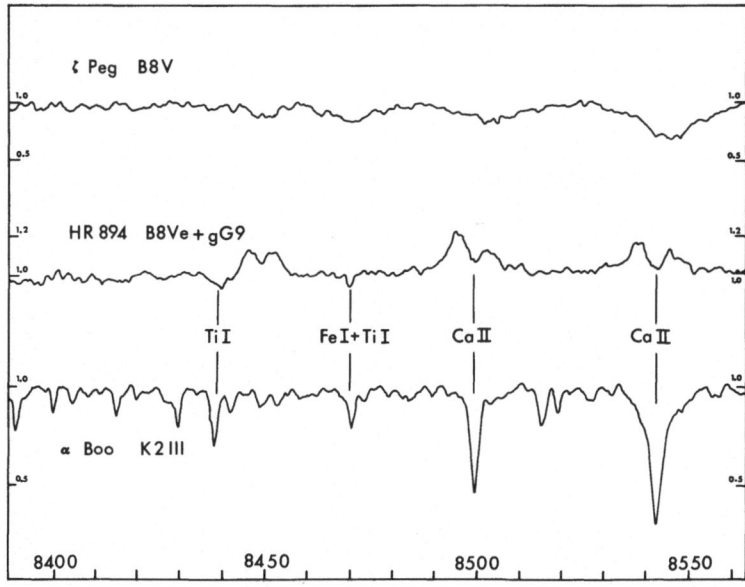

Fig. 6.   The infrared spectrum of HR 894 compared with that of ζ Peg (B8V) and α Boo (K2III). The most prominent lines of the companion are marked. Ordinate and original dispersion same as for Figure 3.

Acknowledgements

We wish to thank Dr M. Plavec for helpful discussion and for his continued support of the project. We also wish to thank Mr E. A. Harlan of Lick Observatory for assistance in obtaining some of the observations. This research was supported by NSF grant MPS 74-04194A01 (Popper/Plavec), being a part of a project studying close binary stars.

References

Andrillat, Y. and Houziaux, L.: 1967, *J. Obs.* **50**, 107.
Andrillat, Y. and Houziaux, L.: 1972, *Astrophys. Space Sci.* **15**, 240.
Bowen, I. S.: 1947, *Publ. Astron. Soc. Pacific* **59**, 196.
Doazan, V. and Peton, A.: 1970, *Astron. Astrophys.* **9**, 245.
Hiltner, W. A.: 1947, *Astrophys. J.* **105**, 212.
Kitchin, C. R.: 1970, *Monthly Notices Roy. Astron. Soc.* **150**, 455.
Kitchin, C. R. and Meadows, A. J.: 1970, *Astrophys. Space Sci.* **8**, 463.
Osmer, P. S.: 1972, *Astrophys. J. Suppl. Ser.* **24**, 247.
Slettebak, A.: 1951, *Astrophys. J.* **113**, 436.
Zappala, R. R.: 1971, 'Proc. Conf. Late Type Stars', *Contr. Kitt Peak National Obs.* No. 554, 1.

NEW OBSERVATIONAL DATA CONCERNING
4 HER AND ζ TAU

PETR HARMANEC, PAVEL KOUBSKÝ, JIŘÍ KRPATA,
and
FRANTIŠEK ŽĎÁRSKÝ

Astronomical Institute of the Czechoslovak Academy of Sciences, Ondřejov, Czechoslovakia

Abstract. Rectified intensity profiles of the Hα and Hβ Balmer lines of the shell star 4 Her were studied on 86 coudé spectrograms taken during 1969–1974. The central intensities of both lines vary periodically with the period of velocity changes equal to 46.194 days, reaching two maxima and minima each cycle. Radial-velocity curves of individual Balmer lines differ systematically one from another, the amplitude of the Hα variations being largest. The V/R ratio of the Hα emission peaks varies in phase with the velocity changes. The velocity of the Hα emission is found to be almost invariable. A model is suggested to explain the observed variations in which 4 Her is considered to be an interacting binary. The full paper will appear in *Bull. Astron. Inst. Czech.* **27**, No. 1 in 1976.

Radial velocities of high-dispersion spectrograms of ζ Tau, obtained mostly during the last observing season, indicate the following facts: (1) Several distinct velocity systems are present in the Balmer lines, not a continuous progression. (2) The main absorption component of the Hα line does not share the long-term velocity variations, its velocity being always rather close to the systemic velocity $+22$ km s^{-1}. (3) The pseudoperiod of the long-term velocity variations has changed and is probably decreasing.

A. Slettebak (ed.), Be and Shell Stars, 67. *All Rights Reserved*

THE SURFACE GRAVITIES OF Be STARS

GERALDINE J. PETERS

Dept. of Astronomy, University of California, Los Angeles, Calif., U.S.A.

Abstract. In an attempt to shed some light on the origin of the material in the envelopes of Be stars, surface gravities were determined for 30 objects by comparing their observed profiles of Hγ and Hδ with those computed from the Princeton model atmospheres and the VCS theory of hydrogen line broadening. The program stars are predominately well-known Be stars and display a wide range of envelope spectra and $v \sin i$. The mean and range in log g for the Be stars appear to be identical to that obtained from a similar analysis on non-Be stars. No correlation was found between log g and Hα emission strength or the strength and/or presence of emission of Fe II, O I λ 7774 Å, or the infrared Ca II triplet. The suggestion made by Schild (1973) and Schild *et al.* (1974) that the *extreme* Be stars are in the post main sequence phase of rapid core contraction is weakened by the fact that there are several members of the class which have log $g \geq 3.8$. All shell stars considered in the program appear to have low values of log g (≤ 3.5). Some possible explanations for this occurrence are discussed.

1. Introduction

One approach toward gaining an understanding of the origin of the material in Be star envelopes is to quantitatively study the spectra of the underlying stars. In particular, the surface gravity is a useful physical parameter which can help us establish to what extent the Be phenomenon is a normal stage in the post main sequence evolution of rapidly rotating B stars. Accordingly, I have obtained surface gravities for 30 Be and Be-shell stars by comparing the observed wings of their Hγ and Hδ profiles with those computed from the Princeton model atmospheres and the hydrogen line broadening theory of Vidal *et al.* (1973). The program stars range from B0–B7 and display a wide range in $v \sin i$ and envelope spectra. Included in the sample are many objects which have been classified by Schild (1973) and Schild *et al.* (1974) as *extreme* Be stars.

The spectrograms used in the analysis were obtained with the coudé spectrograph of the 120-in (305 cm) telescope at Lick Observatory and ranged in dispersion from 5.5–16 Å mm^{-1}. Two-thirds of the objects were observed at the higher dispersion and more than one plate was available for over one-third of the stars.

2. Predictions

What do competing theories for the origin of Be envelopes predict for the value of log g of the underlying star? Three theories which have been recently considered along with the range in log g which they suggest are briefly discussed below.

2.1. 'ROTATIONALLY UNSTABLE' SINGLE STARS

Computations made by several investigators have suggested that rapidly rotating B-type stars will reach a point of critical rotation at some stage during their post main

A. Slettebak (ed.), Be and Shell Stars, 69–78. All Rights Reserved
Copyright © 1976 by the IAU.

sequence evolution (Sackmann and Anand, 1970; Crampin and Hoyle, 1960). The details of the star's subsequent behavior remain uncertain and a source of controversy. Some researchers feel that when the star reaches this stage, 'rotationally forced ejection' of material via magnetic coupling of envelope to star can occur (Limber and Marlborough, 1968; and Limber, 1970). However, it should be stated that alternatives to the latter idea have been presented (Bodenheimer and Ostriker, 1973 and other papers by these authors referenced therein). Schild (1973) and Schild *et al.* (1974) claim that one can spectroscopically isolate a group of Be stars, called *extreme* Be stars, which are losing mass during the post main sequence phase of rapid core contraction as a result of an instability caused by rapid rotation. These Be stars have strong, permanent Hα emission, Fe II emission, and infrared excesses. If Schild *et al.* are correct, then stars classified as *extreme* Be stars should have a fairly narrow range in log g, $3.5 \leqslant \log g \leqslant 3.7$.

2.2. MASS-EXCHANGE BINARIES

Recently, several researchers have suggested that at least some Be stars gain their emission line envelopes through binary mass exchange (Plavec, 1970; Plavec, 1973; Peters and Polidan, 1973; Kriz and Harmanec, 1975). In fact, later in this symposium arguments will be presented by Dr P. Harmanec that all Be stars may be interacting binary stars. The idea of binary mass exchange as an explanation for the existence of Be stars is attractive since it can explain, at least qualitatively, not only the source of the material in Be envelopes but also the high values of $v \sin i$ associated with Be stars. The expected range in log g for interacting binary Be stars is large, $3.3 \leqslant \log g \leqslant 4.0$, and uncertain on account of our present lack of understanding of the details of mass accretion onto the transformed primary.

2.3. PRE-MAIN SEQUENCE OBJECTS

We must consider the possibility that some stars classified as 'classical Be stars' may actually be pre-main sequence objects which have not completely lost or accreted their 'cocoons'. Prime candidates for such a class of objects would be located in young OB associations such as h and χ Per and the Scorpio-Centaurus association. Lack of nebulosity in the vicinity of such objects may confuse one as to the star's evolutionary state. At this time it is uncertain just how long it will take a rapidly rotating pre-main sequence B-type star to accrete and/or dissipate material in a cocoon. The suspected values of log g for these objects are uncertain; however, we suggest the following range: $3.0 \leqslant \log g \leqslant 3.6$.

3. Assignment of Effective Temperatures

The hydrogen line profiles in early B-type stars are a function both of gravity and temperature. The amount of narrowing of a profile due to a change in log g of 0.1 is nearly identical to that resulting from a change in effective temperature of 1000 K. Therefore, it is essential that one obtain a fairly good estimate of the effective temperature of a B-type star before attempting to determine a surface gravity from a theoretical profile grid.

For this investigation, values of T_{eff} were assigned to the program stars on the basis of : (1) spectral types, (2) UBV colors, and (3) photospheric line strengths and ratios. For most stars, the spectral types determined by Lesh (1968) were adopted. The UBV colors for the program stars were obtained by taking weighted means of values listed in the compilation of Blanco *et al.* (1968). Colors were corrected for interstellar reddening using the two color relationship of Johnson (1966). Temperatures were derived from the writer's calibrations of spectral type and $(U-B)_0$ vs T_{eff} (Peters, 1976). Although continuous Balmer emission (or absorption) contributes significantly to the observed colors in some cases, it was found that, in general, the temperature indicated by the corrected $(U-B)_0$ color was in close agreement with the one suggested by the spectral type.

The preferred method of assigning a temperature to a Be star is to perform a detailed study of the star's spectrum using model atmospheres to interpret the individual line strengths. An investigation of this nature is presently in progress for several of the program stars (i.e. υ Cyg, 31 Peg, χ Oph, μ Cen, HR 2142, and ϕ Per).

4. Analysis of the Hγ and Hδ Profiles

In interpreting the hydrogen line profiles observed in Be stars, we are faced with some additional complications which are not present in those profiles observed in sharp-lined, early B-type stars. Some problems include emission line contamination of the profile, rotational broadening, and broadening due to electron scattering in the star's envelope. In this investigation, I have concluded that the problems are indeed surmountable and that the most useful portion of the hydrogen line profile for analysis is $7 < \Delta\lambda < 14$ Å. For stars in which the effective temperature can be determined to within 1000 K, the uncertainty in $\log g$ obtained by the adopted procedure is less than 0.2.

For the majority of the program stars, emission altered the hydrogen line profile only in the vicinity of the line center. Only in the case of HR 2855 was there any evidence of emission line contamination at a greater distance than 7 Å from the center of the line. Some individual cases will be discussed toward the end of this section.

Rotational broadening has a very small effect on the wing portion of a hydrogen line profile at distances greater than 7 Å from the line center. For a star with $v \sin i > 300$ km s^{-1}, one would obtain an apparent $\log g$ which is about 0.1 too high upon using an 'unrotated' grid of theoretical profiles. Burbidge and Burbidge (1953) also noted that moderate amounts of rotation do not affect the wings of hydrogen line profiles.

In their detailed study of the envelope spectra of six Be stars, Burbidge and Burbidge (1953) presented evidence that the wings of the hydrogen lines in Be stars are significantly broadened by electron scattering in the circumstellar envelopes. They obtained optical depths up to 0.4 due to this process. However, the present investigation does not support their conclusion even though four stars were common to both programs (11 Cam, 48 Per, ω CMa, and β Psc). Whereas Burbidge and Burbidge compared the Be stars' hydrogen line profiles to the profiles observed in two standard stars (ζ Cas and η Aur), the Hγ and Hδ profiles observed in the Be stars considered in this program were compared to a theoretical grid of profiles which have been used successfully to obtain surface gravities for ten normal, sharp-lined, early B-type stars (Peters 1976). The temperature of ζ Cas (about 24 500 K) is significantly higher than the temperatures of the four stars common to the Burbridge program and this investigation. The temperature of η Aur is comparable to that of 48 Per, but Burbidge and Burbidge compared this star to ζ Cas. The fact that the temperatures of the standard stars used by Burbidge and Burbidge were not representative of their program stars is most likely the reason for their conclusion regarding the importance of electron scattering in the envelope as a source of broadening in the wings of hydrogen lines.

The surface gravities which were obtained for the Be and Be-shell stars considered in this program are tabulated in Tables I and II. The value quoted for log g is a mean

TABLE I

Summary of surface gravities for Be stars

Star	Sp. type[a]	$v \sin i$[b]	T_{eff}	log g
HD 7636	B2	$-$ km s^{-1}	20 000 K	4.0
ϕ Per	B0	450	28 000	4.0
48 Per	B3	217	16 000	3.5
11 Cam	B2.5	131	19 000	4.0
HR 2142	B1	350	25 500	3.9
ω CMa	B2	137	20 000	3.5
HR 2825	B3	33	17 000	3.7
HR 2855	B0.5	244	28 000	4.0
μ Cen	B2	191	20 000	3.5
χ Oph	B1.5	123	24 000	3.8
66 Oph	B2	241	23 000	4.0
HR 7249	B2	226	22 000	3.6
25 Cyg	B3	229	16 500	3.3
28 Cyg	B2.5	310	20 000	3.7
25 Vul	B7	250	13 000	3.3
59 Cyg	B1.5	$-$	25 000	4.0
υ Cyg	B1.5	261	24 000	3.8
6 Cep	B3	148	17 500	3.5
16 Peg	B3	152	17 000	3.5
31 Peg	B2	134	20 000	3.5
π Aqr	B1	278	27 000	3.9
β Psc	B6	147	14 500	3.5

[a] From Lesh (1968); a few types are from the author.
[b] From Uesugi and Fukuda (1970); value for HR 2142 from Peters (1972).

from Hγ and Hδ and, in the cases where more than one plate was available for the star, the mean was determined by weighting the observations according to the dispersion and quality of the plate. The values of log g deduced from Hγ and Hδ usually agreed within 0.1. For stars in which variations of emission lines or shell components were observed, the wings of the hydrogen line profiles remained unchanged. A general discussion of the results is presented in the following section. Six individual stars will be discussed below.

TABLE II

Summary of surface gravities for shell stars[a]

Star	Sp. type[b]	$v \sin i$[c]	T_{eff}	$\log g$
ψ Per	B5	398 km s^{-1}	15 500 K	3.3
28 Tau	B7	341	13 000	3.3
ζ Tau	B1	310	25 000	3.5
AX Mon	B1	430	25 000	3.5
HD 173219	B0.5	–	27 000	3.5
ε Cap	B2.5	274	19 000	3.2
o And	B6	330	15 000	3.2
HD 218393	B2	–	22 000	3.4

[a] Stars which show 'metallic' shell lines in the ground-based portion of their spectra.
[b] From Lesh (1968) or author.
[c] From Uesugi and Fukuda (1970).

HR 2825 has the lowest projected rotational velocity of the stars considered in this program (33 km s^{-1}) and, hence, the least emission line contamination in its hydrogen line profiles. As shown in Figure 1 for Hδ, the emission is confined to ± 2 Å from the center of the line while the remainder of the profile fits well to one computed from the Princeton model atmosphere of $T_{eff} = 16\,800$ K, $\log g = 3.5$. The photospheric parameters (T_{eff}, $\log g$, and chemical composition) for HR 2825 appear to be quite

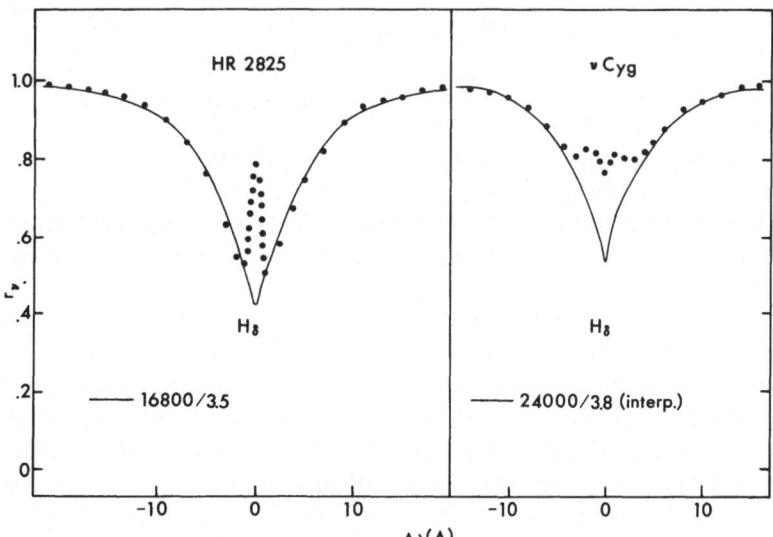

Fig. 1. Comparison between observed profiles of Hδ in HR 2825 and v Cyg and those computed from the Princeton model atmospheres. The model parameters are indicated below the profiles.

close to those for ι Her (Peters 1976). Kodaira and Scholz (1970) reached a similar conclusion from their analyses of these stars.

v Cyg is a good example of a Be star with an intermediate value of $v \sin i$ (~ 250 km s^{-1}). The Balmer line emission appears to be constant in strength and profile and contributes only to the inner ±5 Å of the Hγ and Hδ features. The wings of Hγ and Hδ can be fit to interpolated profiles corresponding to a model of $T_{eff} = 24\,000$ K, $\log g = 3.8$ (see Figure 1).

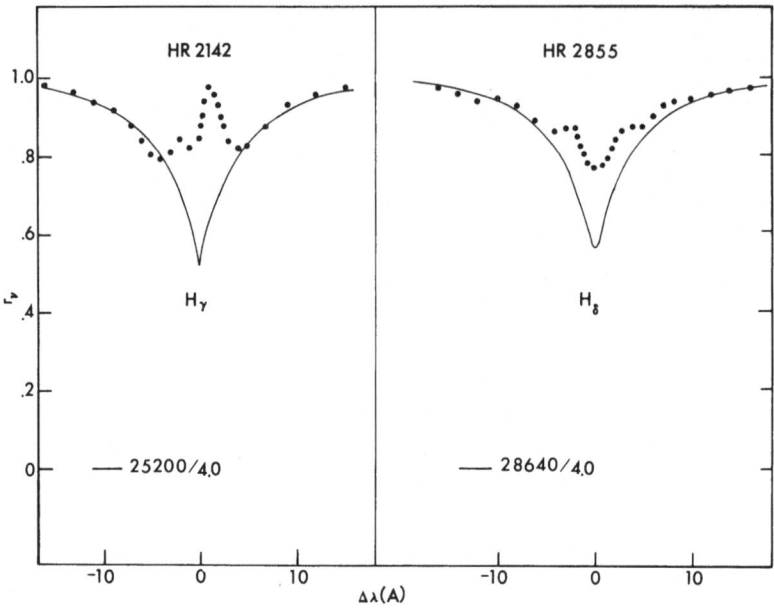

Fig. 2. Comparison between observed and computed hydrogen line profiles for HR 2142 ($\Phi_s = 0.09$) and HR 2855. The model parameters are indicated below the profiles.

Figure 2 shows that the observed profile of Hγ in HR 2142 closely matches one computed from the Princeton model atmosphere of 25 200 K, $\log g = 4.0$. The profiles of the Balmer lines in HR 2142 undergo cyclic variations (V/R type) with a period of $80^{d}86$ (Peters, 1972). The profile which appears in Figure 2 was observed at $\phi_s = 0.09$ when $R \gg V$. Emission contributes at distances less than 7 Å from the line center and the wings of the hydrogen line profiles remain constant with phase. The $v \sin i$ for HR 2142 is about 350 km s^{-1}.

HR 2855 showed the highest degree of emission line contamination to its hydrogen line profiles. As one can see from Figure 2, there appears to be emission in Hδ at 9 Å from the line center! This fact combined with the apparent high temperature and gravity produces a very low contrast hydrogen line profile; the most conspicuous feature is the weak, somewhat broad shell core. The uncertainties in T_{eff} and $\log g$ for HR 2855 ($\Delta T_{eff} = 2000$ K, $\Delta \log g = 0.5$) are considered to be the largest of all the program stars.

The program also included eight Be-shell stars. The values of $\log g$ which were obtained for the shell stars are tabulated in Table II. In this investigation, it was found that shell stars tend to have low gravities. Whereas the mean $\log g$ for the Be stars was 3.7, the shell stars showed values of $\log g \leq 3.5$. In Figure 3, we find the comparison between the observed Hδ profiles of two shell stars, ζ Tau and o And, and computed profiles.

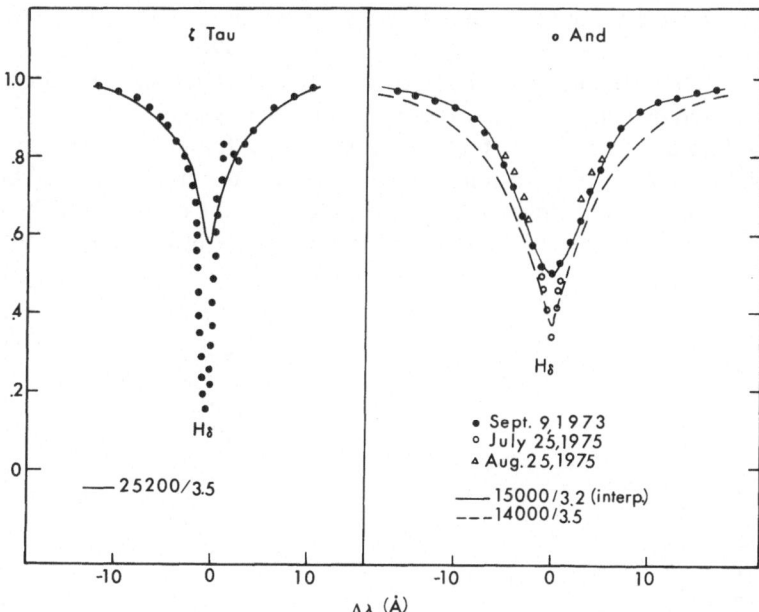

Fig. 3.   Comparison between observed and computed Hδ profiles for ζ Tau (in 1974 and 1975) and
o And (in 1973 and 1975). The observations of o And in 1975 were made during its recent shell phase.

   This investigation suggested that the temperature of ζ Tau is close to 25 000 K and
log g is near 3.5. When transformations between systems of model atmospheres are
considered, the latter atmospheric parameters appear to be in agreement with those
determined by Heap (1975).

   o And is a rapidly rotating B6 star which has shown a recurrent shell phase
(Slettebak, 1952). After remaining inactive for nearly two decades, o And has
recently developed another shell (Koubsky, 1975; Peters and Polidan, 1975). In
Figure 3, the profiles of Hδ observed outside of and during a shell phase are
compared with those computed from the Princeton model atmospheres. Although
the hydrogen line profiles observed during the shell phase show both shell type
absorption and weak emission in the vicinity of the line center, the wings remain
unchanged from their profiles observed outside of shell phase. One would obtain the
same value of log g both inside and outside of shell phase. When the comparisons
between the observed and computed profiles of Hγ and Hδ are made, it becomes
evident that the average surface gravity for o And is quite low. If the effective
temperature is near 15 000 K (a value consistent with the star's spectral type), then
log g is about 3.2. The apparent low gravities of shell stars will be further discussed in
Section 5 of this paper.

5. Discussion of Results

It can be seen from the tabulated values of log g in Table I that the 'classical' Be stars
display a wide range in surface gravity (3.3 ≤ log g ≤ 4.0). The mean value of log g for

the Be stars considered in the program is 3.7. A similar analysis was performed for ten sharp-lined, non-emission B-type stars (Peters, 1976). The mean and range in log *g* was identical to the one obtained in this investigation. The program stars included in each investigation are predominantly field stars.

Can one find a spectroscopic feature whose strength and/or presence correlates with the value of log *g* of a Be star? Possible candidates are considered below.

In Figure 4, we present a plot of the value of log *g* vs the α-index for the star. Both Be stars and shell stars are included. Values of the α index were obtained from

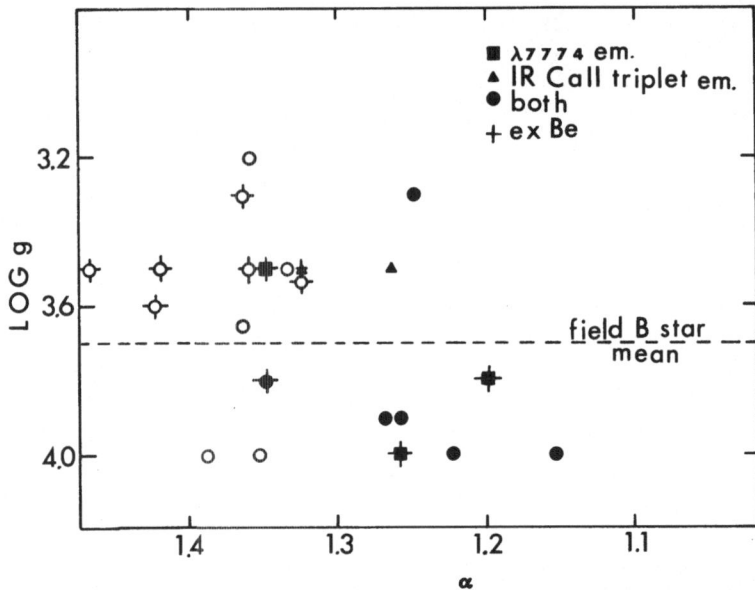

Fig. 4.   The value of log *g* vs Hα emission line strength. Open circles represent objects which do not show O I λ7774 Å emission or emission at the infrared Ca II triplet. *Extreme* Be stars are indicated by crosses superposed on the star's symbol. The dashed line denotes the mean value of log *g* for field, non-Be stars.

Feinstein (1974). Certainly, there is no evidence that the surface gravity of a Be star is correlated with the strength of the Hα emission. It is also evident from the plot in Figure 4 that there is no correlation between the value of log *g* and the presence of emission at λ 7774 Å of O I or the infrared Ca II triplet. In addition, the presence of Fe II emission [which correlates well with the presence of λ 7774 Å emission (Peters, 1974)] does not appear to be a function of log *g*.

The stars classified as *extreme* Be stars by Schild (1973) and Schild *et al.* (1974) are noted in Figure 4 by crosses superposed on the stars' symbols which indicate whether emission is present at λ 7774 or the infrared Ca II triplet. It can be seen that the *extreme* Be stars are not restricted to a narrow range in log *g* ($3.5 \leq$ log *g* ≤ 3.7) which one would expect if every member of this class is in the post main sequence phase of rapid core contraction. Although several members of the group have values of log *g* close to 3.5, there are certain *extreme* Be stars which definitely have log *g* ≥ 3.8 (i.e. 11 Cam, υ Cyg). We cannot dismiss entirely the possibility that some Be stars may indeed be losing mass due to an instability which may occur during the phase of rapid core contraction; however, it appears that such objects cannot be isolated by spectroscopic criteria alone. For an individual Be star which has an intermediate value of log g and for which there is no evidence of binary mass exchange or pre-main sequence contraction, at present one cannot determine the source of the material in its envelope.

This study has also revealed that there is no correlation between the surface gravity of the Be star and $v \sin i$. From the values of $v \sin i$ tabulated in Table I, it can be seen that both low and high surface gravities characterize 'sharp-lined' stars (i.e. β Psc and 11 Cam) and 'broad-lined' stars (i.e. 28 Cyg and ϕ Per).

At the beginning of this investigation, it was not anticipated that the Be-shell stars would have low surface gravities. Yet the values of log g found in this investigation for these objects range from 3.2–3.5. Although the reason for the low gravities is not immediately evident, we can offer the following three suggestions.

(1) The stars may simply be in the early stages of their post main sequence, hydrogen shell burning phase. If this is the case, then it is not clear why they should have emission line envelopes and 'metallic shells'. It should be mentioned that 28 Tau, a well-known member of the Pleiades, appears to be located near the main sequence (Johnson and Mitchell 1958).

(2) Shell stars could have extremely high rotational velocities and, hence, low mean surface gravities. However, rotational velocities in excess of 90% of the critical value are required to reduce the mean log g to the range of values obtained in this investigation. For example, if $v_{eq} = 350$ km s^{-1} and log $g_p = 3.9$, then log $g_{eq} = 3.7$. The mean log g for such a star is 3.8.

(3) The model atmospheres which were used for the analysis may not be representative of the photospheres of shell stars. However, at this time it is not possible to decide appropriateness of the models since we are as yet uncertain about the rotational velocities and surface geometries of shell stars.

It is my feeling that there are several mechanisms for the formation of Be envelopes. This investigation has shown that the surface gravities for Be stars are comparable to those determined for non-Be stars. The Be stars apparently do not display a narrow range in log g. Actually, the wide range in luminosity classes (III–V) which have been assigned to Be stars allows one to anticipate this conclusion. On the other hand, the Be-shell stars seem to have low surface gravities. These objects invariably are classified as Bp! Perhaps by making use of new observations in several spectral regions combined with new line transfer computations for circumstellar envelopes we can eventually isolate the various groups of Be stars.

Acknowledgements

I wish to thank Dr M. Plavec for the support that he has given this project in the form of generous amounts of his time on the Lick Observatory 120-in. telescope. The project was part of an extensive program at UCLA aimed at studying interacting binary systems and was supported in part by NSF MPS 74-04194A01 (Popper/Plavec). I also wish to acknowledge numerous interesting and helpful discussions with Dr M. Plavec and R. S. Polidan.

References

Blanco, V. M., Demers, S., Douglass, G. G., and Fitzgerald, M. P.: 1968, *Publ. U.S. Naval Obs.* **21**.
Bodenheimer, P. and Ostriker, J. P.: 1973, *Astrophys. J.* **180**, 159.
Burbidge, G. R. and Burbidge, E. M.: 1953, *Astrophys. J.* **117**, 407.
Crampin, J. and Hoyle, F.: 1960, *Monthly Notices Roy. Astron. Soc.* **120**, 33.
Feinstein, A.: 1974, *Monthly Notices Roy. Astron. Soc.* **169**, 171.
Heap, S. R.: 1975, personal communication.
Johnson, H. L.: 1966, *Ann. Rev. Astron. Astrophys.* **4**, 193.
Johnson, H. L. and Mitchell, R. I.: 1958, *Astrophys. J.* **128**, 31.
Kodaira, K. and Scholz, M.: 1970, *Astron. Astrophys.* **6**, 93.
Koubsky, P.: 1975, *IAU Circ.* No. 2802.
Kriz, S. and Harmanec, P.: 1975, *Bull. Astron. Inst. Czech.* **26**, 65.
Lesh, J. R.: 1968, *Astrophys. J. Suppl.* **17**, 371.
Limber, D. N.: 1970, in A. Slettebak (ed.), *Stellar Rotation*, D. Reidel Publ. Corp., Dordrecht-Holland, p. 274.
Limber, D. N. and Marlborough, J. M.: 1968, *Astrophys. J.* **152**, 181.
Peters, G. J.: 1972, *Publ. Astron. Soc. Pacific* **84**, 334.
Peters, G. J.: 1974, *Bull. Am. Astron. Soc.* **6**, 456.
Peters, G. J.: 1976, in preparation.
Peters, G. J. and Polidan, R. S.: 1973, in A. H. Batten (ed.), 'Extended Atmospheres and Circumstellar Matter in Spectroscopic Binary Systems', *IAU Symp.* **51**, p. 174.
Peters, G. J. and Polidan, R. S.: 1975, *IAU Circ.* No. 2814.
Plavec, M.: 1970, *Publ. Astron. Soc. Pacific* **82**, 957.
Plavec, M.: 1973, in A. H. Batten (ed.), 'Extended Atmospheres and Circumstellar Matter in Spectroscopic Binary Systems', *IAU Symp.* **51**, p. 216.
Sackmann, I. J. and Anand, S. P. S.: 1970, *Astrophys. J.* **162**, 105.
Schild, R.: 1973, *Astrophys. J.* **179**, 221.
Schild, R., Chaffee, F., Frogel, J. A., and Persson, S. E.: 1974, *Astrophys. J.* **190**, 73
Slettebak, A.: 1952, *Astrophys. J.* **115**, 573.
Uesugi, A. and Fukuda, I.: 1970, *Contrib. Inst. Astrophys. Kwasan Obs. Univ. Kyoto*, No. 189.
Vidal, C. R., Cooper, J., and Smith, E. W.: 1973, *Astrophys. J. Suppl.* **25**, 37.

A COMPARATIVE STUDY OF THE SHELL OF
ζ TAU AND 48 LIB

A. M. DELPLACE and M. Th. CHAMBON

Observatoire de Meudon, France

1. Introduction

Two unstable shell stars, ζ Tau and 48 Lib, are compared. In the Herman-Rojas (1974) and Rountree-Lesh (1968) classifications, both stars have luminosity class III. These stars show similar features in the variation of the intensities, the profiles, and the radial velocities of the shell lines. An attempt has been made to explain the evolution of these unstable features.

2. ζ Tauri

This Be star is a binary system with period 132.91 days. The shell was moving in 1914–17 (Losh, 1932), then quiescent from 1920 to 1952, but since 1955 it has again been in motion. In 1958–59, an expanding shell was found by Hack (1962); in 1958 the radial velocities of the shell lines were -60 km s^{-1} and in 1959 -80 km s^{-1}.

This star has been observed at the Haute Provence Observatory with the 193 and 152 cm telescopes for fifteen years. The dispersions of the spectra are 9.67 Å mm^{-1} and 12.27 Å mm^{-1}.

2.1. Outer layers of the shell

The radial velocities of the Hγ, Hδ, Hε shell lines are shown in Figure 1. The short period is that of the binary system. The long variation is a representation of the radial velocities of the outer layers, but we shall see that the behaviour of the inner layers is more complex. A possible explanation of these variations is the following.

In 1956–59 an expanding shell was driven by rapid internal motions, probably resulting from the instability of the photosphere of the star; when the internal motions stopped, the outer shell fell down on the star until another expanding phase occurred. In our diagram, therefore, the outer shell is oscillating around the 25 km s^{-1} value, which is probably the mean radial velocity value of the star, as it was shown by Losh (1932), Hynek and Struve (1942) and Underhill (1952). We can see that the period of variation of the radial velocities increases with increasing velocity of expansion.

2.2. Inner layers of the shell

The differences between the radial velocities of the inner layers and the outer layers are plotted in Figure 2. For metallic lines, the total excitation energy is defined as the sum of the ionization potential and the excitation potential.

A. Slettebak (ed.), Be and Shell Stars, 79–85. All Rights Reserved

80

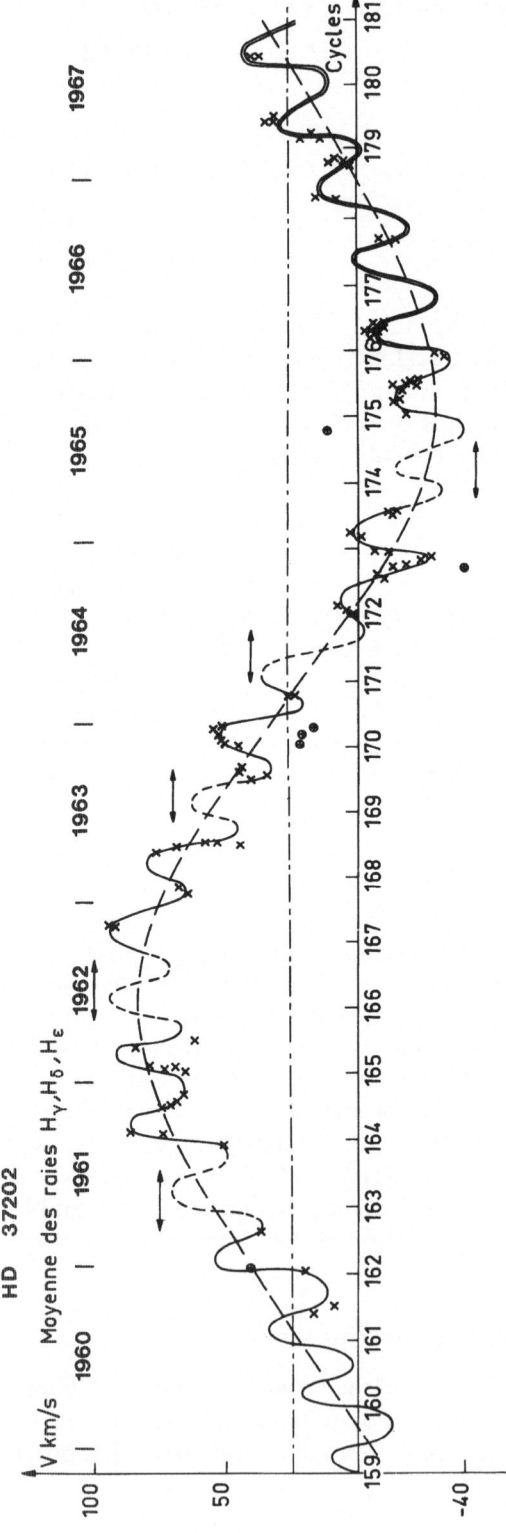

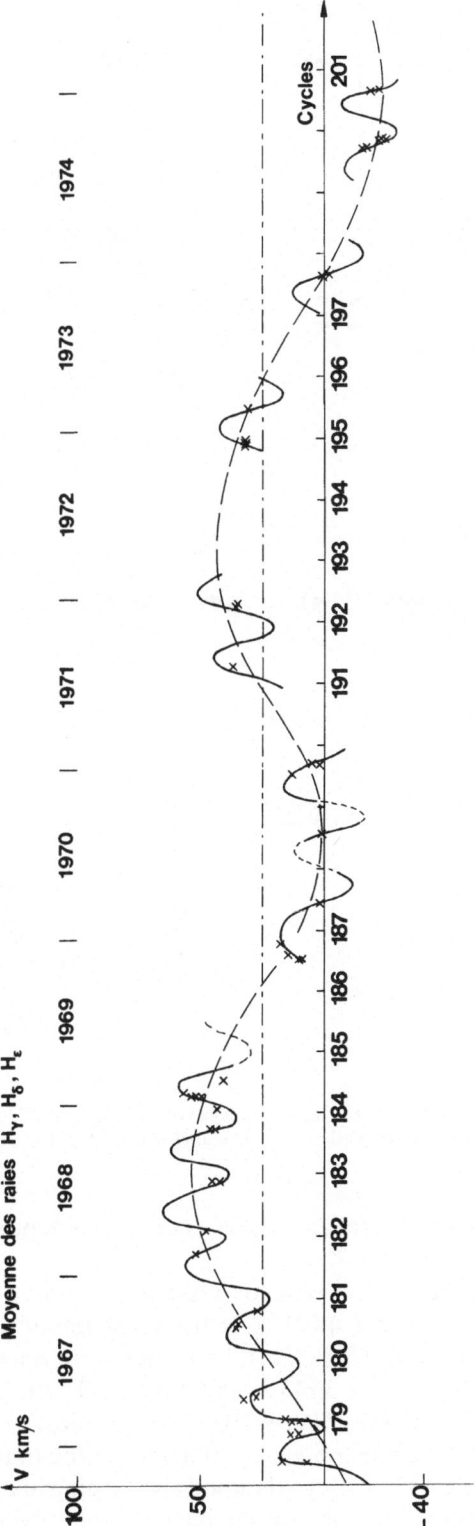

Fig. 1. Variation of the radial velocity of the Hγ, Hδ, Hε lines of ζ Tauri during the period 1960–1975. ⊕ Perkins plates 1949–57. ↔ no observations possible at the Haute Provence Observatory.

81

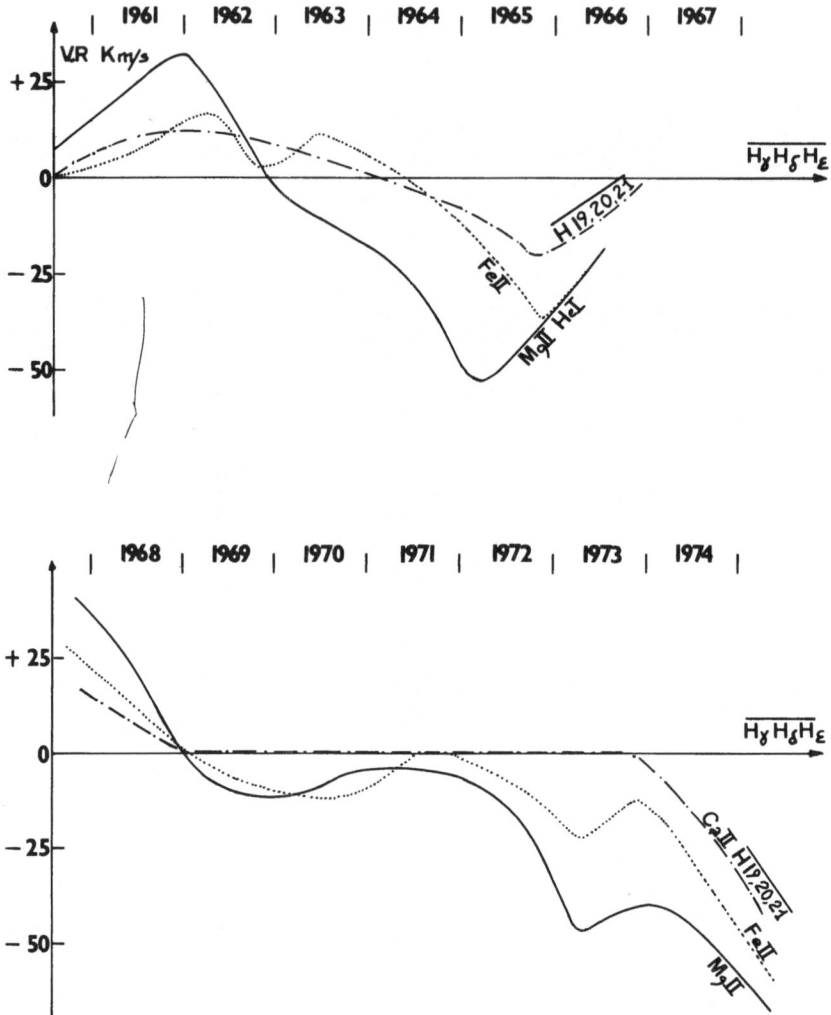

Fig. 2.   A schematic representation of the radial velocity curves of the Mg II and He I, Fe II, Ca II and
H19, H20, and H21 lines in comparison with the radial velocity of the Hγ, Hδ, and Hε lines.

We assume that with increasing total excitation energy, the depth in the shell at
which the line is formed increases.

(a) The radial velocities of the line increase for higher total excitation energy. The
radial velocity of the Mg II line at λ 4481 Å is the most important. The radial
velocities of the metallic lines (Fe II, Cr II, Ti II, Ni II) are more important than the
radial velocities of the Ca II lines at λλ 3933 Å and 3968 Å. These Ca II lines have
about the same radial velocities as the H19, H20, and H21 lines.

(b) The expanding motions are much bigger and longer in the inner layers. They
are transmitted to the outer layers but with a damped and lagging manner (the radial
velocity curve of the hydrogen lines is shifted back, about several months, in

comparison to the radial velocity curve of the Mg II line). Thus at the end of 1962, the outer layers were falling back down on the star but some expanding motions occurred in the inner layers because a satellite component, shifted to shorter wavelengths, appears in the He I and Mg II lines at about -60 km s^{-1}. In 1964, the expanding motions of the inner layers were increasing and at the end of 1964 the whole shell was moving out (Delplace, 1970). At the end of 1972, the outer shell was stationary but the radial velocities of He I and Mg II became negative. In 1974, the expanding motions in the inner layers were more important and the whole shell moved out again.

3. 48 Librae

This shell star has been studied particularly by Underhill (1966), Faraggiana (1969, 1971), and Geuverink (1970). From 1904–31 it was considered as a fast rotator surrounded by a faint hydrogen shell. Since 1932–35, the shell has become variable. The radial velocities of the shell lines given by Underhill, Faraggiana, and our own measurements on the spectra taken at the Haute Provence Observatory since 1968 are considered and the schematic curve of these variations shown in Figure 3. The

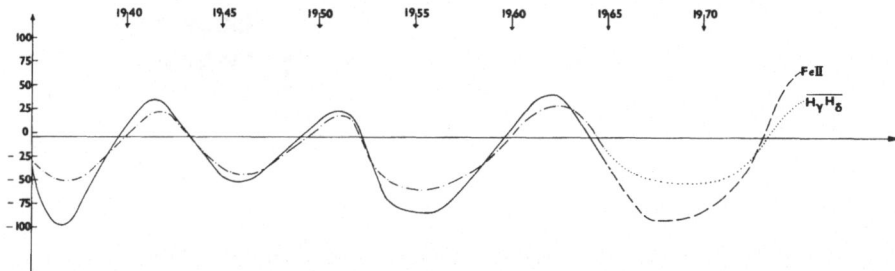

Fig. 3.   A schematic representation of the radial velocity curve of the Hγ, Hδ and Fe II lines of 48 Librae during the period 1935–75.

radial velocity of the star is about -5 km s^{-1}. As for ζ Tauri, the radial velocity variations can be interpreted as a sequence of the expanding phases followed by the infalling of material. The time of the oscillation increases with increasing amplitude of the variation. For 48 Librae the infalling material phases are short (about 3 or 4 years).

In the inner layers the motions are much more important than in the outer layers. Faraggiana (1969) showed that the radial velocity of the Hα shell line was -30 km s^{-1} in 1957–58 when the radial velocity of the He I and Mg II lines was -100 km s^{-1}.

The difference between the radial velocities of the inner and outer layers is shown in Figure 4. The amplitude of the motions increases for higher total excitation potential. The radial velocity curve of the Ca II lines is shifted several months back in comparison to the Mg II radial velocity curve. In 1974, when the outer shell was

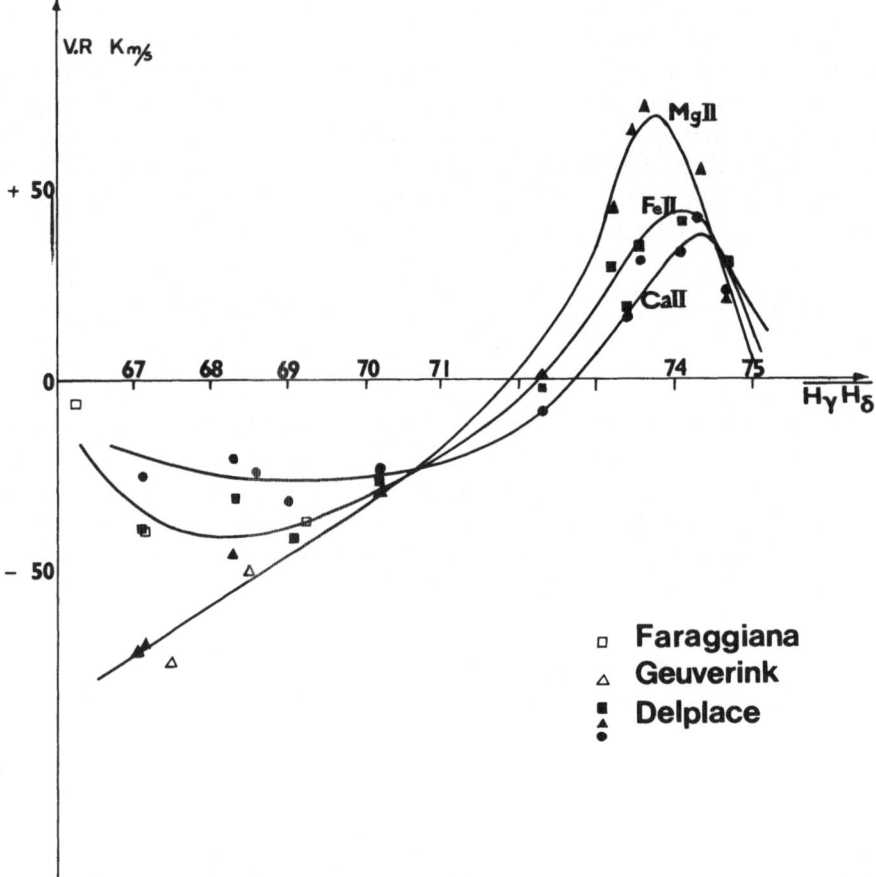

Fig. 4.   A schematic representation of the radial velocity curve of the Mg II, Fe II, and Ca II lines in comparison with the radial velocity of the Hγ and Hδ lines.

falling back on the star, a new expanding phase began because the lines which are formed in the inner layers are double, one component being shifted to shorter wavelengths.

4. Conclusion

The behaviour of the shells of ζ Tauri and 48 Librae is similar. The amplitude of the motion increases for higher total excitation potential. The period of the oscillation of the outer layers increases with increasing velocity of expansion. The internal motions in the shell give rise to a damped oscillation of the outer layers.

References

Delplace, A. M.: 1970, *Astron. Astrophys.* **7**, 68.
Faraggiana, R.: 1969, *Astron. Astrophys.* **2**, 162.

Faraggiana, R.: 1971, *Astrophys. Letters* **8**, 45.
Geuverink, H. G.: 1970, *Astron. Astrophys.* **5**, 341.
Hack, M.: 1962, *Publ. Astron. Soc. Pacific* **74**, 78.
Herman, R.: 1974, private communication.
Hynek, J. A. and Struve, O.: 1942, *Astrophys. J.* **96**, 425.
Losh, H. M.: 1932, *Publ. Obs. Univ. Michigan* **4**, 1.
Rountree-Lesh, J.: 1968, *Astrophys. J. Suppl.* **16**, 371.
Underhill, A. B.: 1952, *Publ. Dominion Astrophys. Obs.* **9**, 138.
Underhill, A. B.: 1966, *The Early Type Stars*, D. Reidel Publ. Co., Dordrecht, p. 238.

A MODEL FOR THE SHELL OF HD 50138*

LÉO HOUZIAUX

Université de Mons, Département d'Astrophysique, Belgium

and

YVETTE ANDRILLAT

Observatoire de Haute-Provence, France

(Paper read by J. P. Swings)

Abstract. A model for the shell star HD 50138 is inferred from the observation of the continuous spectrum from 0.14 to 10 μ.

1. The Observations

HD 50138 was discovered to be a Be star by Humason in 1920. Merrill mentioned in 1931 the variability of its spectrum, which has been studied by Doazan (1965) and ourselves (1972). Further examples of this variability are shown in Figures 1, 2, and 3. The spectra (dispersion 39 Å mm^{-1}) cover the region from Hα to P12 and show striking changes in line profiles and intensities over periods of 24 h. The HD spectral type is B8e but, on the basis of its Balmer discontinuity, the star has been classified B5 IV–V by one of us (Houziaux, 1960). New measurements of the strength of the photospheric He I lines (λ 4471 and λ 4026) lead to an effective temperature of 12 000° for the central star. In order to resolve the apparent contradiction between the spectral types as given by these two criteria, we have gathered as much information as possible on the continuous spectrum. The S2/68 orbiting telescope (Boksenberg *et al.*, 1973) provided absolute fluxes between λ 1400 and λ 2500 Å. We observed the continuous spectrum photographically in the region λ 3200–λ 5000 (at 66 Å mm^{-1}) and in the λ 5500–λ 9200 wavelength interval (at 230 Å mm^{-1}). *UBV* colors have been published by Haupt and Schroll (1974), while *R, I, J, K*, and *L* magnitudes are given by Allen (1973). All these data are summarized in Figures 4 and 5. *V, R*, and *I* magnitudes are in good agreement with our photometric measurements. The *B* filter however indicates a flux 10% higher than our photographic determination.

2. Interstellar Reddening

The color indices $B - V = 0.0$ and $U - B = -0.37$ indicate that the star is abnormally reddened; the $U - B$ excess is -0.35 with respect to normal B-type stars, which corresponds to a difference of 0.13 in the Balmer discontinuity. From the He I lines, a

---

* The observations have been carried out at the 193 cm and 120 cm telescopes of the Observatoire de Haute-Provence (CNRS).

A. Slettebak (ed.), Be and Shell Stars, 87–93. All Rights Reserved

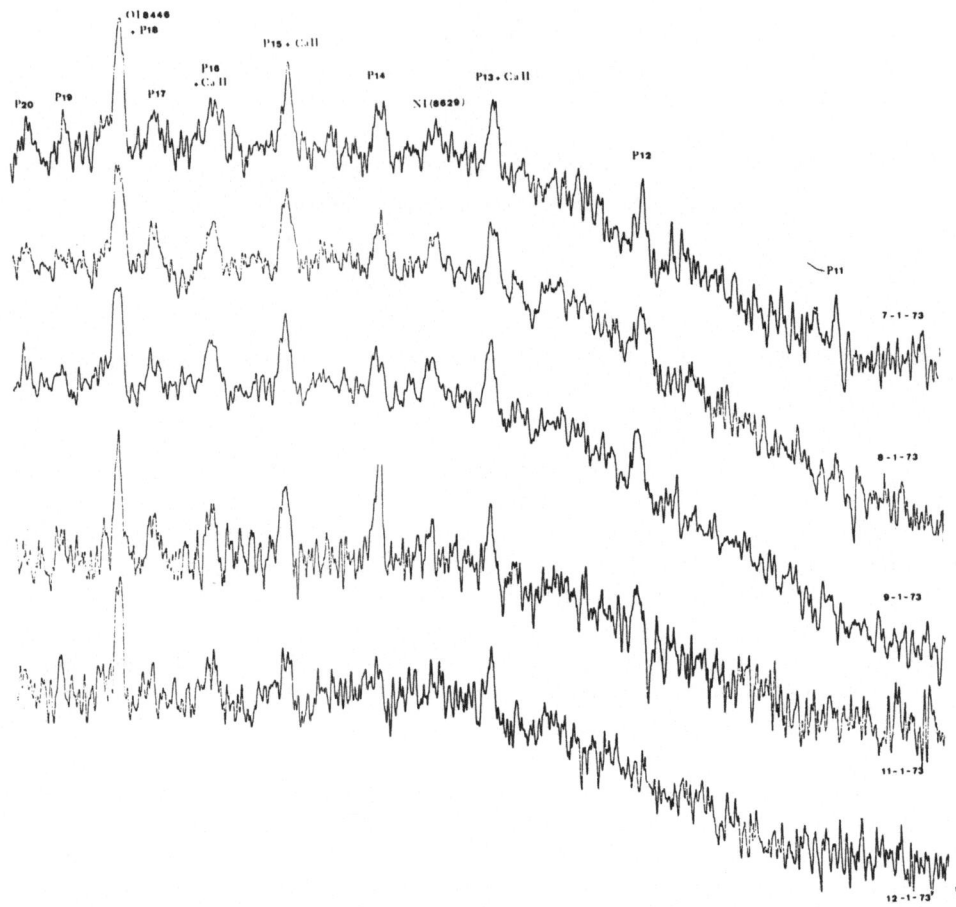

Fig. 1.   Microphotometer tracings of the near infrared spectrum of HD 50138, showing rapid changes in the line profiles and intensities during the nights from Jan. 7 to Jan. 12, 1973.

temperature of 12 000° has been assigned to the stellar photosphere. Comparison of the far wings of Hγ, Hδ and H8 with computed profiles indicates a surface gravity of 10^4 cm s^{-2}. Hence, the star's photosphere can be classified B8 V. The normal $(B - V)_0$ color for such a spectral type is −0.12. Hence, the color excess is 0.12. If it is entirely due to interstellar reddening, it would correspond to an absorption of 0.8 magnitude at λ 2200 Å, according to Nandy (1974), if a normal $A_V/E(B - V)$ ratio is adopted. Such an absorption would imply the existence of a hump in the ultraviolet continuum around λ 2200 Å, which is not observed (see Figure 4). On the other hand, an absorption $A_V = 0.36$ locates the star at about 400 pc. Its absolute visual magnitude would then be −1.7, in disagreement with the spectral type B8 V. In order to obtain an adequate absolute visual magnitude (−0.1), the distance should be 200 pc, corresponding to an interstellar color excess $E(B - V)$ of 0.06.

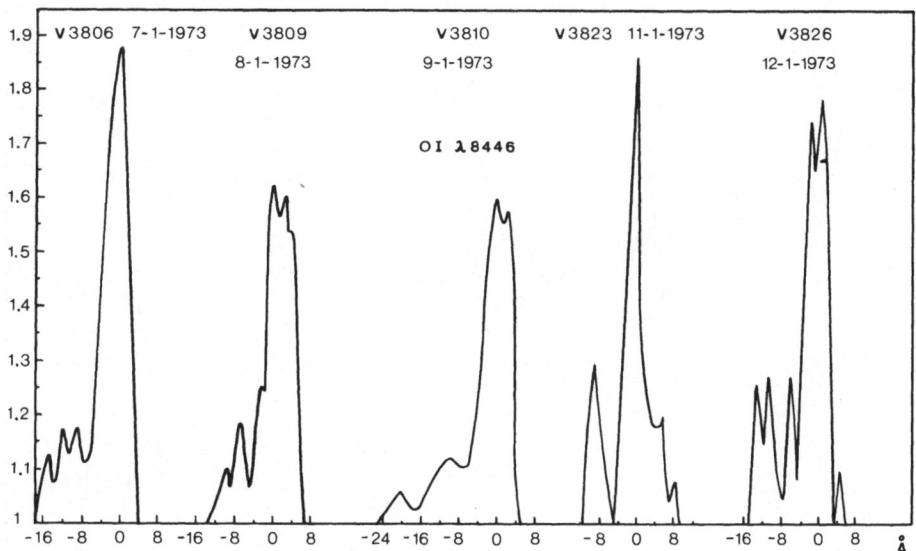

Fig. 2. Changes in the O I line at λ 8446 Å.

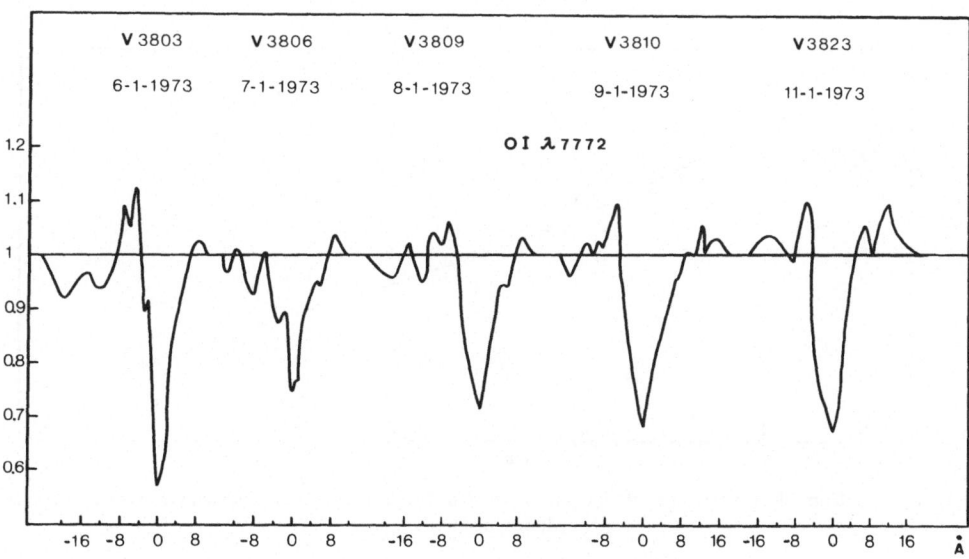

Fig. 3. Changes in the O I line at λ 7772 Å.

3. The Model

Hence, the proposed model should account for:
  (a) an 'intrinsic' color excess $E(B - V) = 0.06$;
  (b) a Balmer discontinuity smaller by 0.13 than the discontinuity assigned to a B8 V star;

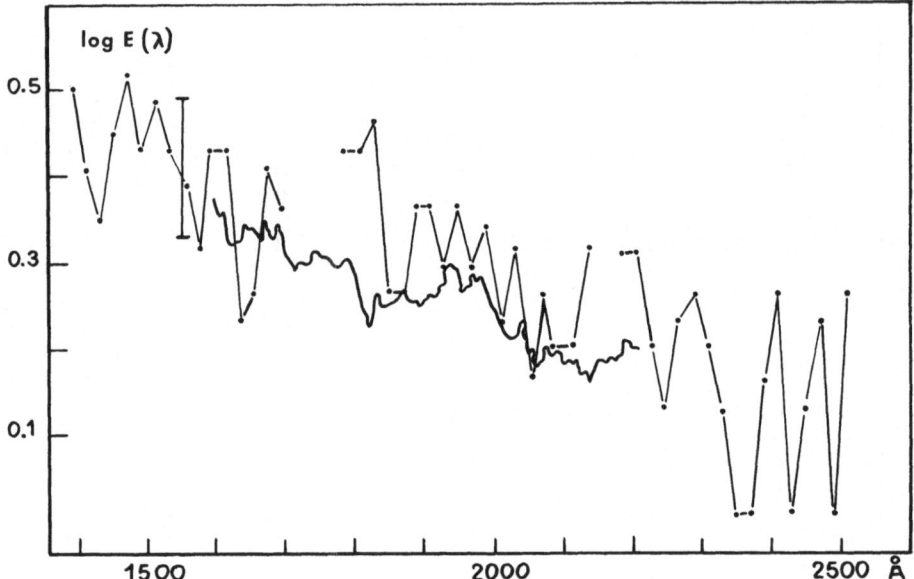

Fig. 4.   Ultraviolet spectrum of HD 50138. Full line: computed spectrum. Dots: observed fluxes. See text.

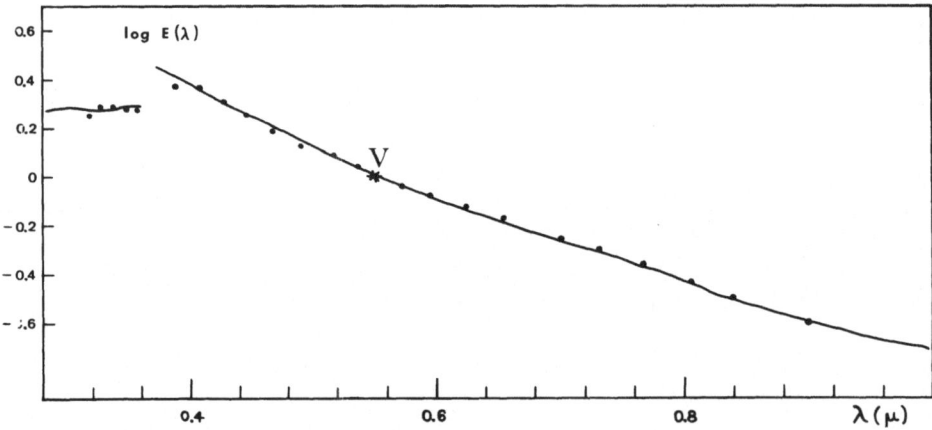

Fig. 5.   Spectrum of HD 50138 in the region 0.3 to 1 μ. Full line: computed spectrum. Dots: observed fluxes. See text.

(c) the observed intensity of emission lines;
(d) the infrared excess;
(e) the presence of forbidden lines (namely O I).

Starting with a B8 V central star with $T_{\text{eff}} = 12\,000°$, $\log g = 4$, and $R = 3.2\,R_{\odot}$, we shall compute how the photospheric spectrum can be altered by a surrounding shell. Previous analysis of the shell absorption spectrum (Houziaux, 1960) led to an electron temperature of $10^{4°}$ and an electron density of 3×10^{11} cm^{-3}. Because of the stellar rotation ($v \sin i = 150$ km s^{-1}), we shall admit that the shell is flattened with a flattening parameter α of 0.1, although this value is by no means critical.

Assuming that the envelope is transparent to its own continuous radiation, the observed spectrum will result in the superposition of:

$$(\text{central star radiation}) \cdot e^{-\tau} + \text{shell emission} + \text{interst. abs.}$$

The central emission is $4\pi^2 R_*^2 F_\lambda$, where F_λ is computed from a model atmosphere by Kurucz *et al.* (1974). The shell emission $E_{\text{shell}}(\lambda)$ consists of bound-free ($E_{bf}(\lambda)$) and free-free emission ($E_{ff}(\lambda)$) of a hydrogen plasma, which we assume to be fully ionized at the conditions prevailing in the shell. Using well-known notations, we shall write

$$E_{\text{shell}}(\lambda) = (R_{\text{shell}}^3 \alpha - R_*^3)$$

$$\times \left[2.71 \times 10^{-21} N_e^2 T^{-3/2} \sum_n \lambda^2 n^{-3} 10^{68.544/n^2 T} e^{-1.439/\lambda T} \bar{g}_u \right.$$

$$\left. + 8.54 \times 10^{-27} \lambda^{-2} N_e^2 T^{-1/2} e^{-1.439/\lambda T} \bar{g} \right]$$

the sum being extended to appropriate n quantum numbers, depending on the wavelength.

It is evident that $E_{\text{shell}}(\lambda)$ will show maxima at short wavelength edges of the hydrogen discontinuities, the maxima decreasing with increasing n. Precisely in our case, we note that there is a discrepancy of the Balmer discontinuity of 0.13. In order to account for the 'filling-in' of the Balmer edge absorption, we determined from the above expressions R_{shell} to be 5 R_*, as all other parameters are known. If the star is seen edge-on (as may be inferred from the aspect of the shell absorption spectrum), we see that the opacity $\tau(\lambda)$ at the violet edge of the Balmer discontinuity is rather small:

$$\tau(3650 \text{ Å}) = \tau_{bf}(3650 \text{ Å}) + \tau_{ff}(3650 \text{ Å}),$$

where

$$\tau_{bf}(3650 \text{ Å}) \sim (R_{\text{shell}} - R_*) 1.6 \times 10^{-17} N_2$$

$$\tau_{ff}(3650 \text{ Å}) \sim (R_{\text{shell}} - R_*) 3.69 \times 10^8 [1 - \exp(-1.439/T)] \bar{g} T^{-1/2} \nu^{-3} N_e^2$$

N_2 can be computed with the usual hydrogen recombination theory and is found to be 2×10^3 cm^{-3}. The resulting value of τ (λ 3650 Å) is 0.7, which indicates that the opacity is indeed very low in the Paschen continuum.

4. Comparison between Observed and Computed Fluxes

The observed flux between λ 1400 Å and 1 μ is given in Figures 4 and 5. Satellite data are fairly uncertain, as indicated by the error bar. Absolute ultraviolet fluxes have been normalized to a V magnitude 0.0, assuming a $V = 0.0$, $B - V = 0$ star produces a flux of 3.64×10^{-9} erg cm^{-2} s^{-1} Å^{-1} at the effective wavelength of the V filter at the top of the Earth's atmosphere. The dots in Figure 5 result from our photographic observations and from the photometric observations mentioned above. The normalization is the same as in Figure 4.

The computed fluxes $E(\lambda)$ are also reported on the same figures, with the same normalization:

$$\log E(\lambda) = \log [4\pi^2 R_*^2 F(\lambda) e^{-\tau_\lambda} + E_{\text{shell}}(\lambda)] - 0.4 A(\lambda) + C,$$

C being a normalization factor and $A(\lambda)$ the interstellar absorption in magnitudes. Between 0.36 and 1 μ the fluxes $F(\lambda)$ in the continuum have been computed in a straightforward manner from the above mentioned model. However, between λ 1590 and λ 2200 Å, because of the extreme crowding of absorption lines, we have taken into account the effects of the lines, and folded the resulting fluxes with the instrumental profile of the S2/68 spectrometer (see Boksenberg et al., 1973). Several thousands of absorption lines have been considered, using a table established by Kurucz and Peytremann (1975). The interstellar extinction curve has been provided by Nandy (1974). It can be seen that the agreement is satisfactory throughout the spectrum, except between λ 3700 and λ 4000 Å. This region is, however, crowded with many high Balmer lines.

5. Hydrogen Shell Lines

In order to check the validity of the model, we have computed the strength of the higher Paschen emission lines, where the opacity is very small. As P13, P15 and P16 are blended with the Ca II triplet, and P17 and P18 are seriously perturbed by the strong O I emission at λ 8446 Å (see Figure 1), we have used only P14.

The energy $E(14-3)$ emitted in the transition 14 to 3 may be written

$$E(14-3) = \frac{4\pi}{3}(R_{\text{shell}}^3 \alpha - R_*^3)N_{14}A(14-3)h\nu_{14-3},$$

where

$$N_{14} = b_{14}N_e^2 \frac{h^3}{(2\pi mkT_e)^{3/2}} \frac{\varpi_{14}}{2} \exp(hR/14^2 T_e)$$

with the usual notations. Taking $b_{14} \sim 1$ as a mean for the envelope, we find an equivalent width W_λ of 4 Å for P14, whereas we observe 3.2 Å. Taking into account the adopted approximations, this agreement may be considered as satisfactory.

On the other hand, the shell is opaque to Balmer lines and, for the members of the Balmer series where no emission is detected, it is easy to separate the shell absorption component from the photospheric line. We have measured the lines from H8 to H16, and, using the isothermal thin layer approximation, as improved by Huang and Struve (1956), we find

$$N_2(R_{\text{shell}} - R_*) = 1.129 \times 10^{20} \frac{W_\lambda}{\lambda^2 f}\left(\frac{R_c}{R_c - R_0/2}\right),$$

where R_c is the limiting depth (0.8) and R_0 the central depth of the shell line. Taking the value $R_{\text{shell}} - R_*$ as above, we find a mean value of $N_2 = 1.5 \times 10^3$ cm^{-3}, in reasonable agreement with the value found in using the hydrogen recombination theory.

6. Conclusions

The proposed model is thus consistent with several observational facts:

  (a) the B8 V central star accounts for the strength of the He I lines. It fits the observed ultraviolet flux in a spectral region very sensitive to the photospheric temperature;

  (b) the shell reemission explains the apparently low value of the Balmer discontinuity (0.27) for a B8 V star. It explains also a moderate $E(B-V)$ intrinsic reddening;

  (c) the H II shell model is compatible both with the strength of the Paschen emission lines and with the intensity of the Balmer absorption components.

It is clear however that the infrared excess mentioned by Allen (1973) must find its origin outside the H II shell. An H I region certainly exists around this H II shell, where low excitation forbidden lines may be formed. The infrared radiation may come from H^- free-free and/or dust radiation in this region. So far, no evidence of molecular absorption has been detected in the spectrum of HD 50138; hence a binary nature for this object is not to be considered at present.

Acknowledgements

We would like to thank A. Delcroix (Mons) for computing the ultraviolet spectrum at our request. Thanks are due to the S2/68 reduction team in Liège for providing us with the ultraviolet spectral counts of HD 50138, and to E. Roques (Montpellier) and G. Houziaux (Mons) for help in the reduction of the various spectrometric data.

References

Allen, D. A.: 1973, *Monthly Notices Roy. Astron. Soc.* **161**, 145.
Andrillat, Y. and Houziaux, L.: 1972, *Astrophys. Space Sci.* **18**, 324.
Boksenberg, A., Evans, R. G., Fowler, R. G., Gardner, I. S. K., Houziaux, L., Humphries, C. M., Jamar, C., Macau, D., Malaise, D., Monfils, A., Nandy, K., Thompson, G. I., Wilson, R., and Wroe, H.: 1973, *Monthly Notices Roy. Astron. Soc.* **163**, 391.
Doazan, V.: 1965, *Ann. Astrophys.* **28**, 1.
Haupt, H. F. and Schroll, A.: 1974, *Astron. Astrophys. Suppl. Ser.* **15**, 311.
Houziaux, L.: 1960, *Publ. Obs. Haute Provence* **5**, No. 24.
Huang, S. S. and Struve, O.: 1956, *Astrophys. J.* **123**, 231.
Kurucz, R. L., Peytremann, E., and Avrett, E. H.: 1974, Smithsonian Astrophysical Observatory, Washington D.C.
Kurucz, R. L. and Peytremann, E.: 1975, Special Report, Smithsonian Astrophysical Observatory.
Merrill, P. W.: 1931, *Astrophys. J.* **73**, 348.
Nandy, K.: 1974, private communication.

OBSERVATIONS RÉCENTES DE HD 200120

R. HERMAN et H. HUBERT

Observatoire de Paris, France

HD 20012 (f_1 Cyg) fait partie des étoiles des premiers types B. Nous l'avons classée O9 V?, Nina Morguleff l'a classée B1 IV et Janet Lesh B1,5 V). Elle fait partie de notre programme d'observation qui concerne les étoiles Be de magnitude ≤7 et de déclinaison ≥ −15°. Parmi les étoiles les plus chaudes de ce programme, on peut signaler au moins deux catégories:

(1) celles qui présentent, à certains moments, des raies d'émission de H et He I et Fe II

TABLEAU I

		R-H	Morguleff	Lesh	Chal-Div	
HD	5394	B0 II??	B0 IV	B0,5 IV	B0	
	10516	O9,5 V?	O9 Ia	B2 V P		
	11606	B1V?				
	24534	B0–O9??	B0	O9,5 P	B0	
	200120	O9 V?	B1 IV	B1,5 V	B0 I	
	206773					B0nnek

(2) celles qui ne présentent que des raies d'émission de H et Fe II

TABLEAU II

HD	19243	B1 II?				
	30076	B1 V??	B2 V			
	32343	B3 V	B2 V	B2,5 V	B2 II–III	
	32991	B3,5 V	B2 V			
	36576	B2 V-IV		B2 IV-V		
	37967	B	B5 V	B2,5 V	B2–3 V	
	38010					B3ne
	41335	B3 V?				
	44458			B1 V		
	199356	O III?				
	202904	B2 V–IV	B2 V	B2 V	B1–2	
	212571	B0,5 III–II		B1 V		

HD 200120 vient de subir une explosion d'hydrogène en Décembre 1973 et nous nous proposons de montrer les diverses phases de ce phénomène.

HD 200120 est une étoile brillante dont la déclinaison (+47°) permet l'observation quasi-continue pour les observateurs de l'hémisphère nord. Elle a donc été étudiée particulièrement par nos aînés tels que Curtiss, McLaughlin et autres.

McLaughlin l'avait considérée comme une étoile présentant de longues périodes de calme suivies de brèves périodes d'activité. En fait, lorsqu'on étudie son spectre à grande dispersion on s'aperçoit qu'il y a une masse importante d'hydrogène

A. Slettebak (ed.), Be and Shell Stars, 95–97. All Rights Reserved

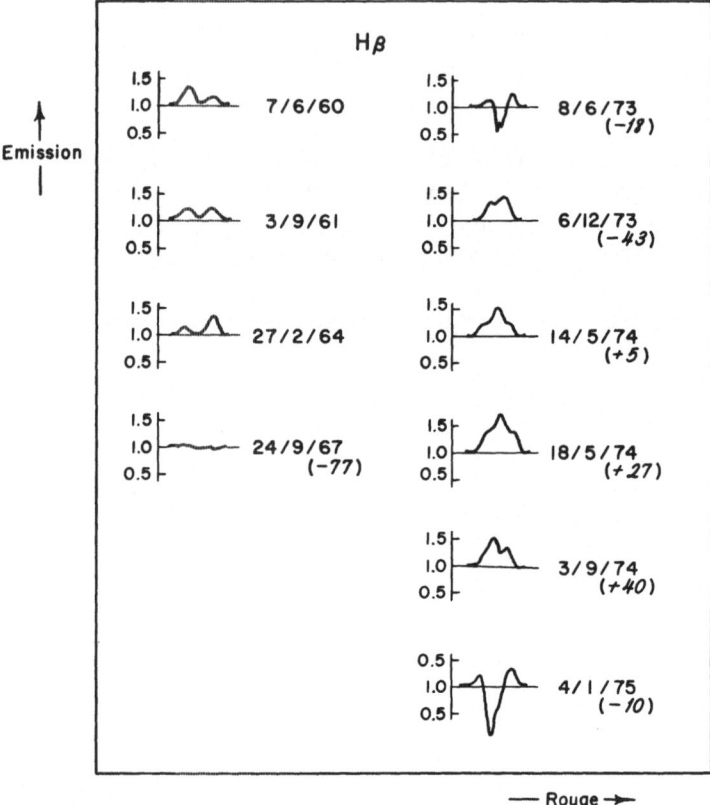

Fig. 1.   Profils de la raie Hβ de HD 200120 de 1960 à 1975.

gravitant autout de l'étoile et que cette atmosphère absorbante est très fluctuante
(Figure 1). Indépendamment de ces fluctuations, plus ou moins erratiques, la
bibliographie permet de prévoir, plus ou moins, des périodes d'activité maximales.
Miczaîka avait observé, en Août, 1948, les raies de l'hydrogène en émission jusqu'à
H15. Cette observation date de 25 ans par rapport à l'explosion de décembre 1973.

   Les prémisses de ce phénomène ont été indiquées probablement avant le mois de
Juin 1973. En effet, le 8 Juin 1973, avec Danièle Briot, nous avons observé cette
étoile avec une dispersion de 12,5 Å mm^{-1} à l'aide du spectrographe Coudé du
150 cm de l'Observatoire de Haute-Provence, sur plaque II-aO chauffées. Le
spectre montrait une enveloppe absorbante importante. En Décembre 1973, les
profils des raies H indiquaient nettement que l'atmosphère extérieure retombait sur
l'étoile d'où un réchauffement considérable des couches basses provoquant
l'émission. Toute la série de Balmer est en émission ainsi que le continu Balmer, au
moins jusqu'à la fin de Septembre 1974. Malheureusement, nous n'avons pas
d'observations entre Septembre et Décembre 1974. En Décembre 1974,
l'atmosphère de HD 200120 ressemble à une géante ou même à une supergéante.

   Ce n'est pas le lieu, ici, d'indiquer tous les processus possibles et je vous indique
seulement les faits observés. Un point intéressant concernant le cliché à grande

dispersion du 4 Janvier 1975, pris fort aimablement par nos collègues Nina Mor-guleff et Agop Terzan: Les raies d'enveloppe H sont toujours prononcées mais les profils de ces raies sont déformés, indiquant une légère émission dans les ailes rouges; de plus, on observe la présence d'une atmosphère très étendue d'hélium montrant toutes les séries: $2\ ^3P^0-n\ ^3S$ et $2\ ^3P^0-n\ ^3D$ jusqu'à λ 3471,8 Å; la série $2\ ^1S-n\ ^1P^0$ jusqu'à λ 3613,6 Å; et les séries $2\ ^1P^0-n\ ^1S$ jusqu'à λ 3872; $2\ ^3S-3\ ^3P^0$ (λ 3889 Å).

Un tel phénomène a été observé et commenté largement par Struve et Wurm (1938) au sujet d'une comparaison des enveloppes de φ Per, ζ Tau.

Il faut noter que les grandes enveloppes de Juin 1973 et de Janvier 1975 sont essentiellement différentes. Avant l'explosion, c'est-à-dire en Juin 1973, les raies d'enveloppe de He I et de Fe II sont de mêmes intensités et la densité de l'enveloppe est assez grande. Au contraire, l'enveloppe de Janvier 1975 ne concerne guère que He I qui est très fort comme nous l'avons indiqué. Les métaux comme Fe II et Si II sont à peu près de même intensité dans les deux enveloppes (Figure 2).

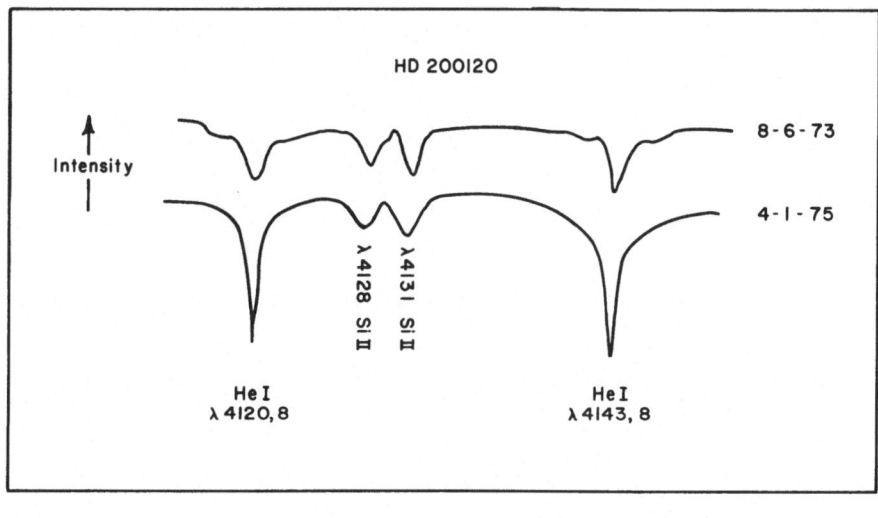

Fig. 2.   Enregistrements, au microphotomètre, de la région λ 4120 Å–4145 Å du spectre de HD 200120, mettant en lumière la différence des enveloppes d'hélium et des métaux *avant* l'explosion d'hydrogène (8-6-73) et *après* cette explosion (4-1-1975).

Enfin, en Juin 1975, des clichés à petite dispersion indiquaient que Hα, Hβ, Hγ(?) et He I étaient en émission et que l'enveloppe d'hydrogène restait importante.

Références

Lesh, J. R.: 1968, *Astrophys. J. Suppl.* **17**, 371.
Morguleff, N.: 1970, communication privée.
Chalonge, D. et Divan, L.: 1971, communication privée.
Struve, O. et Wurm, K.: 1938, *Astrophys. J.* **88**, 84.

SINGLY-IONIZED IRON EMISSION LINES IN THE SPECTRA OF EARLY TYPE STARS

R. VIOTTI*

Laboratorio di Astrofisica Spaziale, Frascati, Italy

and

P. KOUBSKÝ

*Stellar Department, Astronomical Institute of the Czechoslovak Academy of Sciences,
Ondřejov Observatory, Czechoslovakia*

(Paper read by D. G. Hummer)

Abstract. The appearance of singly ionized iron emission lines in the spectra of early type stars is studied, and the results of a spectroscopic investigation of EW Lac and other Be stars are given. We also discuss the atomic processes of excitation of Fe II in the stellar envelopes using a two-parameter diagram $W, N_e T_e^{-1/2}$.

1. Introduction

This paper is concerned with a systematic investigation of singly ionized iron emission lines in the spectra of early type stars. This study is justified by some important facts. First, in the spectra of Be and shell stars the strongest lines of singly ionized iron – notably λ 4233, 4583, 6238 Å, etc. – often have emission components, such as in the 1937–39 spectra of γ Cas (Wellmann 1952), and in EW Lac (see later). Lesh (1968) classified the extreme Be stars, characterized by spectra with prominent Fe II lines, as Be_3–Be_4. These stars are most probably surrounded by very extended atmospheric envelopes. Actually, as we shall show later, infrared observations confirm that stars with more extended envelopes have stronger emission lines, as also suggested by simple theoretical computations.

Forbidden lines of [Fe II] have been observed in HD 200775, the central star of NGC 7023 (Weston, 1949; Viotti, 1969), and in some peculiar B-type stars like XX Oph, BD +61°154, HK Ori and HD 45677. It is worth remembering that ionized metal emission lines frequently appear in superluminous B- and A-type stars, including several stars in the Magellanic Clouds, and in many peculiar objects (η Car, symbiotic stars, etc.). A further argument for this study is that the presence of Fe II in emission in early type stars provides evidence for cool extended atmospheric envelopes. Finally, the conditions of excitation of the ionized metal lines in stellar envelopes are different from those of hydrogen and helium. In addition, the line opacity is obviously much smaller so that in many cases the emitting regions are optically thin to ionized metal lines. Thus their study could be faced in a different and a somewhat simpler way.

In the following we describe (Section 2) the appearance of Fe II emission lines in the spectra of some Be stars, in particular of EW Lac. In Section 3 we discuss the

* Presently ESRO fellow at the Astronomical Institute, Utrecht.

A. Slettebak (ed.), Be and Shell Stars, 99–103. All Rights Reserved
Copyright © 1976 by the IAU.

physical processes populating the Fe II levels and compare spectroscopic observations with infrared photometry.

2. Fe II Emission Lines in Be Stars

EW Lac (HD 217050) is a typical shell star with many sharp Fe II absorption lines. P. Koubsky analysed a spectrogram of this star taken at the Ondřejov Observatory (Plate number 1287, October 8, 1972; original dispersion 8.5 Å mm^{-1}). The following ionized iron lines show, or probably show, emission components: λ 4178.85, 4233.17 (s), 4351.76, 4385.38 (?), 4472.92 (?), 4491.40, 4508.28 (?), 4515.34 (?), 4520.22, 4522.63, 4549.47, 4555.89 (?), 4583.83 (s; see Figure 1), 4629.34, 4923.92, 5018.43 (s), 5169.03 (?), 5275.99, 5316.61 (s), 5534.86 (?), 6238.38, 6247.56 (?), 6318.0 (s) and 6385.5 Å (s). The identification of this last line is questionable, but it is also present in ζ Tau, 48 Lib, in MV Sgr (Herbig, 1975), in η Car and probably in RR Tel.

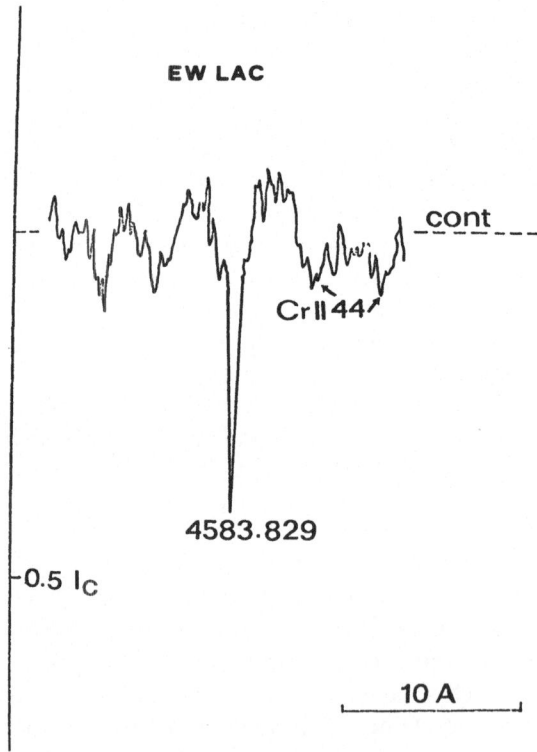

Fig. 1.   The spectrum of EW Lac near the Fe II line at λ 4584 Å showing the shell absorption and the broad emission components (plate N.1287, original dispersion 8.5 Å mm^{-1}).

The intensity tracing near λ 4584 Å is shown in Figure 1. The Fe II emission is double with a strong central shell absorption. In a few cases the Fe II lines present a very broad absorption.

The following equivalent widths have been estimated from the spectrograms for the emission components of seven lines: λ 4233, 0.34 Å; 4520, 0.26 Å; 4583, 0.44 Å; 4628, 0.32 Å; 6238, 0.59 Å; 6318, 0.68 Å; 6385, 0.60 Å. The intensities are comparable to those of γ Cas given by Wellmann (1952). In both stars and in MV Sgr we found that their intensity, normalized to gf λ^{-3} exp $(-E_u/kT_{ex})$ $(E_u =$ energy of the upper level of the transition, $T_{ex} =$ mean excitation temperature of Fe II lines), increases with wavelength as a consequence of the decreasing stellar continuum. γ Cas and EW Lac are classified by Lesh (1968) as Be_1, that is with weak H emission and no Fe II in emission, and this gives an example of the large variability of the intensity of the emission lines in Be stars. In AX Mon according to the profiles published by Peton (1974) the emission at λ 4233 Å is variable in time with an equivalent width ranging from 0.3 to 0.52 Å. A search for emission at this line failed for the stars φ Per, ψ Per, Pleione (28 Tau) and 1 Del, while the line has a V emission peak in the 1968 spectrum of ζ Tau and at all times in the spectrum of HD 218393 during the years 1972–74.

3. Atomic Processes

The problem is how to relate the observed line intensities with the physical properties of the emitting envelopes. An earlier attempt to study the processes of formation of Fe II and [Fe II] emission lines in the stellar spectra was made by Wurm in 1937. Wellmann (1952) applied the curve-of-growth method to hydrogen, He I and Fe II emission lines in the spectrum of γ Cas. Recently Viotti (1976) analysed the problem of excitation of Fe II and [Fe II] in the envelopes of hot stars, and discussed the dominant processes of excitation in some early type stars mostly on the basis of the infrared observations of Gehrz et al. (1974). The physical parameters for the envelopes of some early type stars are illustrated in Figure 2, where W is the geometrical dilution factor, and N_e and T_e the electron density and temperature. Actually, in the figure the abscissa is a measure of the rate of electron collisions, while the ordinate is a measure of the stellar radiation in the envelope. By comparing the rate of collisional excitation of the upper levels of the Fe II transitions (E.P. about 5–6 eV) with photoexcitation through the ultraviolet resonance lines, one finds that the latter dominates when $W \gg 10^{-12} N_e T_e^{-1/2}$ ('radiative' regime of excitation). This boundary is represented in Figure 2 by a dashed line. A competitive process populating the upper levels of Fe II is radiative recombination from Fe III followed by cascade. A precise estimate of its rate is at present impossible, but it seems to be negligible with respect to photoexcitation from the ground level as long as the photospheric temperature is below 30 000 K. These considerations regarding the excitation processes of Fe II are obviously valid for all the ions having similar Grotrian diagrams.

Let us now compare the above considerations with observations. Figure 2 shows the representative points (derived from the infrared observations of Gehrz et al. 1974) for the envelopes of 9 Be stars (γ Cas, π Aqr, χ Oph, ω Ori, φ Per, ζ Tau, EW Lac, 48 Lib and β Psc), of P Cyg and the O 7.5 supergiant ξ Per. For ζ Ori we used the results of the recent study of its Hα profile made by Hearn (1975). All the

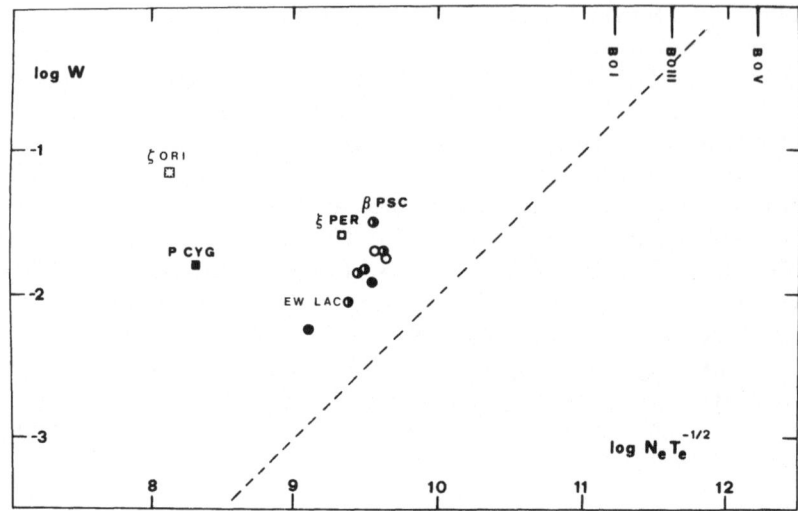

Fig. 2.   Physical conditions in the envelopes of some Be stars (circles), of P Cyg and ξ Per (Viotti 1976), and of ζ Ori (Hearn, 1975). The emission character of each Be star is indicated according to Lesh (1968). Open circles: no emission. Half-filled circles: stars with weak emission (Be_1). Filled circles: stars with strong emission (Be_4). EW Lac classified Be_1 by Lesh in 1972 showed many Fe II emission lines. The dashed line ($N_e T_e^{-1/2} = 10^{12}\ W$) is the boundary between the 'radiative' region (upper left) and the region where electron collision is the dominating process of excitation of the emission lines of Fe II. The value of $N_e T_e^{-1/2}$ in the photospheres of B-type stars is also indicated.

representative points of the Be stars, except perhaps β Psc, apparently form a linear sequence with $N_e T_e^{-1/2}$ proportional to the dilution factor W, that is to R_E^{-2}, where R_E is the radius of the envelope. This suggests that the same physical process of mass loss is acting in these stars and that probably the rate of mass loss is similar. In Figure 2 the representative points of the Be stars fall on the left side of the dashed line, implying that in their envelopes the rate of collisional excitation of the upper levels of metallic ions is very low. Under these conditions the lower (metastable) Fe II levels are in LTE with the ground level, while the upper levels of the permitted transitions are depopulated with respect to LTE roughly by a factor W (Viotti, 1976), so that the intensity of the Fe II emission lines is proportional to the number density of the Fe^+ ions and to R_E. As a matter of fact, in Figure 2 the stars with more extended envelopes have the stronger emission character (filled circles and EW Lac). This result is of particular importance for the study of some peculiar variables, namely of the spectral variations of the novae. Taking $R_E = 2 \times 10^{12}$ cm, $N(Fe^+) = 10^7$ cm^{-3}, and $T_{star} = 20\,000$ K we find that for an optically thin envelope the strongest Fe II lines should have an emission equivalent width of a few 0.1 Å, in agreement with the intensities observed in EW Lac and γ Cas.

Many Be stars have an infrared excess produced by free-free emission in an extended envelope, and we have shown that in these stars the Fe II lines are, or should be according to elementary theoretical computations, in emission with an intensity increasing with the radius of the envelope. Thus, *there is a physical connection between the presence of singly ionized iron emission lines and an infrared excess in Be stars.* These lines are excited by the ultraviolet stellar radiation, and it would be of great importance to make a systematic investigation of the ultraviolet flux of the Be stars with Fe II in emission near the ionized iron resonance lines. Forbidden ionized iron emission lines should appear when the envelopes are much more extended (*W* smaller than 10^{-3}, Viotti, 1976), in a more favourable condition for dust grain formation. *We therefore expect that* [Fe II] *emission lines are more likely related to infrared excesses by dust emission.*

Acknowledgement

R.V. is grateful to ESRO for the allowance of a fellowship at the Astronomical Institute of Utrecht where this investigation was completed.

References

Gehrz, R. D., Hackwell, J. A., and Jones, T. W.: 1974, *Astrophys. J.* **191**, 675.
Hearn, A. G.: 1975, *Astron. Astrophys.* **40**, 277.
Herbig, G. H.: 1975, *Astrophys. J.* **199**, 702.
Lesh, J. R.: 1968, *Astrophys. J. Suppl.* **17**, No. 151, 371.
Peton, A.: 1974, *Astrophys. Space Sci.* **30**, 481.
Viotti, R.: 1969, *Mem. Soc. Astron. It.* **40**, 75.
Viotti, R.: 1976, *Astrophys. J.* **204**, 293.
Wellmann, P.: 1952, *Z. Astrophys.* **30**, 96.
Weston, E. B.: 1949, *Publ. Astron. Soc. Pacific* **61**, 256.

PLEIONE AS A VARIABLE STAR

A. S. SHAROV and V. M. LYUTY

Moscow University, Moscow, U.S.S.R.

The well-known shell star Pleione is also an interesting variable star, BU Tau. Its variability was first discovered by Calder (1937) from photoelectric observations in the 1930's. The light curve of this star from the end of the 19th century to 1954 was studied by Binnendijk (1949, 1955). Botsula and Sharov (1959) and Sharov (1961) added some old and new observations and continued the light curve up to 1959.

Variations of Pleione are connected with the absorption of the starlight in the shell material. Possibly this phenomenon also took place in 1883–1884, when the star was faint according to the observations of Lindemann (1884). Near that time hydrogen emission lines were observed in the spectrum of the star. Between the two decreases the brightness of the star was practically constant. The well observed ejection of the shell in 1936 led to the decrease of light during the period from 1937 to 1958. Some observations of Pleione were published after 1958. In autumn 1971 we started systematic photoelectric *UBV* observations of Pleione at the Crimean station of the Sternberg Astronomical Institute (Sharov and Lyuty, 1972a, 1972b, 1973, 1975). It is interesting to note that the decrease of Pleione was found during the second night of observations! Now these observations are continuing. The light curve from all our observations up to April 1975 is presented in Figure 1. The decrease in brightness lasted for two years, and the minimum took place at the end of 1973. The brightness

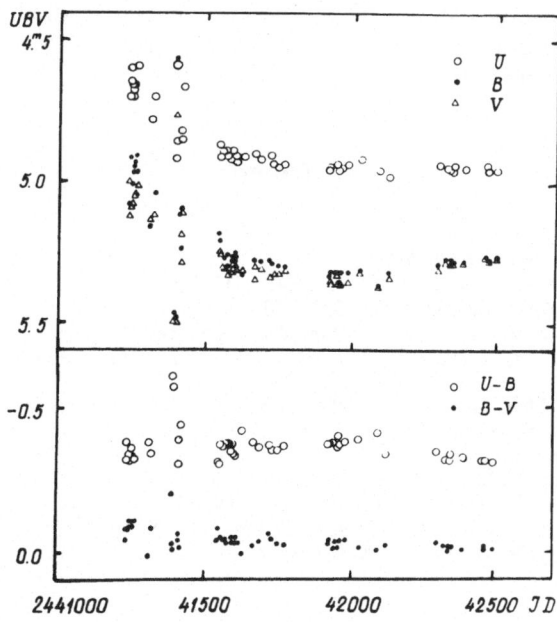

Fig. 1.   Light curve of Pleione from photoelectric observations from 1971 to 1975.

A. Slettebak (ed.), Be and Shell Stars, 105–106. *All Rights Reserved* in *B* and *V* then started to increase slowly, but in *U* the star is still at minimum. We have an opportunity to observe a new shell phase of Pleione photometrically from the beginning in 1971. The appearance of the absorption shell spectrum of Pleione was observed a year later, in December 1972 (Morgan *et al.*, 1973), and thereafter.

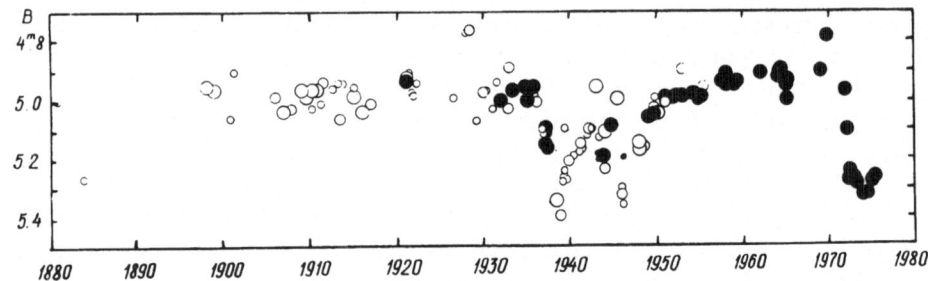

Fig. 2.   Total light curve of Pleione from 1880 to 1975. Filled circles represent photoelectric observations.

The total light curve of the star is presented in Figure 2. We used all the published material (Botsula and Sharov, 1959; Sharov, 1961; Sharov and Lyuty, 1972b, 1973, 1975; Haupt and Schroll, 1974; Johnson *et al.*, 1966). Here we give average magnitudes from our observations. The observations of Johnson *et al.* need some systematic corrections.

During a period of less than one century, astronomers have observed two (or three) shell outbursts of Pleione. The previous case was studied in detail. We hope also that new photometric and spectral observations will give much interesting information about the nature of Pleione.

References

Binnendijk, L.: 1949, *Astron. J.* **54**, 117.
Binnendijk, L.: 1955, *Astron. J.* **60**, 364.
Botsula, R. A. and Sharov, A. S.: 1959, *Perem. Zvezdy* **12**, 398.
Calder, W. A.: 1937, *Ann. Harv. Coll. Obs.* **105**, 453.
Haupt, H. F. and Schroll, A.: 1974, *Astron. Astrophys. Suppl. Ser.* **15**, 311.
Johnson, H. L., Mitchell, R. I., Iriarte, B., and Wisniewski, W. K.: 1966, *Commun. Lunar Planet. Lab.* **4**, No. 63, 99.
Lindemann, E.: 1884, *Mém. Acad. Imp. Sci. St-Petersb.*, ser. VII, **32**, No. 6.
Morgan, W. W., White, R. A., and Tapscott, J. W.: 1973, *Astron. J.* **78**, 302.
Sharov, A. S.: 1961, *Perem. Zvezdy* **13**, 443.
Sharov, A. S. and Lyuty, V. M.: 1972a, *Inform. Bull. Var. Stars*, No. 698.
Sharov, A. S. and Lyuty, V. M.: 1972b, *Perem. Zvezdy* **18**, 377.
Sharov, A. S. and Lyuty, V. M.: 1973, *Inform. Bull. Var. Stars*, No. 814.
Sharov, A. S. and Lyuty, V. M.: 1975, *Astron. Circ.*, No. 872.

ENERGY DISTRIBUTIONS OF Be STARS

(Review Paper)

R. E. SCHILD

Center for Astrophysics, Harvard College Observatory and Smithsonian Astrophysical Observatory, Cambridge, Mass., U.S.A.

Abstract. Energy distributions from 0.32 to 1.06 μ for field Be stars having p designations in their spectral types show that stars with Balmer discontinuities in emission do exist. Furthermore, the filling in of the Balmer continuum near the discontinuity is usually observed. The continuum emission showing a peak near 1 μ is a common feature in the energy distributions presented here.

Energy distributions for a number of bright field Be stars are also presented and discussed.

1. Introduction

Why should we study Be star continuum energy distributions? Do they promise to tell us anything about the photospheres or shells around Be stars that we do not already know from 75 years of observations of line spectra? What promise do they offer for new insight into contemporary problems of shell support and acceleration, shell geometry, and stellar evolution?

While it would surely be ambitious to attempt to solve all of the outstanding problems with a single type of observation, I hope to suggest that the study of continuum energy distributions can potentially improve our understanding greatly. This comes about principally because the continuous energy distribution originates in different regions of the star than does the line spectrum. As Woolf *et al.* (1970) have shown, the continuum in the infrared can tell us about the hydrogen density in the shell, while, as we shall see below, the Balmer and Paschen discontinuities may teach us about the photosphere itself.

To summarize what is already known about energy distributions of Be stars, I show in Figure 1 data for stars in the vicinity of h and χ Persei from photoelectric and PbS measurements by Schild *et al.* (1974). It was particularly important to study stars in a region where the interstellar extinction is known, because, as can be seen, the energy distributions are very peculiar and offer, at the outset, little clue of the color excesses, $E(B - V)$, which might be attributed to the interstellar medium. However, comparison of these dereddened energy distributions with stellar models, fitted as shown in Figure 1, shows that the slope of the energy distribution in the λ 4000–5000 Å region probably follows rather well the model continuum, and might profitably be used to deredden stars of unknown $E(B - V)$. Note that the excess emission which appears to peak in the 1 μ region affects the continuum flux at wavelengths as short as V of the UBV system, so that these Be stars have an intrinsic $B - V$ color excess relative to normal stars of the same spectral type.

The excess radiation at visual and near-infrared wavelengths has been interpreted by Schild *et al.* (1974) in terms of the H^- free-bound emission predicted by Milkey and Dyck (1973). The identification is based upon the observation that the excess emission appears to become prominent in the λ 5000 Å region, reach a peak in the 1 μ region, and decline in the 2 μ region. The excess relative to the stellar continuum

A. Slettebak (ed.), Be and Shell Stars, 107–119. All Rights Reserved

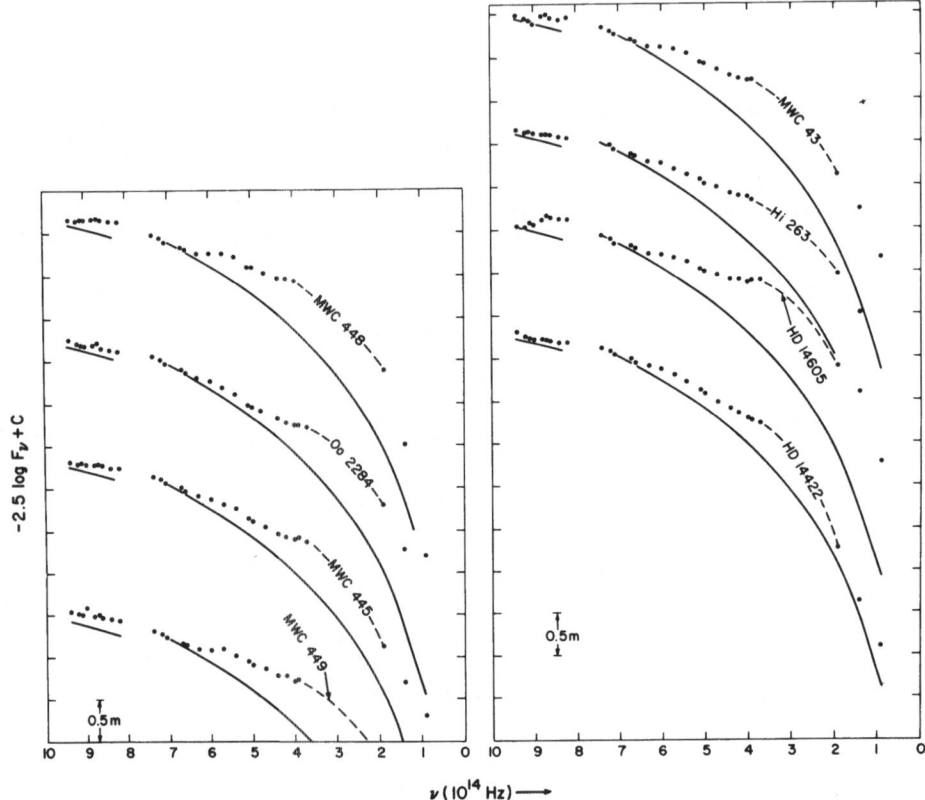

Fig. 1.   Energy distributions of Be stars in the h and χ Persei association, corrected for interstellar reddening with $E(B-V)$ estimated from a map of color excesses for stars having UBV photometry.

is shown in Figure 2, where the theoretical H⁻ free-bound emission spectrum is also shown for comparison. I underscore here the problem with the H⁻ free-bound interpretation, namely, that the peak of the excess appears to occur at significantly shorter wavelengths than the theoretical curve. The latter results entirely from the physics of the electron–hydrogen interaction and must be considered fixed. As I shall show shortly, newer observations confirm the existence of this discrepancy.

Also evident in Figure 1 is an excess of continuum emission below the Balmer discontinuity. This excess causes the energy distribution of the stars in Figure 1 to be nearly continuous across the Balmer jump. Schild *et al.* (1974) also presented energy distributions of 10 field stars and found that several had Balmer jumps too small for their spectral types. While the interpretation of this fact, in terms of the temperature gradient in the Be star atmosphere, is not clear, it is probably safe to say that the U color of the UBV system will be affected, so that the relationship between spectral type and $B-V$ or $U-B$ for normal stars will be invalid for Be stars. A similar conclusion has been reached, on the basis of UBV photometry, by Feinstein (1968), who also noted variability in $(U-B)$.

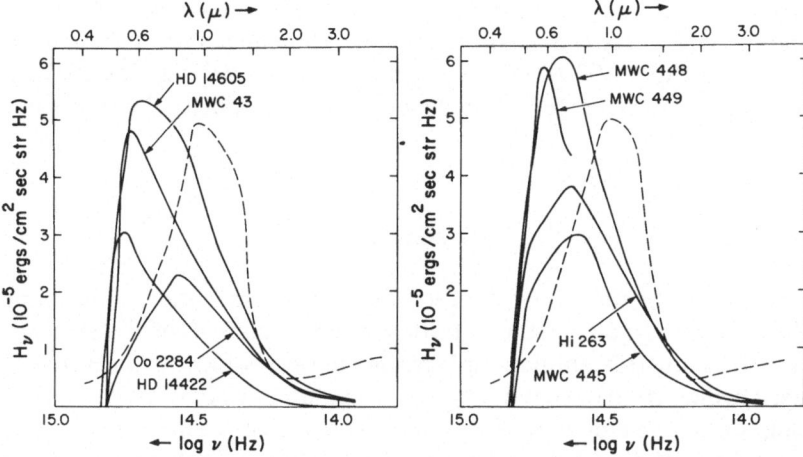

Fig. 2.   The $1\,\mu$ feature, shown from heavily smoothed data for the stars in Figure 1, but with the continuum of a normal B1 star, HD 13900, subtracted off. The normalized curve for an isothermal shell of 3000 K is shown as a dashed line.

The effects of the excess continuum radiation in the visual to near-infrared and below the Balmer jump are shown schematically in the two-color diagram, Figure 3. The excesses in both colors, $(B-V)$ and $(U-B)$, cooperate to cause reddening to be overestimated when conventional dereddening procedures are followed. Because no correlation of the intrinsic $(B-V)$ or $(U-B)$ excesses with other measurable

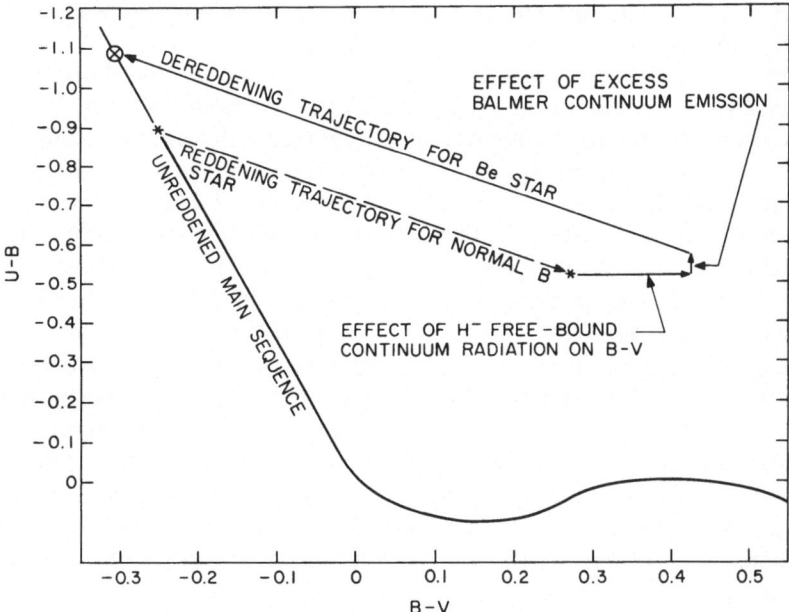

Fig. 3.   Schematic two-color diagram, to show the effects of continuum emission below the Balmer jump and in the near infrared. Relative to a normal reddened B star, the Be star would be dereddened to a color too blue for its photospheric temperature.

parameters, such as Hα emission line strength, are yet known, the only way to determine the color excess due to interstellar reddening is to fit the continuum to stellar models in the λ 4000–5000 Å region. Infrared observations are affected by free-free emission to the extent that they do not help specify the interstellar color excess.

2. New Data

To provide a basis for further exploration of energy distributions, I have measured energy distributions for 60 Be stars. The observing list consists of all the stars north of declination 0° having BD numbers and having the *pe* classification in Hiltner's (1956) catalogue. This selection of spectral type is expected to yield a large number of stars with strong continuum emission. Not all stars were measured in the near infrared (0.8 to 1.0 μ), and I present at this time energy distributions of only those stars for which full energy distributions from 0.32 to 1.06 μ are available. Some questions which we might ask of these energy distributions are:

  (i) how common are the infrared excesses which peak at 1 μ?
  (ii) do stars with negative Balmer jumps exist?

Before these questions can be answered, the energy distributions must be corrected for interstellar reddening. Three special problems arise in this connection:

  (i)   the problem of determining the color excesses, $E(B-V)$;
  (ii) the variation of the shape of the interstellar reddening curve with galactic longitude;
  (iii) the shape of the interstellar reddening curve for our shortest wavelengths, 0.32 to 0.34 μ.

Problem (i) has been dealt with by plotting the energy distributions for eight choices of $E(B-V)$, and fitting a Kurucz *et al.* (1974) model in the λ 4000–5000 Å region to achieve a best fit. Problem (ii) has arisen recently with the recognition from OAO II data (Bless and Savage, 1972) that the far ultraviolet reddening law has important longitude variations, and from the results of Hayes *et al.* (1973) that smaller differences also affect visible energy distributions, especially in the 0.4 to 0.5 μ region. Because many of our stars are heavily reddened, with several $(B-V)$ color excesses of one magnitude, it is important to allow for these longitude variations. We have done this by using an average of Radick's (1973) curves 1 and 2 for $18^h \leq \alpha < 1^h$, curve 3 for $1^h \leq \alpha \leq 5^h$, and curve 4 for $\alpha > 5^h$. Problem (iii), the shape of the interstellar extinction curve at λ 3200 Å remains vexing because of the difficulty of getting accurate energy distributions of heavily reddened stars at the shortest wavelengths. The Radick's (1973) data which show the longitude dependence of interstellar reddening extend to only 0.3448 μ, but are simply extrapolated to 0.32 μ for the present work. It seems likely that the intrinsic Be star energy distributions are too peculiar to derive the interstellar reddening law from the present data.

In Figure 4a, b, c, d, and e we show the new energy distributions corrected for interstellar extinction. They are arranged in a sequence of increasing Balmer discontinuity. All energy distributions shown here are on the calibration system of

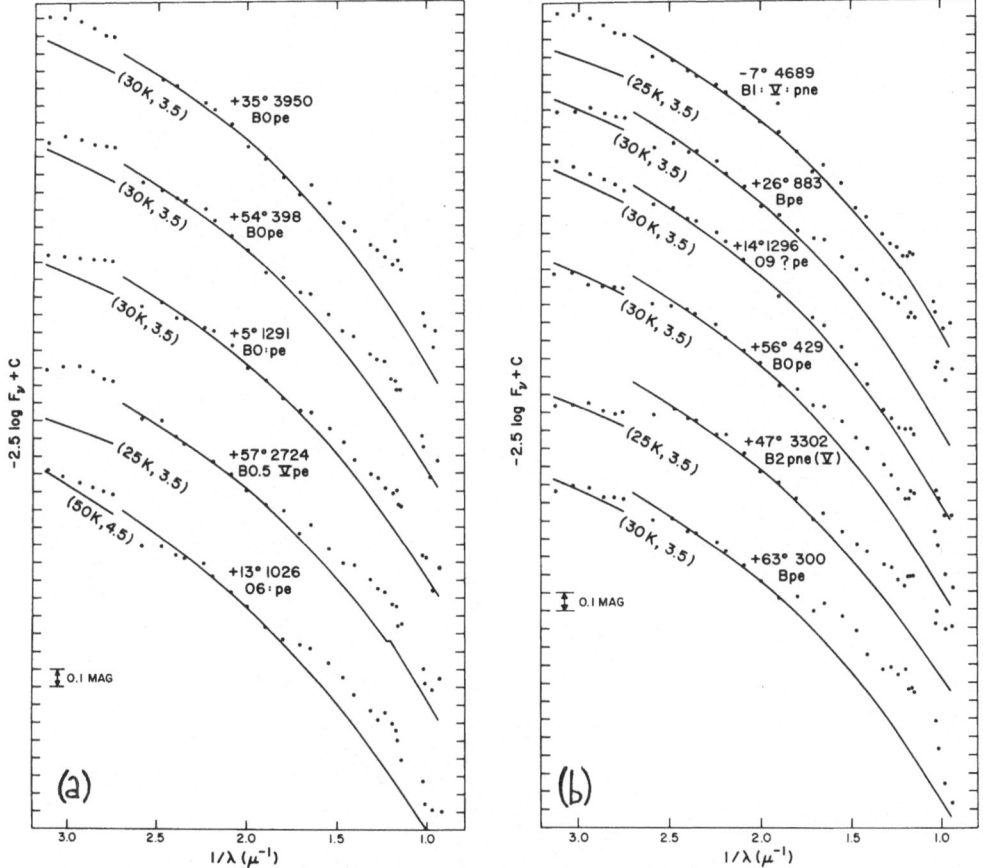

Fig. 4a–e. Field Be star energy distributions, corrected for reddening and fitted to models with temperatures appropriate to the spectral types. (For Figure 4c–e see next page.)

Hayes and Latham (1975), because this has a more positive Balmer discontinuity than the calibration of Oke and Schild (1970). Thus, our conclusions regarding the existence of Balmer discontinuities in emission are conservative.

In Figure 4a, we show energy distributions of five stars in which the Balmer discontinuities are in emission by 0.05 mag. or more. These stars tend to be the Be stars of earliest spectral type, B0pe, and are not confined to any one region of the sky. The existence of emission Balmer discontinuities had already been noted by Barbier and Chalonge (1949), for the bright star γ Cas, which is also of spectral type B0. The existence of Balmer discontinuities in emission is of interest because it implies that scattering is not the dominant source of opacity in the Be star envelope, and that continuum emission from the envelope is also likely to be important on the long wavelength side of the Balmer discontinuity. This in turn implies that line profiles are diluted by continuous emission from the Be star envelope, and cannot be safely used to estimate surface gravities or $v \sin i$.

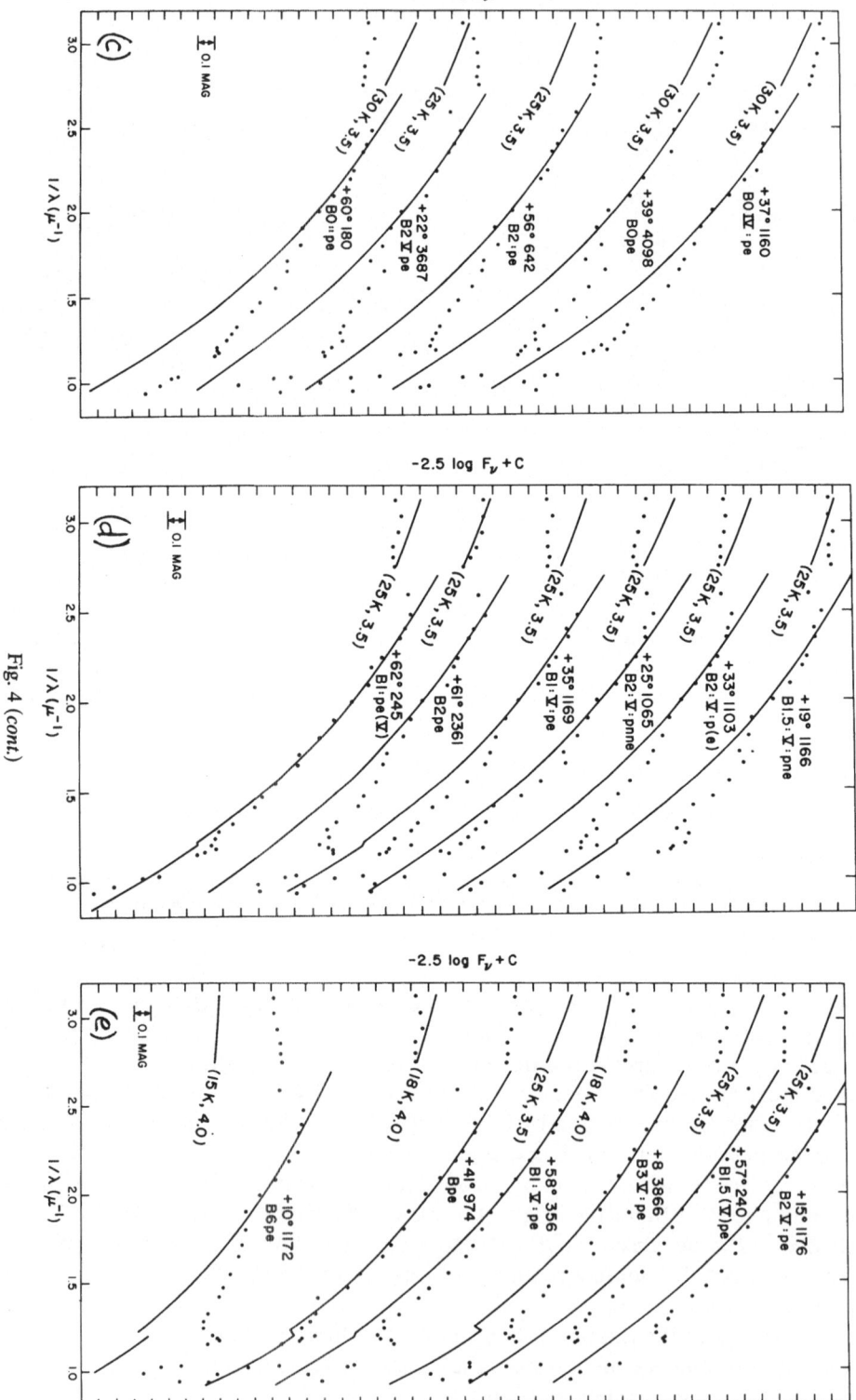

Fig. 4 (cont.)

At the shortest wavelengths below the Balmer discontinuity, data for all of our stars often show a slope inconsistent with the model continuum. At present we must consider this an effect of the uncertainty in the interstellar reddening law at these wavelengths.

Turning next to the near infrared, we note that all of the stars in Figure 4a have moderate to large excesses, of the order of 0.2 mag., which peak in the 0.8 to 0.9 μ region and which are significantly diminished at 1.06 μ. We also note the lack of evidence for a Paschen discontinuity in emission or absorption for these very hot Be stars.

As we pass to increasingly cool stars, in Figures 4b, c, d, e, we see the Balmer discontinuity becoming longer, the spectral type becoming later, and, possibly, evidence for emergence of the Paschen discontinuity. Although the data in the infrared are noisy, I have the impression that the Paschen discontinuities are larger than the model prediction.

The existence of infrared continuum emission has been recognized in many southern Be stars by Feinstein (1968) from *UBVRI* photometry. Feinstein (1970) also has reported variability in *V* magnitude and *B* − *V* color, which suggest that the near infrared emission is probably variable in at least some stars.

3. I.A.U. Symposium 70 Stars

I wish now to pass on to the IAU Symposium 70 stars. As most of you know, the organizers of this symposium announced a consensus of early symposium participants that intensive observations of four stars, γ Cas, ϕ Per, ζ Tau, and 28 Tau, be made by all available observational techniques to permit a synthesis of their many properties. I show in Figure 5 the energy distributions from 0.32 to 1.06 μ for these four stars.

The B0 p star γ Cas is seen to have a continuous energy distribution across the Balmer jump, and a moderate excess in the near infrared. Chalonge and collaborators (Arnulf *et al.*, 1938) have found the Balmer discontinuity strongly in emission in 1937, and significantly variable on a time scale of six months. Thus we conclude that the shell in γ Cas was relatively thin when our observations were made in October 1974.

The B1 p star ϕ Per is also nearly continuous across the Balmer jump, and also has a moderate excess in the near infrared. We note here that the energy distributions follow the extrapolated Paschen continuum and parallel the Balmer continuum of the Kurucz *et al.* (1973) models quite well, which suggests that the fainter stars in Figure 4 still have systematic errors in the dereddening at the shortest wavelengths, as discussed in Section 2. The Balmer discontinuity of ϕ Per given by Chalonge and collaborators appears to be constant and near zero, in agreement with the October 1974 measurement.

For the B2 IV p star ζ Tau I was unable to decide on the correct color excess, and have plotted energy distributions for two choices. The first choice fits well in the λ 4000 Å region and in the infrared, but has a moderate trough in the λ 5000 Å region, which is usually an indication that the star has been overcorrected for

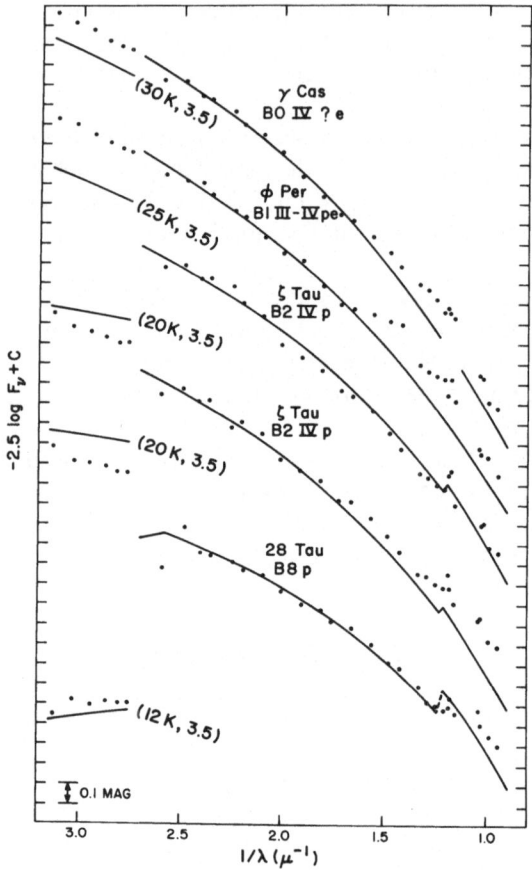

Fig. 5. Energy distributions of IAU Symposium 70 stars, fitted to models appropriate to the spectral types. The data for ζ Tau are shown for two choices of color excess, $E(B - V)$.

reddening. A better fit at the λ 4000–5000 Å region is obtained for the second choice of color excess, which gives a mild excess in the near infrared. Note, however, that the infrared excess does not peak in the 0.8 μ region, but instead appears to increase monotonically with wavelength, unlike the stars in Figures 4a, b, c, d.

Finally, we note that the fit of 28 Tau to the (12 K, 3.5) model is quite good, except that the observed Paschen discontinuity appears to be somewhat too large, as was noted for the cooler stars in Figure 4.

4. Energy Distributions of 'Pole-on' Stars

In this section we discuss the two groups of 'pole-on' stars which were investigated by means of infrared data and other techniques by Schild (1973). At the time I made the present observations, I expected all of the original pole-on stars to show the 1 μ excess.

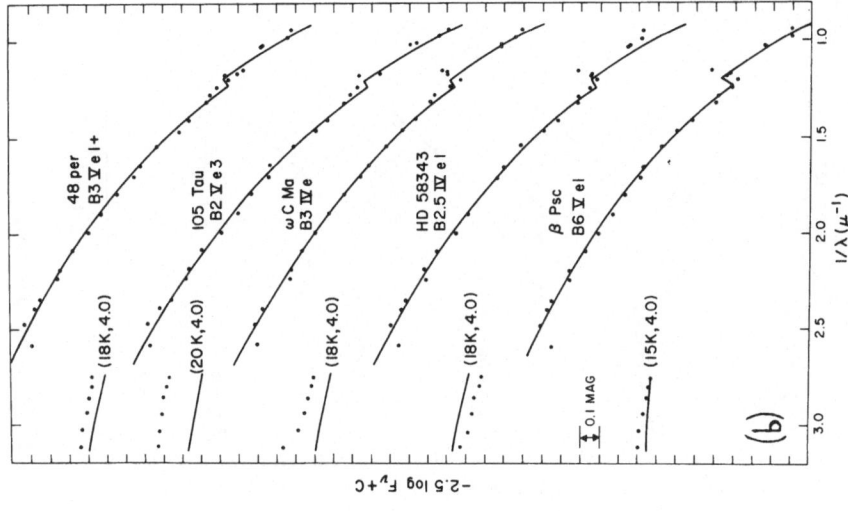

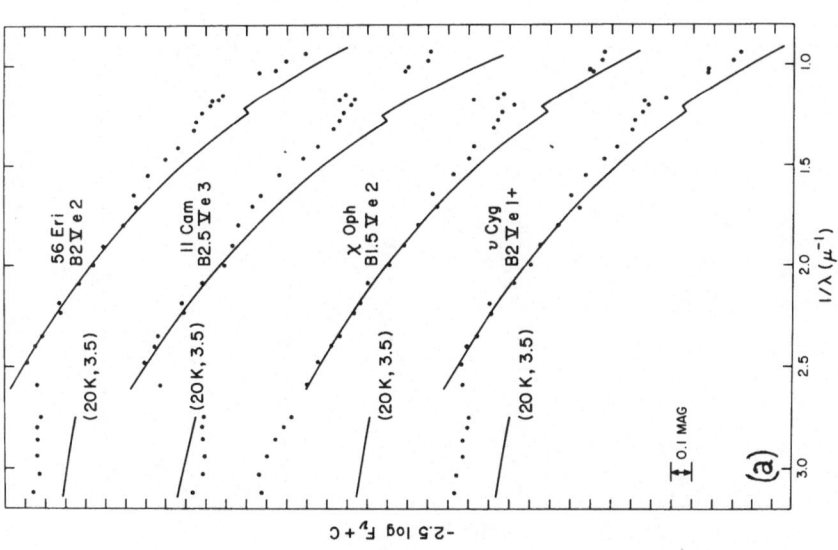

Fig. 6.   Energy distributions for the classic 'pole-on' stars.

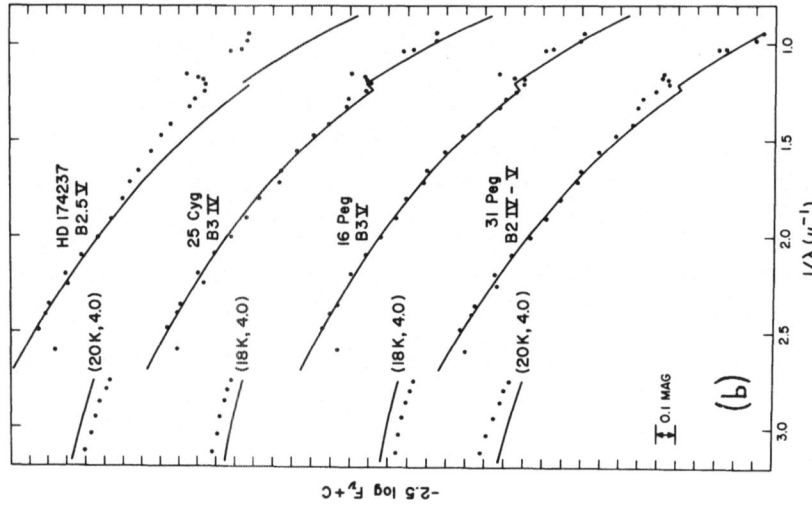

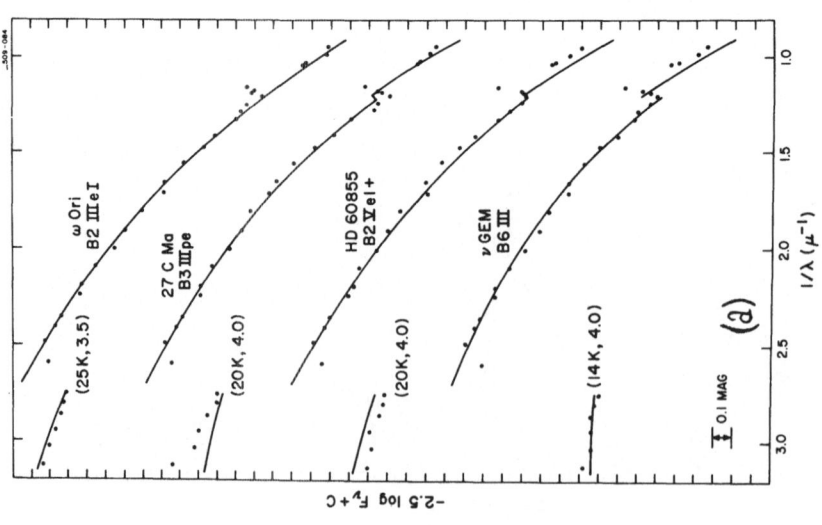

Fig. 7.   Energy distributions for the group of stars identified by Schild (1973) as true pole-on stars.

Figure 6a shows energy distributions of the classic pole-on stars which do show the 1 μ excess. We note that χ Oph appears to be nearly constant across the Balmer discontinuity, much as was found by Barbier and Chalonge (1941). In Figure 6b we have energy distributions for the five remaining pole-on stars. These appear to be quite normal for their spectral types, which are on the average only slightly later than the stars in Figure 6a. We may note that the Balmer jumps for stars having an infrared excess are systematically smaller than for those lacking the 1 μ excess.

One of the stars in Figure 6b, 105 Tau, may have a weak excess in the near infrared. When stars having such an excess are fitted to continuum models by overestimating the color excess $E(B - V)$, the fitted energy distribution shows a slight excess in the λ 4000 Å region and a moderate deficiency in the λ 5000 Å region. This is faintly seen for 105 Tau as seen in Figure 6b and has also been discussed in the preceding section with regard to ζ Tau, where two choices of the reddening are given. We have adopted here the conservative value of $E(B - V)$ in plotting our data for 105 Tau, so a good fit to the model is obtained in the near infrared.

It is not yet clear why some of the stars in Figures 6a and b have the near infrared excess whereas others lack it. We note, however, that the stars in Figure 6b which lack the excess appear systematically to be cooler than the stars in Figure 6a. This is particularly true if the ambiguous case 105 Tau is allowed for.

In Figures 7a, b we show energy distributions of stars identified by Schild (1973) as 'true pole-on stars'. They have been fitted to Kurucz *et al.* (1973) models appropriate to their spectral types. In general, the fit to the continuum is good except that the Balmer discontinuities are, on the average, too small. One star, HD 174237, shows evidence for the 1 μ infrared excess. Since there is no tendency for the spectral types of the true pole-on stars to be different from those of the classical pole-on stars, there may be some slight evidence for a greater incidence of 1 μ excess emission among the latter group of stars.

References

Arnulf, A., Barbier, D., Chalonge, D., and Jafir, H.: 1938, *Ann. Astrophys.* **1**, 293.
Barbier, D. and Chalonge, D.: 1941, *Ann. Astrophys.* **4**, 30.
Bless, R. and Savage, B.: 1972, *Astrophys. J.* **171**, 293.
Feinstein, A.: 1968, *Z. Astrophys.* **68**, 29.
Feinstein, A.: 1970, *Publ. Astron. Soc. Pacific* **82**, 132.
Hayes, D. and Latham, D.: 1975, *Astrophys. J.* **197**, 593.
Hayes, D., Mavko, R., Radick, K., Rex, K., and Greenberg, J. M.: 1973, in J. M. Greenberg and H. C. van de Hulst (eds.), 'Interstellar Dust and Related Topics', *IAU Symp.* **52**, p. 83.
Hiltner, W. A.: 1956, *Astrophys. J. Suppl.* **2**, 389.
Kurucz, R., Peytremann, E., and Avrett, E.: 1974, *Blanketed Model Atmospheres for Early Type Stars*, Smithsonian Institution Press, Washington, in press.
Milkey, R. and Dyck, M.: 1973, *Astrophys. J.* **181**, 833.
Oke, J. B. and Schild, R.: 1970, *Astrophys. J.* **161**, 1015.
Radick, R. R.: 1973, Master's Thesis, Rensselaer Polytechnic Institute.
Schild, R.: 1973, *Astrophys. J.* **179**, 221.
Schild, R., Chaffee, F., Frogel, J. and Persson, E.: 1974, *Astrophys. J.* **190**, 73.
Woolf, N. J., Stein, W. A. and Strittmatter, P. A.: 1970, *Astron. Astrophys.* **9**, 252.

DISCUSSION

Peters: (1) I doubt whether continuous emission from the envelope contributes significantly in the blue portion of the spectrum in most cases. The maximum measured equivalent widths of the He I features in Be stars are comparable to the maximum values measured for non-Be stars. Specifically, in many cases we measure equivalent widths of 1.3 Å or higher for the stronger He I features (i.e. $\lambda\lambda$ 3819, 4026, 4471).

Schild: Our models do not yet completely reproduce the continuum so I cannot yet be sure of the contribution of the shell. My stars were selected to have strong emission, and are not typical.

Peters: Can you be sure that your infrared magnitudes are free from Paschen line emission?

Schild: The infrared bands were chosen to be free of the Paschen lines, except for one which was chosen to measure emission from a single Paschen line.

Peters: What T_{eff} and log g did you obtain for ϕ Per?

Schild: The energy distribution of ϕ Per is so nearly continuous across the Balmer jump that it would imply a temperature in excess of 50 000 K if fitted to a plane parallel model in static equilibrium. Such a temperature is completely incompatible with the absorption line spectrum.

Snow: If the infrared excesses in these stars are due to H^- free-bound emission, then the amount of excess flux is an indication of the formation rate of bound H^- in the circumstellar shell. H^- has predicted discrete transitions in the far-ultraviolet, and *Copernicus* data have been used to show that H^- is probably not present in the amount inferred from the infrared excess in at least two cases (χ Oph: Snow, *Astrophys. J.* **198**, 361, 1975, and υ Cyg, unpublished) of Be stars which show the kind of infrared excess which you attribute to H^- free-bound emission. This implies that these infrared excesses are not due to H^- free-bound emission, although it should be pointed out that, because it is a very difficult laboratory experiment, the expected discrete transitions in bound H^- have yet to be observed, and at present nearly all that is known about them is based on theoretical calculations.

Schild: There is a strong line just shortward of the wavelength of the predicted H^- feature (λ 1129.6 Å) in the spectrum of χ Oph, and I suggest that this is the H^- feature, shifted due to expansion in the circumstellar shell.

Snow: The feature you refer to is identified as Si IV λ 1128.34, and is strong in all early B-stars observed with *Copernicus*. It may be possible that the H^- feature is present and is shifted so that it is superposed on the Si IV line and therefore is not seen, but this seems unlikely for two reasons: (1) it is somewhat ad hoc to assume that the line is present but shifted by just the right amount so that it can't be seen, especially since more than one star is under consideration, and (2) in χ Oph the numerous shell lines due to Fe II and Fe III in the vicinity of λ 1130 Å are not noticeably more shifted than the stellar rest frame, so it is difficult to see why the H^- feature should be.

Slettebak: You stated that you chose stars having a 'p', or peculiar, designation. In what sense were these stars peculiar?

Schild: These stars were taken from Hiltner's 1956 list (*Astrophys. J. Suppl. Ser.* **2**, 389) and the peculiarity was not stated. My selection inclined to favor stars with strong infrared excesses, however.

Feinstein: With respect to the near infrared excess of the Be stars, I showed some years ago (*Z. Astrophys.* **68**, 29, 1968) that when one makes two-color diagrams with the *UBVRI* measures (for example, $V-I$ vs $B-V$), the Be stars appear above the main sequence. This gives evidence of the near infrared excess of these stars, in agreement with Schild's paper.

Cowley: Have you considered the effects of cool companions in your work?

Schild: I would not expect the energy distributions to reach a maximum and then go down again in the cool companion model. I once tried to add cool energy distributions to hot energy distributions in an attempt to synthesize the observed ones, but the synthesized energy distributions always level off to longer wavelengths instead of receding as observed.

Feinstein: In several bright southern Be stars, we found no infrared excess in *JHKL* photometry.

Haight: Cassinelli and I constructed very extended model atmospheres taking into account the radiative transfer and spherical geometry, and we found that practically all of our models have the Balmer discontinuity in emission. Only the very coolest stars with effective temperatures of 16 000° or 17 000° have Balmer discontinuities in absorption.

Schild: Have these results been published?

Haight: No. These are essentially scaled down versions of a Wolf-Rayet model which was constructed in connection with work on polarization.

Hutchings: Have you concluded that the pole-on stars show infrared excesses?

Schild: I've concluded that half the pole-on stars do and half don't. But if we are talking about the stars defined by Slettebak in 1949, they tend to be somewhat cooler than the stars selected from the Hiltner catalogue which I observed. Therefore it may be that when you get to the cooler stars the frequency of occurrence of this phenomenon is lower. If indeed the interpretation of infrared excess as H⁻ free-bound continuum emission is correct it takes an enormous big shell at a temperature of about 3000°, of the order of 1000 stellar radii, to generate the amount of emission that we found here. It may be that the stars of later spectral type do not sustain such a large shell at the required temperature.

PART II

Be STARS AS ROTATING STARS

Be STARS AS ROTATING STARS: OBSERVATIONS

(Review Paper)

ARNE SLETTEBAK

Perkins Observatory, Ohio State and Ohio Wesleyan Universities, Delaware, Ohio, U.S.A.

Abstract. The classical rotational hypothesis for Be stars is reviewed and discussed. Methods of measuring rotational velocities are considered and the particular difficulties that Be stars pose in this respect are pointed out. Several tests of the rotational hypothesis are discussed and shown to support this idea. New observations of Balmer emission-line widths in the spectra of 46 Be stars show a correlation with rotational velocity, also in support of the rotational hypothesis. The very large emission widths found for some Be stars are discussed and possible reasons for the broadening are presented.

1. Introduction

Over 40 years ago, Otto Struve wrote his classical paper, 'On the Origin of Bright Lines in Spectra of Stars of Class B' (Struve, 1931), in which he stated in the Abstract:

It is found that stars of class B, having widely separated double bright lines are characterized by extremely flat and broad absorption lines suggestive of rapid axial rotation, of the order of several hundred km s^{-1}. Stars having narrow, single emission lines, few in number, show little rotation.

The suggestion is now offered that rapidly rotating single stars of spectral class B are unstable, and form lens-shaped bodies which eject matter at the equator, thus forming a nebulous ring which revolves around the star and gives rise to emission lines. The inclination of the star's axis would then be responsible for the observed range in width of the emission lines.

In view of a considerable number of observations plus some new ideas about Be stars (including the binary hypothesis being discussed at this symposium) which have been developed since the above words were written, it seems appropriate to review and discuss the observational evidence for the classical rotational hypothesis for Be stars at this time.

2. Rotational Velocities of Be Stars

2.1. PROFILE ANALYSIS

If we exclude Galileo's observations of sunspots (1613), profile analysis is the oldest method for estimating stellar rotation. Abney (1877) first demonstrated that stellar rotation would result in a broadening of spectrum lines and suggested that "... other conditions being known, the mean velocity of rotation might be calculated". The first list of rotational velocities, $v \sin i$, was given by Elvey (1930), following a graphical method suggested by Shajn and Struve (1929).

Rotational velocity determinations based on the Shajn-Struve graphical method assume that the flux profile of a sharp-lined star can be used to approximate the non-rotating intensity profile at each place on the disk of a rotating star. Most of the rotational velocities in the literature have been determined using this method; references may be found in the catalog of $v \sin i$'s by Uesugi and Fukuda (1970). With large computers now available, it has become feasible to calculate rotationally

A. Slettebak (ed.), Be and Shell Stars, 123–136. All Rights Reserved
Copyright © 1976 by the IAU.

broadened line profiles directly from model atmosphere results, without recourse to the assumptions implicit in the graphical method. Such calculations will be discussed later in this section.

A different approach was first suggested by Carroll (1933), in which the integral equation relating the observed line profile to the true line profile is solved using Fourier analysis, thereby yielding the equatorial velocity in the line of sight. More recently, Gray (1973) has used a Fourier transform technique to distinguish between microturbulence, macroturbulence, and rotation. The difficulty with such profile analysis methods is that the observed line profile must be known with a very high degree of accuracy in order to extract the required information.

The Be stars as a class represent a special problem with respect to rotational velocity measurements. The very large line broadening in Be stars presumably corresponds to large axial rotation, which suggests that they must be flattened objects. This in turn suggests that the effective gravity must vary across the stellar surface, being high at the poles and low in the distended equatorial regions. If von Zeipel's theorem, or a variation thereof, is valid, the surface brightness, and therefore the effective temperature, is related to the local effective gravity. Thus, the temperature and pressure will vary across the stellar surface. Corrections for such shape distortion and gravity darkening have been applied to line profiles by Slettebak (1949), Collins and Harrington (1966), Friedjung (1968), Stoeckley (1968a), Hardorp and Strittmatter (1968), and others.

Collins (1974) has recently computed a large set of line profiles from rotating model atmospheres, using the ATLAS program (Kurucz, 1970). He assumed a Roche model (Harrington and Collins, 1968) with polar radius and luminosity specified by the interior models of Sackman and Anand (1970), taking shape distortion and gravity darkening (assuming von Zeipel's theorem to hold) into account. Collins computed line profiles for He I λ 4471 Å, Mg II λ 4481 Å, and Fe I λ 4476 Å for models of spectral type ranging from O9 to F8 for various values of the fractional angular velocity and the inclination of the rotation axis. He found that the helium and iron lines are very sensitive to rotation whereas the magnesium line is not, and pointed out that these effects can lead to ambiguities in spectral classification.

The work of Collins was used to establish a system of standard rotational velocity stars (Slettebak et al., 1975), in which high-resolution line profiles in the spectra of 217 bright, early-type stars were compared with his theoretical profiles to estimate rotational velocities. The new $v \sin i$'s obtained in this way are systematically smaller than those derived using the graphical method: about 5% for the A-type and F-type stars and approximately 15% for the B-type stars (cf. Slettebak et al., 1975, Figure 7).

Unfortunately, the Be stars present the biggest problems. Observationally, the broad and shallow line profiles from the underlying star are difficult to measure accurately and may be distorted by both absorption and emission effects from the shell. On the theoretical side, the computed line profiles for gravity-darkened, distorted stars may show considerable changes in equivalent width (and therefore to a lesser degree also in the half-intensity width) depending upon aspect angle, which leads to additional ambiguities in assigning the proper $v \sin i$ to the star (cf. Slettebak *et al.*, 1975, Figure 2). All $v \sin i$'s derived from line profiles for Be stars must therefore be considered rather uncertain.

2.2. PERIODICITIES IN LIGHT OR SPECTRUM

A more direct way of measuring stellar rotation is by detecting a periodicity in the total light or color or line profile from a nonuniform stellar surface. The idea of starspots to explain photometric variability was first proposed by Pickering in 1880 but not confirmed until more recent times (see Vogt, 1975 for references). Thus, Krzeminski (1969), using UBV photometry, found equatorial velocities of 10 to 15 km s^{-1} for dMe stars, and Dr Butler, who reported to us on his Hα photometry yesterday, finds a 40-day periodicity which may be due to rotation in a K3 III star (Butler, 1976). Deutsch (1954) first suggested that for the peculiar A-type spectrum variable stars "we observe the rotation of A stars that exhibit intensely magnetic areas, within which the peculiar line strengths are produced".

In view of the ambiguities associated with the determination of rotational velocities from line profiles in the Be stars, it would be of great importance to measure their rotations more directly. Hutchings (1970) reported a periodicity of 0.7 days in the peak separation and V/R ratio of the double emission profiles of Hγ and Hβ in γ Cas, which he attributed to rotation. Additional measurements of this type for a number of Be stars, using scanners or narrow-band photometry, would be very desirable to help answer the question of how nearly the rotation of Be stars approximates the critical velocity at which the centrifugal force at the equator balances the gravitational force.

2.3. TESTS OF THE ROTATIONAL HYPOTHESIS

We can now raise the question: do the observed $v \sin i$'s for Be stars confirm Struve's rotational hypothesis? Several tests are possible.

Assuming that all Be stars of a given type rotate with the same equatorial velocity v_0 and that their axes of rotation are randomly distributed in space, Struve (1945) showed that the frequency distribution would be expected to show a pronounced maximum at v_0. A sample of 42 Be stars of spectral types B6–B9 (Slettebak, 1966) does indeed show such a distribution. Furthermore, the percentage of stars that show shell absorption increases with increasing $v \sin i$, as would be expected if the shells are equatorial features and the stars with small observed $v \sin i$ are viewed nearly pole-on while those with large $v \sin i$ are seen essentially equatorially. Studies of the distribution of rotational velocities by Stoeckley (1968b) and Hardorp and Strittmatter (1970) reach similar conclusions. Statistical studies by Bernacca (1970), Balona (1975), and Massa (1975) suggest, however, that Be star rotational velocities do not reach the critical velocity.

I recently made visual estimates of $v \sin i$ on the new system (Slettebak *et al.*, 1975) for all the Be stars for which I have spectrograms and plotted the frequency vs $v \sin i$. The stars and estimated rotational velocities are listed in Table I, where standard rotational velocity stars are identified with an asterisk after their $v \sin i$. It should be emphasized that the values listed are of varying quality, since a number of different dispersions were used. Values of $v \sin i$ followed by a colon are considered

TABLE I
Be-star rotational velocities and velocities corresponding to one-half the emission widths

Star	HD	Hβ			Hγ			$v \sin i$
		v_T (km s^{-1})	v_E (km s^{-1})	v_S (km s^{-1})	v_T (km s^{-1})	v_E (km s^{-1})	v_S (km s^{-1})	(km s^{-1})
10 Cas	144							150
o Cas	4 180							220
γ Cas	5 394	340	180	85	315	180	95	230:
φ And	6 811							60
	9 709							320
α Eri	10 144							225*
φ Per	10 516	360	205	120	320	185	125	400
	13 867	–	–	–	105	65	–	60
HR 894	18 552							270
HR 985	20 336	–	–	–	205	170	140	320
	21 641							140
ψ Per	22 192	300	150	85	205	145	100	350
17 Tau	23 302							190
23 Tau	23 480							260
HR 1160	23 552							210
28 Tau	23 862	310	175	115	220	175	135	320
HR 1204	24 479							100
48 Per	25 940	200	100	45:	145	95	55:	200
	26 398							160
11 Cam	32 343	180	90	–	165	65	–	100
25 Ori	35 439							260
120 Tau	36 576	325	190	90	215	165	120	280
ζ Tau	37 202	305	215	145	225	180	140	300
ω Ori	37 490	295	210	115	245	200	155	160*
α Col	37 795	200	140	85	160	135	115	180*
HR 2142	41 335	405	190	65	305	190	110	400
HR 2309	44 996							100
ν Gem	45 542							180
β Mon A	45 725	345	175	110	240	175	130	260*
HR 2418	47 054							240
κ CMa	50 013	325	175	90	265	180	105	150*
ψ^9 Aur	50 658							230
ω CMa	56 139	170	75	20:	150	85	35	80*
β CMi	58 715	215	175	130	160	–	–	245*
HR 2932	61 224							230
HR 3034	63 462	390	250	145	335	235	150	320*
HR 3135	65 875	290	115	30	215	115	50	170
HR 3237	68 980	240	85	30	190	85	35	115*
HR 3498	75 311							240*
HR 3642	78 764	240	150	90	170	140	110	120*
HR 3858	83 953	305	215	140	210	185	160	260*
ω Car	89 080	220	170	120	155	145	135	200
	89 884	285	190	115	190	150	120	300
HR 4123	91 120							270
HR 4140	91 465	305	175	90	200	145	105	250*
HR 4460	100 673							125*
HR 4537	102 776							205*
HR 4618	105 382							65*
δ Cen	105 435	270	125	55	200	120	60	220*
κ Dra	109 387	275	165	95	180	–	125	200
λ Cru	112 078							280*
μ Cen	120 324	160	80	40	120	80	45	155*
	127 617	155	70	–	–	–	–	150:
η Cen	127 972							260*

Table I (*cont.*)

Star	HD	v_T (km s^{-1})	v_E (km s^{-1})	v_S (km s^{-1})	v_T (km s^{-1})	v_E (km s^{-1})	v_S (km s^{-1})	$v \sin i$ (km s^{-1})
			Hβ			Hγ		
θ CrB	138 749							320*
4 Her	142 926							300
48 Lib	142 983	310	230	170	205	175	145	400*
χ Oph	148 184	255	75	30:	220	75	35:	140*
ζ Oph	149 757							320*
α Ara	158 427	370	180	100	235	170	115	230*
	162 428							350
88 Her	162 732							300
	163 848							300:
66 Oph	164 284	370	200	85	335	230	120	280
HR 6720	164 447							220
HR 6873	168 797							240:
	168 957							50:
HR 6881	169 033							200
	171 219							260
	173 371							350
λ Pav	173 948	265	200	150	215	195	170	170*
	174 105	–	–	–	170	150	125	220
HR 7084	174 237							170:
HR 7249	178 175							150
	179 343	–	–	–	210	170	–	320:
HR 7415	183 656							300:
β^2 Cyg	183 914							240
11 Cyg	185 037							340
12 Vul	187 811							260
25 Cyg	189 687							200
	189 689							130
	190 150							270:
28 Cyg	191 610	355	240	160	255	215	175	280
20 Vul	192 044							300
	192 954							330:
25 Vul	193 911	275:	110:	–	–	–	–	210
HR 7843	195 554							220
HR 7890	196 712							220
λ Cyg	198 183							120
59 Cyg	200 120							350:
60 Cyg	200 310							320
υ Cyg	202 904	285	135	55	235	140	75	200
	203 356							300:
6 Cep	203 467	280	145	70	225	170	125	150
HR 8259	205 551							150
ε Cap	205 637	320:	240:	155:	–	–	–	260
	207 232							300
16 Peg	208 057							120
o Aqr	209 409	305	190	115	200	160	130	320
25 Peg	210 129							170
31 Peg	212 076	230	115	50	170	120	80	110
π Aqr	212 571	365:	200:	85:	300:	200:	130:	300:
HR 8682	216 057							320
HR 8731	217 050	335	200	120	235	175	125	300
o And	217 675							280
β Psc	217 891	175	85	–	135	75	35	100
	220 300							350:
	220 582							300

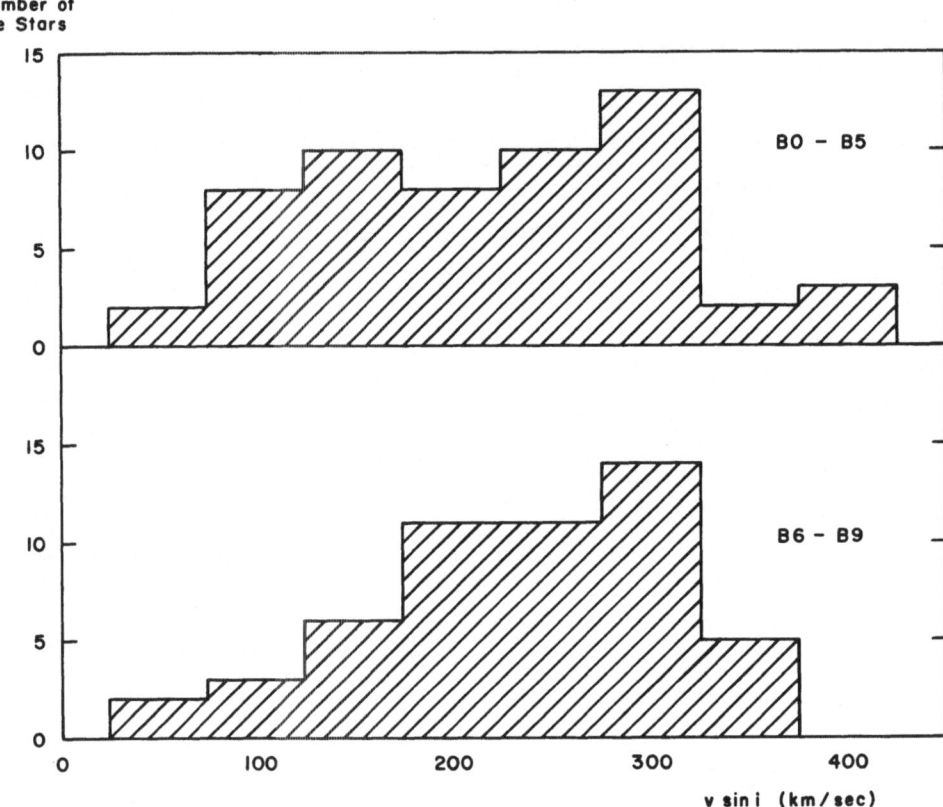

Fig. 1.   Histograms showing the frequency distribution of observed rotational velocities for 56 Be stars of spectral type B0–B5 and 52 Be stars of spectral type B6–B9.

to be rather uncertain, usually because of inferior plate quality. The 56 B0–B5 and 52 B6–B9 stars are plotted in Figure 1. Again, the B6-B9 distribution shows the expected maximum at large $v \sin i$. The B0–B5 distribution shows a tendency in that direction but it is not as pronounced. There are several reasons for not expecting the observed distributions to look exactly as predicted by Struve. First, the observational material is not strictly homogeneous and may contain an observational bias. Secondly, the critical rotational velocity v_0 varies with spectral type, particularly for the B0–B5 stars (cf. Figure 2); therefore the histograms in Figure 1 cannot be expected to come to a sharp maximum at a single v_0. Thirdly, a number of Be stars with strong absorption shell spectra are not included in Figure 1 because the underlying stellar absorption lines were obscured on the spectrogram and no $v \sin i$ estimate was possible. But these are precisely the stars with large $v \sin i$, since they are being viewed essentially equatorially. Finally, as has already been stated, ambiguities in the line profile method of estimating $v \sin i$ make values of 300 km s^{-1} and higher quite uncertain. In summary, it seems to me that the observed frequency of Be stars having various $v \sin i$'s is consistent with the rotational hypothesis.

As another test, the largest observed rotational velocities for stars of various spectral types (assumed to be viewed equatorially) may be compared with the computed equatorial critical velocities for stars of corresponding type. If the rotational hypothesis for Be stars is correct, the curves should intersect for those spectral types which include the Be stars. Such a comparison was made some years ago (Slettebak, 1966b) and while the curves representing the computed equatorial breakup velocities and the largest observed rotational velocities did not actually touch, the region of minimum difference was approximately that occupied by the Be stars. Recently, Collins (1974) recomputed the equatorial critical velocities for main sequence Roche models and found values which are considerably lower than in the earlier work. This would tend to bring the curves closer together, but the observed rotational velocities on the new scale should also be reduced by about 15%. Figure 2

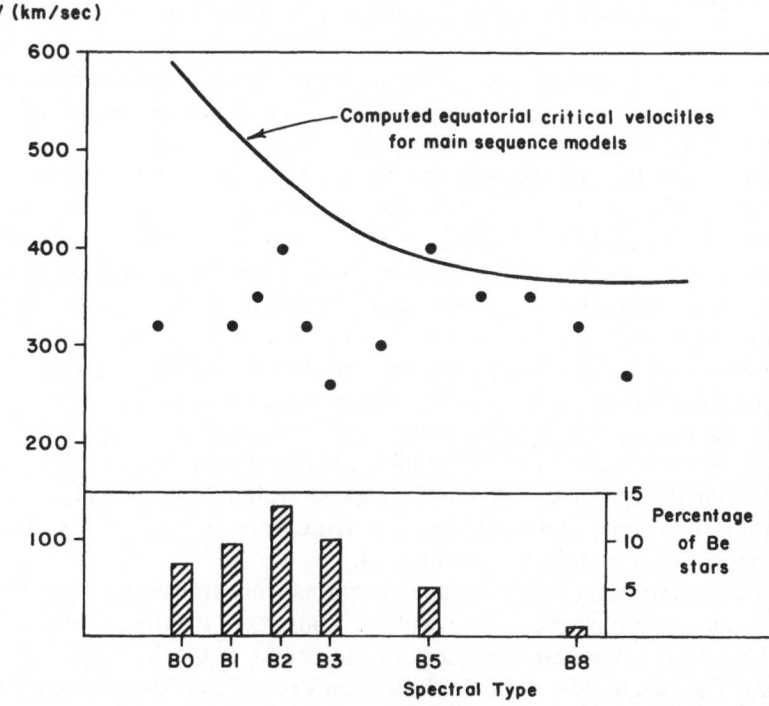

Fig. 2.   Comparison of the largest observed rotational velocities for Be stars of spectral types O9.5 to B9 with computed equatorial critical velocities (at which the centrifugal force balances the gravitational force) for main sequence models of the same spectral type range. The relative frequency of Be stars is also shown.

shows Collins' theoretical curve with the largest observed rotational velocities from Table I and the percentage of Be stars as a function of spectral type also plotted. The observed rotational velocities may lie somewhat closer to the computed curve than in the earlier work but, on the average, are still smaller than the computed critical velocities. Three factors may play a role: (1) The equatorial critical velocities were computed for main sequence models whereas there is evidence that some Be stars at least have evolved off the main sequence into the subgiant and giant regions of the

H-R diagram. Such evolution would increase their radii without a change in mass, thereby lowering the critical rotational velocity. (2) The spectral types for a number of the stars plotted in Figure 2 are rather uncertain because absorption shell features make it difficult to classify the underlying stellar spectrum. (3) The largest observed rotational velocities are also the most uncertain, for reasons already discussed.

In any case, a mechanism in addition to rapid rotation appears to be necessary to transport material from the star into the shell. Ostriker (1970) has shown, for example, that a contracting rotating star in which viscous and magnetic forces may be neglected will never shed mass at the equator. The observed rotational velocity need then only be close to the equatorial critical velocity, with a pulsational instability or radiation pressure or some other mechanism giving the material the final push.

3. Observations and Interpretation of Be Emission-Line Widths

In his 1931 paper, Struve showed that there was a correlation between axial rotation, as shown by the widths of stellar absorption lines, and the widths of the emission lines in Be stars. This conclusion was based on three independent sets of data: (1) spectrograms from the collection of the Yerkes Observatory; (2) the emission-line widths of R. H. Curtiss (1923); and (3) the observations of P. W. Merrill (Merrill *et al.*, 1925). Some years later, Underhill (1953) found her measurements of Hα and Hβ emission in the spectra of five Be stars to be consistent with Struve's rotational hypothesis and, more recently, Gray and Marlborough (1974) obtained a good correlation between their emission widths and $v \sin i$ for 12 Be stars.

But Doazan (1970), from measurements of the Balmer emission lines in 26 Be stars, found that "no correlation exists between the widths of the Hα, Hβ and Hγ emission lines and the velocity due to the rotation of the central stars ($v \sin i$) when this velocity is less than 350 km s^{-1}". Since this conclusion casts considerable doubt on the rotational hypothesis, Doazan's work stimulated me to collect my Be star spectrograms of the past 25 years together and measure emission-line widths to see if my material shows a correlation with $v \sin i$.

It should be emphasized that my spectrograms were not originally taken for the purpose of measuring emission-line widths. Thus, they do not include Hα (which always shows the strongest emission in Be stars) and often had no intensity calibration. The spectrograms also include a variety of dispersions, ranging from 9 to 46 Å mm^{-1}, with the projected slit width at the plate taking values between 0.2 Å and 0.9 Å.

Transmission microphotometer tracings of all Be stars showing emission on my plates were made with the University of Vienna Observatory PDS-1000 microphotometer and PDP-12 computer combination in 1975. The three emission widths measured for each line when a measurement was possible are shown schematically in Figure 3. The quantity $(\Delta\lambda)_E$, measured from the steepest slopes of the emission profile, is essentially the same quantity measured by Curtiss (1923), and a comparison of 9 stars in common shows good agreement. This quantity should also be expected to represent the emission width best, since $(\Delta\lambda)_S$ is strongly affected by self-absorption and $(\Delta\lambda)_T$ is a rather uncertain quantity to measure.

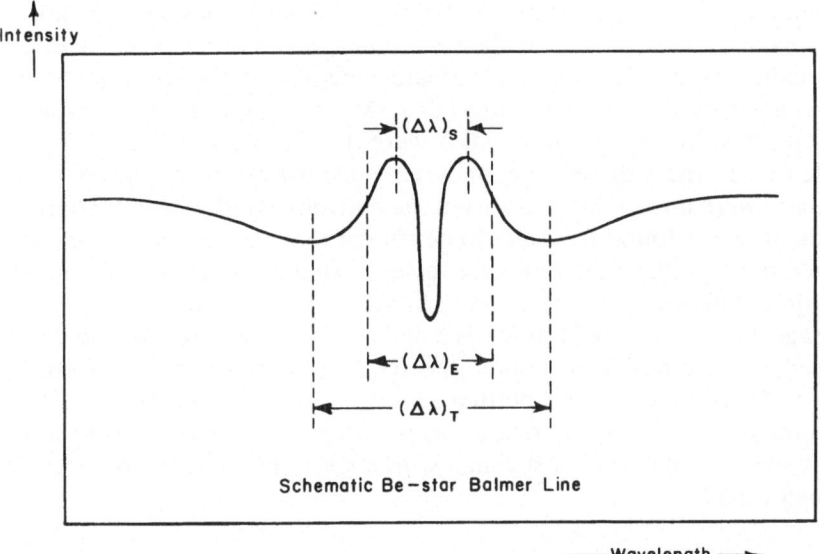

Fig. 3. A schematic Be-star Balmer line, showing the three emission widths measured in this paper.

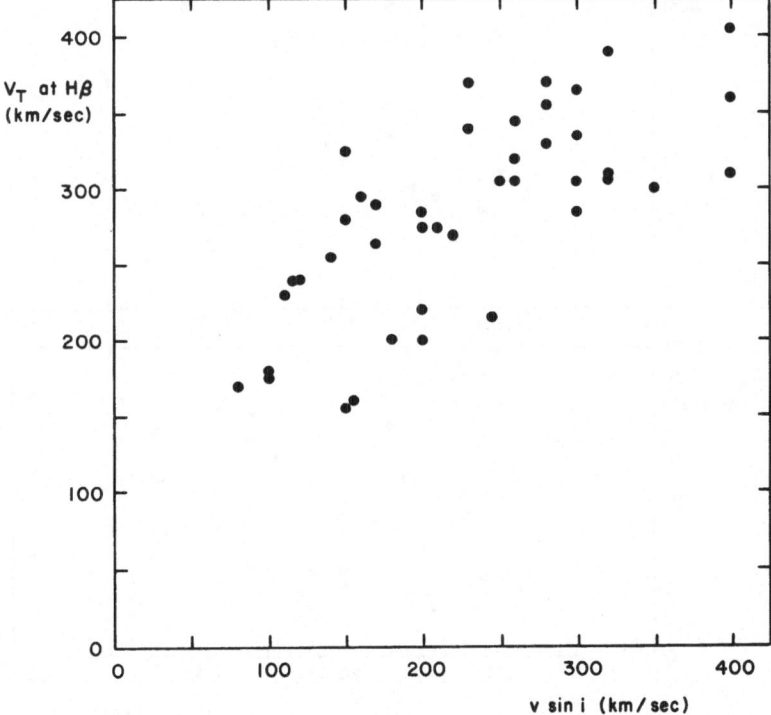

Fig. 4. Comparison of the velocity v_T corresponding to one-half the emission width $(\Delta\lambda)_T$ of $H\beta$ in 42 Be stars.

A comparison of emission-line width measurements from spectrograms of differ-ent dispersions for the same star showed good agreement for plates with projected slit widths in the range 0.2–0.6 Å, but greater emission widths for the plates of poorer resolution, particularly in the quantity $(\Delta\lambda)_T$. Measurements from spectrograms with projected slit widths greater than 0.6 Å were therefore discarded.

In a few cases where the strength of the emission in a given star varied from plate to plate, there were no striking changes in the emission widths. McLaughlin (1962) in his study of π Aqr found the same to be true of that star, for which he states "The total measured width of the emission remains roughly constant". Emission widths from different plates for each star were therefore simply averaged.

Average emission-line widths for Hβ and Hγ in 46 Be stars are listed in Table I, expressed as velocities corresponding to the half-width of the emission, $1/2(\Delta\lambda)$. Figures 4–7 show these values plotted against $v \sin i$. Although there is not a strict correlation, there does appear to be a correlation in the sense that Struve stated: wide emission lines are generally associated with large $v \sin i$ and narrow emission lines with small $v \sin i$.

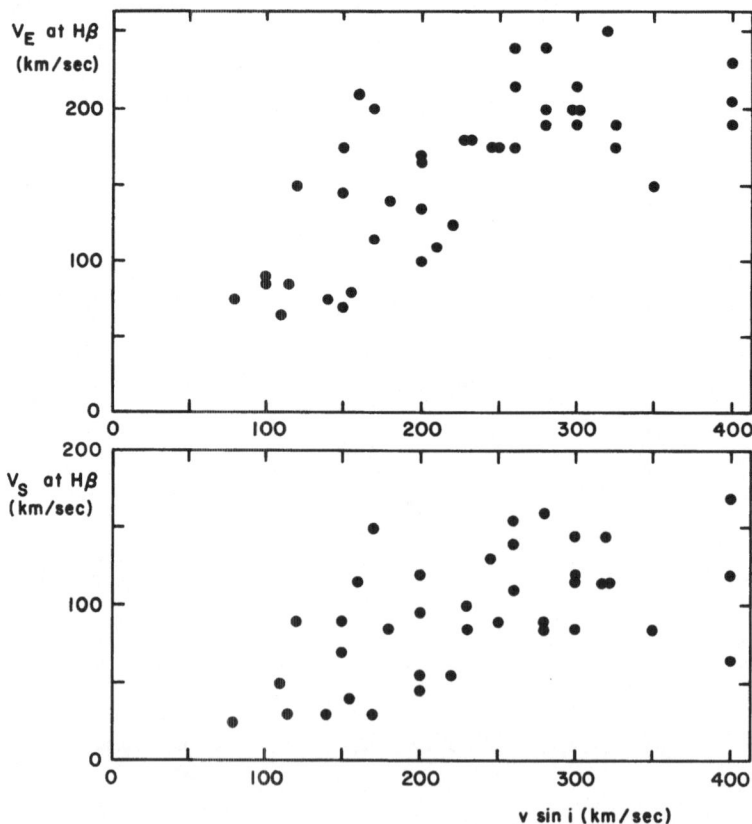

Fig. 5.   Comparisons of the velocities v_E and v_S corresponding to one-half the emission widths $(\Delta\lambda)_E$ and $(\Delta\lambda)_S$ of Hβ in 42 and 38 Be stars, respectively.

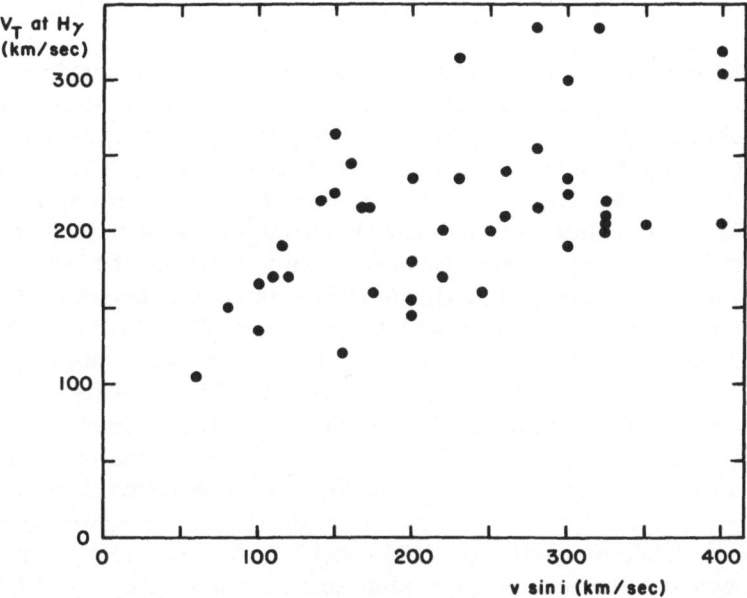

Fig. 6.   Comparison of the velocity v_T corresponding to one-half the emission width $(\Delta\lambda)_T$ of Hγ in 43 Be stars.

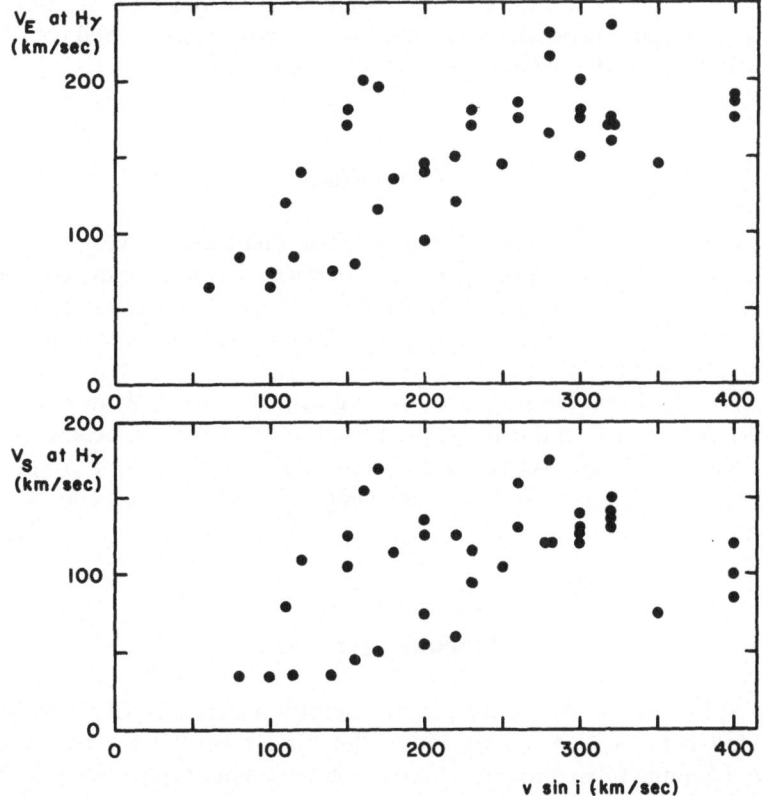

Fig. 7.   Comparisons of the velocities v_E and v_S corresponding to one-half the emission widths $(\Delta\lambda)_E$ and $(\Delta\lambda)_S$ of Hγ in 41 and 39 Be stars, respectively.

This conclusion does support the rotational hypothesis for Be stars but the lack of a strict correlation suggests that other factors must also be involved. This is also evident from Table I, where the velocities corresponding to the emission half-widths are sometimes larger than $v \sin i$, particularly for $H\beta$. Doazan (1970) found a similar effect for her 26 Be stars: "For all the stars studied, the rotational velocities corresponding to the $H\alpha$ emission line width are larger than that due to the rotation of the central star ($v \sin i$). This is also the case for the $H\beta$ line of 17 stars and for the $H\gamma$ line of 12 stars". Indeed, Underhill in 1953 pointed out the very extensive wings of $H\alpha$ in some shell stars and suggested that "it is possible that the extra broadening of the emission at $H\alpha$ is caused by large chaotic motions of the emitting material". But turbulent velocities of 100 km s^{-1} and larger in the shell imply a degree of disorder which does not seem consistent with the Be phenomenon.

Another possible explanation of the excessive emission-line broadening in Be stars may involve electron scattering. Münch (1948) showed that electron scattering in early type supergiants can modify absorption lines by producing very extended and shallow wings, while the Burbidges (1953) held electron scattering responsible for the broad wings of the Balmer absorption lines in pole-on Be stars. Marlborough (1969) also suggests that "electron scattering seems to be a reasonable mechanism to explain the extensive emission wings of $H\alpha$" in Be stars. Gray and Marlborough (1974), from an analysis of photoelectric profile measurements of $H\alpha$ and $H\beta$ in 14 Be stars, suggest that "more likely we observe a combination of shell rotation and a velocity distribution either of mass elements or electrons".

4. Conclusions

The classical picture of a Be star as an object rotating near the verge of instability, with an equatorial shell responsible for the emission characteristics, appears still to be supported by the observations of absorption and emission line widths, at least in a qualitative way. Other factors, including duplicity, undoubtedly modify this simple picture in many individual cases, however.

The observed fact that Be stars are rotating near but probably not at the critical velocity does not contradict the above picture since recent theoretical work suggests that a star will not shed mass equatorially through rotation alone in any case. A number of ejection mechanisms have been suggested but none has been worked out in any detail as yet.

Acknowledgements

I am grateful to George W. Collins, II, for helpful discussions of a number of the topics treated here. I also acknowledge the kind hospitality I received at the University of Vienna Observatory, where the measurements described in this paper were made, and particularly thank R. Albrecht for his help with programming the PDS-1000 microphotometer.

References

Abney, W. de W.: 1877, *Monthly Notices Roy. Astron. Soc.* **37**, 278.

Balona, L. A.: 1975, preprint.

Bernacca, P. L.: 1970, in A. Slettebak (ed.), *Stellar Rotation*, D. Reidel Publ. Co., Dordrecht-Holland, p. 227.

Burbidge, G. R. and Burbidge, E. M.: 1953, *Astrophys. J.* **117**, 407.

Butler, H. E.: 1976, this volume, p. 51.

Carroll, J. A.: 1933, *Monthly Notices Roy. Astron. Soc.* **93**, 478, 508, 680.

Collins, G. W., II: 1974, *Astrophys. J.* **191**, 157.

Collins, G. W., II, and Harrington, J. P.: 1966, *Astrophys. J.* **146**, 152.

Curtiss, R. H.: 1923, *Publ. University of Michigan* **3**, 1.

Deutsch, A. J.: 1954, *Trans. IAU* **8**, 801.

Doazan, V.: 1970, *Astron. Astrophys.* **8**, 148.

Elvey, C. T.: 1930, *Astrophys. J.* **71**, 221.

Friedjung, M.: 1968, *Astrophys. J.* **151**, 779.

Galilei, G.: 1613, Letters to Mark Welser (translation in *Discoveries and Opinions of Galileo* by S. Drake, Doubleday and Co., New York, 1957).

Gray, D. F.: 1973, *Astrophys. J.* **184**, 461.

Gray, D. F. and Marlborough, J. M.: 1974, *Astrophys. J. Suppl. Ser.* **27**, 121.

Hardorp, J. and Strittmatter, P. A.: 1968, *Astrophys. J.* **153**, 465.

Hardrop, J. and Strittmatter, P. A.: 1970, in A. Slettebak (ed.), *Stellar Rotation*, D. Reidel Publ. Co., Dordrecht-Holland, p. 48.

Harrington, J. P. and Collins, G. W., II: 1968, *Astrophys. J.* **151**, 1051.

Hutchings, J. B.: 1970, in A. Slettebak (ed.), *Stellar Rotation*, D. Reidel Publ. Co., Dordrecht, Holland, p. 283.

Krzeminski, W.: 1969, in S. S. Kumar (ed.), *Low Luminosity Stars*, Gordon and Breach, New York, p. 57.

Kurucz, R. L.: 1970, ATLAS: A Computer Program for Calculating Model Stellar Atmospheres (Cambridge: *Smithsonian Astrophys. Obs. Rept.* No. 309).

Marlborough, J. M.: 1969, *Astrophys. J.* **156**, 135.

Massa, D.: 1975, preprint.

McLaughlin, D. B.: 1962, *Astrophys. J. Suppl. Ser.* **7**, 65.

Merrill, P. W., Humason, M. L., and Burwell, C. G.: 1925, *Astrophys. J.* **61**, 389.

Münch, G.: 1948, *Astrophys. J.* **108**, 116.

Ostriker, J. P.: 1970, in A. Slettebak (ed.), *Stellar Rotation*, D. Reidel Publ. Co., Dordrecht-Holland, pages 19 and 47.

Sackman, I.-J. and Anand, S. P.: 1970, *Astrophys. J.* **162**, 105.

Shajn, G. and Struve, O.: 1929, *Monthly Notices Roy. Astron. Soc.* **89**, 222.

Slettebak, A.: 1949, *Astrophys. J.* **110**, 498.

Slettebak, A.: 1966a, *Astrophys. J.* **145**, 121.

Slettebak, A.: 1966b, *Astrophys. J.* **145**, 126.

Slettebak, A., Collins, G. W., II, Boyce, P. B., White, N. M., and Parkinson, T. D.: 1975, *Astrophys. J. Suppl. Ser.* **29**, 137.

Stoeckley, T. R.: 1968a, *Monthly Notices Roy. Astron. Soc.* **140**, 121.

Stoeckley, T. R.: 1968b, *Monthly Notices Roy. Astron. Soc.* **140**, 141.

Struve, O.: 1931, *Astrophys. J.* **73**, 94.

Struve, O.: 1945, *Popular Astronomy* **53**, 202.

Uesugi, A. and Fukuda, I.: 1970, *Contrib. Inst. of Astrophys. and Kwasan Obs., Univ. of Kyoto*, No. 189.

Underhill, A. B.: 1953, *Monthly Notices Roy. Astron. Soc.* **113**, 477.

Vogt, S. S.: 1975, *Astrophys. J.* **199**, 418.

DISCUSSION

Doazan: I have two remarks to make: (1) I would like to emphasize that Merrill's and Struve's results are based on line widths estimated by visual inspection of the plates, and that the only previous quantitative results are those of Curtiss, who measured line widths of only 11 stars, I think. (2) The way you measure the line widths of the emission lines underestimates these widths. It is necessary to trace the photospheric absorption lines to determine the position of the beginning of the depression caused by the emission. If you do so your line widths will be greater.

Slettebak: That is correct. I was less interested in the absolute values of the emission widths, however, than in whether or not they show a correlation with rotational velocity.

Doazan: Our spectra indicate that a certain number of Be stars show emission in the helium lines as, for example, γ Cas, which was also observed by Hutchings. My question is: have you observed many Be stars of this kind and can the observed $v \sin i$'s really be interpreted then as rotation?

Slettebak: Certainly there are stars like γ Cas, where the helium profile shapes are peculiar. I did not find such line profiles as a common thing, however, nor did I find emission in the helium lines (at least, not in He I 4471). By and large, the profiles looked like the dish-shaped profiles that are predicted under the rotational hypothesis.

Peters: Concerning He I emission, we have been observing Be stars at λ 7065 of He I and, although the survey is far from complete, the only star which appears to have significant He I emission is ϕ Per.

Plavec: As a matter of curiosity, I know that high velocities of rotation are somewhat uncertain, but which star is rotating fastest?

Slettebak: We have assigned several stars velocities of $400 \, \mathrm{km \, s^{-1}}$, including 48 Lib, ϕ Per, and HR 2142. It looks from Collins' models as though that is the largest numerical value we can assign.

Bidelman: I wonder whether your emission-line measurements depend at all on the emission intensity. Have you measured the same star at times when the emission strengths were different?

Slettebak: Yes, in a few cases, and there seems to be little difference in the emission widths. McLaughlin noted the same thing: for stars showing V/R variations he noted that the total measured width of the emission remains roughly constant. But I do not have a lot of data to support that; just a few plates.

Coyne: When one goes to the catalogue material on $v \sin i$, one finds large differences in the various catalogues. Is there a recommendation among the people who use $v \sin i$'s as to which catalogue gives a homogeneous set of values?

Slettebak: As you know there are several. Originally there was the Boyarchuk-Kopylov catalogue of $v \sin i$'s, and more recently the Bernacca catalogue and the Uesugi and Fukuda catalogue. In each case they weight the various $v \sin i$ values and try to put them all on one system. Our hope in the paper which I just presented is to establish a system which people would use much like a system of standards for spectral classification and that if everyone stuck with these stars we would have a consistent system. At the present time I think that the Bernacca and the Uesugi-Fukuda catalogues are the best.

Coyne: How does one relate your recent work to those two catalogues?

Slettebak: The new scale, which is based on Collins' theoretical work and our new measures, is about 15% lower for the B stars and about 5% lower for the A and F stars. So if you believe this work you could take values from those catalogues and apply corrections to them of about that amount.

Heap: The ultraviolet region of the spectrum has many strong lines which can be used both for spectral classification and for rotational-velocity determinations. The profiles of these lines are easily measured even in rapidly rotating stars. Some of these lines are lines of high ionization state, like Si IV, so you do not have the problem of shell emission or shell absorption superimposed on them. I would like to persuade anyone who is theoretically inclined to compute gravity-darkened profiles for these ultraviolet lines.

Hutchings: Can you distinguish between pole-on rotating stars and non-rotating stars?

Slettebak: Normally no, but in the case of Be stars, if you accept the hypothesis that they are all rotating near a certain critical velocity, then the sharp-line stars are, by definition, pole-on stars.

Hutchings: Do you believe that Be stars are rotating at breakup velocity?

Slettebak: No. I think the best evidence now is that they are not. They still seem to rotate below the critical velocity but I do not think this is a problem since there now seem to be a number of mechanisms to get material out into the shell, even if the star's rotation does not reach the critical velocity.

ULTRAVIOLET OBSERVATIONS OF RAPIDLY ROTATING B-TYPE STARS

SARA R. HEAP

Goddard Space Flight Center, Greenbelt, Md., U.S.A.

Abstract. Evidence of anomalous C−N abundances in the atmospheres of six rapidly rotating B-type stars has been found from an LTE analysis of ultraviolet line spectra obtained by *Copernicus*. Such anomalies are in the direction (carbon depletion, nitrogen enhancement) predicted by Paczyński (1971), but the extent of the anomaly grows toward earlier spectral type, contrary to Paczyński's predictions.

DISCUSSION

Garrison: I would like to comment that θ Car is very obviously very peculiar in the visual region as well. It obviously is a very nitrogen rich star. There are some O-type stars which are nitrogen rich and carbon poor. However, none of them are emission-line stars, as far as I know, and not all are rapid rotators. Perhaps the effect you are seeing here is somehow related to those stars and is not a rotation effect.

Heap: Yes, I agree.

Conti: Interestingly enough, there is a small group of main sequence O-type stars that also have a property of enhanced nitrogen and deficient carbon. However, none are emission-line stars, nor are they among the most rapid rotators. The origin of the anomaly is not certain.

Henize: My spectra in the ultraviolet for θ Car show that the C IV line at λ 1548 is very weak for a B0 star. Another B0 star with quite weak C IV is HR 1887, but I know nothing about its rotational characteristics. It is not a Be star.

Peters: HR 1887 forms a visual binary with HR 1886. I have analyzed HR 1887 using the Princeton model atmospheres and find that the carbon and nitrogen abundances are quite comparable to those in the sun.

Snijders: The results by Kamp show that for the Si III–Si IV ionization balance which you used as an effective temperature criterion, LTE is a good approximation. However for carbon and nitrogen, the importance of non-LTE effects is unknown so far. The increase of the abundance deviations you find with increasing effective temperature resembles the behavior of the LTE abundances derived from the Mg II 4481 line as discussed by Mihalas. Especially the different abundances found from different nitrogen ions is worrisome. Extending your work to standard slow rotators like γ Peg and τ Sco might clarify some of these problems.

Heap: Yes, I agree.

Latham: There is another mechanism that may be able to explain CNO abundance anomalies. It requires a close binary companion which was more massive and has already evolved enough to expand and transfer CNO processed mass onto the surface of the star we see now. Can you rule out this mechanism?

Heap: As far as I know, only one of the six stars studied, ζ Cen, is a binary. Perhaps, λ Sco is also.

Hummer: I would be worried about using strong lines for this kind of analysis because of the possibility of getting effects from atmospheric motion.

Heap: Except for the case of θ Car, the ultraviolet line spectra did not show any blue shifts or asymmetries.

Peters: How sensitive are your abundances to your adopted Stark broadening parameters?

Heap: In the calculations for the strengths of the carbon and nitrogen lines, I used Sahal-Brechot and Segre's values for the electron damping constant. Had I excluded electron damping, the derived abundances would have been raised by up to a factor of two.

THE RELATIONSHIP OF THE Oe TO THE Be STARS

STEWART A. FROST* and PETER S. CONTI**

Joint Institute for Laboratory Astrophysics
University of Colorado and National Bureau of Standards
and
Dept. of Physics and Astrophysics
University of Colorado, Boulder, Color., U.S.A.

Abstract. Oe stars are earlier type analogues of the better known Be class. These stars have relatively narrow emission in the hydrogen lines and sometimes in the helium lines, which often appears to be double. In several Oe stars, the emission is intermittent. Other properties of the class include relatively broad absorption lines and luminosities near the main sequence. Line profiles in three representative Oe stars are presented and discussed. The similarity of the spectra of Oe and Be stars suggests that the formation mechanism for the emitting region, whatever it is, must extend to the mid O-type stars, at least.

Recently, Conti and Leep (1974) introduced the spectral classification type Oe for those O-type stars showing emission due to the Balmer lines of hydrogen and *not* showing emission due to N III λλ 4634, 40, 41 Å and He II λ 4686 Å. The spectral types of the stars that they included in this class are given in Table I. One additional

TABLE I

Spectral types of the Oe stars

	MWC	Conti and Leep	Walborn	Morgan *et al.*
X Per	78	OBe		
HD 39680	783		O6 V : [n]pe var	
HD 45314	140	OBe		
HD 46056	808	O8 V(e)	O8 Vn	
HD 60848	184			O8 V : pe
ζ Oph		O9 V(e)		
HD 155806	252	O7.5 IIIe	O7.5 V[n]pe	
68 Cyg		O8 V	O7.5 III : n((f))	

star, 68 Cyg, has been included for reasons which will be mentioned later. The spectral types are taken from Walborn (1972, 1973) and Conti and Leep (1974), except for that of HD 60848 which is taken from Morgan *et al.* (1955). It is important to note that all of these stars have spectra which indicate that they are on or near the main sequence. The appearance of the line profiles of He II λ 4686 Å, some of which will be illustrated later, are consistent with these classifications. In addition, two of these stars, HD 46056 and ζ Oph, have well established absolute magnitudes, and these magnitudes indicate that they are on the main sequence. The Mount Wilson Catalogue numbers (Merrill and Burwell, 1949) are also listed in Table I. All of these

* Also Visiting Student, Kitt Peak National Observatory, which is operated by AURA, Inc., under contract with the National Science Foundation.
** Also Department of Astrogeophysics, University of Colorado; Guest Investigator, Hale Observatories; and Visiting Astronomer, Kitt Peak National Observatory, which is operated by AURA, Inc., under contract with the National Science Foundation.

stars, except ζ Oph and 68 Cyg were observed to have hydrogen Balmer line emission more than twenty-five years ago. However, Merrill and Burwell noted that the emission in HD 46056 was present intermittently. We will present line profiles for HD 39680, HD 60848, and HD 155806.

The spectrograms from which the line profiles were obtained were taken during the years 1972–1974 at the Kitt Peak National Observatory, Palomar Mountain Observatory, and Mount Wilson Observatory. The coudé spectrographs were employed so as to give a dispersion of about 18 Å mm^{-1}, on nitrogen baked IIa-O, in the λ 3700–4900 Å, blue region, and a dispersion of 25 Å mm^{-1}, on unbaked and nitrogen baked 098–04, in the λ 5000–6800 Å, yellow-red region. The smoothed line profiles were produced from the spectrograms by the computer method described by Conti and Frost (1974). The spectra were traced on a microphotometer. The digital density information was filtered by a fast Fourier transform method (Brault and White, 1971) and converted to intensity through the use of a polynomial calibration curve. The line profiles were calculated and plotted on microfilm by computer, and later copied by hand from prints of the film. The zero point of the velocity scale was determined for each spectrum from a number of absorption lines. The line profiles were sometimes shifted 50–100 km s^{-1} in order to obtain the 'best' agreement between a number of them.

The Balmer line profiles, shown in Figures 1–4, are similar to the double peaked or central emission line profiles characteristic of the Be stars. The emission is strongest at Hα and decreases in strength as one goes to the shorter wavelength lines. For the

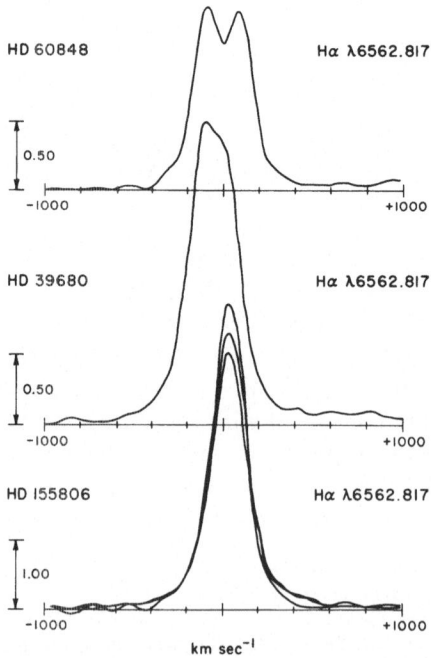

Fig. 1.   Normalized, filtered Hα line profiles in three Oe stars. In this figure, and all subsequent ones, the *zero point* of the velocity scale is only approximately correct. Shifts in line center from 0 km s^{-1} may not be real.

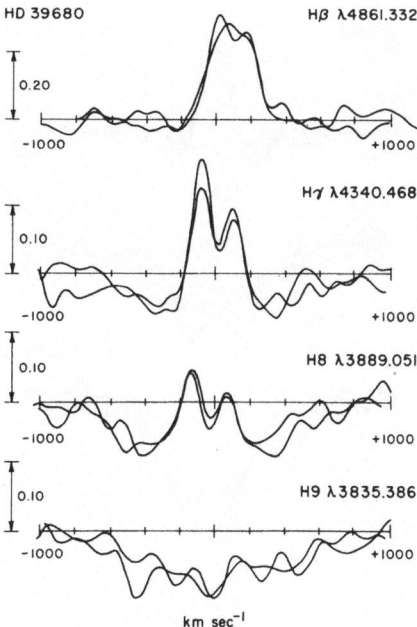

Fig. 2.   Normalized, filtered Balmer line profiles in HD 39680. The emission has disappeared at H9 and only the underlying photospheric feature remains.

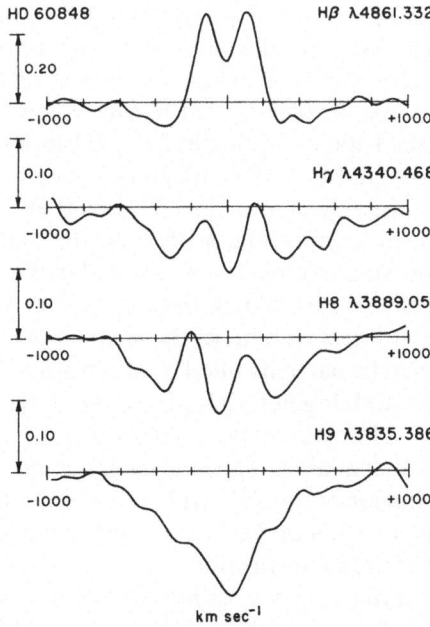

Fig. 3.   Normalized, filtered Balmer line profiles in HD 60848. The emission has disappeared at H9 and only the underlying photospheric feature remains.

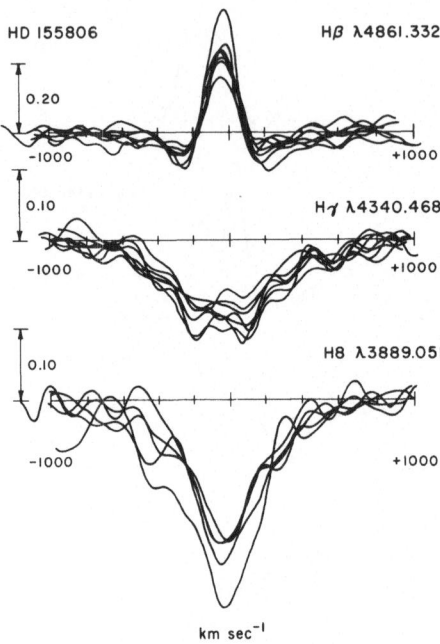

Fig. 4. Normalized, filtered Balmer line profiles in HD 155806. The emission has essentially disappeared by Hγ and only the underlying photospheric feature remains. Note, however, that Hα was much stronger in this star than in the other two as shown by Figure 1.

higher members of the Balmer series, Hγ and H8 in HD 39680, Hβ, Hγ, and H8 in HD 60848, and Hβ in 155086, one clearly sees the profiles as the superposition of emission on a broader absorption. During the period of time covered by these observations, the Balmer line profiles appear not to have varied very much.

Figures 5–7 illustrate the triplet and singlet $2P^0-nD$ series of neutral helium. The velocity scales for λ 5876 Å and λ 6678 Å are compressed by a factor of two relative to the velocity scales for the other lines. In HD 39680 and HD 60848, λ 5876 Å, the first member of the triplet series, and λ 6678 Å, the first member of the singlet series, appear in emission. Similar emission was noted by Hutchings (1970) in γ Cas and Bahng and Hendry (1975) in κ CMa, both Be stars. The appearance of the triplet line, He I λ 4471 Å, is of interest in both of these stars. Relative to the next triplet line, λ 4026 Å, it appears to be partially filled in by emission. Thus, in addition to the first members of the triplet and singlet series, He I λ 4471 Å seems to be formed in an extended atmosphere. Unfortunately, observations of the singlet line, He I λ 4921 Å are not yet available. The higher series members appear as broad absorption lines. The behavior of the $2P^0-nD$ series of neutral helium in these two stars is in some ways similar to that of the Balmer series of hydrogen. In HD 155806, however, all the He I lines are in absorption.

Spectral classification employing the helium ionization balance as given by the ratio of the equivalent widths of He I λ 4471 Å to He II λ 4541 Å will give spectral types which are too early for some Oe stars. In HD 39680, the He II lines indicate a spectral type near O9, in contrast to the earlier spectral type found by Walborn

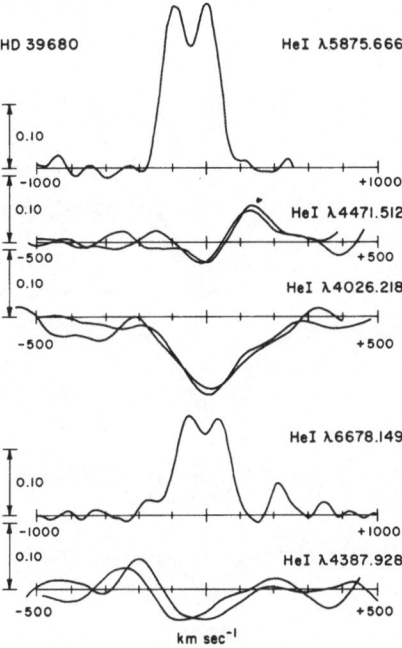

Fig. 5.   Normalized, filtered neutral helium lines in HD 39680. The emission has disappeared at the third
member of the series.

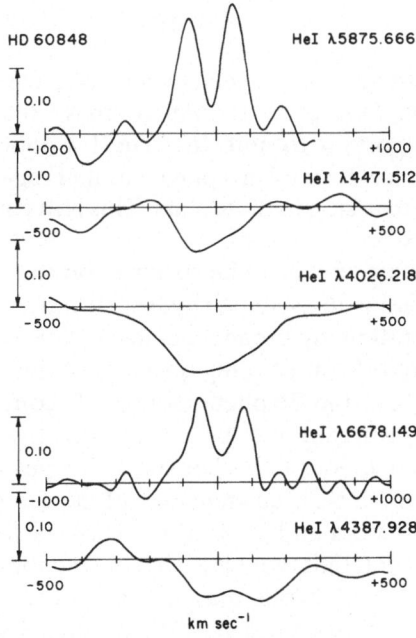

Fig. 6.   Normalized, filtered neutral helium lines in HD 60848. The emission has disappeared at the third
member of the series.

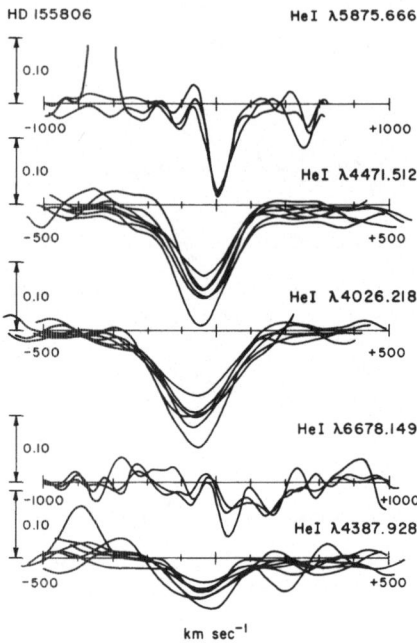

Fig. 7.   Normalized, filtered neutral helium lines in HD 155806. No emission is present.

(1973). Thus, HD 155806, with a spectral type of O7.5, is the earliest Oe star of which we are aware.

These three stars are O-type stars, as seen by the presence of the lines due to singly ionized helium shown in Figures 8–10. All of these features appear as broad absorption lines. It is important to note that He II λ 4686 Å appears strongly in absorption, implying that it is formed in a plane parallel region. In addition, the N III complex is not present in emission in these stars. This is in contrast to the behavior of these lines in the Of stars.

Thus, these three Oe stars appear to be spectroscopically similar to the Be stars. Their spectra contain emission lines, due to hydrogen and neutral helium, arising in a rotating envelope and rotationally broadened absorption lines.

Massa (1975) considered the interesting question of the significance of radiation pressure in the appearance of the Be phenomenon. He concluded that the presence of this force aids in the formation of a disk. That is, as the magnitude of the force increases, presumably as one goes to earlier spectral types, the amount of rotation necessary for the formation of a circumstellar envelope decreases. Massa found that roughly 20% of the B1–5 stars and 12% of the stars earlier than B1 have shown emission at some time. He felt that there is a smaller fraction of Be stars at the earlier spectral types, but did state that this may be an observational effect due to the intermittent nature of the emission in the hotter stars. He suggested that at the earliest spectral types the relationship between the magnitude of the radiation pressure and the appearance of Be type emission may have broken down.

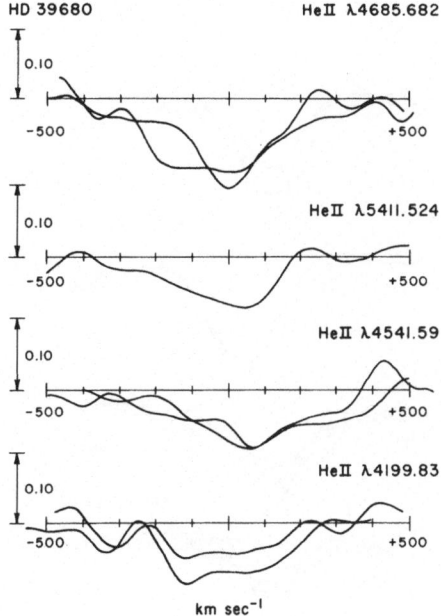

Fig. 8.   Normalized, filtered ionized helium lines in HD 39680. No emission is present.

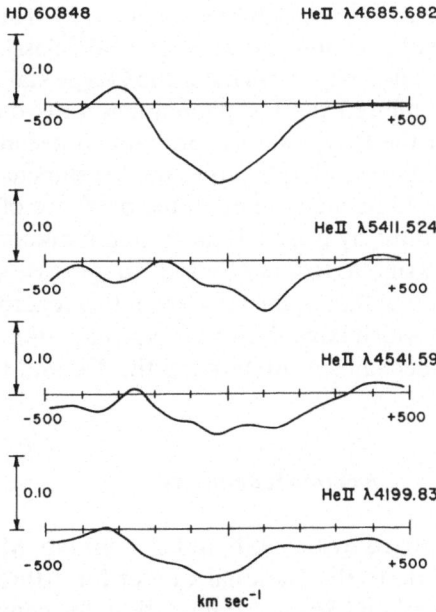

Fig. 9.   Normalized, filtered ionized helium lines in HD 60848. No emission is present.

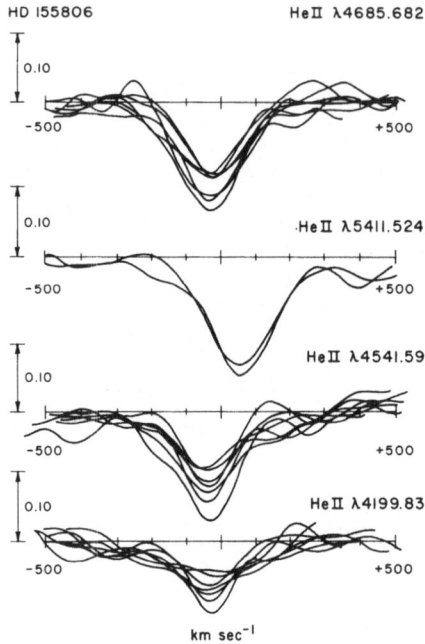

Fig. 10.   Normalized, filtered ionized helium lines in HD 155806. No emission is present.

There are 28 stars in the O star survey of Conti and Alschuler (1971) which meet the following requirements: (1) later than O7.5 (2) on the main sequence (3) $m_v \leq 8.0$ and (4) $\delta > -20°$. Of these, two, X Per and HD 45314, are Oe stars. Two others, ζ Oph (Niemelä and Méndez, 1974; Barker and Brown, 1974) and 68 Cyg (Wilson, 1958) have shown intermittent emission only at Hα, and are classified O(e) according to the spectral types of Conti and Leep (1974). For the O-type stars of this statistically small sample, 14% have exhibited the Oe phenomenon at some time. That this fraction is similar to that for the B-type stars is probably not significant considering the paucity of the statistics. One wonders whether the Oe stars have large winds and how a circumstellar disk would be maintained in the presence of such a wind.

Some of the Oe stars exhibit hydrogen Balmer line emission only at irregular intervals. This activity is very similar to that observed in some Be stars. The behavior of the Oe and Be stars is not like that of the Of stars in this regard. That is, there are no O-type stars known to us which have shown Of type emission *intermittently*. This indicates that the physical mechanisms underlying the Oe and Of phenomena are very different indeed.

Acknowledgments

This research has been supported by the NSF under grant No. MPS72-05062 A02. Acknowledgment is also made to the National Center for Atmospheric Research, which is sponsored by the National Science Foundation, for computer time used in this research.

References

Bahng, J. D. R. and Hendry, E.: 1975, *Publ. Astron. Soc. Pacific* **87**, 137.
Barker, P. K. and Brown, T.: 1974, *Astrophys. J. Letters* **192**, L11.
Brault, J. W. and White, O. R.: 1971, *Astron. Astrophys.* **13**, 169.
Conti, P. S. and Alschuler, W. R.: 1971, *Astrophys. J.* **170**, 325.
Conti, P. S. and Frost, S. A.: 1974, *Astrophys. J. Letters* **190**, L137.
Conti, P. S. and Leep, E. M.: 1974, *Astrophys. J.* **193**, 113.
Hutchings, J. B.: 1970, *Monthly Notices Roy. Astron. Soc.* **150**, 55.
Massa, D.: 1975, *Publ. Astron. Soc. Pacific* **87**, 777.
Merrill, P. W. and Burwell, C. G.: 1949, *Astrophys. J.* **110**, 387.
Morgan, W. W., Code, A. D., and Whitford, A. E.: 1955, *Astrophys. J. Suppl. Ser.* **2**, 41.
Niemelä, V. S. and Méndez, R. H.: 1974, *Astrophys. J. Letters* **187**, L23.
Walborn, N.: 1972, *Astron. J.* **77**, 312.
Walborn, N.: 1973, *Astron. J.* **78**, 1067.
Wilson, R.: 1958, *Publ. Roy. Obs. Edinburgh* **2**, 61.

DISCUSSION

Hutchings: Are there variable radial velocities for any of these stars?

Conti: There is very little radial velocity information in the catalogues on these Oe stars. However, both ζ Oph and 68 Cyg do seem to have non-negligible ranges in radial velocity.

Hutchings: I suggest you look at the Balmer decrements and peak separation progressions in Oe stars to see if X Per is typical of Oe types or still different from all the rest.

Hα AND Hβ MEASURES AS RELATED TO
Be STAR ROTATION

ALEJANDRO FEINSTEIN*

Observatorio Astronómico, Universidad Nacional de La Plata, Argentina

Abstract. Photometric data for the hydrogen lines in 80 southern and 55 northern Be stars give evidence of a constant ratio for the emission in α and β in stars of similar spectral type. In an (α, β) diagram the intersection of this emission-line ratio with the standard relation for non-emission stars gives lower values than those corresponding to standard stars of the same spectral type. This may be due to rotational effects. A correlation of the variation in the β index with the V magnitude is also given.

Photometric observations of some 80 bright southern Be stars were obtained in the *UBV* system during a time interval of 12 yr (1963–1974). In a previous paper (Feinstein, 1975), it was shown that 19 of these 80 stars display changes in the *V* magnitude ($\Delta V > 0^{m}15$). Some RI measures were also added.

In April 1970, February 1972 and December 1974, the Balmer lines Hα, Hβ and Hγ of the same stars were measured. (See Feinstein, 1974.) Previously, in December 1968, the Hβ line in 33 stars of the same group had been measured. Fifty-five northern hemisphere Be stars were also included in the same program during February and October 1970. The comparison of all these measures shows that a few

TABLE I

Be stars variable in β and in V
(difference: 1974 − 1968)

HD	$\Delta\beta$	ΔV
45 910	$+0^{m}090$	$+0^{m}05$
48 917	-0.002	-0.24
54 309	$+0.222$	$+0.15$
56 014	-0.029	$+0.03$
56 139	-0.046	$+0.13$
57 150	-0.127	-0.12
58 343	-0.019	-0.15
58 978	$+0.024$	$+0.02$
60 606	$+0.047$	$+0.16$
66 194	$+0.132$	$+0.01$
67 888	$+0.123$	-0.04
68 980	$+0.114$	-0.23
75 311	$+0.086$	0.00
88 661	$+0.022$	-0.10
131 492	$-0.064*$	$+0.45*$
148 184	$-$	-0.12
178 175	$-$	$-0.17*$
212 571	$+0.108*$	$-$

Notes: 131 492, $\Delta\beta$ (1972 − 1970), ΔV (1975 − 1968)
178 175, ΔV (1973 − 1968).
212 571, $\Delta\beta$ (1974 − 1970).

* Visiting Astronomer, Cerro Tololo Inter-American Observatory, which is operated by the Association of Universities for Research in Astronomy Inc., under contract with the National Science Foundation.

stars display variations in the Balmer lines and those with significant changes in β ($\Delta\beta > 0\overset{m}{.}060$) are related to their variations in V (Table I). The differences in β and in V (obtained nearly simultaneously) for the observations between 1968 and 1974 are plotted in Figure 1, (ΔV, $\Delta\beta$) diagram. The position of the two stars, HD 54309 and 57150, which have the largest variation in β, indicates that a decrease or increase in V (greater or lesser brightness, respectively) is correlated with a decrease or increase in β (larger or smaller emission, respectively). All the other stars are distributed in a band which may be correlated in a different way. In conclusion, the (ΔV, $\Delta\beta$) diagram does not give a clear relation between both values.

It is not possible to find a similar correlation with the α index, since we did not measure it in 1968. However, if we compare the measures between 1970 and 1974, significant variations in the α index are shown by the stars HD 45910, 54309, 66194, 148184 and 212571. The sense of the variations is always the same as in the β index.

The (α, β) diagram for the standard stars is plotted in Figure 2. In Figure 3 a similar diagram for all the observed Be dwarf stars is given; it shows the influence of the emission in both spectral lines, since most of them have smaller values of α and β than the standard stars. To investigate their emission effects the same (α, β) array is plotted for separate groups of spectral types: O8–B1.5, B2, B2.5–B5, and B6–B8 (Figures 4 to 7). These four diagrams suggest that each group has a different α/β emission-line ratio. With the exception of the B2 stars, the low percentage of stars in each spectral type does not allow us to obtain a more precise relation.

The points where each emission-line ratio crosses the standard relation do not coincide with the corresponding $\bar{\beta}$ of each group of spectral types. Since the indices are smaller (column 6 of Table II), an effect due to the star's rotation may be apparent in α and β. The β values for stars with the break-up velocities were computed by means of the theoretical computations of Collins and Harrington (1966). This was

TABLE II

Emission-line relation for Be stars

Sp. T.	$\beta = a\alpha + b$		β_i	$\bar{\beta}$ (Crawford)	$\Delta\beta$	Sp. T.	$\delta\beta_R$
	a	b					
O8–B1.5	0.906	1.266	2.595	2.608	0.013		
B2	0.898	1.303	2.628	2.645	0.017	B2	0.015
B2.5–B5	0.607	1.740	2.634	2.678	0.044	B6	0.034
B6–B8	0.839	1.465	2.721	2.719	−0.002	B8	0.059

also applied for stars with smaller velocities, taking into account the random distribution in the inclinations. The difference of these two values, for three spectral types, multiplied by an appropriate coefficient (1.6), is tabulated in the last column of Table II. The differences are similar to those observed (column 6). The effects of binarity in β are smaller than $0\overset{m}{.}01$.

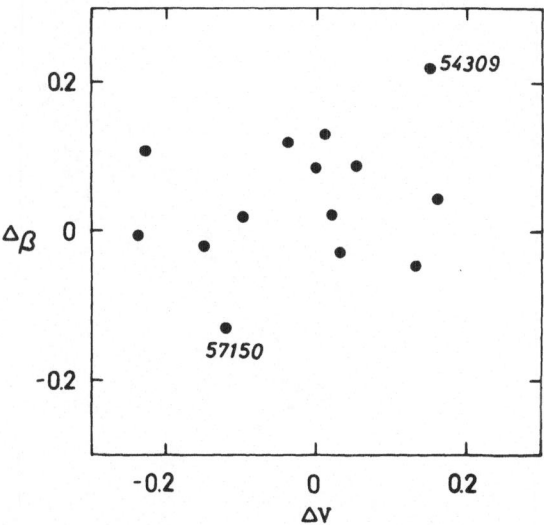

Fig. 1.   The variation in ΔV vs the variation in $\Delta\beta$ from observations between 1974 and 1968. The two stars with large $\Delta\beta$ are indicated by their HD numbers.

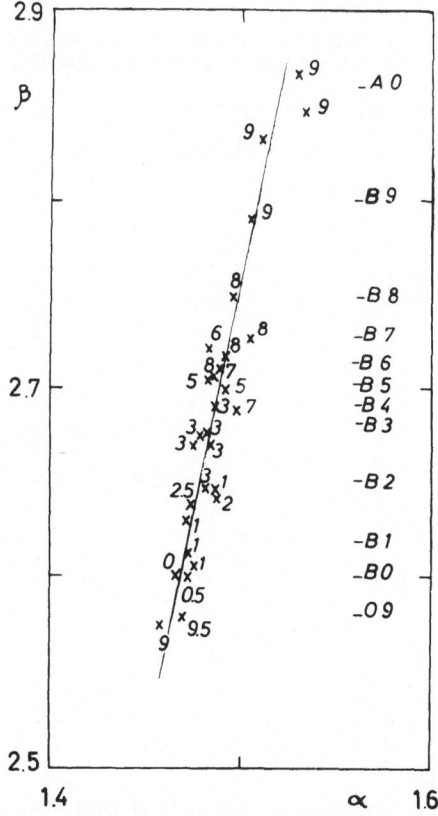

Fig. 2.   The (α, β) diagram for the standard stars. The spectral types according to the calibration for β obtained by Crawford (1975) are given in the right margin.

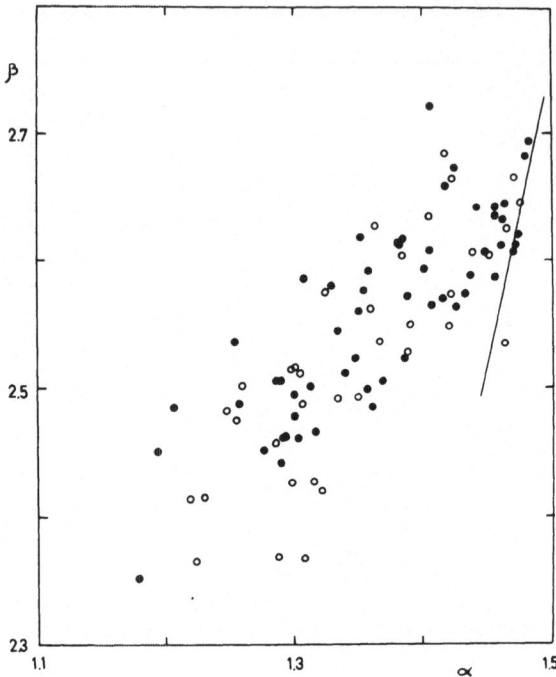

Fig. 3. The (α, β) diagram for all the Be stars. Open circles denote northern stars and dots southern stars. The position of the standard relation taken from Figure 2 is also shown.

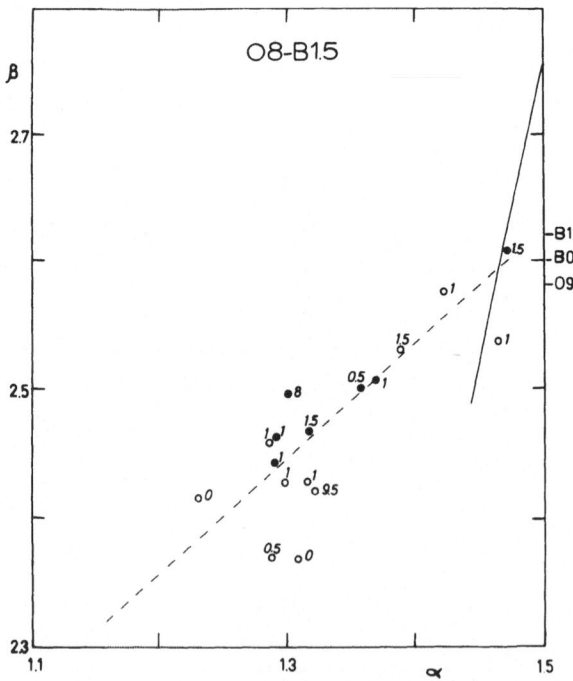

Fig. 4. The (α, β) diagram for the Be stars of spectral types O8–B1.5. The relation of the β values to the spectral types is given in the right margin. The standard relation (full line), and the weighted mean of the measured points (dashed line) are also indicated.

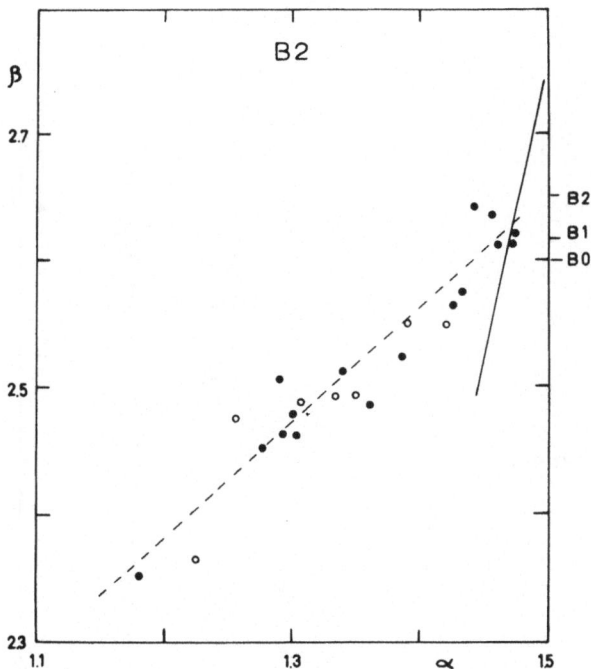

Fig. 5. The (α, β) diagram for the Be stars of spectral type B2. The meaning of the symbols is the same as in Figure 4.

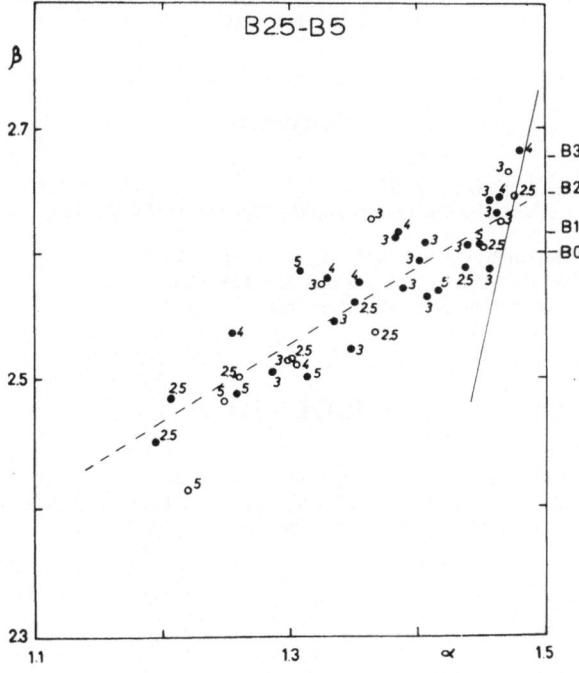

Fig. 6. The (α, β) diagram for the Be stars of spectral types B2.5–B5. The meaning of the symbols is the same as in Figure 4.

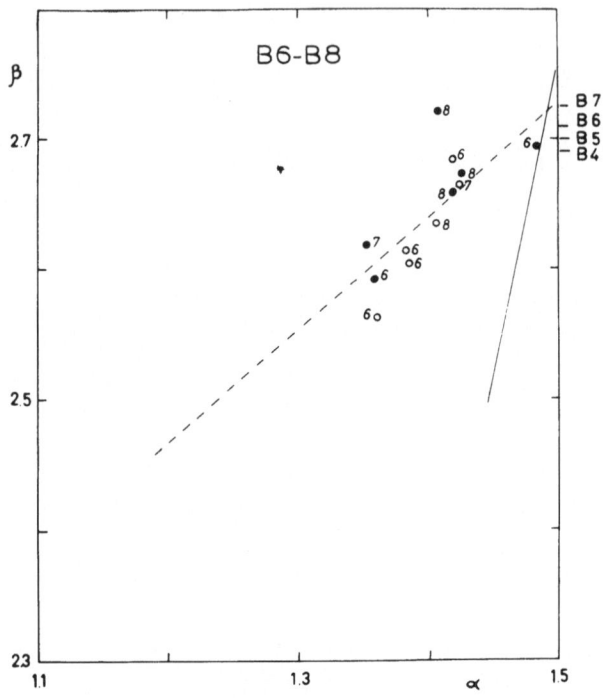

Fig. 7.   The (α, β) diagram for the Be stars of spectral types B6–B8. The meaning of the symbols is the same as in Figure 4.

References

Briot, D.: 1971, *Astron. Astrophys.* **11**, 57.
Crawford, D. L.: 1975, *Multicolor Photometry and the Theoretical HR Diagram*, Dudley Obs. Report No. 9, p. 17.
Collins, G. W. II and Harrington, J. P.: 1966, *Astrophys. J.* **146**, 152.
Feinstein, A.: 1974, *Monthly Notices Roy. Astron. Soc.* **169**, 171.
Feinstein, A.: 1975, *Publ. Astron. Soc. Pacific*, in press.

DISCUSSION

Harmanec: What are the standard errors in your measurements?

Feinstein: For the standard stars, about 0.02 mag. in α and β, but for the Be's, I am not certain. Since we must extrapolate in the reductions of the measures, they may be larger.

PART III

NEW OBSERVATIONAL TECHNIQUES

RADIO OBSERVATIONS OF Be STARS

(Review Paper)

C. R. PURTON

York University, Downsview, Ontario, Canada

1. Introduction

During the past few years the search for radio stars has been both intensive and extensive, and Be stars are among the various types of objects which have been examined. A large number of Be stars have *not* been detected, and to date no classical Be star has been found to emit detectable amounts of radio emission. However, following the tradition of radio astronomy to emphasise the abnormal or extreme cases, radio emission has been observed which is associated with a few peculiar Be stars.

The fact that only peculiar Be stars are seen at radio wavelengths is not surprising. If thermal radio emission from ionised gas is considered, then the same argument can be applied as was invoked a long time ago to demonstrate the very poor prospects for detecting radio emission from stars in general. In terms of modern radio telescopes, an angular diameter for the emitting gas in excess of $0''.1$ is required if the gas is optically thick in the cm. wavelength range, and larger if it is optically thin. If non-thermal radio emission is considered (synchrotron radiation, or plasma oscillations) then clearly an unusual object is required which involves intense flaring, or mass exchange* as in the radio-emitting binaries, or some other energetic phenomenon.

The two different types of radio stars mentioned above can be distinguished on the basis of their radio characteristics, and represent physical objects which are quite different. Those which have a thermal radio spectrum and whose intensity does not vary with time are associated with very large circumstellar envelopes, whereas those with a non-thermal spectrum and/or fairly rapid time variations are associated with mass-transfer in binary systems. As far as the Be stars are concerned the radio emission is, with one exception, attributed to thermal emission from circumstellar envelopes which have dimensions of many hundreds of astronomical units.

2. Instrumental Considerations

To observe such objects the basic requirements of the radio telescope are sensitivity, because of the low flux levels involved, and resolution, because of the incidence of confusing sources – particularly in the galactic plane. Table I summarises the telescopes which are actively used for radio-star work. The list is a simplification, as all major radio telescopes have been used to observe radio stars, but those shown are

---

* During this symposium the hypothesis that mass-exchange binaries might be quite common among Be stars was vigorously defended.

A. Slettebak (ed.), Be and Shell Stars, 157–164. All Rights Reserved

involved in active 'programmes' of such observations. The wavelength range covered is 21 cm to 3 mm (noting that only continuum observations have been attempted for Be stars). It should also be noted that the sensitivities of these instruments, considering the collecting areas and the receiver bandwidths used, are comparable, within an order of magnitude, with the exception of the 11 m dish which in any case is limited by atmospheric effects at 90 GHz.

TABLE I

TELESCOPE (Commonly used name)	FREQUENCY (GHZ)
BONN 100-m DISH (WEST GERMANY)	10.6
WESTERBORK RADIO SYNTHESIS TELESCOPE (NETHERLANDS)	1.4 and 5.0
CAMBRIDGE 5-Km RADIO TELESCOPE (ENGLAND)	5.0 and 14.5
ALGONQUIN 46-m DISH (CANADA)	10.6 and 22.2
NRAO 3-ELEMENT INTERFEROMETER (USA)	2.7 and 8.1
KITT PEAK 11-m DISH (USA)	90.0
PARKES 64-m DISH (AUSTRALIA)	6.2 and 8.9

MAJOR INSTRUMENTS USED IN DETECTION AND OBSERVATION OF RADIO STARS, AND THE FREQUENCIES AT WHICH THEY ARE MOST COMMONLY USED FOR THIS PURPOSE.

The single-dish instruments listed in Table I operate at high frequencies, with resolutions in the 1' to 3' range, and hence are useful for radio star observations. However, confusion effects are appreciable with such resolutions, particularly in the galactic plane or in the Cygnus or Carina regions, and measurements made with single dishes should be approached with considerable caution unless amply confirmed, preferably by one of the interferometer/synthesis arrays. The three arrays currently in operation are listed separately in Table II, with references to published material which provides detailed technical information. The Westerbork and NRAO instruments have also been described and compared by Braes and Miley (1973) at the *IAU Symp.* **55**. It is interesting to note that the resolution which can be obtained with these instruments is comparable to that of conventional optical telescopes.

Some caution is also required in the interpretation of interferometer data, in the cases where only a few Fourier components of the angular structure are measured. If a full synthesis observation is taken, then a reasonably unambiguous radio map is obtained. However, it is often the case that only a selection of spacings is used, in the interests of economy: if then the radio source is resolved, or partly resolved, the angular structure is often derived by what is essentially a model-fitting process. In these cases it is not common practice to publish the complete visibility curve, but only

TABLE II

TELESCOPE	FREQUENCY (GHZ)	MAXIMUM RESOLUTION (ARC secs.)	REFERENCE
WESTERBORK	1.4	22	BAARS & HOOGHOUDT, ASTRON. & ASTROPHYS. 31, 323, 1974
	5.0	6	CASSE & MULLER, ASTRON. & ASTROPHYS. 31, 333, 1974
CAMBRIDGE	5.0	2	RYLE, NATURE 239, 435, 1972
	15.4	0.7	
N R A O	2.7	8	HOGG et al, ASTRON. J, 74, 1206, 1969
	8.1	2	

SYNTHESIS INSTRUMENTS USED IN STUDY OF RADIO STARS.

the final result based on the assumed model. To interpret such information, one should be aware of the spacings that are used, and of the assumptions inherent in the model.

3. Observations

3.1. NEGATIVE RESULTS

The long lists of non-detections for Be stars are, for the most part, unpublished. I will mention only the four 'Symposium 70' stars (ζ Tau, γ Cas, ϕ Per and Pleione) for which we have established upper limits of ~20 mJy.

3.2. RADIO OBJECTS WITH A CONTINUUM SPECTRUM HAVING THE APPROXIMATE FORM OF FLUX DENSITY PROPORTIONAL TO FREQUENCY

The prime example of this type of radio source is MWC 349, classified optically as Bep. In 1942 Swings and Struve (1942) noted the similarity of its spectrum to that of RY Scuti, which is also a radio object. More recent spectra have been obtained by Ackermann (1970), Herbig (1972), Kuhi (1973), Greenstein (1973), and by Ciatti and Mammano (1975). Absorption bands were found by Ciatti and Mammano and by Ackermann in the near infrared, but none have been reported at shorter wavelengths. Greenstein, however, suggested a spectral class in the B0–B5 range, based on the dereddened optical continuum. Extinction is large, giving ~9 magnitudes of absorption in the V band, the Hα line is very strong, and variable, and many other species of lines are observed, including the nebular lines. The object is sometimes referred to as a small H II region, but the radio data indicates that the emitting gas is intimately associated with the star, and I prefer to call it an extended circumstellar envelope. The infrared excess is large (Geisel (1970), Allen (1973)) and has been attributed to reradiation by dust in this envelope.

The radio observations do not indicate variability. The data defining the continuum spectrum is shown in Figure 1, where it is obvious that the spectrum is not the standard form for Bremsstrahlung radiation from a homogeneous slab of gas (i.e.

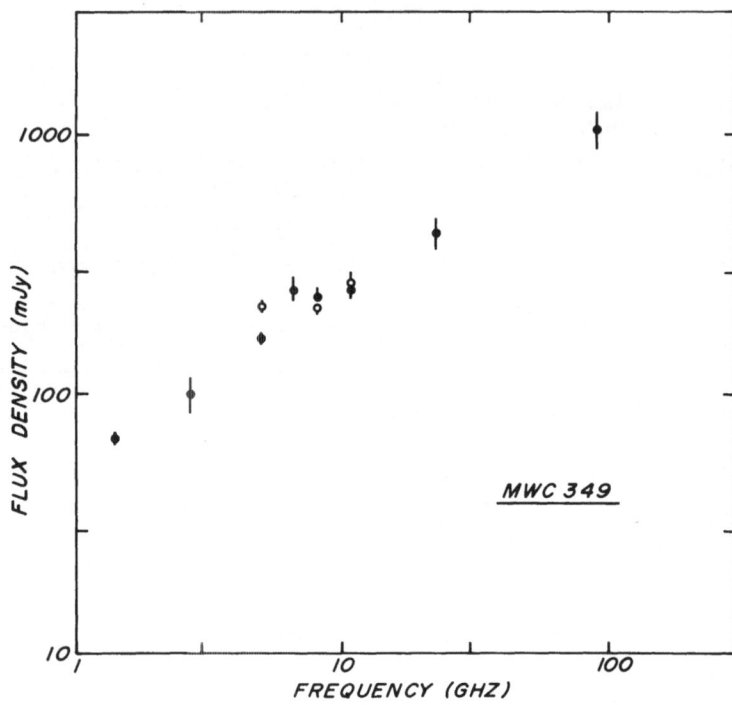

Fig. 1. Flux density measurements of MWC 349. The data have been taken from Olnon (1975) at 1.4, 5.0 (solid circle) and 10.6 (open circle) GHz; from Hjellming *et al.* (1973) at 2.7 and 8.1 (solid circle) GHz; from Baldwin *et al.* (1973) at 5.0 (open circle) GHz; from Gregory and Seaquist at 6.6 and 10.6 (solid circle) GHz; from Greenstein (1973) at 8.1 (open circle) GHz; and from Marsh, Purton and Feldman (unpublished) at 22.2 and 90 GHz.

slope ~0 if optically thin, ~2 if optically thick). In the usual notation of $S \propto \nu^{\alpha}$, the value obtained for the spectral index (α) for MWC 349 is 0.7. This type of radio spectrum is produced by a spherically-symmetric shell of gas with an inverse-square density distribution, such as would result from a steady and continuous mass outflow from the central star. Such a density distribution for radio-emitting gas was first considered by Weymann and Chapman (1965) for α Orionis, and the associated theory has been applied more recently to observations of V1016 Cygni by Seaquist and Gregory (1973). Since then the theory has been refined by three groups: Wright and Barlow (1975), Panagia and Felli (1975), who applied their results to P Cygni, and by Olnon (1975) who considered MWC 349 itself.

The physical size of such a circumstellar shell would be appreciable, and in fact the MWC 349 radio object has been resolved by the Cambridge interferometer. Olnon (1975) has analysed both the Cambridge and Westerbork data in detail, taking into consideration the complete visibility curve (although assuming circular symmetry), and has shown that the data is compatible with an inverse-square density distribution. The exact value of the radio spectral index implies a distribution of the form $n_e = kr^{-2.1}$, and the value of k was found, from a knowledge of the distance to

MWC 349 (2.1 kpc. as suggested by Reddish (1967) on the basis of its membership in the Cyg OB II Association) and an assumed temperature of the emitting gas (the results are rather insensitive to temperature), to be $\sim 6 \times 10^{38}$ in cgs units.

A number of similar objects have been considered by Marsh (1975a), who shows that in each case the spectrum is steeper than expected from an unbounded inverse-square density distribution, but that the data for each agrees well with the spectrum that is produced by an exact r^{-2} distribution truncated at some outer radius. If the same model is taken for MWC 349, then the observed spectrum indicates a value for the outer radius of $r_{out} \sim 10^{17}$ cm. Constraints can also be placed on the inner radius of the envelope. As the continuum spectrum is still optically thick at 90 GHz, then $r_{in} \gtrsim 10^{15}$ cm. However, the density at the inner edge cannot exceed $\sim 3 \times 10^8$ cm^{-3}, otherwise the short recombination times prohibit the ionising photons from penetrating to the remainder of the envelope; hence $r_{in} \gtrsim 10^{14}$ cm. This lower limit to r_{in} applies at all times, and the fact that emission lines have been observed for the past 40 years implies that $r_{in} \gtrsim 10^{14}$ cm for at least that long. Presumably, if the envelope is expanding, as indicated by the r^{-2} density and the P Cygni profiles, then r_{in} was less than that value in the not-too-distant past, and the entire envelope bcame ionised in a relatively short time as the inner radius passed through the critical value.

If you prefer to regard envelopes of this size as planetary nebulae, then note that the total mass of the gas is only $\sim 0.03\ M_{\odot}$ (Marsh (1975b)), which is a bit low. However, MWC 349 may well represent the low-mass equivalent of a planetary nebula.

Other objects are known which have the same type of radio spectrum as MWC 349. Of these, good data is available for V1016 Cygni, Vy2-2 and Hb12, two of which have been observed with the Cambridge interferometer (Ryle, 1975). These latter observations show that the intensity distribution of V1016 Cygni is roughly as expected for an inverse-square density distribution, but indicate a somewhat more complex structure for Vy2-2. Reasonable data is also available for Hen 1044 (Wright et al., 1974) and HD 167362. Finally, the data available for P Cygni and RY Scuti suggest that these objects may also be of this radio type. All these objects form a homogeneous class of radio objects, but appear with the various optical classifications of planetary nebulae, Wolf-Rayet stars, symbiotic stars, and Bep stars. In the above list only the last two have been classified as Bep, and only these will be discussed further.

P Cygni: this well-known star is a weak-radio source (Wendker et al., 1973), and flux-density information is available at only two frequencies. However, the optical spectrum indicates mass outflow, and the radio spectral index of ~ 0.7 is compatible with the inverse-square model. The densities in the circumstellar envelope would be considerably lower than for MWC 349, by a factor ~ 300 at the same radius, but the velocity of the outflowing gas is relatively high. Both Panagia and Felli (1975) and Wright and Barlow (1975) obtained mass-loss rates of $\sim 10^{-5}\ M_{\odot}$ yr^{-1}.

RY Scuti: this object was described in the HD catalogue as having faint dark absorption lines, but in 1922 (Merrill) was classed as Pec, with some P Cygni features, and it still exhibited emission lines in 1928 (Merrill). If the same model is adopted as has been suggested for MWC 349, then the epoch at which emission lines appeared would correspond to the inner radius of the envelope expanding beyond the critical value at which the central star is able to ionise the envelope.

In 1937 Gaposhkin deduced, on the basis of the light curve, that RY Scuti was an eclipsing binary system similar to β Lyrae. In 1943 Popper found that some emission lines varied as for a spectroscopic binary, and interpreted his results as a binary system of large mass ($>100\,M_\odot$) but accompanied by a shell surrounding both components which was responsible for most of the emission lines. A detailed spectrum by Swings and Struve (1940) contained absorption lines which suggested that the brighter component was of spectral class O or B0. The infrared measurements of Geisel (1970) again indicate the presense of circumstellar dust.

The radio data is sparse. RY Scuti was initially detected at Algonquin by Hughes and Woodsworth (1973), but the region is confused, and the measured flux density unreliable. Measurements at two frequencies with the NRAO interferometer (Hjellming *et al.*, 1973) indicate that the radio emission is not variable, and suggest a spectral index of ~1.0. Both the radio and the optical measurements indicate the presence of a significant circumstellar shell, and the radio object is probably similar to MWC 349 rather than to β Lyrae, i.e. the binary nature of the object is not directly related to its radio emission.

3.3. RADIO OBJECTS WITH A CONTINUUM SPECTRUM OF OPTICALLY THIN GAS

An example of this type of object is MWC 957, classified as Be by Vyssotsky *et al.* in 1945, and independently in the Mount Wilson Catalogue (Merrill and Burwell,

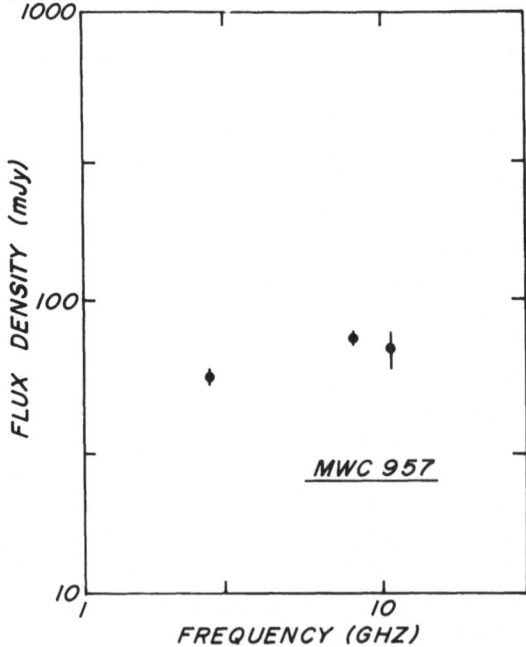

Fig. 2.   Flux density measurements of MWC 957. The data have been taken from Marsh, Purton and Feldman (unpublished) at 2.7 and 8.1 GHz, and from Feldman *et al.* (1973) at 10.6 GHz.

1949). However, both Perek and Kohoutek (1967) and Henize (1967) take it to be a planetary nebula. Measurements by Allen (1973) in the H and K bands show an infrared excess, but do not indicate the presence of circumstellar dust.

The radio data is again sparse (see Figure 2), partly because it is a weak radio source – particularly for the higher frequencies – and partly because less interest is taken in an object of this type. Measurements with the NRAO interferometer show the region to be free of confusing sources, hence the single-dish measurement of Feldman *et al.* (1973) is probably reliable. The data suggest a standard Bremsstrahlung spectrum from gas which is optically thick at frequencies below 3 GHz. The radio spectrum in the optically thick region is not known, and could be of the $\alpha = +2$ or the $\alpha = +1$ type: if the latter is found to be true then the cirumstellar shell could be of the type associated with MWC 349, although of lower density and/or with a larger value of r_{in}. The radio object was unresolved with the NRAO interferometer, hence the angular size is $\gtrsim 1''$, suggesting, independently of the rather dubious spectrum shown in Figure 2, a turnover frequency as high as 3 GHz. The emission measure is then $\sim 3 \times 10^7$ cm^{-6} pc, and for any reasonable distance at all (say a few kpc. within a factor of 3) the mass of the cirumstellar shell is even less than that of MWC 349.

There are many objects which have a similar radio spectrum, but most of them are classified as planetary nebulae, although the radio spectra indicate a shell mass which is relatively small. One of these objects which has been classified as Bep is MWC 247, for which only single-dish measurements are available (Wright, 1975), but these suggest that it is optically thin in the 5–10 GHz range.

3.4. β LYRAE

This is a special case. It is classified as Bep, but is also an eclipsing binary of large mass. The radio emission is weak and variable (Wade and Hjellming, 1972), hence the radio spectrum is difficult to determine. However, the radio luminosity is several orders of magnitude lower than that of the objects discussed above (Woodsworth, 1975), and it is most likely that the radio emission is associated with mass transfer in the system, as has been suggested for Algol (Hjellming, 1973; Jones and Woolf, 1973), and for the radio-emitting X-ray stars (see, for example, Gursky and Schreier, 1975), i.e. the Be classification of β Lyrae is not directly related to its radio emission.

4. Conclusions

Although normal Be stars do not produce detectable amounts of radio emission, a few peculiar Be stars are associated with radio objects by virtue of their very extensive circumstellar shells. The masses of the shells discussed in this review are probably $\gtrsim 10^{-2} M_\odot$, and some appear to be the result of a prolonged mass outflow at a steady rate of $10^{-5} M_\odot$ yr^{-1} or less (Marsh, 1975b). The connection between these objects and Be stars as a class is not clear. However, the connection with any other class of emission-line objects, such as planetary nebulae or Wolf-Rayet stars, is not clear either. It should be noted that the mass of the shell is quite small compared to the mass of a main-sequence early-type star, leaving open the possibility that these objects could represent a transient phase in the life of an early-type star.

References

Ackermann, G.: 1970, *Astron. Astrophys.* **8**, 315.
Allen, D. A.: 1973, *Monthly Notices Roy. Astron. Soc.* **161**, 145.
Baldwin, J. E., Harris, C. S., and Ryle, M.: 1973, *Nature* **141**, 37.
Braes, L. L. E. and Miley, G. K.: 1973, in H. Bradl and R. Giacconi (eds.), 'X- and γ-ray Astronomy', *IAU Symp.* **55**, 86.
Ciatti, F. and Mammano, A.: 1975, *Astron. Astrophys.* **38**, 435.
Feldman, P. A., Purton, C. R., and Marsh, K. A.: 1973, *Nature Phys. Sci.* **245**, 39.
Gaposhkin, S.: 1937, *Harvard Annals* **105**, 509.
Geisel, S. L.: 1970, *Astrophys. J. Letters* **161**, L105.
Greenstein, J. L.: 1973, *Astrophys. J. Letters* **184**, L23.
Gregory, P. C. and Seaquist, E. R.: 1973, *Nature Phys. Sci.* **242**, 101.
Gursky, H. and Schreier, E.: 1975, in V. E. Sherwood and L. Plaut (eds.), 'Variable Stars and Stellar Evolution', *IAU Symp.* **67**, 413.
Henize, K. G.: 1967, *Astrophys. J. Suppl.* **14**, 125.
Herbig, G. H.: 1972, *IAU Circ.* **2457**.
Hjellming, R. M.: 1973, *Nature Phys. Sci.* **238**, 52.
Hjellming, R. M., Blankenship, L. C., and Balick, B.: 1973, *Nature Phys. Sci.* **242**, 84.
Hughes, V. A. and Woodsworth, A. W.: 1973, *Nature Phys. Sci.* **242**, 116.
Jones, T. W. and Woolf, N. J.: 1973, *Astrophys. J.* **179**, 869.
Kuhi, L. V.: 1973, *Astrophys. Letters* **14**, 141.
Marsh, K. A.: 1975a, *Astrophys. J.*, in press.
Marsh, K. A.: 1975b, Ph.D. Thesis, York University, Toronto, Ontario.
Merrill, P. W.: 1922, *Publ. Astron. Soc. Pacific* **34**, 295.
Merrill, P. W.: 1928, *Astrophys. J.* **67**, 179.
Merrill, P. W. and Burwell, C. G.: 1949, *Astrophys. J.* **110**, 387.
Olnon, F. M.: 1975, *Astron. Astrophys.* **39**, 217.
Panagia, N. and Felli, M.: 1975, *Astron. Astrophys.* **39**, 1.
Perek, L. and Kohoutek, L.: 1967, *Catalogue of Galactic Planetary Nebulae*, Prague.
Popper, D. M.: 1943, *Astrophys. J.* **97**, 394.
Reddish, V. C: 1967, *Monthly Notices Roy. Astron. Soc.* **135**, 251.
Ryle, M.: 1975, private communication.
Seaquist, E. R. and Gregory, P. C.: 1973, *Nature Phys. Sci.* **245**, 85.
Swings, P. and Struve, O.: 1940, *Astrophys. J.* **91**, 546.
Swings, P. and Struve, O.: 1942, *Astrophys. J.* **95**, 152.
Vyssotsky, A. N., Miller, W. J., Walther, S. J., and Walther, M. E.: 1945, *Publ. Astron. Soc. Pacific* **57**, 314.
Wade, C. M and Hjellming, R. M.: 1972, *Nature* **235**, 270.
Wendker, H. J., Baars, J. W. M., and Altenhoff, W. J.: 1973, *Nature Phys. Sci.* **245**, 118.
Weymann, R. and Chapman, G.: 1965, *Astrophys. J.* **142**, 1268.
Woodsworth, A. W.: 1975, Ph.D. Thesis, Queen's University, Kingston, Ontario.
Wright, A. E.: 1975, private communication.
Wright, A. E. and Barlow, M. J.: 1975, *Monthly Notices Roy. Astron. Soc.* **170**, 41.
Wright, A. E., Fourikis, N., Purton, C. R., and Feldman, P. A.: 1974, *Nature* **250**, 715.

DISCUSSION

Hutchings: The limits you quoted for the four *Symp.* **70** stars seem rather high. I thought it was possible to do better than that.

Purton: Yes, the limiting flux is determined by the integration time and the estimated confusion. The synthesis arrays can work to ~5 mJy. The limits of 20 mJy come from observations with the Algonquin dish and represent a modest amount of integration time.

Because of the pressure for observing time on the synthesis arrays the practice is growing of taking preliminary observations with dishes, then checking out the possible detections with an interferometer.

ULTRAVIOLET OBSERVATIONS OF Be STARS

(*Review Paper*)

SARA R. HEAP

*Laboratory for Optical Astronomy, NASA Goddard Space Flight Center,
Greenbelt, Md., U.S.A.*

1. Introduction

The history of ultraviolet studies of Be stars is barely ten years old. However, in the last decade, twelve major space experiments have observed Be stars in the ultraviolet region of the spectrum, and ultraviolet data for over 30 Be stars are now available in the literature. Table I shows some of the characteristics of the experiments. They include two rockets, five astronomical satellites, three manned satellites, and one planetary probe. Except for the rocket experiments, they are primarily survey instruments, which have provided ultraviolet data on early-type stars in general, and these data have proved to be extremely useful as standards of comparison for Be stars. Of the twelve spacecraft, two are presently operating: *Copernicus* and ANS. The two experiments complement one another very nicely in that the ANS experiment can obtain absolute continuous flux distributions, while the Princeton experiment can obtain high-resolution line spectra for the brighter Be stars.

Table II shows those bright Be stars, which have been observed in the ultraviolet, and for which there are explicit published data. For most stars, the coverage has been spotty, but for a few stars such as ζ Tau and γ Cas, the ultraviolet data are complete in the sense that absolute continuous fluxes as well as high-resolution line spectra are available, often from more than one source, so that cross-checks can be made and the extent of variability, if any, determined.

Rather than organize this review, experiment by experiment, or star by star, I would like to divide this review into four parts, each one describing an area of study in which ultraviolet data have proved useful (or should prove useful) in coming to an understanding of Be stars. The four areas are: (1) spectral classification and the determination of the stellar atmospheric parameters, (2) evidence for mass loss and the formation of circumstellar shells around Be stars, (3) the effects of very rapid rotation on the properties of Be stars, and (4) the constituents and physical properties of the shells surrounding Be stars. In each of these areas, I shall use one star as a primary example, in the belief that the accumulation and comparison of many data on one star lead to a more certain understanding of the nature of a Be star than only one datum per star for many stars. The example I shall use is ζ Tau because it has been studied most intensively and because it has many characteristics in common with other Be stars: its emission lines are strong, its shell spectrum is very rich, and its infrared excess is sizeable. I do not wish to imply that it typifies Be stars in general. In all probability, it does not. This is something we can assess after hearing all the papers at this Symposium.

A. Slettebak (ed.), Be and Shell Stars, 165–178. All Rights Reserved
Copyright © 1976 by the IAU.

TABLE I

Space experiments which have observed Be stars

Experiment	(Satellite)	Launch date	Range	Resolution	Reference
A. Filter Photometry					
Celescope	(1964 83C)	1964	1376 Å		Smith (1967)
	(OAO-2)	1968	4 filters		Davis et al. (1972)
U. Wisconsin	(OAO-2)	1968	12 filters		Code et al. (1970)
B. Low Resolution Spectrophotometry					
U. Wisconsin	(OAO-2)	1968	1800–3800 Å	20 Å	Code et al. (1970)
			1050–2000	10	
UVS	(Mariner 9)	1971	1150–3400	15	Lillie et al. (1972)
			1150–1650	7.5	
S2/68	(TD1)	1972	1350–2500	35	Boksenberg et al. (1975)
S-169	(Apollo 17)	1972	1180–1680	10	Henry et al. (1975)
Orion-2	(Soyez 13)	1973	2000–4000	variable	Gurzadyan et al. (1974)
UV Exp.	(ANS)	1974	1500–3300	150	van Duinen et al. (1973)
S-019	(Skylab)	1974	1300–5000	variable	Henize et al. (1976)
C. High Resolution Spectroscopy					
S-59	(TD1)	1972	2060–2160 Å	1.9 Å	de Jager et al. (1974)
			2490–2590	2.3	
			2770–2870	1.8	
Princeton U.	(Copernicus)	1972	~750–1500	0.2, 0.05	Rogerson et al. (1973)
			~1500–3200	0.4, 0.10	Bohlin (1970)
———	(Rocket)	1968	1060–2130	2.0	Bohlin (1970)
———	(Rocket)	1972	1100–2040	0.1	Heap (1975)

TABLE II

Bright Be stars observed in the ultraviolet

Star	Spectral type	Reference				
		OAO-2	*Copernicus*	TD 1	Mariner 9	Other
o And	B6p	3		10, 18	12	
π Aqr	B1 Ve	13			12	
ι Ara	B2 IIIe					9
α Ara	B2 Vne	13				
ρ Car	B4 Vne	13				
f Car	B3 Vne					15
γ Cas	B0.5 IVe	5, 13	17	11	12	2, 8, 15
ε Cas	B3 Vp	3			12	
δ Cen	B2 IVne	13		11		
η Cen	B1.5 Vn	13	4, 17			9, 15
μ Cen	B2 IV–Ve	3, 13	14, 17			
α Col	B8 Ve	13		11, 18		
ω CMa	B2 IV–Ve			11		
κ CMa	B2 Ve	13		11		9
υ Cyg	B2 Ve	3	14		12	
59 Cyg	B1.5 Ve	13				
60 Cyg	B1 Vnep	5				9
κ Dra	B7 e	3		11, 18		15
α Eri	B3 Vp	13		11		15
ν Gem	B7 IVe			18		
8 Lac	B1 Ve					9
48 Lib	B5 IIIp	3				
β Mon A	B3 Ve			11		
χ Oph	B1.5 Ve		16			
ω Ori	B2 III	3, 13				
25 Ori	B1 Vn	13				
λ Pav	B1 Ve			11		9, 15
31 Peg	B2 IV–Ve		17			
φ Per	B1 pe	13	17		12	9
ψ Per	B5 e				12	
48 Per	B3 Ve		17			
o Pup	B0 V:pe					9
ζ Tau	B4 IIIp	3, 13	17	11, 1, 6	12	7, 8, 15
η Tau	B7 III	3				
HD 28497	B1.5 Ve		17			
HD 58978	B0 VI?pe					9
HD 120991	B2 IIIe					9
HD 135160	B0.5 Ve					9
HD 200775	B3 IV–Ve			19		
HD 217050	B4 IIIpe	3				

1. Beeckmans (1975).
2. Bohlin (1970).
3. Bottemiller (1971).
4. Burton and Evans (1976).
5. Coyne (1972).
6. Delplace and van der Hucht (1976).
7. Heap (1975).
8. Heap and Stecher (1974).
9. Henize *et al.* (1976).
10. Kondo *et al.* (1975).

11. Lamers and Snijders (1975).
12. Molnar (1975).
13. Panek and Savage (1975).
14. Peters (1976).
15. Smith (1967).
16. Snow (1975).
17. Snow and Marlborough (1976).
18. Underhill and van der Hucht (1976).
19. Viotti (1975).

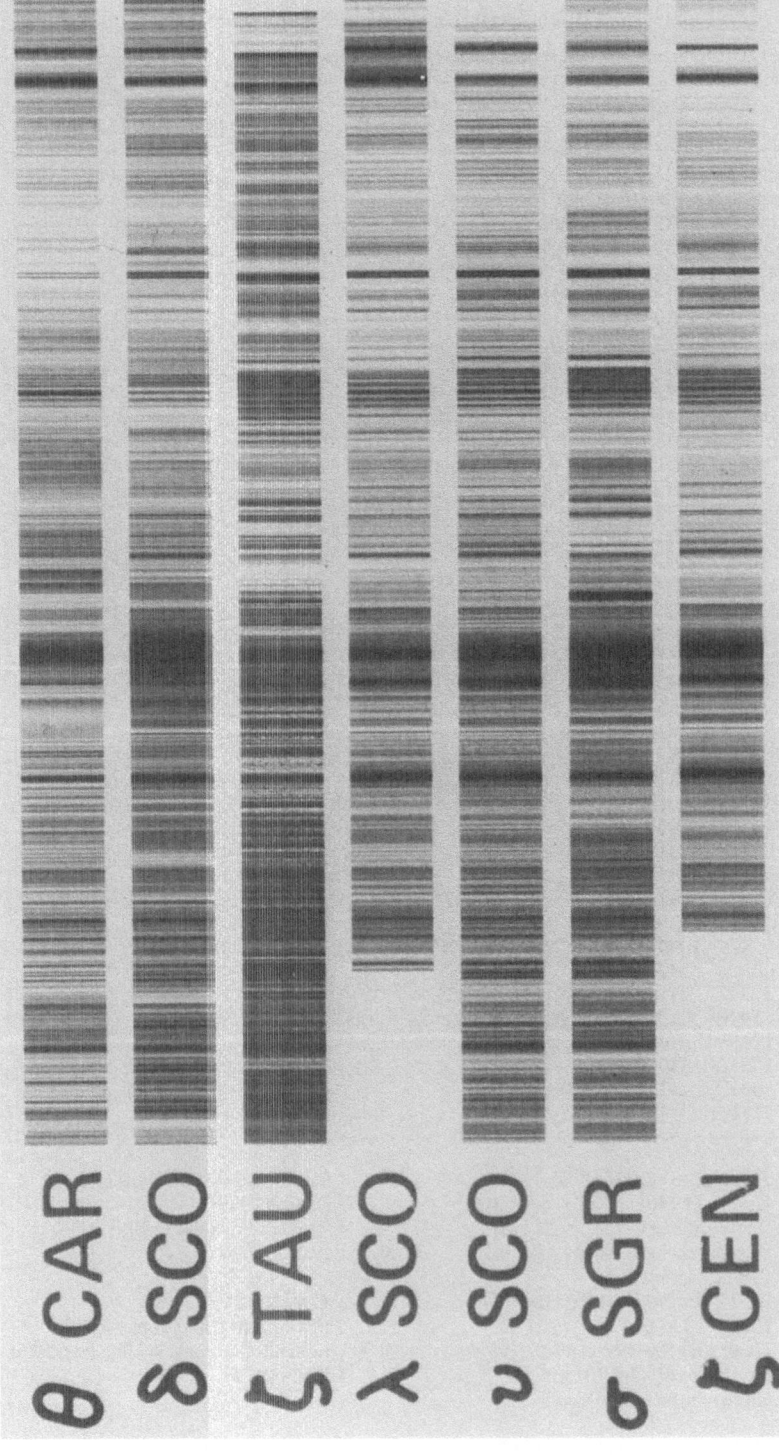

Fig. 1. Photographic Reconstruction of U2 Spectrograms from *Copernicus*. The spectra have been arbitrarily normalized, and noise has been added for simulation purposes.

2. Spectral Classification

A system for ultraviolet spectral classification has not been fully worked out as yet, but it may prove a good deal easier and more accurate – at least for Be stars – than that worked out for the visual region of the spectrum. Figure 1 shows a reconstruction of several U2 spectra of 7 B-type stars that were obtained by the Princeton experiment on *Copernicus*. The spectral interval covered here goes from λ 1050 Å on the left to λ 1420 Å on the right. All the stars have published projected rotational velocities of around 200 km s^{-1} except for υ Sco whose $v \sin i \cong 30$ km s^{-1}, and ζ Tau whose published $v \sin i \cong 300$ km s^{-1}. The spectral types of the stars range from B0.5 at the top to B2.5 at the bottom. One Be star, ζ Tau, is also shown here for comparison. The most prominent line, of course, is Lyman alpha at λ 1216 Å. The strength of this line is a measure of the interstellar column density of hydrogen. This quantity should be a great help in the study of Be stars because Bohlin (1975) has found a tight correlation between the interstellar hydrogen column density and the color excess, $E(B - V)$. This means that it should now be possible to estimate the intrinsic color of a Be star (at least those with early B spectral types) with corrections for interstellar extinction based on the strength of Lyman alpha line.

One of the main advantages of the ultraviolet is that it contains strong lines of abundant elements, e.g., C III λ 1175 Å, N III λ 1183–5 Å, Si III λ 1206 Å, C II λ 1334–5 Å, Si IV λ 1394, 1403 Å. These lines are especially useful in determining the atmospheric properties of Be stars for several reasons. First, since they are strong lines, they are easily and accurately measured, even in stars with high projected rotational velocities. Most of these strong lines are resonance lines of abundant elements so that their associated atomic data are relatively reliable. Secondly, sometimes more than one stage of ionization of an element is present (as is the case here with Silicon, where both Si III λ 1206 Å and Si IV λ 1394, 1403 Å are present) so it is possible to use them for ionization balance tests. Figure 1 shows that Si III λ 1206 Å is very sensitive to spectral type: it is quite strong in the B2.5 stars and very weak in the B0.5 stars. Thirdly, some of these lines, like the Si IV resonance lines, belong to high ionization states so they are not contaminated by a shell component. Fourthly, the very strong lines are useful in determining the velocity field in the atmospheres of Be stars. Since the lines are very strong, they are formed in the outermost layers of the atmosphere, and so they are good indicators of mass motions in these outermost layers. Finally, some of these lines are good measures of projected rotational velocity.

Zeta Tau is a case in point, illustrating – sometimes with surprising results – some of those benefits that I have just outlined. One result is that the observed ionization-balance of Si III vs Si IV and C III vs C IV is best matched by Mihalas' (1972) model with the parameters $T = 27\,500°$ and $\log g = 4.0$. This value for the temperature is much higher than what would be expected from the most recent estimate of its spectral type, which is B4 III. The line profiles computed from this model are compared to the observed profiles in Figure 2. It is evident that the observed lines best fit the profiles for deficient C and Si. Another surprising result is that the stellar lines are all shifted to shorter wavelengths. The exact amount is shown under the word, 'Shift' in Figure 2. These shifts correspond to outward velocities of up to

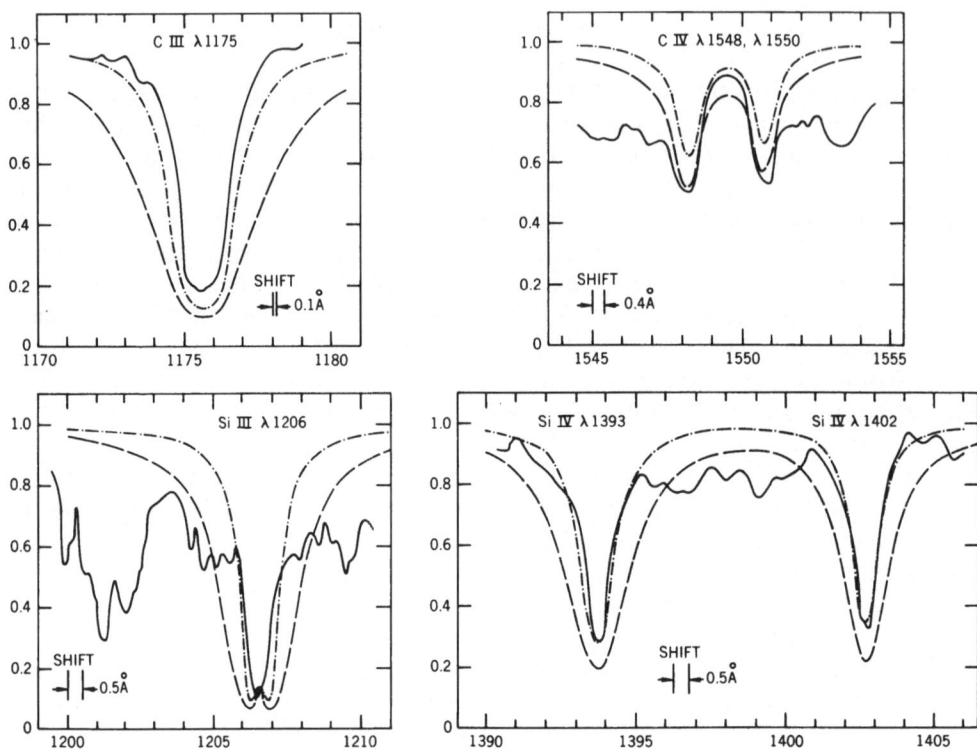

Fig. 2.   Observed and Computed Line Profiles in the ultraviolet Spectrum of ζ Tau (Heap, 1975).
—— = observed profile, — — = predicted profiles for a normal abundance (C = 8.57, Si = 7.42), —·—
predicted profiles for 1/5th normal abundance. All theoretical profiles are based on Mihalas' (1972)
·model with the parameters, $T_{\text{eff}} = 27\,500$ K, $\log g = 4.0$.

$120\,\text{km s}^{-1}$. A third surprising feature of the ultraviolet spectrum is the relative
sharpness of the stellar lines. The rotational velocity of ζ Tau, as derived from the
visual line spectrum, is about $300\,\text{km s}^{-1}$ (Underhill, 1952; Slettebak, 1949). However, the computed line profiles in Figure 2 are all artificially rotated to correspond to
a projected rotational velocity of only $100\,\text{km s}^{-1}$, and their widths are comparable
to the observed line widths. In the case of the C IV doublet, theoretical profiles where
$v \sin i = 200\,\text{km s}^{-1}$ tend to blend together, so we know the rotational velocity
cannot be that high. Right away, we have found an inconsistency in our picture of
ζ Tau. Usually, we assume that a Be star with a *low* projected rotational velocity
must be seen pole-on, or near pole-on. We also assume that *shell* stars must be seen
equator-on, or near equator-on. The ultraviolet spectrum of ζ Tau indicates both a
low projected rotational velocity and a rich shell spectrum. Either the shell around
ζ Tau is not so highly concentrated toward the equatorial plane as we had thought,
and projection effects are important, or the ultraviolet line spectrum yields a grossly
underestimated rotational velocity.

The finding that the effective temperature of ζ Tau is higher than its currently-used
spectral type would suggest, is something also found for other Be stars. Panek and
Savage (1975) have shown that C IV λ 1550 Å is stronger in the spectra of Be and
B-shell stars than in normal B stars of the same spectral type. In most cases, assigning

a spectral type to a Be or B-shell star which is one subclass earlier would resolve the discrepancy. Henize *et al.* (1976) have also found that about half of the Be stars observed by Skylab had C IV lines which were abnormally strong for their spectral types. The discovery of blue-shifted photospheric lines in the UV spectrum of ζ Tau is also something which has recently been found in other (but not all) Be stars studied by Snow and Marlborough (1976).

Some space experiments have looked repeatedly at the same object and have thereby provided information concerning the variability of the ultraviolet spectra of Be stars. These data have not been fully assessed as yet, but it is already clear that both the photospheric and shell spectrum of γ Cas is variable. Both Molnar (1975) and Panek and Savage (1975) have reported changes by a factor of two in the strength of the Si IV and C IV resonance lines – lines which are presumably formed in the photosphere of γ Cas. In addition, a comparison of the strengths of the Mg II resonance lines at λ 2800 Å as obtained for γ Cas by different experiments (*cf.* Table II) also reveals changes which are larger than can be explained by systematic errors of the various experiments. Beeckmans (1975) has found that the ultraviolet absolute flux of ζ Tau as measured by TD-1 varies by about ten percent. Because the spectral resolution of the TD-1 experiment is rather low, it is not clear whether this variability is due solely to changes in the shell absorption lines or whether the photospheric spectrum is also variable.

3. Mass Loss

The presence of emission lines in the spectra of Be stars is generally interpreted as evidence for a shell surrounding the star, and the high average rotational velocity is generally interpreted as evidence that rotation plays an important role in forming the shell. One problem with these interpretations has been that the visible spectra of Be stars showed no indications that the stars were actually losing mass or that Be stars were really all rotating at their critical velocities (Slettebak, 1966). There was hope that the ultraviolet spectrum might provide the needed evidence, since it contains the strong resonance lines whose cores are formed at the outermost layers in the atmosphere, perhaps even the interface between the star and shell. These hopes have been fulfilled, at least for the early B-emission stars, with the discovery of three types of anomalous ultraviolet line profiles which may be interpreted as evidence for outward mass motions, perhaps leading to the formation of a shell. First, Bohlin (1970) showed that the C IV resonance doublet in the spectra of γ Cas had P Cygni profiles. The absorption components of each line were displaced shortward by about 450 km s^{-1}. Second, Heap (1975) showed that the strong stellar ultraviolet resonance lines in ζ Tau, although totally in absorption, were all displaced to shorter wavelengths by up to 120 km s^{-1}. Since the visual lines do not show such displacements, she interpreted these net shifts of the ultraviolet lines as evidence for the radial acceleration of material in the photosphere that should lead to mass loss. Finally, Snow and Marlborough (1976) have shown that some Be stars with the earliest spectral types – B0 and B1 – show asymmetrical profiles, with the blueward wings extending to up to 900 km s^{-1}. Some also show the net shifts that Heap found in ζ Tau.

Two notes of caution are appropriate here. One is that there is no evidence for mass loss in the *late*-Be stars, and the other is that there is no proof that the indications of mass loss in the early-Be stars have anything to do with the formation or maintenance of the shell. Certainly, the way in which a shell is formed around a Be star still warrants further investigation.

4. Effects of Rapid Rotation

It is generally believed that the root, underlying cause of the peculiarities of Be spectra is rotation, although its role has not been fully assessed. Various predictions concerning the effects of rapid rotation have been made. These effects include: (1) the alteration of the effective gravity (through the influence of centrifugal force) in such a way as to allow radiation-driven mass loss to occur (Marlborough and Zamir, 1975), (2) the alteration of the continuous and line fluxes due to gravitational darkening, and (3) the alteration of the atmospheric chemical composition in the direction of carbon depletion and nitrogen enhancement (Paczyński, 1971). The Be stars are a very appropriate class of objects to serve as tests to these theories, since as a class they are rapid rotators. It also happens that most of the observable consequences of these theories are most prominent in the ultraviolet region of the spectrum. I will, therefore, review the relevant ultraviolet observations. Since the preceding section dealt with the first effect – mass loss – this section will deal with the second and third predicted effects of rapid rotation.

4.1. Gravitational darkening

Collins (1974a) and others have shown that rapid rotation produces a change in the apparent color temperature of a star and that this change is most obvious in the ultraviolet, where it can amount to a magnitude or more for stars seen equator-on. Using OAO-2 data, Bottemiller (1972) compared the ultraviolet colors of B-shell stars with those of slowly rotating stars, in the belief that shell stars as a class should be rotating near critical velocity and should be seen nearly equator-on. The result of this comparison was that 'if you push eyeball analysis near its limits', you find that shell stars have lower ultraviolet fluxes than normal B stars, which is in accordance with predictions. However, Bless and Code (1972), who also used OAO-2 data, did not find any firm evidence that rapidly rotating stars showed lower ultraviolet fluxes at λ 1700 Å, even though their sample of rapidly rotating stars included Be stars like ϕ Per.

Part of the problem with these comparisons is that the observational scatter amounts to several tenths of a magnitude, which is comparable with the predicted effects of gravitational darkening. There is only one star, ζ Tau, whose absolute ultraviolet flux has been measured by several experiments with high accuracy. This star, which shows a very rich shell spectrum, is a very likely candidate to show the effects of gravitational darkening, and the observations *do*, in fact, suggest that the radiation from this star is strongly affected by gravitational darkening. Figure 3 shows the observed absolute ultraviolet flux of ζ Tau as obtained by Smith (1967),

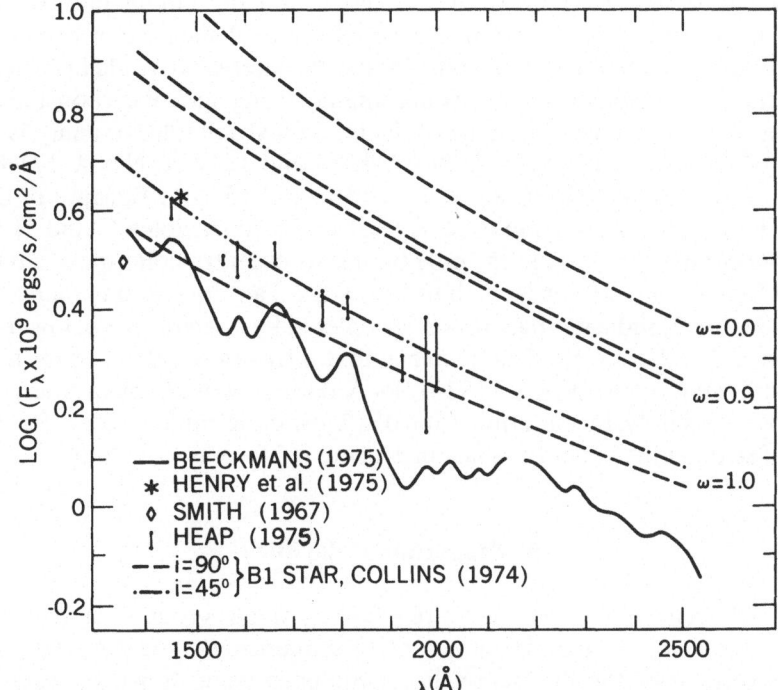

Fig. 3. Ultraviolet Flux of ζ Tau. The observed flux labeled 'Beeckmans' is an eye-estimate of the average of her scans (No. 2, 3, 4 and 5). Collins' theoretical fluxes were scaled such that the visual magnitude corresponds to that of ζ Tau ($V = 2.99$).

the APL group on Apollo-17 (Henry *et al.*, 1975), and by the S2/68 experiment on the satellite, TD-1 (Beeckmans, 1975). The measurements are in substantial agreement with each other. All these fluxes were obtained at low resolution, so they must be corrected for the effects of line-blocking by photospheric and shell lines before comparison with theoretical continuous fluxes can be made. The extent of the line-blocking is considerable (Heap, 1975), and correction for it raises the ultraviolet flux as much as 40% in some regions. However, even when such corrections are made, and allowance for interstellar extinction is made [$E(B-V) \approx 0.01$], the estimated continuous flux of ζ Tau is far below the predicted continuous flux of a non-rotating star. Collins' (1974b) predicted fluxes are shown in Figure 3 for the cases of ω = rotational velocity/critical velocity = 0.0, 0.9, and 1.0, and for angles of inclination of the rotational axis to the line of sight of 45° and 90°. It is evident that the estimated continuous flux of ζ Tau is compatible with a B1-star only if rotating near its critical velocity.

4.2. ALTERATION OF PHOTOSPHERIC CHEMICAL COMPOSITION

Paczyński (1971) suggested that for rapidly rotating stars, circulation currents induced by rotation might be sufficient to bring up to the surface material processed by the CNO cycle near the edge of the core. The result of this transfer would be that carbon would be depleted and nitrogen would be enhanced in the photosphere. The extent of the alteration would be most severe for late-B stars, but it would taper off toward earlier spectral types, until at the O stars the alteration would be insignificant.

The ultraviolet region of the spectrum contains strong lines of carbon and nitrogen in varying stages of ionization. Some of the lines most suitable for analysis are: N ɪɪ λ 1083–5 Å, N ɪɪɪ λ 1183, 1184 Å, N ᴠ λ 1238 and 1242 Å, N ɪᴠ λ 1718 Å, C ɪɪɪ λ 1175 Å, C ɪɪ λ 1334–1335 Å, and C ɪᴠ λ 1548 and 1550 Å. Because of this wide range in ionization, there should be several usable ultraviolet C and N lines for virtually all B-type stars. So far, the only Be star whose ultraviolet spectrum has been analyzed for carbon and nitrogen abundances is ζ Tau. Ionization balance tests, as shown in Figure 2, indicate that carbon is depleted by a factor five below its normal value. However, there is no firm evidence that nitrogen is enhanced in the atmosphere of this star, or the N ɪɪɪ λ 1183–1184 Å doublet would appear more strongly than it does. Certainly, the determination of atmospheric abundances in Be stars with the use of strong ultraviolet lines is a promising field worthy of study.

5. Properties of the Shell

It is generally believed that Be stars are a homogeneous group of stars in the sense that they are all main-sequence (or near main-sequence) B-type stars surrounded by shells. Nevertheless, the distinction has sometimes been drawn between Be and B-shell stars: B-shell stars show absorption·lines in the visual region of the spectrum, while regular Be stars do not. (Of course, according to this definition, a given Be star might become a B-shell star for a time and then revert back to a Be star.) I do not know if the same generalization holds true in the ultraviolet – that B-shell stars, and only B-shell stars, show shell absorption lines in the ultraviolet region of the spectrum. However, Lamers and Snijders (1975) have shown that Be and B-shell stars *can* be distinguished from one another on the basis of the strength of the resonance lines of Mg ɪɪ near λ 2800 Å. Figure 4 shows the observed strengths of Mg ɪɪ as a function of spectral type. The solid line, labeled 'V', represents their mean observed relation for normal main-sequence B-type stars, and the dashed curve shows their mean observed relation for B-type supergiants. The data show that the Mg ɪɪ lines are weaker in the spectra of Be stars than in the spectra of normal main-sequence B stars, and in a few cases, they are even in emission. However, the Mg ɪɪ lines have normal or greater-than-normal strengths in the B-shell stars. The one really discrepant case in this figure is ζ Tau. The abnormally strong absorption at λ 2800 Å is undoubtedly due to absorption by Mg ɪɪ in the circumstellar shell. In fact, enhanced ultraviolet absorption by the shell appears to be the rule for ζ Tau: no ultraviolet emission lines appear in the spectrum of this object.

The visual spectrum of a B-shell star has often been described as appearing composite, with a late-B supergiant spectrum superimposed on an early-B main-sequence spectrum. The same description holds true in the ultraviolet, at least in its gross spectral features. Low-resolution ultraviolet experiments have discovered two features which are characteristic of extended atmospheres: one at λ 1720 Å, first discovered in supergiants by Underhill *et al.* (1972) using OAO-2 spectrometer data;

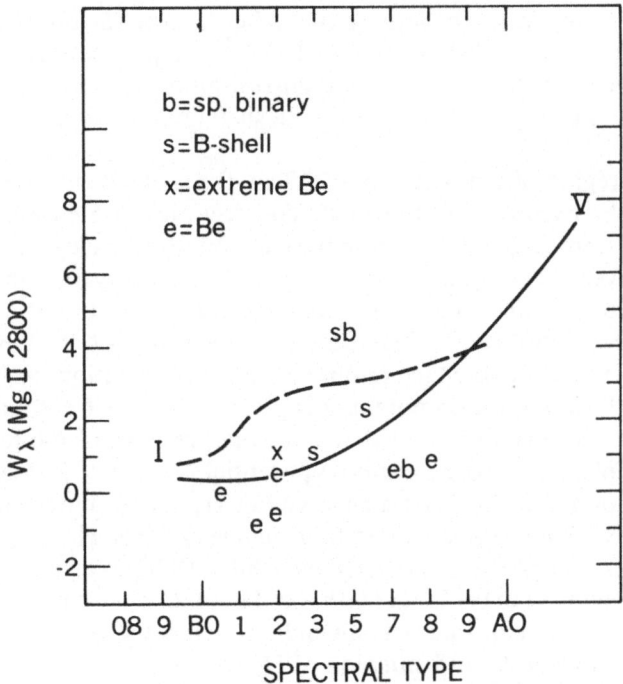

Fig. 4.   Equivalent widths for Mg II λ 2800 Å (Lamers and Snijders, 1975).

and another, very broad feature centered on λ 1920 Å, first discovered in super-giants by Thompson *et al.* (1974) using spectrometer data from TD-1. Both of these features are present in the spectrum of ζ Tau, with strengths comparable, or even superseding those of supergiants (Heap, 1975). Figure 5 shows a portion of the λ 1920 Å feature in the spectrum of ζ Tau. The bottom half of the figure shows the actual spectrum of ζ Tau while the top half shows the wavelengths and intensities of Fe III lines as listed by Kelly and Palumbo (1973). As Thompson *et al.* surmised, the

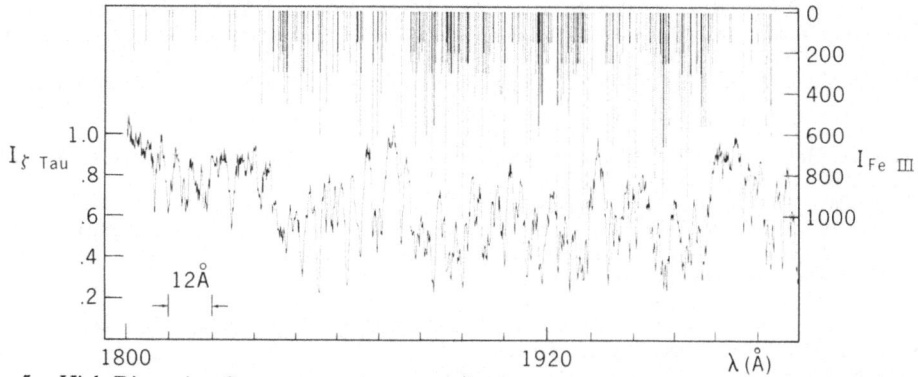

Fig. 5.   High-Dispersion Spectrum of the λ 1920 Å feature in the spectrum of ζ Tau (Heap, 1975).

λ 1920 Å feature appears to be largely due to Fe III, since the only high points in the observed spectrum at λ 1864, 1935 and 1972 Å are precisely those places where there are no Fe III lines. Similarly, a high-resolution spectrum of the λ 1720 Å feature in ζ Tau shows that it is largely due to shell lines of Al II, as well as Fe and Ni lines.

In the visual region of the spectrum of ζ Tau, the shell absorption lines indicate a wide range in the excitation of the shell. The presence of He I lines, for example, indicates that H-ionizing radiation from the star penetrates at least the inner layers of the shell. However, the presence of Mg I lines indicates a very low ionization state, no doubt in the outer region of the shell. The ultraviolet spectrum of ζ Tau also shows some indication of stratification. The ultraviolet spectrum is rich with lines from ions like Fe III and Ti III, both of which require H-ionizing radiation for their formation. Ions like Si II whose ionization potential is greater than 13.6 eV, have very strong lines. However, there is no firm evidence for the presence of the lowest excitation lines. For example, C I, whose ionization potential is 11.2 eV is definitely absent.

Neither H⁻ nor molecular lines such as CO or H_2 appear in the shell spectrum of ζ Tau. Snow (1975) has made a most thorough search for H⁻ in χ Oph, a Be star which has an infrared excess and is listed by Schild (1973) as an extreme-Be star. He found an upper limit to the column density of H⁻ of 2×10^{14} cm⁻², a value nearly of one magnitude lower than that which would be expected if the infrared excess of χ Oph were due to free-bound emission of H⁻.

I would like to close by returning to the introduction of this review and recalling that ultraviolet observations of Be stars are still being obtained and reduced. In these fortunate circumstances, this review can be neither complete nor up-to-date, but I hope that it will provide a starting point for later papers and discussion in this Symposium.

References

Beeckmans, F.: 1975, *Astron. Astrophys.*, in press.
Bless, R. C. and Code, A. D.: 1972, *Ann. Rev. Astron. Astrophys.* **10**, 197.
Bohlin, R.: 1970, *Astrophys. J.* **162**, 571.
Bohlin, R.: 1975, *Astrophys. J.* **200**, 402.
Boksenberg, A., Evans, R. G., Fowler, R. G., Gardner, I. S. K., Houziaux, L., Humphries, C. M., Jamar, C., Macau, D., Macau, J. P., Malaise, D., Monfils, A., Nandy, K., Thompson, G. I., Wilson, R., and Wroe, H.: 1973, *Monthly Notices Roy. Astron. Soc.* **163**, 291.
Bottemiller, R.: 1972, in A. D. Code (ed.), *Symposium on Scientific Results from OAO-2*, Amherst, p. 505.
Burton, W. and Evans, R. G.: 1976, this volume, p. 199.
Code, A. D., Houck, T. E., McNall, J. F., Bless, R. C., and Lillie, C. F.: 1970, *Astrophys. J.* **161**, 377.
Collins, G. W.: 1974a, *Astrophys. J.* **191**, 157.
Collins, G. W.: 1974b, private communication.
Coyne, G.: 1972, in A. D. Code (ed.), *Symposium on Scientific Results from OAO-2*, Amherst, p. 495.
Davis, R. J., Deutschman, W. A., Lundquist, C. A., Nozawa, Y., and Bass, S. B.: 1972, in A. D. Code (ed.), *Symposium on Scientific Results from OAO-2*, Amherst, p. 495.
Delplace, A. M. and Van der Hucht, K. A.: 1976, this volume, p. 197.
Gurzadyan, G. A., Kashin, A. L., Krmoyan, M. N., and Ohanesyan, J. B.: 1974, *Astrofizika* **10**, 177.
Heap, S. R.: 1975, *Phil. Trans. Roy. Soc. London A.* **279**, 371.
Heap, S. R. and Stecher, T. P.: 1974, *Astrophys. J.* **187**, L27.
Henry, R. C., Weinstein, A., Feldman, P. D., Fastie, W. G., and Moos, H. W.: 1975, *Astrophys. J.*, in press.

Henize, K. G., Wray, J. D., Parsons, S. B., and Benedict, G. F.: 1976, this volume, p. 191.
de Jager, C. Hoekstra, R., Van der Hucht, K. A., Kamperman, T. M., Lamers, H. J., Hammerschlag, A., Werner, W., and Emming, J. G.: 1974, *Astrophys. Space Sci.* **26**, 207.
Kelly, R. and Palumbo, L. J.: 1973, *Atomic and Ionic Emission Lines Below 2000 Angstroms*, NRL Report 7599, Naval Research Lab., Washington, D.C.
Kondo, Y., Modisette, J. L., and Wolf, G. W.: 1975, *Astrophys. J.* **199**, 100.
Lamars, H. J. G. L. M. and Snijders, M. A. J.: 1975, *Astron. Astrophys.* **41**, 259.
Lillie, C. F., Bohlin, R. C., Molnar, M. R., Barth, C. A., and Lane, A. L.: 1972, *Science* **175**, 321.
Marlborough, J. M. and Snow, T. P.: 1976, this volume, p. 179.
Marlborough, J. M. and Zamir, M.: 1975, *Astrophys. J.* **195**, 145.
Mihalas, D. M.: 1972, *Non-LTE Model Atmospheres for B and O Stars*, NCAR-TN/STR-76, NCAR, Boulder, Colorado.
Molnar, M. R.: 1975, *Astrophys. J.* **200**, 106.
Paczyński, B.: 1971, *Acta Astron.* **23**, 191.
Panek, R. J. and Savage, B. D.: 1976, *Astrophys. J.*, in press.
Peters, G.: 1976, this volume, p. 209.
Rogerson, J. B., Spitzer, L., Drake, J. F., Dressler, K. Jenkins, E. B., Morton, D. C., and York, D. G.: 1973, *Astrophys. J.* **181**, L97.
Schild, R. 1973: *Astrophys. J.* **179**, 221.
Slettebak, A.: 1949, *Astrophys. J.* **110**, 498.
Slettebak, A.: 1966, *Astrophys. J.* **145**, 126.
Smith, A. M.: 1967, *Astrophys. J.* **147**, 158.
Snow, T. P.: 1975, *Astrophys. J.* **198**, 361.
Thompson, G. I., Humphries, C. M., and Nandy, K.: 1974, *Astrophys. J.* **187**, L81.
Underhill, A. B.: 1952, *Publ. Dominion Astrophys. Obs.* **9**, 139.
Underhill, A. B., Leckrone, D. S., and West, D. K.: 1972, *Astrophys. J.* **171**, 63.
Underhill, A. B. and Van der Hucht, K. A.: 1975, submitted to *Astron. Astrophys.*
Van Duinen, R. J., Aalders, J. W. G., and Wesselius, P. R.: 1973, 'Announcement of Opportunity', Dept. of Space Research, Groningen.
Viotti, R.: 1975, 'Ultraviolet Energy Distribution of the B3e Star HD 200775 in NGC 7023 From TD1-S2/68 Observations', submitted to *Astron. Astrophys.*

DISCUSSION

Bidelman: Were there adequate ground-based observations of ζ Tau at the time that your observations were made?

Heap: Yes. Helmut Abt used the coudé spectrograph of the Kitt Peak 84-in. telescope to get spectra of ζ Tau the night before the rocket launch, and these plates have been a great help, because they allow an estimate of the electron density in the shell (from the Balmer shell lines), an estimate of the surface gravity of the star (from the wings of the photospheric Balmer lines), and things like that.

Swings: (1) I wish to point out that the identification of the λ 1940 absorption feature as Fe III was suggested, and published in 1973, by Swings, Jamar, and Vreux (*Astron. Astrophys.* **29**, 207) in a paper devoted to the reduction of S2/S68 spectrophotometric tracings of B stars. (2) The confirmation came on the basis of *Copernicus* data (note added in proof in the same paper). (3) The strength of the λ 1940 absorption feature is greater in the earlier B stars (earlier than B3); among these λ 1940 is much stronger for the supergiants than for the giants or main sequence stars. A comparison of *Copernicus* data to models and to laboratory intensities of Kelly and Palumbo is given in a paper by Swings, Klutz, Vreux, and Peytremann (*Astron. Astrophys. Suppl.* 1975, in press). (4) I believe it is shown in the paper by Beeckmans that the ultraviolet continuum of ζ Tau between λ 1400 and λ 2500 Å has changed appreciably between 1972 and 1974, both in the slope and in the level in absolute energy of the continuum.

Heap: The variability of the ultraviolet continuous flux of ζ Tau found by Beeckmans is small compared to the ultraviolet deficiency as indicated in Figure 3. I believe that Figure 5 represents the only high-dispersion spectrum of the λ 1920 feature in a Be star obtained so far.

Conti: This talk has been a very nice summary of our knowledge about Be stars derived from ultraviolet data. I am particularly impressed by what appears to be a fundamental difficulty in interpretation. We see evidence of a shell in ζ Tau which is stationary or may even be falling into the star. On the other hand, the same star has an outflowing wind of over 100 km s^{-1}. How can a shell be maintained in the pressure of a wind? Perhaps the wind begins outside the shell, but my impression is that dilution effect arguments, etc., always place the shell appreciably away from the star. It seems to me there is a considerable geometric problem in accounting for these observations.

Heap: I agree. There is no guarantee that the mass outflow indicated by the ultraviolet spectra of a Be star has anything to do with the maintenance of its shell.

Hummer: In response to Peter Conti's question, I can imagine two ways in which a reasonably stationary shell can feed a wind flowing from it. (1) If radiation pressure builds up in the shell, say because of recombination, its outer layers could be blown off, while the shell itself remains; (2) a disturbance moving up from below could shock when it reaches a sufficiently low density region, and carry off the outer layers of the shell.

Hutchings; Can I ask about the rotational velocities? You mentioned that in the visual region you get 250 or 300 km s^{-1} while the ultraviolet lines gave you 100 km s^{-1}. Are they the shell lines? Are they also photospheric? Why are they different?

Heap: The lines whose profiles I showed in Figure 2 – Si III λ 1206, Si IV λ 1393, λ 1403, C III λ 1175, C IV λ 1548, λ 1550 – are all *photospheric* lines, and *all* indicate a projected rotational velocity of around 100–150 km s^{-1}. I don't know why the ultraviolet lines suggest a lower rotational velocity than do the visual lines.

Snijders: I should like to comment on the results for the Mg II lines at λ 2800 Å obtained by H. Lamers and me and discussed by you. For some Be stars these lines were observed by many different experiments and comparison of results in various studies shows that the lines are often strongly variable; γ Cas and ζ Tau are good examples. Before 1973 the Mg II lines in γ Cas were far too weak for a B0.5 star. In 1973 they were at least a factor 20 too strong. In ζ Tau the Mg II lines changed by about a factor 2 between 1972 and 1973. If variations of this size occur in other shell lines in ζ Tau they could be responsible for some of the narrow band (35 Å) 'continuum' variations observed by Beeckmans with the S2/68 experiment which were discussed earlier by Swings.

A SURVEY OF MASS LOSS FROM Be AND SHELL STARS USING ULTRAVIOLET DATA FROM *COPERNICUS*

J. M. MARLBOROUGH

University of Western Ontario, London, Ontario, Canada

and

THEODORE P. SNOW, JR.

Princeton University Observatory, Princeton, N.J., U.S.A.

Abstract. Ultraviolet spectra of intermediate resolution have been obtained with *Copernicus* of twelve objects classified as Be or shell stars, and an additional 19 dwarfs of spectral classes B0–B4. Some of these spectra show marked asymmetries in certain resonance lines, especially the Si IV doublet at λ 1400 Å, indicating the presence of outflowing material with maximum velocities of nearly 1000 km s^{-1}. Direct evidence for mass loss at these velocities is seen for the first time in dwarf stars as late as B1.5. Later than B0.5, the only survey objects showing this phenomenon are Be stars. Among the stars considered there is a correlation between the presence of mass-loss effects and projected rotational velocity, suggesting that the UV flux from B1–B3 dwarfs is sufficient to drive high-velocity stellar winds only if rotation reduces the effective gravity near the equator. The role of mass-loss in producing the Be star phenomenon and the effects of rotation on mass loss are discussed.

1. Introduction

It is generally accepted that the shell and/or emission lines observed in the spectra of Be and shell stars arise from a low density extended atmosphere or circumstellar envelope which exhibits both differential rotation and strong concentration to the equatorial plane of the star. From the spectroscopic observations one infers that this circumstellar matter is in motion, most of the motion being rotational, with typical velocities of a few hundred km s^{-1} at a distance of several stellar radii above the star's surface. At such distances the velocity of escape from the star is of the same order of magnitude as the observed velocities. Consequently it seems probable that some or all of the observed gas ultimately escapes from the star. General discussions of these and other properties of Be and shell stars are given by McLaughlin (1961) and Underhill (1966), and reviewed by Hack and Struve (1970).

Nevertheless, little direct evidence in support of this widely accepted picture exists. In the optical region the radial component of velocity of the shell material, obtained from the central reversals of hydrogen lines and/or shell lines, is small, generally less than 10–20 km s^{-1}. Only during the active phases of some Be and shell stars does this radial component increase and then it normally does not exceed 100 km s^{-1}. Perhaps radial components of velocity of this order of magnitude indicate mass loss during the active phases, although the interpretation of what these velocities mean during these phases is not clear. Ultraviolet data obtained by Heap (1975) using a rocket-borne spectrograph show displaced resonance lines in the spectrum of ζ Tau corresponding to an outflow velocity of approximately 120 km s^{-1}, but whether or not this truly represents mass loss depends on the location of the absorbing material. Similar displacements have been seen by Marlborough (unpublished) in *Copernicus* spectra of α Eri, 59 Cyg, and HD 28497.

A. Slettebak (ed.), Be and Shell Stars, 179–189. All Rights Reserved
Copyright © 1976 by the IAU.

In the present study we are more concerned with evidence for mass outflow at velocities of several hundred km s^{-1}, which requires substantial acceleration of the material beyond the photosphere, and which almost certainly represents true mass loss. Bohlin (1970) noted that the C IV resonance doublet in γ Cas has a P Cygni profile with the absorption component displaced by at least 450 km s^{-1} from the stellar rest frame, but additional evidence for high-velocity mass loss from Be and shell stars has not been forthcoming. It is the purpose of this paper to report the results of a survey of such effects, based on *Copernicus* spectra of Be and shell stars. A partial description of the *Copernicus* results has been published elsewhere (Snow and Marlborough 1976).

2. The Data

All of the spectra used in this study were acquired with photomultiplier tube U2 (Rogerson *et al.*, 1973) at a nominal resolution of 0.2 Å. Backgrounds due to charged particles were subtracted using standard procedures developed at Princeton (Snow *et al.*, 1975) and the stray light contributions were removed following the algorithm of Bohlin (1975). The overall photometric accuracy, with errors dominated by drifts in the spacecraft guidance, is thought to be roughly 10% of the continuum level. Wavelengths are accurate to about half the U2 bandpass, i.e. $\pm\sim$0.1 Å.

For comparison with the Be and shell stars, all objects of luminosity class IV or V in the spectral class range B0 to B4 for which a complete U2 spectrum is available have been considered. This includes several stars from the mass-loss survey of Snow and Morton (1976). The distribution of these normal stars as well as the Be and shell stars is depicted in Figure 1, where it is seen that each subclass from B0 to B4 is

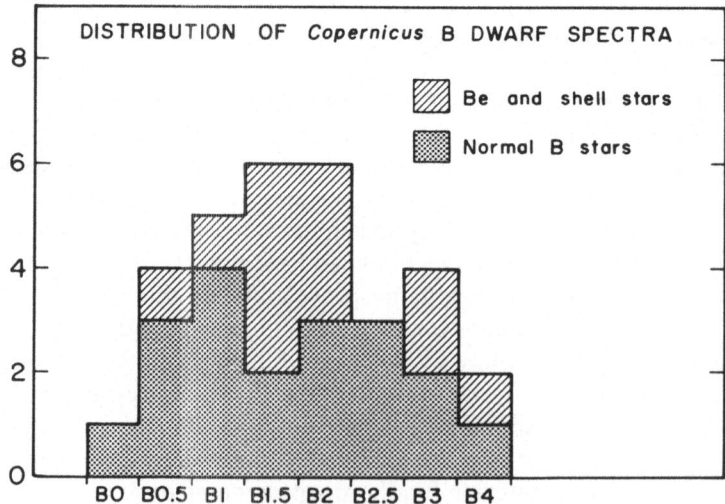

Fig. 1.   The distribution by spectral type of the B-type dwarfs and Be stars considered in this survey. Included are only objects for which complete (or nearly so) U2 spectra are available, and for which the signal-to-noise ratio is reasonably high as far longward as the Si IV doublet at λ 1400 Å.

represented by at least one star. Many more objects in this category have been scanned with *Copernicus*, but only partial spectra were obtained, and these stars are not included here because data are available at most for only a few of the features commonly showing mass-loss effects. Also omitted from consideration were stars for which the quality of the data was too poor to allow a conclusion to be reached regarding the presence or absence of line asymmetries. This criterion primarily applied to the Si IV λ 1400 Å doublet, since this is in a wavelength region where the sensitivity of *Copernicus* is relatively low.

All the spectra were plotted at a scale of 4 Å in.$^{-1}$, and careful comparisons were made in order to determine whether or not any indications of mass ejection were present. The wavelength regions studied included most lines considered in the survey of mass-loss effects in hotter stars (Snow and Morton, 1976); specifically, C II (λ 1334.532 Å), C III (λ 1176 Å), N II (λ 1085 Å), N V ($\lambda\lambda$ 1242.804, 1238.821 Å), O VI ($\lambda\lambda$ 1031.945, 1037.627 Å), Si III (λ 1206.510 Å), Si IV ($\lambda\lambda$ 1393.755, 1402.769 Å), and S IV ($\lambda\lambda$ 1062.672; 1073 Å).

Lines were considered to show mass-loss effects only if asymmetries were present which could not be attributed to blends with other nearby lines, or if large shifts were seen. Checking for blends was accomplished primarily by making comparisons among stars of similar spectral type. In the case of doublets such as Si IV λ 1400 Å, this procedure was simplified by the requirement that any asymmetry due to mass-loss must be present in both lines, whereas blends would not be expected to affect both in the same way.

In Table I are listed the Be and shell stars included in this survey, along with relevant spectroscopic and photometric data. The spectral classifications are primarily taken from Lesh (1968, 1972) or Morgan and Keenan (1973); the photometric data from Lesh (1968, 1972), Johnson *et al.* (1966), or the catalogue of Blanco *et al.* (1968); and the projected rotational velocities from Uesugi and Fukuda (1970) or Boyarchuk and Kopylov (1964). Although systematic inaccuracies may be present in the $v \sin i$ data, they should not affect the following discussion of the correlation of

TABLE I

Be and shell stars surveyed

Star	HD	Spectral class	V	$(B-V)$	$v \sin i$ (km s^{-1})	Observation date Year	Day
γ Cas	5 394	B0.5 IVel	2.58	−0.20	300	1974	317–318
α Eri	10 144	B3 IV ev	0.48	−0.16	411	1972	280–285
ϕ Per	10 516	B1pe (III, V)	4.06	−0.04	450	1973	242–272
48 Per	25 940	B3 Ve	4.01	−0.03	217	1974	352–354
	28 497	B1.5 Ve	5.60	−0.23	340	1973	280–285
ζ Tau	37 202	B4 IIIp	2.95	−0.19	310	1972	319–323
μ Cen	120 324	B2 IV–Ve	3.2:	−0.14	191	1975	142–143
η Cen	127 972	B1.5 Vn	2.3:	−0.20	300	1975	60–63
χ Oph	148 184	B1.5 Ve	4.43	+0.28	123	1974	242–243
59 Cyg	200 120	B1.5 Ve2nn	4.79	−0.07	450	1972	289–293
υ Cyg	202 904	B2 Ve$^+$	4.28	−0.08	261	1974	306–308
31 Peg	212 076	B2 IV–Ve	5.04	−0.16	134	1974	295–296

mass-loss effects with projected rotational velocity. Also included is the date of observation of each star.

3. Results

Effects of mass ejection at speeds of several hundred km s^{-1} were seen in a few stars in the survey. Of the normal B-type stars, only τ Sco (already reported by Rogerson and Lamers 1975) and θ Car (Snow and Morton, 1976) showed these effects. In both cases strong asymmetries were seen in Si IV λ 1400 Å and N V λ 1240 Å, and in the case of τ Sco, also O VI $\lambda\lambda$ 1031, 1037 Å were broadened and asymmetric. The latter is of particular interest because, as pointed out by Rogerson and Lamers, O VI requires for its formation temperatures higher than normally expected in a B0 star photosphere; hence its presence in the spectrum may indicate that heating is taking place above the photosphere. Furthermore, it is suggested that the presence of highly-ionized species such as O VI helps to drive the high-velocity mass loss which is observed, via photon absorption in resonance lines. The general O-B star mass-loss survey of Snow and Morton (1976) provides some evidence that the high-velocity ejection of material may be linked to the presence of highly-ionized species.

Among the Be and shell stars, 59 Cyg, η Cen, and ϕ Per show marked evidence of mass loss. Si IV λ 1400 Å is strongly asymmetric in both 59 Cyg and ϕ Per, and somewhat asymmetric in η Cen. In addition, both 59 Cyg and ϕ Per show strong NV absorption, with extended short wavelength wings. Since most B1.5 stars show no NV absorption at all in U2 spectra, the fact that it is strongly present in these two objects is indicative that some unusual ionization mechanism must be acting. In the spectrum of 59 Cyg, the Si III λ 1206 Å line is also seen to be very asymmetric, further indication of mass flow. ϕ Per is discussed in much more detail by Plavec (this Symposium), as is η Cen by Burton and Evans (also this Symposium).

Possible evidence for high-velocity mass loss is seen in the spectra of γ Cas and HD 28497, in the form of Si IV asymmetries in the former case, and large shifts ($\sim$1 Å) of the Si IV lines in the latter. As noted earlier, small displacements corresponding to expansion velocities of about 100 km s^{-1} or less are not being considered in this survey, but the apparent shifts in the case of HD 28497 are significantly larger, representing an outflow velocity of slightly more than 200 km s^{-1}.

The fact that Bohlin (1970) has observed the C IV doublet in γ Cas to show a P Cygni profile tends to support the suggestion here that mass loss is occurring in this star. Since C IV requires more energy for its formation than does Si IV, the two species may exist in different levels in (or above) the stellar atmosphere, so it may be that the rather different character of the C IV and Si IV profiles is due to a stratification of the velocity field. It is also quite possible that γ Cas was more active at the time of Bohlin's (1970) observations than it was some five years later, when the *Copernicus* scan was made.

Si IV λ 1400 Å profiles of γ Cas, 59 Cyg, HD 28497, η Cen, and α Eri are shown in Figure 2. It is interesting to note that asymmetries may be marginally present in α Eri, a B3 star. Quite probably the λ 1393 Å line in this star is affected by blends,

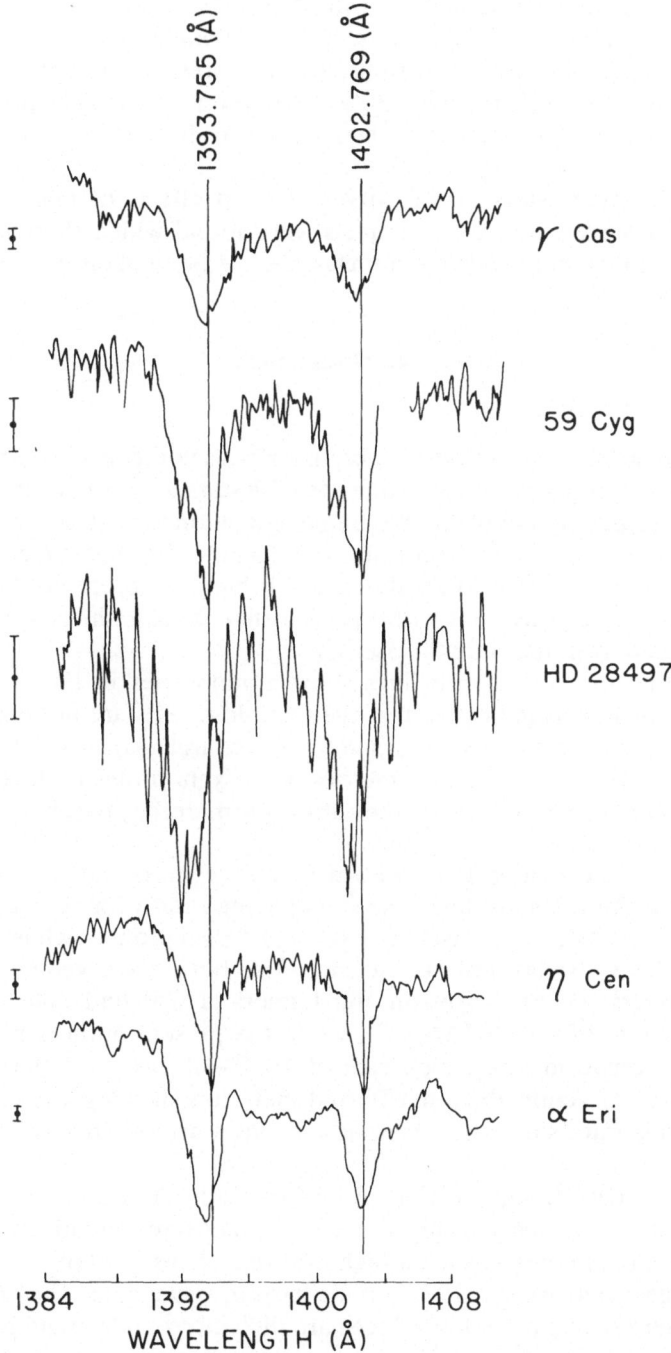

Fig. 2. Si IV λ 1400 Å doublet profiles in five of the Be or shell stars surveyed. The rest positions of the two lines are indicated. The error bars at the left represent the 2σ rms noise at the continuum level due to photon statistics. Strong asymmetries are seen in 59 Cyg and η Cen, and a slight asymmetry appears to be present in γ Cas.

since it is more asymmetric than the λ 1402 Å line. However, the longer wavelength line itself may be asymmetric; if it is, its short wavelength edge represents a velocity of outflow of about 400 km s^{-1}. It would be of interest to study these profiles with higher resolution to clarify the possibility that asymmetries may be present, because if so, this is certainly the latest dwarf star known to show evidence of high-velocity mass loss.*

None of the other stars in the survey showed effects of mass ejection. It is noteworthy, however, that a few objects were found which show apparent shell absorption features, but which are not classified as Be or shell stars on the basis of visual spectra.

4. Discussion

4.1. Mass loss rates

Since normal dwarfs of spectral type later than B1 are not seen to have strong enough stellar winds to produce mass loss, while a few Be stars of type later than B1 do show its effects, it seems very likely that the Be phenomenon is related to the existence of mass ejection. As noted in Section 1, this is not a new idea, but evidence in the form of directly-observed high-velocity stellar winds has heretofore not been available. *Copernicus* data are consistent with the idea that Be star shells are formed from material ejected from the stars themselves.

The fact that only resonance lines show asymmetries in the Be stars having detectable mass flow is indicative that the mass-loss rates are not high, because in cases of high-volume mass loss, often some features arising from excited lower states (such as C III λ 1175 Å) show asymmetries or P Cygni profiles. The lack of a large population of these excited levels in the outflowing material probably implies that the density in this region is relatively low.

Furthermore, the comparative weakness of the mass-loss profiles is another indication that the rates are low. The stars having rates of mass ejection of order 10^{-6} $M_\odot$ yr^{-1} (e.g. Morton, 1967) show strong P Cygni profiles, while those of lower mass-loss volume display little or no emission, but only asymmetric absorption features (Lamers, 1976). Rogerson and Lamers (1975) find that τ Sco is losing material at a rate of 4×10^{-9} $M_\odot$ yr^{-1}, while the present authors find for the Be star 59 Cyg a crude mass ejection rate of 10^{-10}–10^{-9} $M_\odot$ yr^{-1} (Snow and Marlborough, 1976). Certainly the other Be and shell stars showing line asymmetries do not exceed this rate, since their mass-loss profiles are no stronger than those of 59 Cyg.

Since τ Sco (B0 V) and θ Car (B0.5 Vp) both show mass loss but no Be phenomena, it may be that stars hotter than B1 which have radiatively-driven stellar winds are ejecting matter at such a high rate that shells are not able to remain in place, as suggested by Massa (1975) on the basis of visual data. The Oe stars, which can display very strong mass-loss effects, are difficult to understand in this context, however, since the presence of nearly stationary emission lines indicates that

---

* **Note added in proof.** New high-resolution (U1) scans of α Eri show that the Si IV λ 1400 lines have extended short-wavelength wings, indicating high-velocity mass flow in this star and further confirming the correlation shown in Figure 3.

cirumstellar material must survive the rapidly streaming gas. Perhaps in these cases the visual emission occurs in a relatively dense region just above the photosphere, while the strong acceleration producing the high-velocity mass flow becomes effective at greater distances from the star.

Although the *Copernicus* data are consistent with the possibility that Be star shells are formed by radiatively-driven mass loss from single stars, it is also possible that the observed gas motions result from some other mechanism. One suggestion is that all Be stars are binaries with late-type secondary components, and that the shells form as a result of mass exchange (see Kriz and Harmanec, 1975, and references cited therein). It is known from *Copernicus* data that P Cygni profiles in cases of mass-exchange binaries can be very similar to those seen in single stars (e.g. McClusky *et al.*, 1975).

4.2. EFFECTS OF ROTATION

It was suggested long ago (Struve, 1931) that the rapid rotation rates often observed in Be stars may play a role in the development of their extended atmospheres. The early concept of rotational break-up has in recent years been replaced by the development of models for radiatively-driven stellar winds (Lucy and Solomon, 1970; Cassinelli and Castor, 1973; Marlborough and Zamir, 1975; Rogerson and Lamers, 1975; Lamers, 1976). Since the existence of a wind in these models depends on the dominance of radiation pressure over gravitational forces for at least some ions or atoms, it is clear that a reduction in the effective gravity by rotation can enhance the formation of strong winds. Hence it might be expected that the presence or absence of high-velocity mass flow effects in Be stars is a function of their rotational velocity. That this is apparently so has already been pointed out in an earlier discussion of *Copernicus* data (Snow and Marlborough, 1976).

In Figure 3 is shown the distribution in the spectral class-$v \sin i$ plane of all the stars considered in the present survey, with filled-in circles or triangles to indicate respectively normal B dwarfs or Be stars not showing mass-loss effects, and open circles or triangles representing those which do. It is worth reiterating that evidence of atmospheric or shell expansion at velocities of about $100 \mathrm{~km~s}^{-1}$ or less is not considered to represent mass-loss effects in the present context.

It is seen in Figure 3 that all the stars showing evidence for high-velocity mass loss lie in the upper left-hand portion of the diagram, consistent with the expectation that such effects should depend on rotational velocity and, of course, effective temperature. This more complete survey is consistent with the earlier suggestion (based on fewer data) that dwarfs in the spectral class range B1 to B3 or so lack sufficient radiative flux to drive strong stellar winds unless rotation sufficiently reduces the effective gravity (Snow and Marlborough, 1976). Hence Be stars may fall into a transition region where the degree of rotation determines the presence or absence of an extended atmosphere. It would be of very great interest to obtain high-resolution data on many of the Be and normal B stars which do not show mass loss effects in the U2 data, to determine whether milder stellar winds can be detected in stars below the apparent cut-off in Figure 3. It is possible, of course, that the lack of strong mass-loss effects in Be stars of relatively low $v \sin i$ is due to a geometrical effect, if these

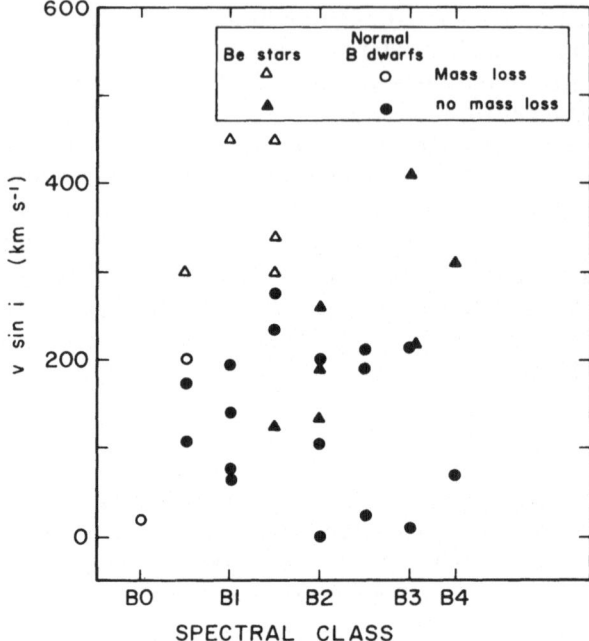

Fig. 3. Correlation of high-velocity mass loss with $v \sin i$ for all survey stars. The tendency of objects showing mass-loss profiles to lie in the upper left-hand corner is consistent with theories of radiatively driven stellar winds, as discussed in the text. All survey objects later than B0.5 which show high-velocity mass loss are Be or shell stars, supporting suggestions that rotation and mass loss play roles in producing the Be star characteristics.

objects are actually rapid rotators seen pole-on, because it may be that the outward mass flow is confined to the equatorial plane so that its effects on absorption lines can only be seen by observers in or near this plane. In any case, however, it is likely that the mass-loss rate is affected by the true rotational velocity, for reasons outlined above.

It will be interesting in future studies not only to see whether higher resolution data may reveal mass-loss effects in stars of lower effective temperature and $v \sin i$, but also to see whether many 'normal' early B stars with moderately high values of $v \sin i$ show shell absorption lines in the ultraviolet, as has already been noticed in a few cases.

4.3. CORRELATION OF MASS-LOSS EFFECTS WITH PROPERTIES
OF VISUAL SPECTRA

It is of interest to consider whether or not any correlations exist between mass-loss effects and other properties of the visual spectrum. In particular is there any correlation of mass loss with Balmer emission line strength? In view of the fact that the emission lines in Be and shell stars are known to vary on a variety of time scales from hours to days and even disappear and reappear over periods of years, observations of the $H\alpha$ line strength, or some other Balmer line, should have been obtained at the same time as the *Copernicus* data if meaningful correlations are to be found. Unfortunately no such results appear to be available for the stars in Table I. However, data on the Hα line strengths for some of the stars in Table I do exist corresponding to observations obtained by various workers over the period 1961 to 1975. In this section these observational results are examined to see if any trend exists between mass loss and Hα line strength.

Andrews (1968) observed many early-type stars during 1961 to 1963 using a spectrometer to isolate the region near Hα. An estimate of the Hα line strength for those Be and shell stars in Table I in common with Andrews can be obtained by subtracting the observed Hα measure from the mean value for a star of the same spectral type and luminosity class obtained from Figure 5 of Andrews' paper. The estimates of the Hα line strength so obtained are given in column 2 of Table II and are such that the larger the numerical value the stronger is the Hα emission. The stars are arranged in order of decreasing strength of the Hα emission.

TABLE II

Hα line strength estimates

Star	Andrews (1961–1963)	Feinstein (1970–1972)	Poeckert and Marlborough (1974–1975)
χ Oph	0.696	0.303	5.76
φ Per	0.688	0.249	4.50
HD 28497	–	0.161	2.77
μ Cen	–	0.160	–
υ Cyg	0.627	0.135	3.67
ζ Tau	0.583	–	2.53
48 Per	0.497	0.156	2.94
γ Cas	0.335	0.185	2.94
59 Cyg	0.329	0.092	1.66
31 Peg	–	0.064	2.00
α Eri	0.058	–	–
η Cen	–	0.009	–

Feinstein (1974) used interference filters to isolate hydrogen lines and measured line strengths photoelectrically for several hundred early-type stars. The observations were obtained during 1970 to 1972. An estimate of the Hα line strength was obtained in a manner similar to that above by comparing the measured Hα index to that of a standard star of similar spectral type and luminosity class given in Table III of Feinstein's paper. Where more than one observation was available simple means were taken. The results are listed in column 3 of Table II. Stars observed by Feinstein but not by Andrews have been inserted in the appropriate place according to the Hα line strength, larger numbers again indicating stronger lines.

Finally estimates of the Hα line strength are available from the data of Poeckert and Marlborough (1976) and are listed in column 4 of Table II, larger numbers again indicating stronger lines. The data were obtained during 1974 and 1975. The Hα line strengths were derived from measures of linear polarization obtained with a polarimeter using narrow interference filters with a half power band-width of 5 Å. Because of the narrowness of the filter used the line strength estimates for stars with broad strong lines may be underestimated.

It is clear from inspection of columns 2, 3 and 4 of Table II that the relative positions of some stars as regards line strength are not the same. Some of the rearrangement in column 4 relative to columns 2 and 3 may be due to underestimates of the Hα emission strength for cases where the emission is strong and v sin i large. Apart from this the rearrangement of some stars is probably an indication that the Hα emission strength varied over the period of observation. Indeed variations of the Hα index over the period 1970–1972 were reported by Feinstein for some of the stars he observed more than once. However for the majority of cases in Table II, large changes such as complete disappearance of emission lines do not seem to have occurred. Consequently some appropriate average will be taken as representative of the Hα emission strength at the time of the *Copernicus* observations.

The stars which showed marked evidence of mass loss from *Copernicus* data were 59 Cyg, η Cen and φ Per and possibly γ Cas and HD 28497. In Table II, these stars cover the range of strong to weak Hα emission but there does not seem to be any particular trend of the Hα strength compared to the degree of line asymmetry of Si IV λ 1400 Å. Specifically 59 Cyg has one of the weakest Hα emission lines while φ Per shows one of the strongest. However both stars possess markedly asymmetric Si IV lines while γ Cas, with lines which are not as asymmetric, has a Hα line of intermediate strength. Present data, therefore, do not appear to support any correlation of Hα line strength with mass-loss effects.

Acknowledgements

The portion of this work which has been carried on at Princeton has been supported by U.S. National Aeronautics and Space Administration contract NAS5-1810. The research at the University of Western Ontario has been supported by the National Research Council of Canada.

References

Andrews, P. J.: 1968, *Mem. Roy. Astron. Soc.* **72**, 35.
Blanco, V. M., Demers, S., Douglas, G. G., and FitzGerald, M. P.: 1968, *Publ. U.S. Naval Obs.* **21**.
Bohlin, R. C.: 1970, *Astrophys. J.* **162**, 571.
Bohlin, R. C.: 1975, *Astrophys. J.* **200**, 402.
Boyarchuk, A. A. and Koplyov, I. M.: 1964, *Publ. Crimean Astrophys. Obs.* **31**, 44.
Cassinelli, J. P. and Castor, J. I.: 1973, *Astrophys. J.* **179**, 189.
Feinstein, A.: 1974, *Monthly Notices Roy. Astron. Soc.* **169**, 171.
Hack, M. and Struve, O.: 1970, *Stellar Spectroscopy of Peculiar Stars*, Osservatorio Astronomico di Trieste, Trieste.
Heap, S. R.: 1975, *Phil. Trans. Roy. Soc. London* **279**, 371.
Johnson, H. L., Mitchell, R. I., Iriarte, B., Wisniewski, W. Z.: 1966, *Comm. Lunar Planetary Lab.* **4**, 99.
Kriz, S. and Harmanec, P.: 1975, *Bull. Astron. Inst. Czech.* **26**, 65.
Lamers, H. J. G. L. M.: 1976, in preparation.
Lesh, J. R.: 1968, *Astrophys. J. Suppl.* **17**, 371.
Lesh, J. R.: 1972, *Astron. Astrophys. Suppl.* **5**, 129.
Lucy, L. B. and Solomon, P. M.: 1970, *Astrophys. J.* **159**, 879.
Marlborough, J. M. and Zamir, M.: 1975, *Astrophys. J.* **195**, 145.

Massa, D.: 1975, *Publ. Astron. Soc. Pacific* **87**, 777.
McLaughlin, D. B.: 1961, *J. Roy. Astron. Soc. Can.* **55**, 13 and 73.
McCluskey, G. E., Kondo, Y., and Morton, D. C.: 1975, *Astrophys. J.* **201**, 607.
Morgan, W. W. and Keenan, P. C.: 1973, *Ann. Rev. Astron. Astrophys.* **11**, 29.
Morton, D. C.: 1967, *Astrophys. J.* **150**, 535.
Plavec, M.: 1975, in preparation.
Poeckert, R. and Marlborough, J. M.: 1976, *Astrophys. J.*, in press.
Rogerson, J. B. and Lamers, H. J. G. L. M.: 1975, *Nature*, **256**, 190.
Rogerson, J. B., Spitzer, L., Drake, J. F., Dressler, K., Jenkins, E. B., Morton, D. C., and York, D. G.: 1973, *Astrophys. J. Letters* **181**, L97.
Snow, T. P. and Marlborough, J. M.: 1976, *Astrophys. J. Letters* **203**, L87.
Snow, T. P. and Morton, D. C.: 1976, in preparation.
Snow, T. P., York, D. G., Welty, D., and Hornack, P.: 1975, in preparation.
Struve, O.: 1931, *Astrophys. J.* **73**, 94.
Uesugi, A. and Fukuda, I.: 1970, *Contr. Inst. Astrophys. Kwasan Obs. Kyoto*, No. 189.
Underhill, A. B.: 1966, *The Early Type Stars*, D. Reidel Publ. Co., Dordrecht-Holland, p. 226.

DISCUSSION

Heap: You say that you have found evidence for mass flow outwards at a speed near 1000 km s^{-1}, and you found this by comparing the observed profile against some theoretical profile. What does the theoretical profile look like?

Snow: We did not assume any theoretical line profile. We simply assumed that the observed asymmetries are due to the presence of an optically thin, low-density stream of gas, so that no natural or collisional broadening takes place in the part of the absorption line which is formed in this stream. In that case the extreme short wavelength edge of the absorption profile represents material moving at the velocity given by the total displacement of that edge from rest.

Hutchings: Are your stars all high v sin i objects? That is, are they all seen equator-on?

Snow: All the Be stars which show mass loss have v sin i's of 300 km s^{-1} or higher, and are all considered equator-on stars. However, our sample includes several B and Be stars of low and moderate v sin i, as seen in Figure 3. The normal B0 V star τ Sco has v sin $i = 20$ km s^{-1}, yet shows strong mass-loss effects.

ULTRAVIOLET Si IV/C IV RATIOS FOR Be STARS

KARL G. HENIZE*, JAMES D. WRAY, S. B. PARSONS, and G. F. BENEDICT

Dept. of Astronomy, University of Texas, Austin, Tex., U.S.A.

Abstract. The intensities of the very strong lines of C IV λ 1549 Å and Si IV λλ 1394, 1403 Å observed in spectra obtained with Skylab experiment S019 provide a sensitive discrimination of spectral type between B0 and B2. Eye estimates of the Si IV/C IV ratio are tabulated for 33 B0–B2, class III–V stars of which 11 are emission-line stars. Seven of the emission-line stars show significantly smaller ratios than normal stars of the same MK class. The most outstanding examples are 60 Cyg, *o* Pup, η Cen, and ι Ara.

1. Introduction

During this symposium Dr Slettebak has already reviewed reasons why it is difficult to assign firm MK classes to the rapidly rotating Be stars, and Dr Heap has noted how data on ultraviolet C IV and Si IV line intensities may be useful in providing improved classifications of such stars. This subject has also been on our minds as we carry out the analysis of ultraviolet spectra obtained with Skylab experiment S019. A preliminary survey of the behavior of the C IV and Si IV lines in early type main sequence stars (Henize *et al.*, 1975) shows that the Si IV/C IV ratio varies dramatically from a value of about 10 at B2 to a value of about $\frac{1}{4}$ at B0. This paper presents a more detailed study of the variation of the Si IV/C IV ratio as a function of spectral type for 33 B0–B2 stars of which 11 are Be stars.

2. The Observations

The instrumentation with which these data were obtained is described by Henize *et al.* (1975) and by O'Callaghan *et al.* (1976). The basic instrument is an objective-prism spectrograph with a 15-cm aperture and a 4° prism of CaF_2 giving resolutions of 2 Å and 12 Å at wavelengths of λ 1400 and 2000 Å respectively.

A total of 400 spectra showing measurable fluxes at λ 1500 Å have been obtained and, of these, roughly 120 show evident absorption or emission lines.

Although these spectra are calibrated so that equivalent widths and flux curves may be derived, all factors entering the calibration are not yet completely analyzed and, as a consequence, the data presented here are based on eye estimates of absorption line intensities. These estimates are affected by the differing dispersions and effective exposures at λ 1400 Å vs 1550 Å and it is not to be expected that they will correspond to ratios of the equivalent widths. Nevertheless they represent a self consistent set of data from which stars showing anomalous behavior may be detected.

Initially, spectra of all stars with spectral types B0–B2 and luminosity classes III–V which showed visible flux at λ 1400 Å were examined. This list of 98 stars yielded 34 stars for which both the C IV and Si IV lines are well-exposed and in which reasonably reliable line intensities can be determined. These stars and their Si IV/C IV ratios are

---

* Astronaut Office, Code TE, NASA Johnson Space Center.

A. Slettebak (ed.), Be and Shell Stars, 191–195. All Rights Reserved

TABLE I

Si IV λ 1394/C IV λλ 1548, 1551 ratios for B0–B2 stars

HD	Desig.	Sp.	Si IV/C IV	Remarks
3 360	ζ Cas	B2 IV	$\frac{1}{2}$	C IV very strong for B2
5 394*	γ Cas	B0.5 IVe	1	
10 516*	φ Per	B2 Vep	>2	Si IV very weak
34 816	λ Lep	B0.5 IV	$\frac{3}{4}$	
35 468	γ Ori	B2 III	10	
36 512	υ Ori	B0 V	$\frac{1}{8}$	
36 822	φ^1 Ori	B0.5 IV–V	$\frac{1}{4}$	
50 013*	κ CMa	B1.5 IVne	>4	
58 978*	HR 2855	B0.5 IVnpe	$<\frac{1}{4}$	
63 462*	o Pup	B1 IV:nne	$\frac{1}{3}$	
75 821	HR 3527	B0 III	$\frac{1}{6}$	
79 351	a Car	B2 IV–V	>4	
93 030	θ Car	B0.5 Vp	2	C IV very weak for B0.5
108 248	α^1 Cru	B0.5 IV	4	with α^2 Cru B1 V
116 658	α Vir	B1 IV	4	
120 640	HR 5206	B2 Vp	>6	
120 991*	HR 5223	B2 IIIe		Si IV very weak for B2
127 381	σ Lup	B2 III	5	
127 972*	η Cen	B1.5 Vn	2	
132 058	β Lup	B2 III	>6	
132 200	κ Cen	B2 IV	>6	
135 160*	HR 5661	B0.5 V	$\frac{1}{2}$	
143 018	π Sco	B1 V+B2	10	
147 165	σ Sco	B1 III	4	
149 438	τ Sco	B0 V	$\frac{1}{3}$	
151 890	μ^1 Sco	B1.5 IV	4	
157 042*	ι Ara	B2 IIIne	1:	UV spectrum unwidened
158 408	υ Sco	B2 IV	6	
158 926	λ Sco	B1.5 IV	10	
173 948*	λ Pav	B2 II–III	1:	Si IV very weak; C IV broad, Fe III?
188 439	V819 Cyg	B0.5 IIIp	$\frac{1}{2}$	1600–2000 blends strong
200 310*	60 Cyg	B1 Vne	$\frac{1}{2}$	
214 168*	8 Lac	B1 Ve	1:	with HD 214167 B1.5V; C IV broad, Fe III?
224 572	σ Cas	B1 V	4	

* Emission-line star.

listed in Table I. Since the Si IV doublet is well resolved in our spectra and the C IV doublet is not, the ratio given is defined as Si IV λ 1394/C IV λλ 1548, 1551. The spectral types in Table 1 are derived from Hiltner *et al.* (1969) and Lesh (1968).

3. Discussion

The data of Table I are displayed in Figure 1. For the non-emission line stars there is a clear cut trend for the Si IV/C IV ratio to increase from about $\frac{1}{5}$ at B0, through 4 at B1, to roughly 8 or 10 at B2. The emission-line stars, on the other hand, show

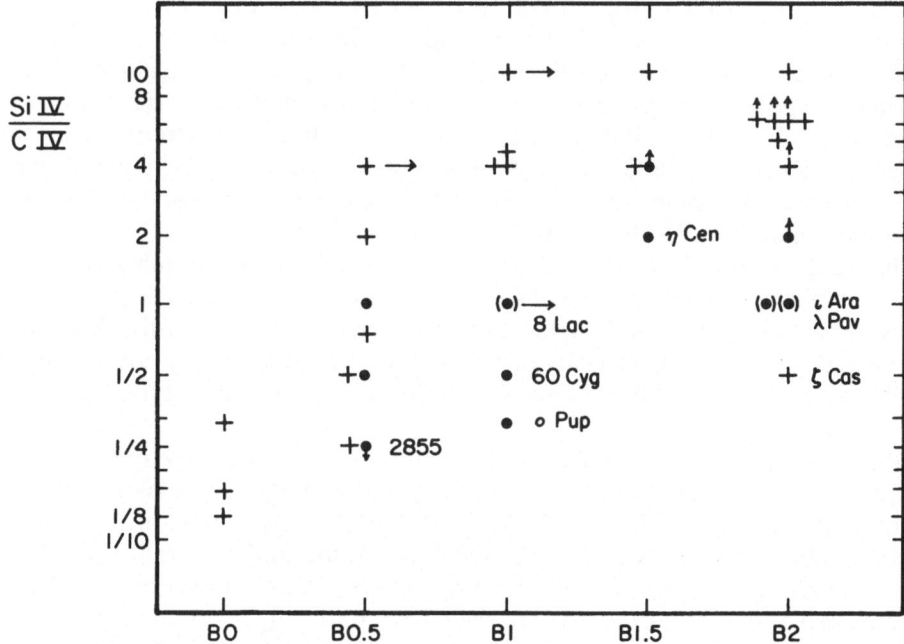

Fig. 1.   The ratio Si IV λ 1394 Å/C IV λλ 1548, 1551 Å plotted as a function of spectral type. Crosses indicate non-emission-line stars; dots indicate emission-line stars. Parentheses indicate uncertainty in the Si/C ratio. ↑ indicates values which are lower limits of the ratio. → indicates that the spectrum is blended with one of slightly later type.

generally smaller ratios than the non-emission-line stars. In particular, seven of the eleven emission-line stars show ratios significantly less than those for normal stars. Of these, four stars (60 Cyg, o Pup, η Cen, and ι Ara) have a C IV intensity conspicuously greater than that of the normal stars of the same spectral class. Two of the remaining stars, λ Pav and HR 2855, are peculiar in that the Si IV lines are unusually weak for their spectral class. This peculiarity is also evident in φ Per and HR 5223.

For the four stars in which C IV is enhanced the data in Figure 1 suggest that they are hotter than the MK classes indicate. The rough calibration of the Si/C ratio provided in Figure 1 suggests spectral classes of B0.2, B0.5, B0.5, and B0.7 for o Pup, 60 Cyg, ι Ara, and η Cen, respectively, assuming them to be main sequence stars. However, the Si/C ratio is also correlated to luminosity and at spectral class B1 the ratio ranges from about 0.1 for main sequence stars to about 1 for supergiant stars (see Figure 3d of Henize *et al.*, 1975). Thus the enhancement of C IV may also be due to a lower than expected surface gravity. The Si/C ratio in 60 Cyg, for example, is also compatible with a classification of B1 II. Since these stars may be expected to show incipient shell absorption and since shell spectra generally show high luminosity characteristics, this is the more attractive of the two possibilities.

The suggestion that the enhancement of C IV absorption may be due to an incipient absorption shell leads to the further question as to whether these four stars show any indication of the extensive blends of weak lines in the λ 1600–2000 Å region which are evident in spectra of the advanced shell stars 48 Lib and ζ Tau and also in many of the O and early B-type supergiants (see Figure 3a, Henize *et al.*, 1975). A weak indication of such lines is present in *o* Pup and 60 Cyg but in *ι* Ara and *η* Cen this region of the spectrum is over-exposed. It should be noted in passing that these blended shell (or supergiant) absorption features are strong in the star V819 Cyg. There is also a suspicion of weak emission present on the redward edge of C IV. Thus the ultraviolet spectrum of V819 Cyg is more like that of a B0 or B1 supergiant than that of a B0.5 III star.

The star ζ Cas also shows enhanced C IV even though it is not an emission-line star. The spectral class suggested by the Si/C ratios in Figure 1 is B0.5. The peculiarity of ζ Cas is further accentuated by the fact that it is an MK standard (Morgan and Keenan, 1973). The C IV anomaly brings into question whether or not this star is a reliable standard star and further investigation of abundances in this star would seem to be warranted.

One further possibility for explaining C IV enhancement should be mentioned; i.e. the possibility that the C IV line is severely blended with Fe III as is suggested by Peytremann (1975) for all stars cooler than 30 000 K. However, in the five stars discussed above the C IV absorption is sharp and distinct and there is little doubt that it is attributable to C IV. A broad blending of weak lines in the λ 1500 to 1600 Å region which probably corresponds to the blend studied by Peytremann is visible in many B1 and B2 stars on our plates, but at the resolution of the S019 spectra, it is not easily confused with the C IV line. The question as to why the empirical data do not agree well with theory is a matter of interest and will be the subject of further study in the S019 spectra.

Two stars in Table I, λ Pav, and 8 Lac, do show a somewhat diffuse C IV line and in this instance it might be suspected that the feature is seriously blended with Fe III. If so, then their Si/C ratio is somewhat greater than is indicated in Figure 1. This would remove both stars from the anomalous group of stars so far as the Si/C ratio is concerned. However, the fact that λ Pav (together with HR 2855, *φ* Per, and HR 5223) shows abnormally weak Si IV bears further consideration. Inspection of Figure 3 of Henize *et al.* (1975) suggests that the ultraviolet spectra of these stars are more like those of B3 stars in which the Si IV and C IV have almost completely disappeared. The reason for such a trend of misclassification is difficult to understand, however, since in rapidly rotating stars the main effect is to obscure the weak lines required to establish a class of B0 or B1 and to enhance the strength of helium. Both effects tend to lead to a classification of B2 and B3 for rapid rotators even though they may be considerably hotter.

References

Henize, K. G., Wray, J. D., Parsons, S. G., Benedict, G. F., Bruhweiler, F. C., Rybski, P. M., and O'Callaghan, F. G.: 1975, *Astrophys. J. Letters* **199**, L119.
Hiltner, W. A., Garrison, R. F., and Schild, R. E.: 1969, *Astrophys. J.* **157**, 313.
Lesh, J. R.: 1968, *Astrophys. J. Suppl. Ser.* **17**, 371.
Morgan, W. W. and Keenan, P. C.: 1973, *Ann. Rev. Astron. Astrophys.* **11**, 29.
O'Callaghan, F. G., Henize, K. G., and Wray, J. D.: 1976, *Appl. Opt.*, in press.
Peytremann, E.: 1975, *Astron. Astrophys.* **39**, 393.

DISCUSSION

Noerdlinger: I can comment on the peculiarity in Si IV you mentioned for ζ Pup. About a year or more ago Tom Hewitt and I published in *Astrophys. J.* **188**, 315 (1974) a theoretical discussion of the formation of P Cyg resonance lines in the case of close doublets. The absence of a red absorption component in ζ Pup can be explained by filling-in by the redward emission from the blue component, provided the density is high enough. In fact, doublet structure can be used in this way to diagnose the density in a mass outflow. For Si IV lines, one needs the greatest expansion velocity to be 2000 km s^{-1} or more. That value may be reasonable for ζ Pup, which has a very rapid wind.

THE NEAR ULTRAVIOLET SPECTRUM OF ζ TAU

A. M. DELPLACE

Observatoire de Meudon, France

and

K. A. VAN DER HUCHT

Space Research Laboratory of the Astronomical Institute at Utrecht, The Netherlands

By using several ultraviolet spectrograms in the λ 2100–2800 Å wavelength range, the absorption features and the flux values of ζ Tau have been determined and compared to different early type stars.

The material was obtained by the Utrecht Orbiting Stellar Spectrophotometer S59 on board the ESRO satellite TD-1A. This instrument is described in detail by de Jager *et al.* (1974).

Three wavelength regions were scanned simultaneously (2060–2160 Å, 2495–2595 Å, 2775–2875 Å) with spectral resolution 1.8 Å. The decrease of the S59 sensitivity between 1972–1973 has been considered (Hoekstra, 1974).

Since ζ Tau is classified, in the visible wavelength range, B2III–B2IV according to various classifications, we have chosen for comparison several stars of about the same spectral type and different luminosity; γ Lup (B2IV), γ Peg (B2IV), β Lup (B2III), ε CMa (B2II), χ Car (B3IV), o^2 CMa (B3Ia), η CMa (B5Ia).

Strong shell features are identified in the ultraviolet spectrum. The same photospheric lines (Fe III, Mn III, Cr III, Ti III, Ni III, Si III, O III, He I, C II ...) are generally found in the S59 spectrum of the B1II-III stars (Van der Hucht *et al.*, 1975) and the shell lines (Fe II, Cr II, Ti II, Ni II, Si II, Mg II, Mg I) generally correspond to the strongest lines observed in the A2 Ia star, α Cyg, *ibid.* ...

Several lines have an interstellar or circumstellar origin.

The flux measurements were calibrated by Stecher's measurements of ζ Pup as standard (1970). In each channel, the wavelength dependent sensitivity was taken into account (Lamers, 1974). The fluxes were measured at λλ 2110 Å, 2553 Å, and 2815 Å; they are normalized to magnitude $V = 0$ and corrected for the interstellar reddening using the law of Bless and Savage (1972).

For the four B2 comparison stars, the flux values at λλ 2110 Å, 2553 Å, and 2815 Å are in sufficiently good agreement.

By comparison to the B2 stars, ζ Tau shows a deficiency in the continuum of about 0.3 magnitude, but the flux ratios (2100 Å/2800 Å, 2500 Å/2800 Å) of ζ Tau and the B2 comparison stars are in agreement. According to a study by Heap (1974), in the 1100–2050 Å wavelength range the comparison of the ratios C III/C IV and Si III/Si IV indicates an effective temperature of about 27 000 K. Therefore, ζ Tau must be considered at least as hot as a B2 star.

The deficiency of the continuum of ζ Tau does not seem to be a rotational effect. In 1973, there was more material in the shell as the result of a new outburst (Delplace, 1975). In the ultraviolet range, the shell looks like an absorbing screen, while the line blocking effect is as high as for an A-type supergiant star.

A. Slettebak (ed.), Be and Shell Stars, 197–198. All Rights Reserved
Copyright © 1976 by the IAU.

References

Bless, R. C. and Savage, B. D.: 1972, *Astrophys. J.* **171**, 293.
Delplace, A. M.: 1975, *Astron. Astrophys.*, in preparation.
Heap, S.: 1974, *Astronomy in Ultraviolet*, The Royal Society, London.
Hoekstra, R.: 1974, *S59 Information Sheet*, No. 7401.
Hucht, K. A. van der, Lamers, H. J., Faraggiana, R., Hack, M., and Stalio, R.: 1975, *Astron. Astrophys.*, in press.
Jager, C. de, Hoekstra, R., Hucht, K. A. van der, Kamperman, T. M., Lamers, H. J., Hammerschlag, A., Werner, W., and Emming, J. G.: 1974, *Astrophys. Space Sci.* **26**, 207.
Lamers, H. J.: 1974, *S59 Information Sheet*, No. 7401.
Stecher, T. P.: 1970, *Astrophys. J.* **159**, 543.

SPECTROSCOPIC OBSERVATIONS OF THE
Be STARS η Cen, γ Cas AND φ Per

W. M. BURTON and R. G. EVANS

Appleton Laboratory (Astrophysics Research Division)
Culham Laboratory, Abingdon, Oxfordshire, England

1. Introduction

The main part of this paper is concerned with observations obtained using the Princeton ultraviolet telescope-spectrometer on the *Copernicus* satellite through the Princeton guest investigator programme. These observations provided high resolution ultraviolet spectra of the bright southern shell star η Cen, which are now being analysed in detail. A short description of the new η Cen spectra and a progress report on the data analysis are presented below. In the second part of the paper we describe visible-region spectra of the northern Be stars γ Cas and φ Per obtained with a cross-dispersed echelle spectrograph on the 36-inch Yapp telescope at the Royal Greenwich Observatory.

2. Copernicus Ultraviolet Spectra of η Cen (HD 127972)

The Princeton telescope-spectrometer on the *Copernicus* satellite (Rogerson *et al.*, 1973) consists of a Cassegrain telescope of 0.8 m aperture with a concave grating spectrometer which has four scanning photomultiplier detectors. Two of these detectors, designated U1 and U2, scan the second order spectrum in the wavelength range λ 750 Å–1500 Å giving spectral resolutions of 0.05 Å and 0.2 Å respectively. Two other detectors, designated V1 and V2, simultaneously scan the first order spectrum in the wavelength range λ 1500 Å–3000 Å with spectral resolutions of 0.1 Å and 0.4 Å. At each discrete point in the wavelength scan, the detectors pause for 14 seconds and record an integrated count for this time interval.

The observations of η Cen were obtained on 1–2 March 1975 and included two scans in the wavelength range λ 1000 Å to 1435 Å made with the U2 detector $(d\lambda \simeq 0.2$ Å$)$. Simultaneously, the U1 detector scanned selected spectral regions at high resolution $(d\lambda = 0.05$ Å$)$ to record individual spectral features, including resonance absorption lines of N I, N II, S II, Fe II and Fe III. Additionally, data were also obtained with the V2 detector in the wavelength range λ 2000 Å to 2870 Å but these observations are not discussed in the present paper.

The new ultraviolet observations of η Cen are generally similar to previously observed spectra of normal stars of comparable spectral type, no emission lines being detected in the ultraviolet region. In the wavelength range λ 1000 Å to 1430 Å the spectrum of η Cen shows considerable line blanketing, the predominant ions being Si III, Cr III and Fe III. Many of the observed spectral features show doppler broadened profiles resulting from the high rotational velocity which is considerably

A. Slettebak (ed.), Be and Shell Stars, 199–207. *All Rights Reserved*
Copyright © 1976 by the IAU.

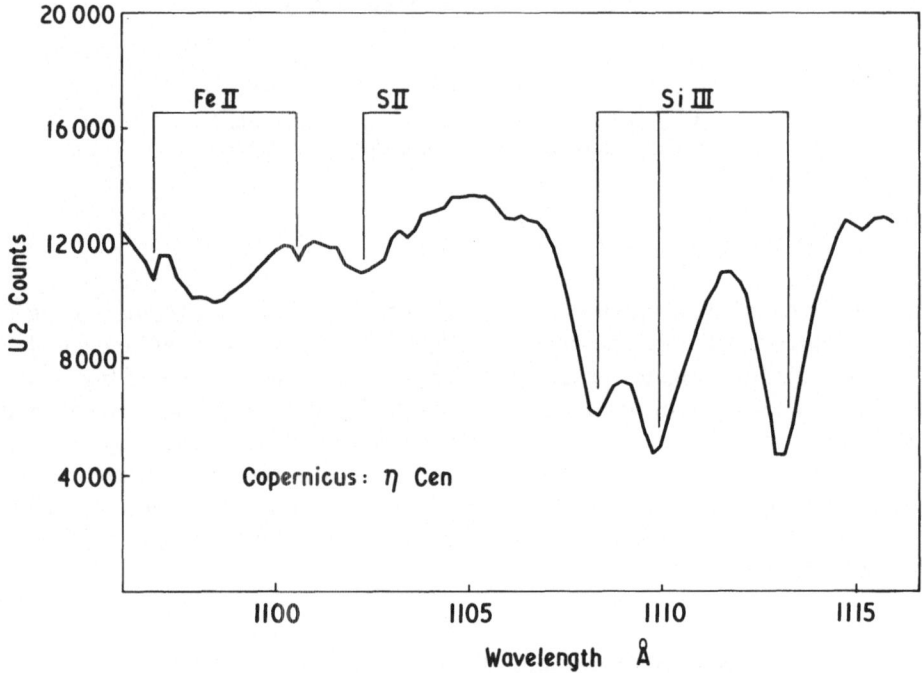

Fig. 1.   *Copernicus* ultraviolet spectrum of η Cen (U2 scan λ 1096 Å–1115 Å) showing absorption lines
of Si III, S II and Fe II.

larger than the instrumental resolution ($\simeq$50 km s^{-1}). Figure 1 shows part of the U2
scan between λ 1096 Å and 1115 Å which includes absorption lines of Si III, S II and
Fe II. The well developed triplet of Si III is almost certainly formed near the stellar
surface and is broadened both by rotation and by radiative damping. The weak S II
line at λ 1102 Å shows the dish-shaped profile characteristic of rotational broadening, giving a projected rotational velocity $v \sin i \simeq 190$ km s^{-1}. Table I gives a
summary of similar measurements on other weak and unblended rotation-
broadened lines in the ultraviolet which provide velocity data for a wide range of
ions.

TABLE I

Measured rotational velocities in the η Cen spectrum

Ion	Wavelength (Å)	$v \sin i$ (km s^{-1})
S II	1102.3	190
S III	1077.1	220
S IV	1062.7	200
P III	1003.6	160
Fe III	1068.2	140
Fe III	1124.9	210
Fe III	1126.7	160
Cr III	1052.9	250
Cr III	1136.7	220

Observations of η Cen in the visible region give a value of $v \sin i \simeq 300$ km s^{-1} for absorption lines formed near the stellar photospheric surface (Uesugi and Fukuda, 1970). This velocity is greater than the measured values for the ultraviolet lines listed in Table I, giving some evidence of a progression from this high value near to the stellar surface through intermediate values (Cr III, S III, S II) to the much smaller value (~ 140 km s^{-1}) observed for those absorption lines (e.g. P III) which are probably formed further out in the differentially rotating extended atmosphere. If it is assumed that angular momentum is conserved in this differentially rotating atmosphere, then these observations suggest that the envelope has an equatorial radius $R_e \simeq 2R_*$.

Visible spectrum observations of η Cen have been discussed by Jaschek *et al.* (1964) who describe it as a shell star of spectral type B1.5 Vne. The Hα line was observed as a single broad emission feature in 1960 but one year later the line became double. This variability was also present in Hβ and Hγ which have changed from double emission lines to wide emission features with sharp central absorption cores during the period 1959–1962.

In contrast to this earlier classification of η Cen as a shell star, the new ultraviolet observations provide little evidence for the presence of a shell spectrum. The lines of Fe II at 1097 Å and 1101 Å which are shown in Figure 1 are very weak and narrow, having widths equal to the limiting instrumental spectral resolution (~ 50 km s^{-1}). These lines would be expected to be much stronger and broader if formed in a circumstellar shell and they are most probably of interstellar origin, being in good agreement with the observations of interstellar absorption lines in normal B-type stars at similar distances (Rogerson *et al.*, 1973). Further evidence for the extreme weakness of possible shell features is obtained from the profiles of the N I triplet at λ 1134 Å and the N II line at λ 1084 Å. Figure 2 shows the N I lines which are extremely narrow ($d\lambda \simeq 0.1$ Å or 25 km s^{-1}), their width and intensity being consistent with purely interstellar absorption. The λ 1084 Å N II line appears as a very narrow feature ($d\lambda \simeq 0.1$ Å) within the much broader stellar N II multiplet and is again consistent with purely interstellar absorption. However, some evidence for the existence of a low density shell is seen in Figure 3 which shows a U2 scan from λ 1188 Å to 1208 Å. The dominant absorption lines are due to a multiplet of S III but the more interesting features are the two pairs of Si II lines, one originating from the ground state (0 cm^{-1}) and the other from a very low-lying excited fine structure state (287 cm^{-1}).

The narrow absorption line at λ 1193 Å corresponds to the transition from the ground state and is probably interstellar but the λ 1198 Å line which originates from the excited 287 cm^{-1} level is much broader. The λ 1198 Å line is not normally seen in the interstellar gas since it requires a density $n_{\mathrm{H}} \simeq 10^3$ atom cm^{-3} to populate this fine structure level (Bahcall and Wolf, 1968). It may be that this absorption occurs in a shell or circumstellar envelope with a gas density $\gtrsim 10^3$ atom cm^{-3}, but the possibility remains that it could be a weak stellar feature, its profile being generally similar to the S II line at λ 1102 Å. The spectrum of Fe III is well developed, but this does not necessarily indicate the presence of a shell, since the profiles of the ultraviolet multiplets of Fe III in the present spectrum are wholly consistent with their being of stellar origin.

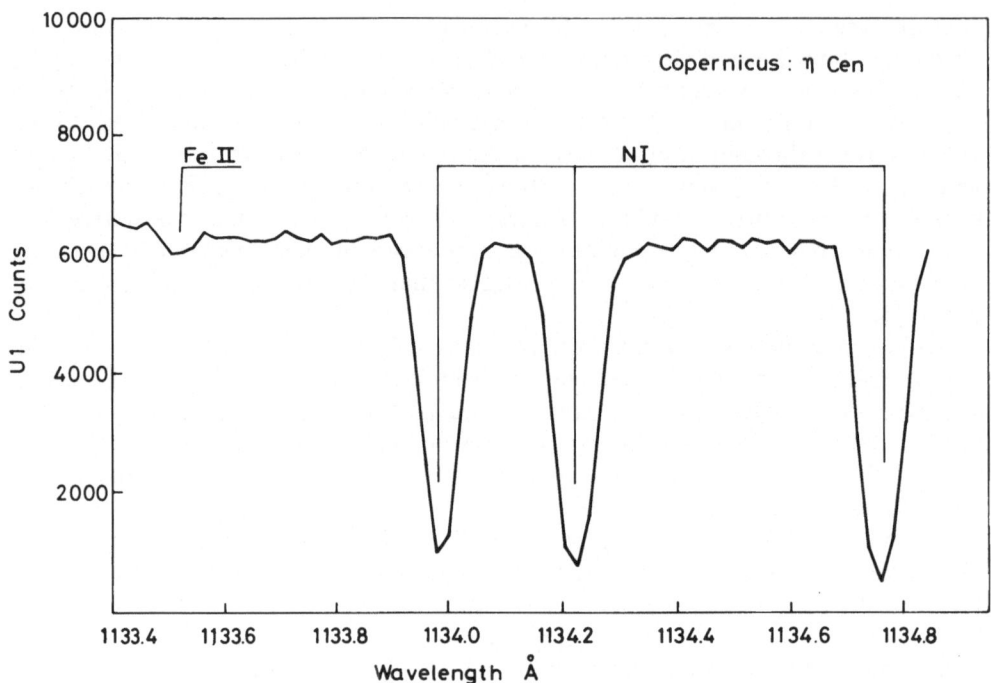

Fig. 2.   *Copernicus* ultraviolet spectrum of η Cen (U1 scan λ 1133.4 Å–1134.8 Å) showing interstellar lines of N I and Fe II.

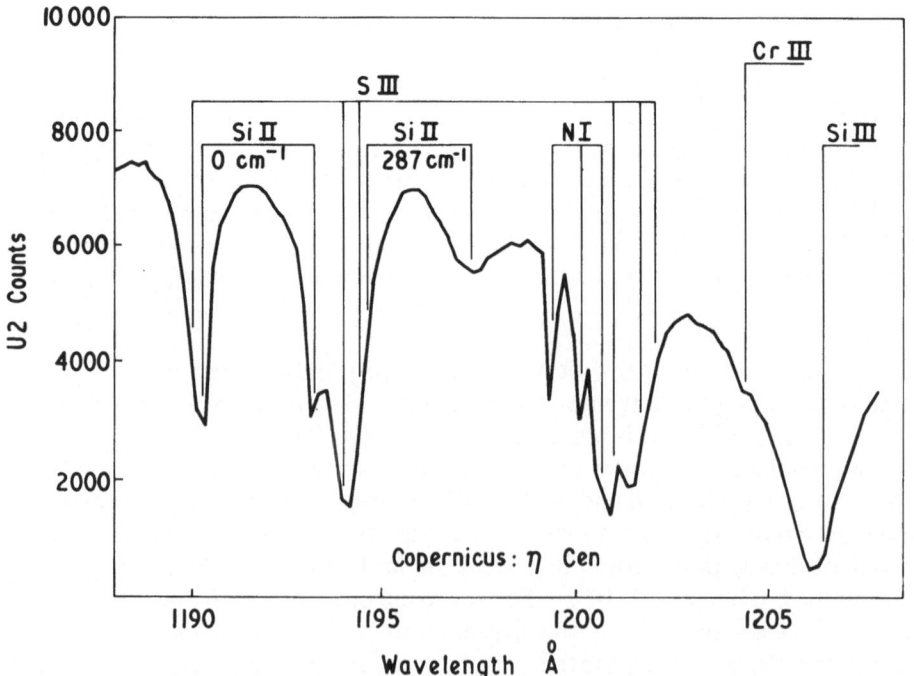

Fig. 3.   *Copernicus* ultraviolet spectrum of η Cen (U2 scan λ 1188 Å–1208 Å) showing absorption lines from the ground level and excited levels in Si II.

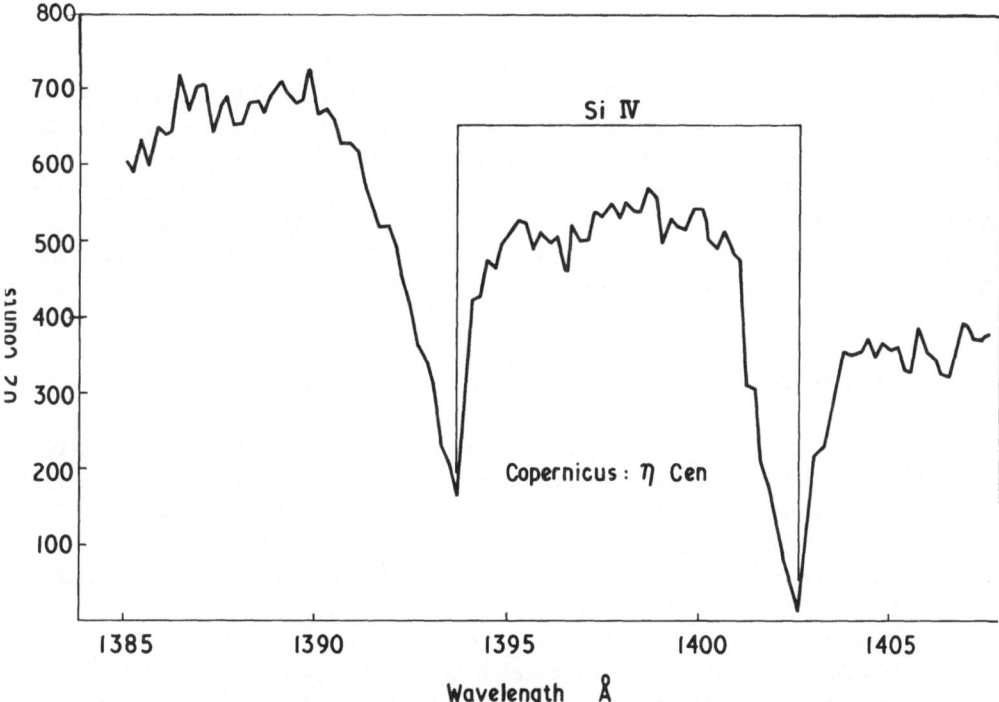

Fig. 4.   *Copernicus* ultraviolet spectrum of η Cen (U2 scan λ 1385 Å-1408 Å) showing the resonance doublet of Si IV with asymmetric absorption line profiles indicating high expansion velocities.

Another feature of interest in the η Cen ultraviolet spectrum is illustrated in Figure 4 which shows the U2 scan in the region of the Si IV resonance lines near λ 1400 Å. The continuum level in this tracing has been distorted by an incomplete correction for stray light in the spectrometer which has resulted in an incorrect stepped continuum level. The absorption lines are in fact saturated and should reach the zero intensity level. Both components of the Si IV multiplet show asymmetric profiles with an enhanced short wavelength wing, giving a limiting velocity value of about -250 km s^{-1}. Blue-shifted absorption lines are normally interpreted in terms of mass outflow from the star, which would result in mass-loss if the flow velocity exceeds the local escape velocity. The predicted escape velocity at the surface of a normal B2 V star is estimated to be $\sim$700 km s^{-1} but for η Cen, which has a rapidly rotating extended atmosphere, the escape velocity from this atmosphere will be significantly lower. If we assume that the observed Si IV expansion velocity in η Cen is in fact resulting in mass loss and that the Si abundance is normal, then the mass loss rate is calculated to be $\sim$2 $\times$ 10^{-11} $M_\odot$ y^{-1}. This value is based on the assumption that the mass outflow is approximately spherical and it would be reduced by about an order of magnitude if equatorial flow predominated. This estimated mass loss rate from η Cen is less by a factor of $\sim$10^{-5} than from Wolf Rayet and Of stars, but if this value is typical, then the larger number of Be stars may make them an important factor in determining the chemical balance of the interstellar medium.

Although the analysis of the new ultraviolet observations is by no means complete at the present time, it is nevertheless possible to draw a few preliminary conclusions.

There is good evidence that the atmosphere of η Cen is extended and rapidly rotating. The Si IV lines provide very strong evidence for the existence of an expanding atmosphere and possibly for mass loss. The existence of a circumstellar shell is less certain, but there does seem to be some indication of a significant density ($\gtrsim 10^3$ cm^{-3}) at a high level in the extended stellar atmosphere. However, the strong absorption lines of ions such as Fe II, which are normally associated with shell spectra, do not seem to occur in the ultraviolet spectrum of η Cen suggesting that there is no strongly developed shell at the present time.

3. Visible Region Spectra of γ Cas (HD 5394) and ϕ Per (HD 10516)

The two bright Be stars γ Cas and ϕ Per were designated as stars of particular interest for the present symposium and it was suggested that new observations would be valuable. For this reason spectra of both stars were recorded on 7–8 June 1975, using the 36-in. Yapp telescope at the Royal Greenwich Observatory. The primary aim during this observing run was to evaluate the performance of a recently developed image tube echelle spectrograph which combines good spectral resolution with wide wavelength coverage. The spectrograph was originally designed for use in rocket vehicles to observe ultraviolet stellar spectra (Burton *et al.*, 1971) but was subsequently converted for use with ground-based telescopes. In its new configuration (CRESTA) this spectrograph provides a dispersion of ~5 Å mm^{-1} (spectral resolution $d\lambda \simeq 1.0$ Å) over a wavelength range extending from the image tube window limit ($\gtrsim \lambda$ 3800 Å) up to the sensitivity limit of the S25 photocathode (~λ 8000 Å). The optical system consists of a concave collimator mirror, echelle grating, plane cross-dispersing grating for order separation and a concave camera mirror which focuses the spectrum on to the input window of a Westinghouse WL30677 image intensifier. Photographic film (Kodak 103aD or 103aF) is contacted against the output fibre optic of this image tube to record the intensified spectra.

An example of the two-dimensional spectrum format produced by this cross-dispersed echelle grating system is shown in Figure 5. This spectrum of ϕ Per (B0e $m_v = 4.2$) was obtained in a 10 minute exposure using Kodak 103aF film. One advantage of the large area echelle format for the study of Be stars is at once apparent. The extended wavelength coverage enables the entire Balmer series to be recorded in a single exposure while the echelle dispersion provides sufficient spectral resolution to record the detailed profiles of the various emission lines, thus permitting simultaneous study of spectrum changes in the different members of the Balmer series. Observations of this type were made for both γ Cas and ϕ Per on 7–8 June 1975 and Figure 6 shows densitometer traces of the profiles of Hα and Hβ emission lines in these two stars. Further data reduction will be necessary to correct for the instrument sensitivity functions but nevertheless the observations show the general shape of the emission line profiles.

The general form of the observed profiles is similar to those obtained by Gray and Marlborough (1974) using a photoelectric line-profile scanner. The Hα emission line in ϕ Per does not show appreciable splitting but the Hβ line has a pronounced

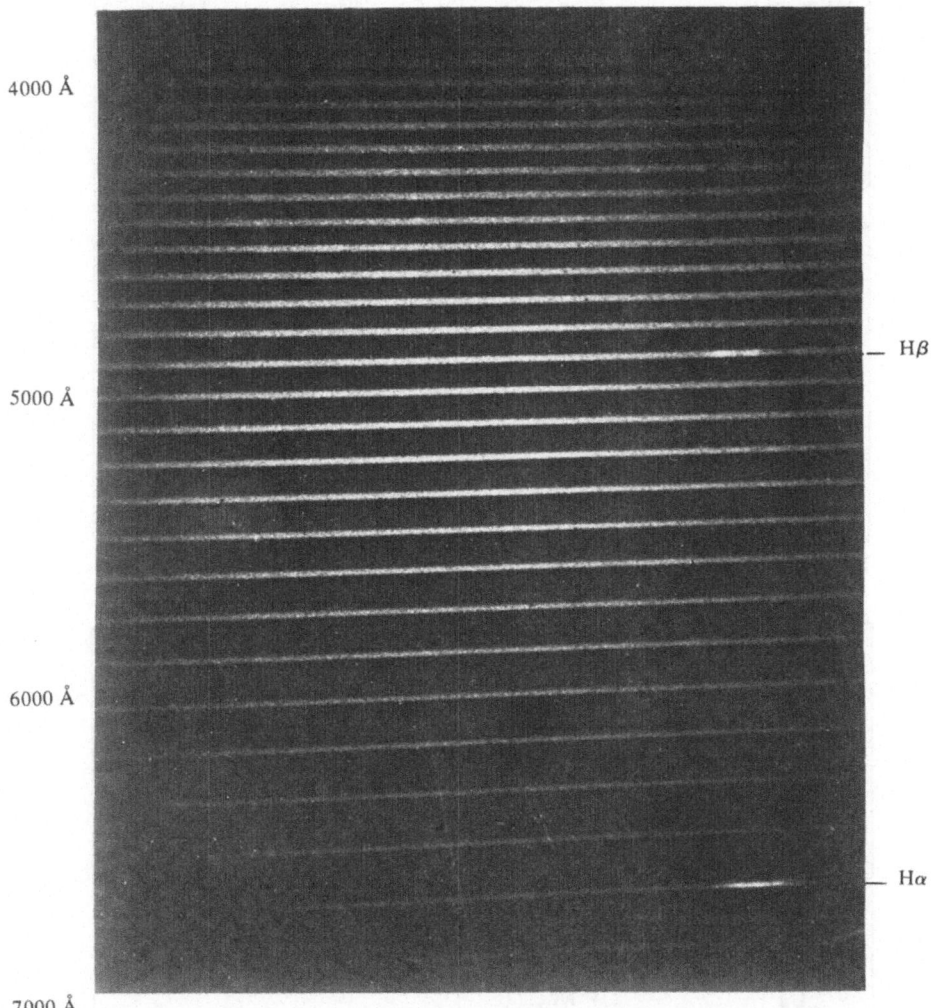

4000 Å

5000 Å

6000 Å

7000 Å

— Hβ

— Hα

Fig. 5.   Echelle spectrum of ϕ Per (B0e $m_v = 4.2$) recorded with the CRESTA spectrograph on the RGO 36-in. telescope (cassegrain focus). 7–8 June 1975. Exposure 10 min on 103aF film.

double peak with $V < R$. In γ Cas the Hα line is double and has the same general shape as the double-peaked emission in Hβ but in this case $V > R$. If the observed V/R ratio is interpreted as the result of a wavelength shift of a central absorption component, then we might conclude that the absorbing atmospheric material in ϕ Per is expanding at the present time while that in γ Cas, being red-shifted, is falling back towards the underlying star.

The profiles illustrate an important point in the common terminology associated with Be star spectra. It was suggested in an earlier paper at this symposium (Bidelman, 1975) that a shell star can be recognised by the presence of Balmer emission lines with central absorption features. In the case of γ Cas and ϕ Per the

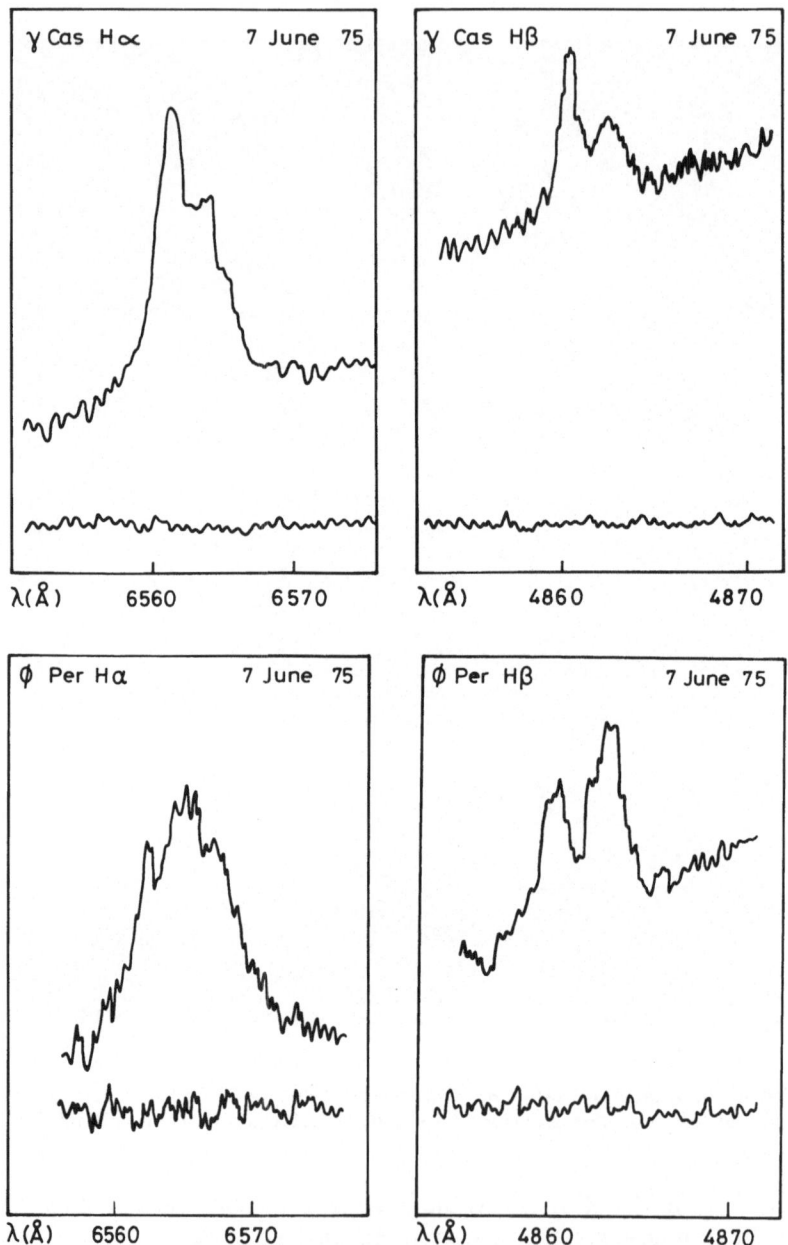

Fig. 6.   Densitometer tracings of Hα and Hβ emission lines in γ Cas and φ Per from echelle spectrograph exposures with the RGO 36-in. telescope on 7–8 June 1975.

Balmer lines indeed show double emission peaks but in neither case does the
absorption minimum extend below the local stellar continuum level, so that in terms
of the stellar continuum this central 'absorption' is still an emission feature. A
possible criterion which could be adopted for consistency would be to designate as

'shell stars' only those stars with distinct shell absorption lines which depress the stellar flux below the local continuum level. The Balmer line profiles obtained in the present observations would then be described as double emission peaks and would not necessarily indicate the presence of a shell.

Acknowledgements

The *Copernicus* observations described in the first part of this paper were obtained through the generosity of Professor L. Spitzer and his colleagues at Princeton University Observatory. In particular we thank Dr D. York and Dr T. Snow for their help in obtaining the observations.

The observations made with the 36-inch Yapp telescope were obtained with the permission of the Director of the Royal Greenwich Observatory. Important contributions made by Mr R. A. Hardcastle and other members of the Appleton Laboratory Astrophysics Research Division during the modification of the echelle spectrograph for the ground-based observations are gratefully acknowledged. This paper is published with the permission of the Director of the Appleton Laboratory.

References

Bahcall, J. N. and Wolf, R. A.: 1968, *Astrophys. J.* **152**, 701.
Burton, W. M., Reay, N. K., Shenton, D. B., and Wilson, R.: 1971, in F. Läbuhn and R. Lüst (eds.), 'New Techniques in Space Astronomy', *IAU Symp.* **41**, 304.
Gray, D. F. and Marlborough, J. M.: 1974, *Astrophys. J. Suppl.* **27**, 121.
Jaschek, C., Jaschek, M., and Kucewicz, B.: 1964, *Z. Astrophys.* **59**, 108.
Rogerson, J. B., York, D. G., Drake, J. F., Jenkins, E. B., Morton, D. C., and Spitzer, L.: 1973, *Astrophys. J. Letters* **181**, L 110.
Uesugi, A. and Fukuda, I.: 1970, *Contrib. Inst. Astrophys and Kwasan Obs., Univ. of Kyoto*, No. 189.

THE FAR ULTRAVIOLET SPECTRA OF υ Cyg AND μ Cen

GERALDINE J. PETERS*

Dept. of Astronomy, University of California, Los Angeles, Calif., U.S.A.

Abstract. The spectra of the 'pole-on' Be stars υ Cyg and μ Cen have been observed in the region λλ 1000–1450 Å with the Princeton Ultraviolet Spectrometer on board the *Copernicus* satellite. The data include scans of intermediate resolution (0.2 Å) with U2 and high resolution scans (0.05 Å) of selected suspected shell lines with U1. The spectra of υ Cyg and μ Cen are compared with a complete intermediate resolution *Copernicus* scan of the equator-on Be star φ Per. The ultraviolet spectra of υ Cyg and μ Cen appear to be identical to those observed in non-Be stars of comparable ground-based spectral type. Neither emission nor shell features are observed. However, numerous strong shell features are seen in φ Per. An upper limit on the column density of hydrogen in our line of sight is computed for υ Cyg from the observed strength of a weak feature at λ 1130.4 Å identified as an Fe III shell line. Estimates for the dimensions of the circumstellar envelope are then obtained. Some implications of the strengths of the Fe III shell lines at λ 1130 Å observed in υ Cyg, μ Cen, φ Per, and other Be stars are discussed.

1. Introduction

The far ultraviolet spectra of the so-called 'pole-on' Be stars υ Cyg (B1.5Ve) and μ Cen (B2IVe) have recently been observed with the Princeton Ultraviolet Spectrometer on board the *Copernicus* satellite. A detailed study of both the ultraviolet and ground-based spectra of these stars is presently in progress. However, some early results from the investigation are interesting and worth reporting at this symposium.

In order to establish whether there are obvious differences between the ultraviolet spectra of the 'pole-on' Be stars and non-emission B-type stars of comparable ground-based spectral type, the *Copernicus* scans of υ Cyg and μ Cen were compared with those of λ Sco (B1V) and σ Sgr (B2.5V). Some possible ways in which the ultraviolet spectra of Be stars can differ from those of non-Be stars include (1) the presence of emission lines, (2) the presence of shell features, (3) veiling of photospheric features due to continuous Balmer emission, and (4) peculiarities in the photospheric line profiles which indicate significant line formation in the circumstellar envelope.

A major goal of this investigation was to search for shell features in the ultraviolet spectra of pole-on Be stars. In the far ultraviolet we might expect to find shell lines of C II, C III, N II, N III, Si II, Si III, S III, S IV, Fe II, and Fe III. The strengths of these shell lines can then be used to obtain an estimate of $N_H h$, the product of the density of hydrogen atoms times the path length, at polar latitudes. Thus, if one can determine a value for the mean density in the envelope from some other means such as the Balmer emission line strengths, the far ultraviolet shell lines can help shed some light on the geometry and/or physical extent of Be star envelopes.

With the above-stated goal in mind, μ Cen ($v \sin i \approx 200$ km s^{-1}) was chosen as a representative 'pole-on' Be star and υ Cyg ($v \sin i \approx 275$ km s^{-1}) as a Be star of intermediate inclination. In collaboration with Dr M. Plavec, the ultraviolet spectra

* Guest Investigator with the Princeton University telescope on the *Copernicus* satellite, which is sponsored and operated by the National Aeronautics and Space Administration.

of these stars were then compared with *Copernicus* scans of ϕ Per, a Be star which can be considered as being viewed nearly equator-on ($v \sin i \approx 400$ km s^{-1}). If the classical model for a Be star envelope is correct (i.e. a thick disk of material positioned about the equator of the star), then one should expect to see strong shell lines in ϕ Per, ones of intermediate strength in v Cyg, and no shell features in μ Cen.

2. Observed Properties of μ Cen, v Cyg, and ϕ Per

v Cyg was designated as a 'pole-on' Be star by Slettebak (1949). Members of this class of objects typically have relatively narrow Balmer emission features with no conspicuous central reversals and relatively sharp photospheric features which suggest values of $v \sin i \le 280$ km s^{-1}. If one assumes that all Be stars have equatorial velocities near 400 km s^{-1}, then the 'pole-on' Be stars have inclinations less than 45° to our line of sight. According to Slettebak (1949) and Slettebak and Howard (1955), $v \sin i = 280$ km s^{-1} for v Cyg.

μ Cen is a member of the Scorpio-Centaurus association and according to Slettebak (1968) has a value of $v \sin i = 190$ km s^{-1}.

The He I photospheric features in ϕ Per are rotationally broadened to a high degree and show evidence of emission contamination. The value of $v \sin i$ is thus uncertain (see the Review paper given by Slettebak at this Symposium for a discussion of the accuracy of rotational velocities for the most rapidly rotating stars). A value of 400 km s^{-1} is tentatively assigned to ϕ Per.

μ Cen, v Cyg, and ϕ Per have strong Hα emission features with no conspicuous structure. The Hα emission line profiles observed for these stars are shown in Figure 1. The spectrograms from which these profiles were obtained were taken with the

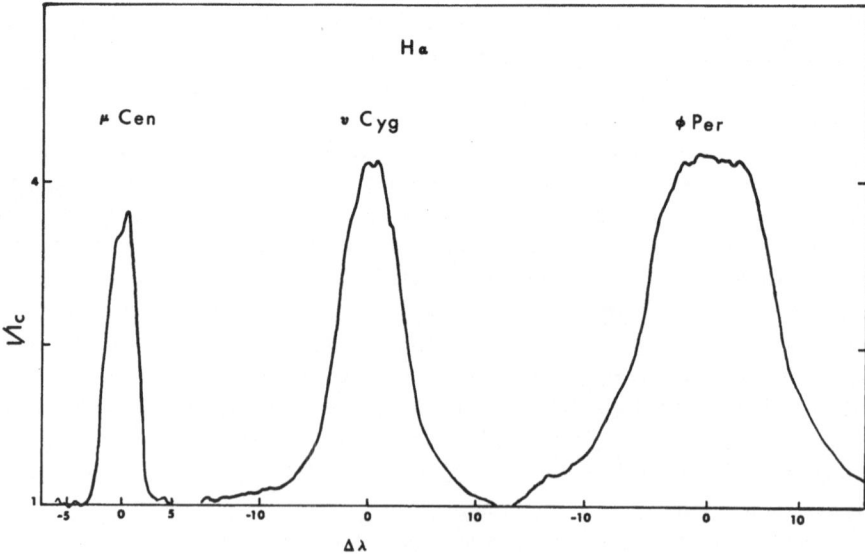

Fig. 1.   Profiles of the Hα emission features seen in μ Cen, v Cyg, and ϕ Per. The original dispersion of the spectrograms is 11 Å mm^{-1}.

Lick Observatory cooled 40 mm Varo image tube and the 24-in. (61 cm) Coudé Auxiliary Telescope (dispersion: 11 Å mm^{-1}). The peak intensities of the Hα emission lines are approximately four times the continuum value for all three stars. However, the widths of the features increase noticeably with the $v \sin i$ of the star. Thus, except for the fact that the Hα profile in φ Per does not show a deep central core, the Hα emission features are suggestive of the 'classical' model for a Be envelope. Variations in the 'fine structure' in the Hα emission profiles have been observed for μ Cen and φ Per. However, ten Hα plates of υ Cyg taken over the course of two years showed a constant profile.

We have also observed μ Cen, υ Cyg, and φ Per in the near infrared with the Lick Observatory Varo tube (dispersion: 23 Å mm^{-1}). All three stars show λ 8446 Å of O I in emission. This feature is formed via a fluorescence involving Lβ (Bowen, 1947) and is discussed later in the paper in connection with the predictions for λ 1302 Å of O I. υ Cyg and φ Per also have emission at the higher order Paschen lines and emission of Fe II, O I λ 7774 Å, and the infrared Ca II triplet.

3. Observational Program

The Princeton Ultraviolet Spectrometer on the *Copernicus* satellite has been described in detail by Rogerson *et al.* (1973). The data for υ Cyg and μ Cen obtained for this project include continuous intermediate resolution (0.2 Å) scans with U2 (λλ 1000–1450 Å), continuous intermediate resolution (0.4 Å) scans with V2 (λλ 1800–3100 Å), and high resolution scans (0.05 Å) of selected features with U1. Although large fluctuations in the dark counts limited the usefulness of the V2 scans (see Rogerson *et al.*, 1973), the observed counting rates for U2 and U1 produced a fairly high signal-to-noise ratio. The observed noise is 1% and 3% of the signal in μ Cen and υ Cyg, respectively.

The features which were scanned with U1 in υ Cyg and μ Cen included lines of C II, N II, S II, S III, S IV, Si II, Si III, Si IV, Fe II, and Fe III which were tentatively identified as 'shell' features in the spectrum of φ Per. In addition, O I λλ 1302, 1305, and 1306 Å were scanned to search for emission and shell structure. Even though substantial interstellar contribution was suspected for some of the lines scanned with U1, they were observed with the anticipation of velocity differences existing between shell and interstellar features which would allow one to separate the components.

The U1 observations were programmed to block stray light from the scans (Snow, 1975). The stray light in U2 was removed with the aid of an algorithm developed by Bohlin (1975). Both stray light and particle background were removed with the aid of computer codes which exist at Princeton University.

4. Results

A comparison between the U2 scans of υ Cyg and μ Cen and those of the standards λ Sco and σ Sgr has revealed that the ultraviolet spectra of these 'pole-on' Be stars are apparently identical to those of non-emission B-type stars of comparable

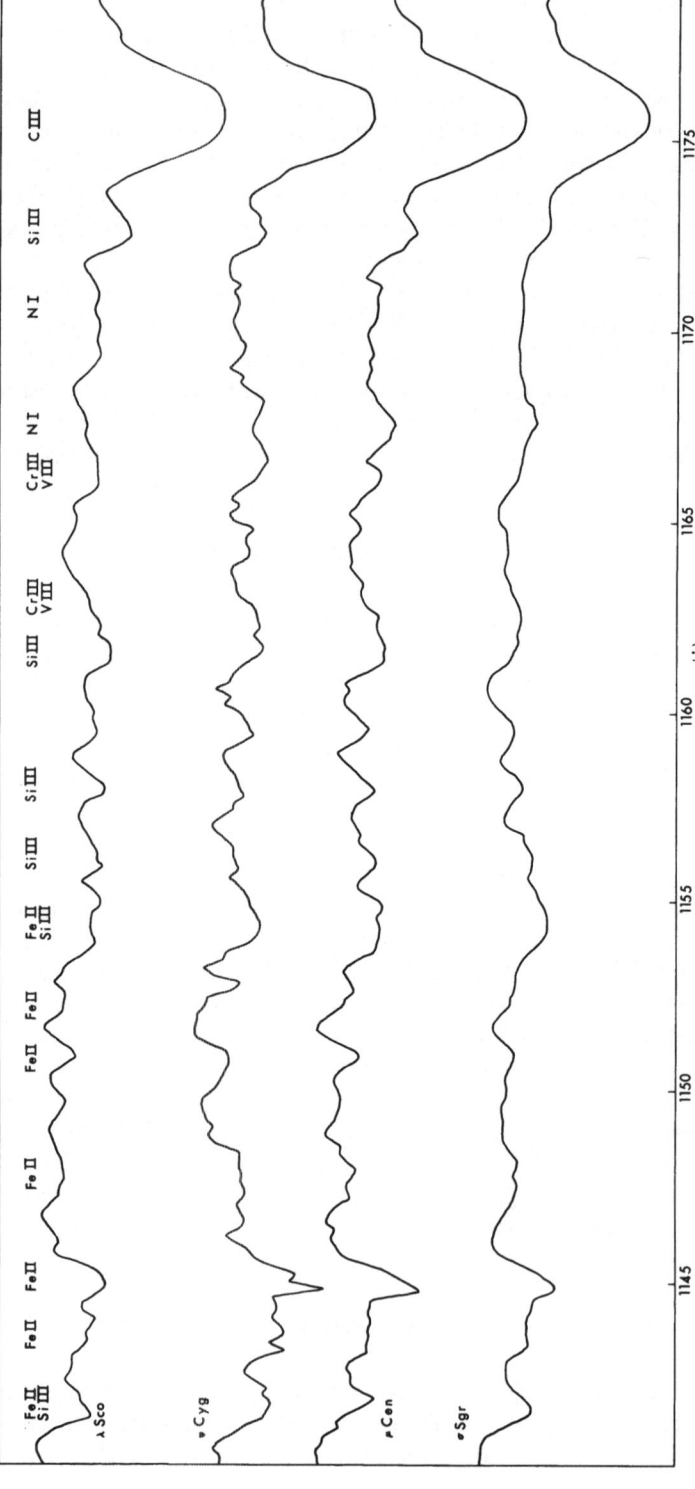

Fig. 2. The ultraviolet spectra of υ Cyg and μ Cen from λλ 1140–1180 Å compared with two standard stars, λ Sco and σ Sgr. The ions which significantly contribute to the stronger absorption features are indicated at the top of the diagram. The strong feature near λ 1175 Å is the C III sextuplet. Sharp features are interstellar.

ground-based spectral type. Emission features are not observed in the far ultraviolet spectra of υ Cyg and μ Cen (not even $L\alpha$ and $L\beta$) and shell lines do not appear to be present. Some of the stronger photospheric features which are seen include: C II λ 1037 Å, N II λ 1084 Å, S IV λ 1073 Å, Fe III λ 1124 Å, C III λ 1176 Å, Si III λ 1206 Å, Si II λ 1265 Å, C II λ 1335 Å, and Si IV λλ 1394 and 1403 Å.

Shown in Figure 2 is an intercomparison between the ultraviolet spectra of μ Cen, υ Cyg, λ Sco, and σ Sgr in the interval λλ 1140–1180 Å. The ions which contribute substantially to the stronger photospheric lines are listed above the features. Certainly, the differences between the spectra are small. The values of $v \sin i$ for λ Sco and σ Sgr (250 and 200 km s^{-1}, respectively) are comparable to those observed for υ Cyg and μ Cen.

The ultraviolet line strengths in υ Cyg compare well with those in λ Sco. The temperature and log g for υ Cyg which are indicated by the ground-based data are 24 000 K and 3.8, respectively (Peters, 1976) while the temperature of λ Sco appears to be near 23 500 K (on the scale of the Princeton model atmospheres). From an inspection of the U2 scans, it appears that the temperature of μ Cen is between that of λ Sco and σ Sgr. An analysis of the ground-based spectrum of μ Cen suggests that $T_{eff} \approx 20\,000$ K and log $g \approx 3.5$. Similarly, the T_{eff} for σ Sgr is near 18 000 K.

The analysis of the ultraviolet spectrum of υ Cyg is complicated by the fact that numerous strong interstellar features (including H_2) are present. These interstellar lines could easily mask weak shell components in the U2 scans (they can be separated in the U1 scans, however). The interstellar features are weak in μ Cen.

The projected rotational velocities for υ Cyg and μ Cen indicated by the weaker photospheric features in the far ultraviolet appear to be consistent with those suggested by the ground-based features. The line widths for the weaker features are one-fourth the values observed for the features near λ 4000 Å. Of course, the ultraviolet features which have substantial Stark broadening have much larger half-widths.

The values for the ultraviolet flux of υ Cyg and μ Cen appear to be comparable to those of non-Be stars of similar ground-based spectral type. The predicted counting rates for U1 and U2, which were based upon those observed for standard stars, were identical to those which were observed.

The O I resonance line λ 1302 Å was observed in υ Cyg and μ Cen both with U1 and U2. This transition is the final step of a set of three which are activated by a fluorescence with $L\beta$ (Bowen, 1947). The sequence proceeds in the following manner: O I λ 1025 Å $(2^3P-3^3D) \to$ λ 11287 Å $(3^3D-3^3P) \to$ λ 8446 Å $(3^3P-3^3S) \to$ λ 1302 Å (3^3S-2^3P). λ 8446 Å is observed to be moderately strong in emission in υ Cyg and weak in emission in μ Cen (1.25 and 1.05 times the local continua, respectively). The U2 and U1 scans of λ 1302 Å in υ Cyg and μ Cen along with the profiles of λ 8446 Å (from Varo image tube spectrograms taken at Lick Observatory) are presented in Figure 3. No obvious emission is present at λ 1302 Å in either star. The strong absorption features are interstellar O I λ 1302 Å.

Based on the strength and width of λ 8446 Å in υ Cyg the predicted intensity and width of λ 1302 Å in this star are 1.03 I_c and 1 Å, respectively, and are, therefore, slightly below the limit of detectability on U2 scans. The predicted strength of λ 1302 Å in μ Cen is lower than that for υ Cyg. However, the strength and width of

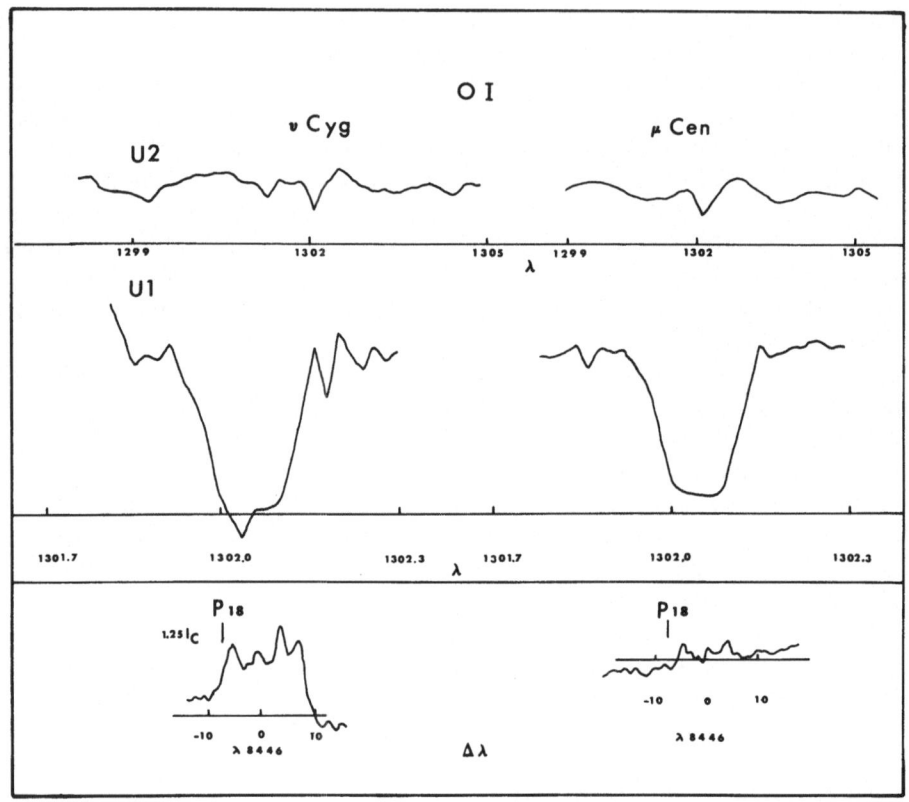

Fig. 3.   U1 and U2 scans of υ Cyg and μ Cen in the region including λ 1302 Å of O I. Shown below these
observations are the features of O I λ 8446 Å which are seen in these stars.

λ 8446 in φ Per (1.7 I_c and 13 Å, respectively) predict a peak intensity of 1.07 I_c and
a width of 2 Å for λ 1302 Å in this star. No emission at λ 1302 Å is observed on a U2
scan of φ Per, however, even though the predicted strength for the feature is well
above the limit of detectability.

The search for shell features in the far ultraviolet spectra of the 'pole-on' Be stars
υ Cyg and μ Cen produced negative results. Whereas numerous shell features are
observed in φ Per, no shell lines have been identified with certainty in the spectra of
the 'pole-on' stars considered in this program. U1 scans reveal the *possible* presence
of weak shell absorptions (≈20 mÅ) of S II λ 1014 Å and Fe III λ 1130 Å in υ Cyg.

U2 scans of μ Cen, υ Cyg, φ Per, and the standard λ Sco in the region of Fe III,
multiplet 1, are intercompared in Figure 4. Note the great strength of the Fe III shell
features in φ Per compared to the conspicuous absence of similar features in υ Cyg
and μ Cen. The Fe III lines in υ Cyg and μ Cen are photospheric features.

An upper limit on the column density of Fe III can be computed for υ Cyg from the
strength of the weak λ 1130 Å feature observed on the U1 scan. The f-value quoted
by Kurucz and Peytremann (1975) was used for the computation. Since $N(\text{Fe III}) \approx$
$N(\text{Fe})$ and $N(\text{Fe}) \approx 10^{-5} N_{\text{H}}$, we obtain a value of 5×10^{19} cm^{-2} for $N_{\text{H}}h$. If $N_{\text{H}} \approx$
10^{10} cm^{-3}, then $h \approx 5 \times 10^9$ cm or about 0.1 $R_{\odot}$! If $N_{\text{H}} \approx 10^8$ cm^{-3}, then $h \approx 10 R_{\odot}$ or

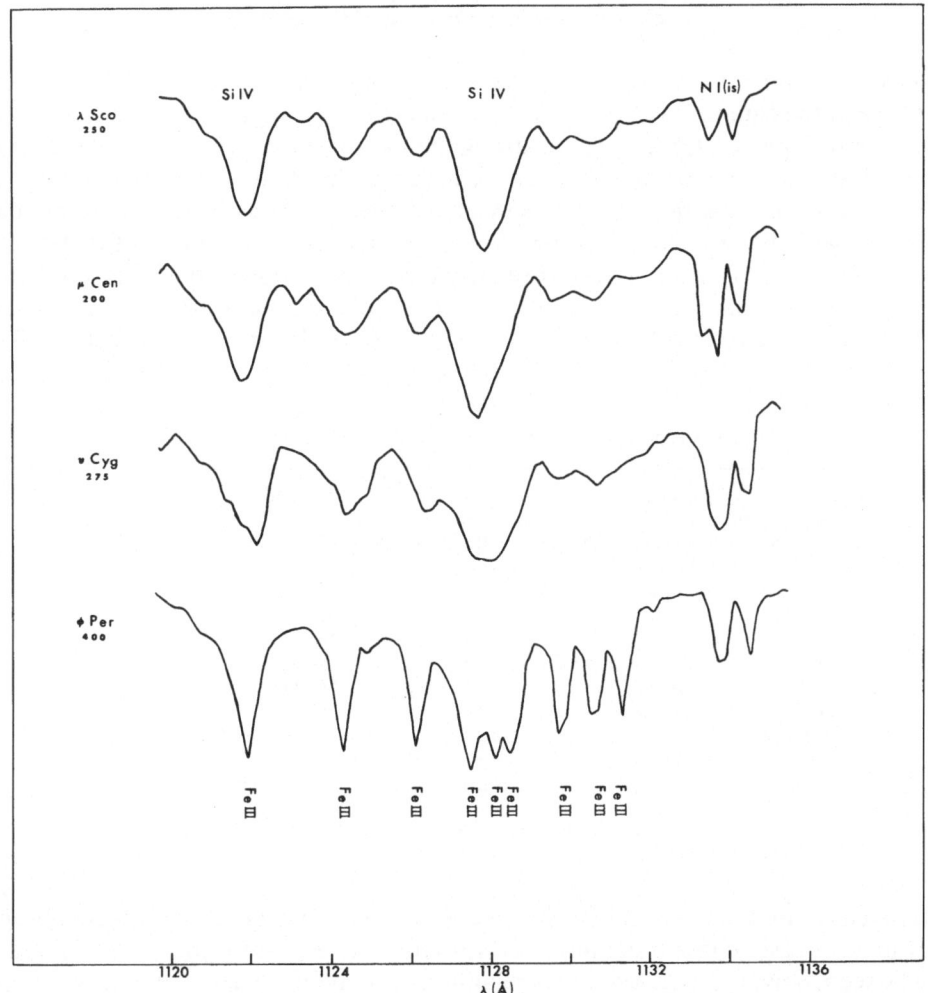

Fig. 4.   The spectra of the Be stars μ Cen, υ Cyg, and φ Per and the standard λ Sco in the vicinity of multiplet 1 of Fe III. Approximate values of $v \sin i$ for each star are quoted below their labels. Shell-type absorptions are seen for Fe III in φ Per. The sharp N I features are interstellar in origin.

$1.0 R_*$! Therefore, the absence of strong Fe III shell absorptions in υ Cyg suggests that virtually all of the material in its circumstellar envelope is located outside of our line of sight.

The volume emission measure for υ Cyg [obtained from the equivalent widths of the Hγ and Hδ emission features with the aid of Wellman's formula quoted in Pagel (1960)] is $N_{ion}N_e V \approx 5 \times 10^{57}$ cm^{-3}. If $N_{ion} = N_e \approx 10^{10}$ cm^{-3} and if the material is confined to a thick disk with a height of $1 R_*$, then $R_{env} \approx 13 R_*$. If the cross-section of the envelope is more wedge-shaped with inner and outer heights of $2 R_*$ and $10 R_*$, respectively, then $R_{env} \approx 5 R_*$. The observations of the ultraviolet Fe III shell lines in υ Cyg are not inconsistent with the model for a Be star envelope suggested by Marlborough (1969).

5. Further Comments on the Fe III Shell

The comparison between the ultraviolet spectra of the 'pole-on' Be stars υ Cyg and μ Cen and the equator-on Be star ϕ Per would seem to support the 'classical' model for the envelope of a Be star in that strong shell absorptions are present in the latter object but absent in the former ones. In order to determine whether it is true, in general, for Be stars that the strength of the shell features in the far ultraviolet correlate with the projected rotational velocity of the star, U2 scans of other Be stars which exist in the Princeton files were examined. The region of the Fe III shell lines (λ 1130 Å) was particularly useful since the Fe III features can be strong and λ 1130 Å is near the peak sensitivity for U2. The results are given in Table I. The

TABLE I

Strength of the ultraviolet Fe III shell in Be stars

Star	Observer	$v \sin i$[a]	Strength
γ Cas	Marlborough[b]	300 km s^{-1}	None
ϕ Per	Plavec[b]	400	Strong
48 Per	Snow	200	Very weak
ζ Tau	Heap[b]	275	Strong
μ Cen	Peters[b]	200	None
η Cen	Burton/Evans[b]	300	Weak
χ Oph	Snow	100	Moderate-weak
υ Cyg	Peters[b]	275	None
31 Peg	Marlborough[b]	100	Moderate-strong

[a] Approximate values based upon those quoted in Uesugi and Fukuda (1970).
[b] Guest Investigator using the Princeton Ultraviolet Spectrometer on the *Copernicus* satellite.

estimated strengths quoted for the Fe III shell features are qualitative values only and may in some cases contain significant contributions from photospheric Fe III lines. It can be seen, however, that a strict correlation between Fe III shell strength and $v \sin i$ does not exist. We find rapidly rotating Be stars ($v \sin i \gtrsim 300$ km s^{-1}) which show no shell (γ Cas), a weak shell (η Cen), and a strong shell (ϕ Per). Whereas the Fe III shell features are effectively absent in υ Cyg, they are strong in ζ Tau even though the projected rotational velocities for both objects are comparable. Even the pole-on Be star 31 Peg appears to have a moderately strong Fe III shell.

Thus, either some Be stars have a substantial amount of material at their polar latitudes or Be stars have a wide range in their intrinsic rotational velocities (the value of $v \sin i$ may not be indicative of the inclination of the star's equator to our line of sight). It is difficult to reconcile a classical model for a Be star envelope which is based upon the concept of forced rotational ejection with the observations of 31 Peg, however. An equatorial velocity of 100 km s^{-1} hardly seems adequate enough to supply a circumstellar disk with material.

This study of the ultraviolet spectra of υ Cyg and μ Cen combined with the observations of the Fe III shell in other Be stars suggests that Be stars are not a homogeneous group of objects. Perhaps the strengths of the Fe III shell lines observed in the far ultraviolet spectra of Be stars can be used in conjunction with the strengths and/or profiles of other envelope features and the photospheric parameters for the star to help shed some light on the geometry, physical structure, and origin of the various types of Be envelopes.

Acknowledgements

I wish to thank the staff of the Department of Astrophysical Sciences at Princeton University for their hospitality during my numerous visits to Princeton. In particular, I am grateful to Drs T. P. Snow and D. G. York and Mr A. E. Miller for their instruction in the use of the *Copernicus* computer codes. I also thank Drs Marlborough, Plavec, Snow, Heap, and Burton for permission to quote the estimates for the strengths of the Fe III shell features in the Be stars which they are studying. This project is supported by NASA grant NSG 5056.

References

Bohlin, R. C.: 1975, preprint.
Bowen, I. S.: 1947, *Publ. Astron. Soc. Pacific* **59**, 196.
Kurucz, R. L. and Peytremann, E.: 1975, *Smithsonian Astrophysical Observatory Special Report No. 362.*
Marlborough, J. M.: 1969, *Astrophys. J.* **156**, 135.
Pagel, B. E. J.: 1960, *Vistas in Astronomy* **3**, 203.
Peters, G. J.: 1976, this volume, p. 69.
Rogerson, J. B., Spitzer, L., Drake, J. F., Dressler, K., Jenkins, E. B., Morton, D. C., and York, D. G.: 1973, *Astrophys. J. Letters* **181**, L97.
Slettebak, A.: 1949, *Astrophys. J.* **110**, 498.
Slettebak, A.: 1968, *Astrophys. J.* **151**, 1043.
Slettebak, A. and Howard, R. F.: 1955, *Astrophys. J.* **121**, 102.
Snow, T. P.: 1975, *A Guide for Guest Investigators Using the Princeton Telescope on the Satellite COPERNICUS.*
Uesugi, A. and Fukuda, I.: 1970, *Contrib. Inst. Astrophys. Kwasan Obs. Univ. Kyoto*, No. 189.

PHOTOGRAPHIC INFRARED SPECTROSCOPY AND NEAR INFRARED PHOTOMETRY OF Be STARS

(*Review Paper*)

J. P. SWINGS

Institut d'Astrophysique, B-4200 Cointe-Ougrée, Belgium

1. Introduction

As the title indicates two topics will be tackled in this presentation: spectroscopy and photometry. Let us therefore define first what 'infrared' means for these two subjects. I chose the following definitions:
– photographic infrared spectroscopy: wavelengths $H\alpha \leq \lambda < 1.2\ \mu$
– near infrared photometry: wavebands: $1.6\ \mu \leq \bar{\lambda} \leq 20\ \mu$

2. Near Infrared Photometry

As pointed out by Allen (1973): "That emission lines in the spectrum of a star are evidence of an extended stellar atmosphere is well established. If the density of material in that atmosphere is sufficiently high, its presence should additionally be manifested by radiation mostly at longer wavelengths than that of the underlying star. We therefore seek to explain the infrared excess of the early type emission-line stars in terms of the circumstellar material we believe to lie in clouds around them". This of course implies that infrared excesses have been observed from Be stars, which indeed is the case. For publications prior to the summer of 1971 I refer to an excellent review paper given that year by J. C. Pecker at the Liège Symposium. What I plan to do here is to show that there is a gradation in the excesses observed in classical Be, less classical Be, and peculiar Be stars. For this matter I rely mainly on the surveys published up to 3.5 μ by Allen (1973) and up to 20 μ by Gehrz *et al.* (1974) (referred to as G-H-J).

2.1. Classical Be stars

It appears that the most likely mechanism for the production of the observed excess infrared radiation in classical Be stars (see e.g. βCMi, Figure 3 of G-H-J) is free-free radiation (remaining optically thin to 20 μ) from proton-electron scattering in a hot ($T_{shell} \geq 10\ 000$ K) ionized circumstellar plasma. The preliminary result of Woolf *et al.* (1970) and the survey by Allen (1973) strongly suggested this mechanism although their data did not include spectral detail in the 4.9–19.5 μ region. The optical hydrogen emission lines, and thus presumably the excess infrared radiation, originate in a circumstellar gas shell, or extended atmosphere, that is formed from material ejected from the equatorial regions of a star near the limit of rotational stability (G-H-J, 1974).

A. Slettebak (ed.), Be and Shell Stars, 219–225. All Rights Reserved

2.2 LESS CLASSICAL Be STARS (SHELL STARS?)

As examples of stars of this class I suggest ϕ Per and ζ Tau. G-H-J indicate in the caption to their Figure 2 that ϕ Per is a typical example of Be stars whose infrared excess is fitted by a pure free-free model in which the shell becomes optically thick at long wavelengths. The infrared energy distribution of ζ Tau (see Figure 6 of G-H-J) can be reproduced by adding either a 1200 K black body spectrum or a 14 000 K free-free spectrum (under some conditions, i.e. $\tau_s = 1$ at 8.7 μ) to the hot stellar continuum. In any case the circumstellar flux makes a negligible contribution to the total emitted energy shortward of 2.3 μ.

For these two subclasses of Be stars, G-H-J derive that "a typical Be star shell, assuming a shell radius-to-thickness ratio of 5, is characterized by an electron density of 3.7×10^{11} cm^{-3}, a radius of $\sim 4\,R_*$ and a Thomson optical depth of 0.16; the infrared excess in Be stars tends to decrease as spectral type becomes late; a number of the survey stars were tested for infrared variability with negative results".

2.3. PECULIAR Be STARS (B[e] STARS)

The prototype for these B[e] stars is, of course, HD 45677 whose energy distribution in the visible and near infrared is given in Swings and Allen (1971), and in Allen (1973) together with a series of objects exhibiting considerable infrared excesses. Starting in 1971 there has been a little debate concerning the mechanisms for producing the color indices of the stars of category D (Allen, 1973), i.e. those with prominent infrared excess. The mechanisms that were suggested and subsequently ruled out were:

  (i) the presence of a cool companion to the Be star;
  (ii) H$^-$ free-free, and free-free radiation in the H II region surrounding the stars;
  (iii) free-free radiation in an ionized metal region around both the star and its H II region.

So, all in all, it is believed that the strong infrared excesses are to be explained by the presence in the circumstellar environment of solid particles (therefore the expression 'dust shell') which absorb the ultraviolet and visible radiations and degrade them to infrared wavelengths. In the case of the prototype HD 45677 one then gets a physical model such as that described by Swings (1973a) where a dust shell of radius about 30 AU optically thick at 5 μ (Swings and Allen, 1971) surrounds the peculiar Be star and its extended atmosphere, ring, and forbidden line regions. Of course, among bright early-type shell stars HD 45677 is an extreme example. In 1952 Paul W. Merrill noted that it "is not a typical shell star, but rather an object intermediate between an ordinary Be star and a planetary nebula. Perhaps it can be thought of as a shell star whose outer atmosphere is extraordinarily extended and brilliant". This possible connection between evolved Be's and young planetary nebulae leads me to introduce an interesting color-color diagram on which one may plot the position of the stars we have been talking about in sections 2.1, 2.2, and 2.3: the $H(1.6\ \mu) - K(2.2\ \mu)$ vs $K - L(3.5\ \mu)$ diagram (see also Allen and Swings, 1972a; Allen, 1973; Swings, 1973b). On such a diagram one sees immediately that the colors of peculiar Be stars and young and dense planetaries are very similar: it is interesting to note that the spectra of most of those objects reveal low excitation emission lines of e.g. [O I], [S II] and [Fe II]. Knowing that such a correlation does exist, Allen and Swings (1972b) suggested that infrared photometry could be a rapid and effective method of making a first order classification of Be stars and in particular of isolating those with unusual spectral properties. First applied to the visible part of the spectrum (cf. a review by Allen and Swings, 1976, and references therein) this method gives equally good results in the photographic infrared (λ 8200–11 200 Å) as we shall see later on the basis of very recent data obtained by Mrs Andrillat and myself.

3. Photographic Infrared Spectroscopy of Be Stars

Since, in the context of this Symposium, we are dealing mainly with new observation techniques I shall not speak of what is done, or has been done, with I-N or I-Z plates except for the work of Andrillat and Houziaux. On the contrary I wish to present a few recent results obtained with the help of some new devices that are becoming more and more used, and useful, in many observatories around the world.

As for part one of this presentation we shall proceed from classical Be stars to peculiar Be stars.

3.1. CLASSICAL AND LESS CLASSICAL Be STARS

In a Bulletin of the American Astronomical Society report written by the 'chair-creature' of U.C.L.A. we read that "G. Peters and R. Polidan have continued a spectroscopic study of Be, Be-shell, and binary stars in the visual and near-infrared portions of the spectrum. They obtained approximately 750 plates of more than 150 different objects brighter than 7^{m}.0 at Lick Observatory's Coudé auxiliary telescope with a Varo image tube . . ."

"As an initial step toward understanding the envelopes of Be stars, G. Peters and R. Polidan are attempting to set up a meaningful classification scheme (Peters, 1974). The spectral line which appears to be the most promising is λ 7774 Å of O I. They have found that the appearance of λ 7774 Å allows them to distinguish three groups of Be stars: (1) Be stars with enhanced λ 7774 Å absorption (the classical shell stars); (2) Be stars with weak λ 7774 Å absorption (Be stars with relatively weak Hα emission); and (3) Be stars with λ 7774 Å in emission (Be stars which as a rule also show Fe II emission and strong Hα emission with no conspicuous structure)". These conclusions are pertinent to Be stars earlier than B6 (Peters, 1975). It is interesting to introduce here a report of the work by Andrillat and Houziaux (1975) who obtained conventional 230 Å mm^{-1} spectrograms in the photographic infrared of 68 'normal' Be stars (of magnitudes between 7 and 9). According to Andrillat and Houziaux, emission appears in the Paschen lines, O I lines and in the Ca II triplet. It is usually confined to spectral types earlier than B5. On the basis of statistics on the HD spectral types (MK types were unavailable for most of the stars), the following percentages of line emission are obtained using the data in Merrill and Burwell

Catalogues of Be and Ae stars:

TABLE I

HD stars ($m_v < 9.0$) (Andrillat and Houziaux, 1975)

HD spectral type	Total number	% Hα emiss.	% Pasch. emiss	% O I emiss. 8446 Å	% Ca II emiss.
B0–B1	364	19	8.6	8.6	3.4
B2	305	21	14	12.6	6.3
B3	902	15	1.8	4.4	1.8
B4–B5–B6	1021	9	1.1	1.1	4
B7–B8	2318	3	–	–	–
B9	4096	1	–	–	–
A0	11184	0.3	–	–	–
A2	8734	0.2	–	–	–

These results are comparable to those of Peters and Polidan (Abell, 1975) who find that for 120 different objects the infrared calcium triplet emission appears to be rare in Be stars: only 10% of their sample shows Ca II triplet emission. Ca II triplet emission does not appear to correlate strongly with any other spectral feature in the region surveyed by Peters and Polidan; however, the incidence of emission is very high among Be stars which are confirmed or suspected interacting binaries. On the basis of Table I however there seems to exist some correlation between the percentages of appearance of Hα, Paschen lines, O I λ 8446 Å and Ca II. According to Briot (1976) the infrared excess is greater for those stars showing emission in the Paschen lines than for those whose spectra exhibit a Paschen series in absorption: it is likely that both the Paschen emissions (and Balmer as well) and the free-free radiation originate from the H II region surrounding the Be star. Concerning emitting zones around Be stars, Peters (1975) feels that 'early-type Be stars which show O I λ 7774 Å emission must have more extended envelopes than early-type shell stars'. However it seems reasonable to her that 'late type shell stars can have very extended envelopes ($>20 R_*$) and not show λ 7774 Å emission due to the scarcity of far ultraviolet radiation needed to ionize the O I atom. All O I emission stars are earlier than B6'. G. Peters thus feels that O I λ 7774 Å (whose lower level is metastable) emission is simply a result of recombination in an extended, low density (10^{10}–10^{11} cm^{-3}) envelope.

Infrared line profiles can be studied in detail when the Be stars are bright enough to give a fair signal to noise ratio: for example a spectrum of γ Cas has been obtained at 4 cm^{-1} resolution over the spectral region 1–1.6 μ with a S/N of 31 in the spectrum at 8050 cm^{-1}. This spectrum was obtained with a Michelson interferometer located at the coudé focus of the 2.72 meter telescope at McDonald Observatory (Texas). The strongest feature in the spectrum is the He I λ 10 830 Å line; in addition, the hydrogen lines P_α, P_β, and P_γ are quite prominent in emission and faint emission is evident in the Brackett series. The line profiles show no reversal at the center of the line (Morgan and Potter, 1975).

3.2. Peculiar Be's

Much less has been done in this field essentially because, on the average, the B[e] stars are much fainter than the other Be stars. Near infrared I-N spectra of objects such as HD 50138 and HD 45677 were obtained more than a decade ago. However the B[e] stars started being frequently observed beyond Hα only very recently after such marvelous gadgets as Varo tubes, image tubes, image intensifiers, and multichannel spectrophotometers were made available to the astronomers. Also those B[e] stars that sounded a bit old-fashioned, that were almost forgotten for 20 years, became fashionable again after their prominent infrared excess radiation was discovered. Spectra from Hα to λ 8600 Å, or even from λ 8000 to 11 000 Å were obtained in various parts of the world: U.S.A. (Lick Observatory for example), Chile, Italy (Asiago), Israel (Mitzpeh Ramon), France (Haute Provence). Astronomers from Asiago have published recently some interesting low dispersion data, up to λ 11 000 Å, of a few BQ[] stars, in which some late type molecular features are detectable (Ciatti and Mammano, 1975a, b; Ciatti *et al.*, 1974). Before describing some very recent, and quite spectacular, spectra it should be mentioned that near infrared data may also be obtained via the use of multichannel spectrometers, such as the one developed by Dr Oke for the Palomar 200-in. telescope. Using that equipment the author gathered data at resolutions between 20 and 80 Å for about 20 B[e] stars that are presently being analysed and that will be joined to the 7 Å resolution spectrograms of August 1975 acquired at the Haute Provence Observatory in the same spectral region. These were obtained by Mrs Andrillat and myself with the Roucas grating spectrograph attached to the Cassegrain focus of the O.H.P. 77-in. telescope. The equipment has its highest efficiency around 1 μ, and covers the spectral region λ 7500–12 000 Å with a dispersion of 230 Å mm^{-1}. The receiver is a cooled two-stage image-tube equipped with an S1 photocathode; 103-aD film is used behind the fiber optics output. The aim of our observations was to detect not only the Paschen series, but mainly the emission lines due to permitted and forbidden transitions. It is indeed well known, as I explained earlier in this presentation, that there exists for peculiar Be stars a correlation between the presence of an infrared excess and the existence in their spectrum of emission lines of e.g. [O I], [S II], [Fe II]. Andrillat and Swings (1976) give a table of the identification of the main emissions although some strong lines remain unidentified (e.g. λ 9999 Å): the strongest features are due essentially to lines of the Paschen series, of the Ca II triplet, and of He I λ 10 830 Å, O I λ 8446 Å, [S III] λλ 9069 and 9532 Å, and [Fe III] λ 10 540 Å. The Ca II triplet is strong when O I λ 8446 Å is weak, and vice-versa (except for Minkowski's footprint M1-92, studied also by Herbig, 1975). It is to be mentioned that spectra of a few stars revealing virtually no infrared excess were obtained with the same equipment: no emission line was detected in 88 Her (B8q; $H-K \sim 0$), BD +61°40 (B2e; $H-K \sim 0.2$) and BD $-$11°4747 (Be[]; $H-K \sim 0.2$). On the contrary it was shown above that the richness in emission lines becomes remarkable for the objects whose $H-K$ index becomes important and especially for those with $H-K$ greater than unity. Therefore the value $H-K \sim 1.0$ which had been considered by Allen and Swings (1972b) as a limit between objects with weak- (free-free) and strong- (dust thermal radiation) infrared excess appears in

Andrillat and Swings (1976) to distinguish the objects with rich emission spectra (groups 2 and 3 in Allen and Swings, 1976) from the others (group 1).

4. Infrared Observations of Be Stars in the Future

Although it is clear that the sophisticated infrared photometer that may be put on board the Large Space Telescope would provide valuable information, it is my impression that the best piece of equipment to rely upon in the coming decade will be the Large Infrared Telescope to be put on Spacelab (L.I.R.T.S.). Thus, to conclude this presentation, I shall simply reproduce an excerpt of the abstract of the report on the mission definition study for the L.I.R.T.S. (Jennings *et al.*, 1974). "Developments in infrared techniques during the last few years have resulted in the discovery of a very high number of astronomical objects which radiate predominantly at infrared wavelengths. These are generally cool objects at temperatures between 3 K and 3000 K, and range from the planets of our own solar system to external galaxies. Infrared observations are providing valuable information also about objects which are intrinsically much hotter (e.g., early type stars) but whose energy is degraded by visibly obscuring dust clouds. The peak emission from many important classes of object, however, lies in the 20 μ–300 μ region which is inaccessible to ground-based observations due to the opacity of the Earth's atmosphere. Information in this region of the spectrum has been limited so far to mainly photometric data obtained with relatively small telescopes flown on aircraft, balloons and rockets. The future development of this very exciting field requires high sensitivity measurements, with high spatial and spectral resolution, and thus the use of large telescopes operating in space, free from the restrictions imposed by the selective atmospheric absorption. The advent of Spacelab in the next decade offers just such a possibility to infrared astronomy. One of the major infrared facilities required on Spacelab is conceived to be a large uncooled telescope which could be used for a wide range of infrared observations. Its unique advantages, however, lie in its capability of carrying out high sensitivity photometry with high spatial resolution and of studying a wide range of astrophysical problems through measurements of atomic and molecular lines in the far infrared. The telescope considered is a 2–3 meter classical Cassegrain-type configuration with uncooled optics. Modulation is provided by rocking the secondary mirror. Its suggested focal plane instrumentation consists of a multi-band photometer, a polarimeter, a Michelson interferometer, and a heterodyne receiver allowing for multi-band mapping and polarimetry, and high and very high resolution spectroscopy. Compared to existing telescopes on ground, balloons and aeroplanes, this telescope will provide more than one order of magnitude greater sensitivity and, most important, a higher spatial resolution in the far infrared".

References

Abell, G.: 1975, *Bull. Am. Astron. Soc.* **7**, 296.
Allen, D. A.: 1973, *Monthly Notices Roy. Astron. Soc.* **161**, 145.

Allen, D. A. and Swings, J. P.: 1972a, *Astrophys. J.* **174**, 583.
Allen, D. A. and Swings, J. P.: 1972b, *Astrophys. Letters* **10**, 83.
Allen, D. A. and Swings, J. P.: 1976, *Astron. Astrophys.* **47**, 293.
Andrillat, Y. and Houziaux, L.: 1975, private communication; in preparation.
Andrillat, Y. and Swings, J. P.: 1976, *Astrophys. J. Letters* **204**, L123.
Briot, D.: 1976, this volume, p. 227.
Ciatti, F., d'Odorico, S., and Mammano, A.: 1974, *Astron. Astrophys.* **34**, 181.
Ciatti, F. and Mammano, A.: 1975a, *Astron. Astrophys.* **38**, 435.
Ciatti, F. and Mammano, A.: 1975b, communication at 20th Liège Symposium.
Gehrz, R. D., Hackwell, J. A., and Jones, T. W.: 1974, *Astrophys. J.* **191**, 675.
Herbig, G. H.: 1975, *Astrophys. J.* **200**, 1.
Jennings, R. E., Laurance, R. J., Lemarchand, A., Lemke, D., Manno, V., Moorwoord, A. F. M., Salinari, P., Swings, J. P., and Winnewisser, G.: 1974, E.S.R.O. report on Large Infrared Telescope on Spacelab, MS(74)24.
Merrill, P. W. and Burwell, C. G.: 1933, *Astrophys. J.* **78**, 87.
Merrill, P. W. and Burwell, C. G.: 1943, *Astrophys. J.* **98**, 153.
Merrill, P. W. and Burwell, C. G.: 1949, *Astrophys. J.* **110**, 387.
Morgan, T. H. and Potter, A. E.: 1975, private communication.
Peters, G. J.: 1974, *Bull. Am. Astron. Soc.* **6**, 456.
Peters, G. J.: 1975, private communication and Abell (1975).
Swings, J. P.: 1973a, *Astron. Astrophys.* **26**, 443.
Swings, J. P.: 1973b, Introductory Report at 1972 Liège Symposium, *Mém. Soc. Roy. Sc. Liège*: Série 6, Tome V, 321.
Swings, J. P. and Allen, D. A.: 1971, *Astrophys. J. Letters* **167**, L41.
Woolf, N. J., Stein, W. A., and Strittmatter, P. A.: 1970, *Astron. Astrophys.* **9**, 252.

DISCUSSION

Polidan: (1) Regarding the *HKL* color diagram and free-free emission: Three stars in Allen's list (AX Mon, HD 218393, and HR 894) are all listed as showing only a free-free excess. Yet in each case we see a K-type secondary at λ 8500. Indeed, in AX Mon the K-type star contributes over half the light at this point. From this one would conclude that one cannot distinguish between a K-type companion and free-free emission. (2) Regarding the Andrillat and Houziaux statistics on Ca II emission, we have extended the survey to later type Be stars and find the same percentages as for the early types. So apparently no correlation exists between Ca II triplet emission and spectral type.

Swings: With regard to your first comment, this fact was noticed by Swings and Allen in their near infrared photometric survey of symbiotic and VV Cephei stars (*Pub.. Astron. Soc. Pacific* **84**, 523, 1972).

PASCHEN DECREMENTS IN Be STARS

D. BRIOT

Observatoire de Paris, France

Abstract. We searched for the general properties of the Be stars with Paschen emission lines. First, we obtained a relation between the infrared excess of the Be stars and the presence of emission in the Paschen lines. Until now, these emission excesses could be related to no physical characteristics of classical Be stars. Then, the measures of Paschen decrements of 12 stars whose spectral types range from B0e to B5e allowed us to check several theoretical calculations about the formation of emission lines in the envelopes of Be stars. Thus we can see the prominent part played by the electronic collisions in the Sobolev theory. Indeed, only with calculations taking the electronic collisions into account, can we obtain theoretical values agreeing with both measured Paschen and Balmer decrements for the hottest stars of our sample. However, no theoretical values agree with the observed decrements for the cooler stars.

A. Slettebak (ed.), Be and Shell Stars, 227. All Rights Reserved
Copyright © 1976 by the IAU.

THE CLASSIFICATION OF FAINT Be STARS

DAVID A. ALLEN

Royal Greenwich Observatory, Anglo-Australian Observatory, Epping, New South Wales, Australia

(Paper read by J. P. Swings)

Abstract. Faint emission-line stars found on Hα surveys are usually classified as planetary nebulae or Be stars on rather arbitrary criteria. A comparison of spectra of such stars with their optical and infrared properties suggests that a next step in their classification can be made quite rapidly by photometry. In particular the more unusual specimens can be readily isolated.

Paul Merrill was the first person to engineer a systematic Hα survey of the northern sky as a means of identifying Be stars. That his technique was successful is beyond question; the three volumes of the MWC (Merrill and Burwell, 1933, 1943, 1949) testify to that.

But Merrill was bedevilled by one problem: how to distinguish his cherished Be stars from the chaff of emission-line objects that litter the sky. Nebulous objects – compact H II regions and large planetary nebulae – he could hope to isolate simply by their extended nature; stellar objects – T Tauri stars, symbiotic stars, Me stars, compact planetary nebulae, Wolf-Rayet stars etc. – could be distinguished only spectroscopically. On the original objective-prism plates this was feasible, and was indeed undertaken, for the brightest sources, but it became increasingly difficult as the plate limit was neared. Merrill's solution was to publish a list of additional stars (the AS; Merrill and Burwell, 1950) about which he could say no more than that they had Hα emission. This is an admirable approach which has subsequently been shunned by the majority of Hα surveyors who seem to prefer risking errors of commission to making admissions which might be construed to imply inadequacy. Thus, rather than hiving off a 'don't know' group *qua* the AS, most recent Hα surveys have turned up only 'Be stars' or 'planetary nebulae'. The reasoning seems to be either: I see a continuum, ergo this is a Be star or: I see no continuum; this object *has* no continuum, ergo it is a planetary nebula. The outcome of this approach is to cram the Be star lists with T Tauri and even later-type stars, and the planetary nebula catalogues with the most bizarre assortment of objects, including even some M stars with pure absorption spectra (Allen and Fosbury, 1975).

Even the impeccable MWC includes a number of stars Merrill chose to classify as Be but which could now better be described as forbidden-line stars. I refer to objects like MWC 17 (Swings and Struve, 1941), MWC 342 (Swings and Struve, 1943), MWC 645 and MWC 819 (Swings and Allen, 1973) in which low-excitation forbidden lines dominate the emission spectrum and there is a strong continuum offering little or no evidence of an underlying B-type star. The study of these enigmatic objects would be greatly benefited by their isolation from Be catalogues, but this was not always possible from the original Mount Wilson plates.

A problem therefore exists. The promulgators of objective prism surveys in the main seem unable or unwilling to classify their discoveries into more than two loosely defined categories. To do so spectroscopically requires an inordinate amount of

A. Slettebak (ed.), Be and Shell Stars, 229–231. All Rights Reserved
Copyright © 1976 by the IAU.

telescope time and offers a low yield of the frontier-pushing astronomy we are supposed to be pursuing. How should we proceed?

A solution offered here is photometry which, by dint of its inherently wider bandpass, is considerably faster than spectroscopy and can be performed on a smaller telescope. Considering first the optical, filter combinations which have already proved successful and which could serve usefully in the classification of emission-line stars include the Hβ/[O III] and the TiO/CaH sets. The first of these combinations was used by Webster (1966) to examine the stellar planetary nebulae of Henize's (1967) survey. On this basis she isolated 15 which had at best only very weak [O III] emission and which therefore merited at least a more thorough examination before they were accepted as genuine planetary nebulae. Two thirds of these have since been shown not to be planetary nebulae, and several of them would more properly be classed as Be stars. An Hβ/[O III] photometric survey of all the objects classified as stellar or compact planetary nebulae would therefore aid in cleaning up this group. Late type stars – M, Me and symbiotic stars – can be identified by the TiO/CaH photometry practiced by Jones (1973), or by similar filter combinations. Many emission-line stars are quite heavily reddened, so working in the red has obvious advantages. Since late-type stars are present in both samples, all faint emission-line stars found by objective prism surveys would need to be examined.

1–4 μm infrared photometry offers another, and perhaps a superior, means of separating the various emission-line objects (Allen and Swings, 1972). At the time of writing, most infrared observations of emission-line stars have been made with PbS cells; the advent of InSb, which is nearly an order of magnitude more sensitive, will considerably speed up this type of work. The standard Johnson filters at 1.25 (J), 1.65 (H), 2.2 (K) and 3.5 μm (L) have been used. At these wavelengths three types of continuum can be identified, and these are listed in Table I.

TABLE I

The three varieties of infrared continua in emission-line stars

Type	Typical colour indices (magnitudes)				Emission-line objects represented
	$J-H$	$H-K$	$K-L$	$V-K$	
i Blue continuum, stellar or free-free	≤0.3	≤0.4	≤0.6	~2	Be stars; planetary nebulae; most Wolf-Rayet stars
ii Late-type star	1.0	0.4	0.4	3–15	M, Me stars; most symbiotic stars; VV cephei stars except that $K-L$ is usually larger. Stars of type (i) if $A_V \simeq 10$ mag.
iii Dust emission at colour temperatures 700–1500 K	Colours depend on dust temperature		>0.6	>0.9	Some dense planetary nebulae; some symbiotic stars; T Tauri stars; forbidden-line stars; compact H II regions; pre-main-sequence Ae and Be stars

The technique so far adopted (Allen, 1974; Allen and Glass, 1974, 1975) was to integrate for 5 minutes at 2.2 μm using a PbS detector on a 1 or 1.2-m telescope. In this time the majority of the late-type and dust emission objects will have been detected. Other wavelengths may then be tackled to derive colours. A one-minute integration with an InSb cell would go deeper, and hence find the fainter late-type and dust emission stars. Five minutes integrating on a similar aperture telescope would reach the continuum of many of the faint Be stars and planetary nebulae. Thus infrared photometry offers a particularly rapid diagnostic for faint emission-line stars and serves to isolate the more unusual and peculiar specimens for further study. The data in Table II indicates the numbers of misclassified objects isolated by the author

TABLE II

Proportion of type (ii) and type (iii) infrared continua amongst emission-line stars fainter than about 13th magnitude

	Total sampled	% (ii)	% (iii)
Classed as Be	85	32	36
Classed as planetary nebulae	376	21	10

in his various infrared surveys of Be stars and compact planetary nebulae, and illustrates the magnitude of the misclassification problem. The use of InSb detectors on the sources undetected with PbS cells would probably lead to an increase in the percentages of misclassified sources, especially amongst planetary nebulae.

References

Allen, D. A.: 1974, *Monthly Notices Roy. Astron. Soc.* **168**, 1.
Allen, D. A. and Fosbury, R. A. E.: 1975, *Observatory* **95**, 15.
Allen, D. A. and Glass, I. S.: 1974, *Monthly Notices Roy. Astron. Soc.* **167**, 331.
Allen, D. A. and Glass, I. S.: 1975, *Monthly Notices Roy. Astron. Soc.* **170**, 579.
Allen, D. A. and Swings, J. P.: 1972, *Astrophys. Letters* **10**, 83.
Henize, K. G.: 1967, *Astrophys. J. Suppl.* **14**, 125.
Jones, D. H. P.: 1973, *Monthly Notices Roy. Astron. Soc.* **161**, 19P.
Merrill, P. W. and Burwell, C. G.: 1933, *Astrophys. J.* **78**, 87.
Merrill, P. W. and Burwell, C. G.: 1943, *Astrophys. J.* **98**, 153.
Merrill, P. W. and Burwell, C. G.: 1949, *Astrophys. J.* **110**, 387.
Merrill, P. W. and Burwell, C. G.: 1950, *Astrophys. J.* **112**, 72.
Swings, J. P. and Allen, D. A.: 1973, *Astrophys. Letters* **14**, 65.
Swings, P. and Struve, O.: 1941, *Astrophys. J.* **93**, 349.
Swings, P. and Struve, O.: 1943, *Astrophys. J.* **97**, 194.
Webster, B. L.: 1966, *Publ. Astron. Soc. Pacific* **78**, 136.

POLARIZATION IN Be STARS

(*Review Paper*)

GEORGE V. COYNE, S.J.

University of Arizona, Tucson, Ariz., U.S.A.

and

Vatican Observatory, Vatican City State

Abstract. A review of polarization produced in the extended circumstellar disks about Be stars is given. While variability of the polarization and its peculiar wavelength dependence in wide-band continuum measurements are discussed, emphasis is placed upon the recently discovered polarization effects in the emission lines and upon a discussion of models. Scattering in the photosphere of a star, even one rigidly rotating near breakup velocity, does not explain the observations. Better fits to the observations are obtained by scattering from electrons in a flattened gaseous disk extending some 10 stellar radii from the star with electron temperatures $T_e \cong 10\,000$ K and electron densities $N_e \cong 10^{12}$ cm^{-3}. In some stars the emission flux in Hα and Hβ appears to be partly polarized.

1. Introduction

One of the ways in which linear polarization can be produced is by the scattering of light from particles which are asymmetrically distributed as seen by the observer. The emission lines observed in the spectra of classical Be stars are said to originate in a flattened gaseous envelope extending some 10 stellar radii out from the photosphere of a rapidly rotating hot star. Thus one might expect to detect linear polarization in these systems as the light from the central star is scattered to the observer from this flattened gaseous disk. One might then employ the polarimetric observations as a means of studying the nature of the extended gaseous envelope about these stars. While we now realize, indeed, that the polarization produced in these envelopes is a powerful means of studying the nature of the envelopes themselves, the original detection of intrinsic polarization in these stars did not come about in such a reasoned, systematic way.

In fact, the first prediction of intrinsic stellar polarization stimulated a search which resulted in the detection of the interstellar polarization. Chandrasekhar (1946) predicted that the linear polarization could be as high as 11% at the limb of a star with a plane-parallel atmosphere in which there was pure electron scattering. In a search for such polarization in O and B-type stars Hall (1949) and Hiltner (1949) independently discovered the interstellar polarization.

It was about ten years later, as a matter of fact, that the first detection of intrinsic stellar polarization in Be stars was made by Behr (1959) who suspected variable polarization in the shell star, γ Cas. On the other hand, the first hint that intrinsic stellar polarization had been observed may be found already in Hall and Mikesell (1950) where, at the end of their catalog, a comment is made about the unexpectedly large polarization in ζ Tau, considering its color excess.

Intrinsic polarization in stars can be detected in four ways: (1) variability of the polarization; (2) variation of the polarization position angle with wavelength; (3)

A. Slettebak (ed.), Be and Shell Stars, 233–260. All Rights Reserved

peculiar $p(\lambda)$ for the continuum radiation; (4) changes in the polarization across discrete absorption or emission features. Historically, the first detections of intrinsic polarization were due to observed variations and observational progress was made more or less in the sequence of the four criteria listed above.

Subsequent to Behr's (1959) discovery of variable polarization in γ Cas, Shakhovskoi (1962) found variable polarization for χ Oph and later (Shakhovskoi, 1964) for other Be stars. Still other early-type stars were found to show variable polarization by Vitrichenko and Efimov (1965) and Coyne and Gehrels (1967).

Serkowski (1968) first indicated the peculiar nature of the wavelength dependence of the polarization for stars with extended atmospheres, although a similar $p(\lambda)$ had already been observed for the recurrent nova T Pyxidis by Eggen *et al.* (1967) and for β Lyrae by Appenzeller and Hiltner (1967).

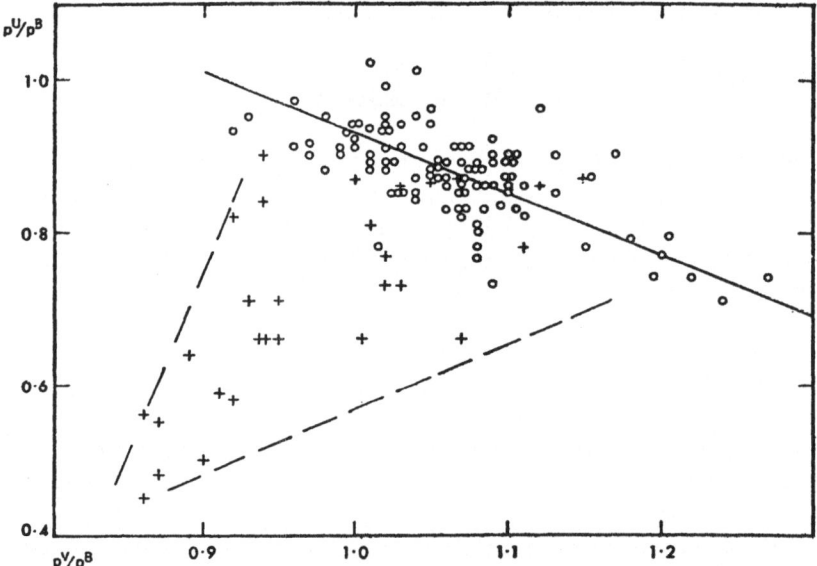

Fig. 1.   A plot of polarization ratios for stars which represent the interstellar polarization (open circles) and for Be stars (crosses).

Figure 1 (from Serkowski, 1970) shows a plot of the polarization ratios, ultraviolet-to-blue vs visual-to-blue, for stars which represent the interstellar polarization (open circles) and for emission-line stars (crosses). The emission-line stars have a systematically lower ultraviolet polarization than that for the interstellar medium. Coyne and Kruszewski (1969) also observed this effect and in addition showed that the polarization in the Be stars, as compared to that for the interstellar medium, increased longwards of the Paschen limit. Serkowski (1970) verified this and in Figure 2 his observations of two Southern Be stars p Car (full circles) and α Ara (open circles) are shown contrasted to $p(\lambda)$ curves for the interstellar medium (dashed lines). Here are seen both the decrease in the ultraviolet polarization and the increase beyond the Paschen limit.

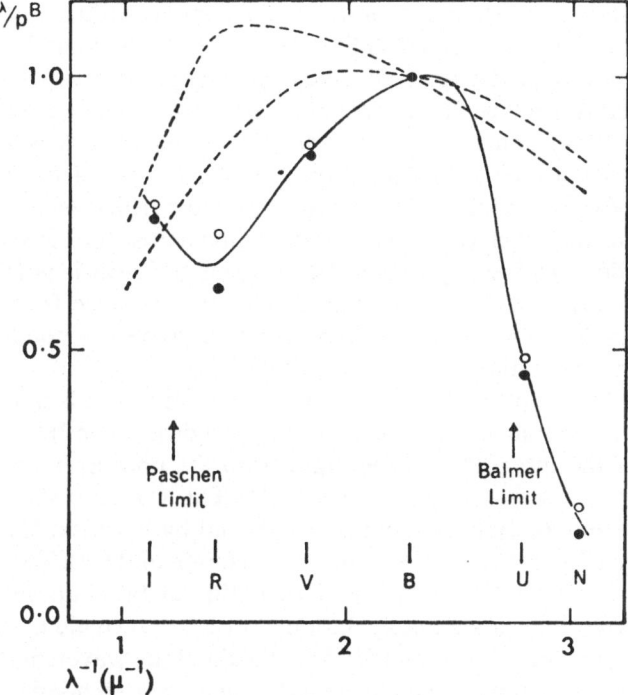

Fig. 2.   Polarimetric observations of two Be stars (p Car and α Ara) are shown by solid line; these are
compared to two representative curves for the interstellar medium shown by dashed lines.

As we shall see subsequently (Section 2) it is not always easy to separate out the
effects of the interstellar and intrinsic polarization. Serkowski (1970) was able to
show the diminishing effects of interstellar polarization on a relatively large group of
early-type stars which he studied. His results are shown in the polarization-ratio plot
of Figure 1 where the apparent apex of the triangular region confining the emission-
line stars represents the intrinsic polarization least contaminated by the interstellar
polarization. The values for α Ara and p Car, plotted in Figure 2, lie near this apex.
Coyne (1971) observed a group of Be stars at high galactic latitudes in seven colors
and his $p(\lambda)$ curve together with that of Serkowski can be considered characteristic of
the $p(\lambda)$ for the intrinsic polarization in emission-line stars. Indeed, there are
differences in the intrinsic $p(\lambda)$ from star to star, but we shall discuss this in Section 2.

To date about 50% of the Be and shell stars observed polarimetrically show
intrinsic polarization detectable either because the polarization is variable or
because the $p(\lambda)$ differs from that for the interstellar medium.

The most recent developments in the polarimetry of Be stars have been observationally, the detection of changes in the polarization across the emission lines and
theoretically, the development of more detailed models to explain the observed
polarizations. Zellner and Serkowski (1972) report an unpublished observation by
Serkowski of a decrease of polarization across the Hβ line of ζ Tau. Actually Öhman
(1934) noted that one wing of Hγ in β Lyr showed polarization dependent on the
phase of the eclipsing binary. Öhman (1946) also predicted that the Chandrasekhar

effect could cause a variation of as much as 0.8% polarization across an absorption line Doppler-broadened by rapid stellar rotation. Later Sen and Lee (1961) made a similar prediction. But the first published systematic observations of polarization effects across emission lines were made recently, first by McLean (1974) and Clarke and McLean (1974) and then by several others whose work will be reviewed in Section 4. Observations of polarization effects in spectral lines by Clarke and Grainger (1966), by Derviz (1970) have not been verified (see Hayes and Illing, 1974). The detection of changes in the polarization across the emission lines of some Be stars provides conclusive proof of the presence of intrinsic polarization in these stars and in many cases allows one to isolate the intrinsic from the interstellar polarization and to estimate the electron distribution in the extended envelope. We shall discuss this in Section 4.

As for models to explain the polarimetric observations, it is unlikely that photospheric effects contribute in any significant way. Nagirner (1962) first showed that the polarization of the integrated visible light from the photosphere of a rotationally distorted early-type star does not exceed 0.1%. This was later shown to be true in a more sophisticated analysis by Collins (1970) and by Ruciński (1970). It has been recently shown (Cassinelli and Haisch, 1974; Haisch and Cassinelli, 1976) that the polarization from early-type stars rigidly rotating at breakup velocities does not exceed 0.3% and that only a highly flattened disk can provide polarization greater than 1.0%. Coyne and Kruszewski (1969) explained the wavelength dependence of the polarization in Be stars by electron scattering in the extended envelope together with hydrogen absorption, a model suggested by Appenzeller and Hiltner (1967) for β Lyr. A more detailed and quantitative model in which free-free emission also is significant has been given by Capps *et al.* (1973). These will be discussed in more detail in Section 5.

Reviews of polarization in circumstellar envelopes have been published by Zellner and Serkowski (1972) and by Kruszewski (1974). Both of these covered a wide variety of stars. While the same physical processes may explain the polarization observed in a wide variety of objects, there are specific characteristics of the polarization found in early-type emission-line stars. The aim of this review is to emphasize both those peculiar characteristics and the more recent developments, specifically the polarization changes across emission lines.

In separate sections we shall treat of the wavelength dependence of the continuum polarization (Section 2), the variability of the polarization (Section 3), the polarization changes across emission lines (Section 4) and models (Section 5). A summary of all available data on the variability of polarization in Be stars is given in Table I of Section 3, and in Section 4, Table II summarizes available information on all stars for which polarimetry in the Hα and Hβ emission lines has been measured.

2. Wavelength Dependence of the Polarization in the Continuum Radiation of Be Stars

The peculiar characteristics of the wide- and intermediate-band polarimetry in the visual regions of the spectrum of Be stars are: (1) a more rapid decrease in the

TABLE I

Summary of data for Be and shell stars which show variable polarization

HD[a] (1)	Star (2)	l^{II} (3)	b^{II} (4)	Sp (5)	$v \sin i$ (km s^{-1}) (6)	$\bar{P}_{vis}$ (%) (7)	n (8)	ΔP_{max} (%) (9)	Δt_{total} (days) (10)	Δt_{max} (days) (11)	Δt_0 (days) (12)	Sources[b] polarimetry (13)	Remarks (14)
5 394*	γ Cas	124°	−2°	B0.5IVe$_1$	300	1.08	17	0.42	2714	1433	30	2, 4, 7	shell
10 516*	φ Per	131	−11	B2Ve$_4$p	493	0.95	32	0.37	3005	2324	20	6	shell, sp. bin.
28 497*	HR 1423	209	−37	B1.5Ve$_2$	340	1.06	12	0.33	1110	608	25	10, 11	shell
37 202*	ζ Tau	186	−6	B4IIIp	310	1.50	11	0.59	2537	1416	1	1, 2, 5, 8	shell, sp. bin.
44 458*	HR 2284	220	−12	B1Ve$_2$	270	1.08	4	0.26	320	134	30	10	var. rad. vel.
45 725*	β Mon A	217	−8	B3Ve$_2$	405	0.49	3	0.19	397	397	10	10	shell, var. rad. vel.
53 179	Z CMa	225	−3	B5neq	–	0.76	14	1.24	2540	338	5	3, 10, 11	shell
91 465	p Car	287	−3	B4Vne$_2$	325	0.60	28	0.73	623	554	5	10, 11	shell
137 387	κ^1 Aps	314	−14	B1pne$_2$	–	0.87	14	0.27	646	329	15	10, 11	shell, var. rad. vel.
142 983*	48 Lib	356	+29	B5IIIp	331	0.84	23	0.27	2815	299	1	3, 4, 5, 10, 11	shell, sp. bin.
148 184*	χ Oph	358	+21	B1.5Ve$_4$	109	0.52	21	0.39	3143	784	20	3, 4, 5, 9, 11	extreme Be, sp. bin.
158 427	α Ara	341	−9	B2Vne$_2$	375	0.63	12	0.26	549	278	15	10, 11	shell, var. rad. vel.
173 948	λ Pav	334	−24	B2II–III	130	0.60	10	0.17	479	393	15	10, 11	
181 615	υ Sgr	22	−14	Apep	38	0.94	32	0.40	1841	484	5	4, 11	ec. bin.
205 637*	ε Cap	32	−45	B2.5Vpe	287	1.23	32	0.38	1967	471	15	4, 5, 9, 10, 11	shell
212 571*	π Aqr	66	−45	B1Ve$_1$	278	0.90	25	0.39	1914	1908	30	4, 5, 9, 10, 11	shell

[a] An asterisk (*) indicates stars for which polarization has been measured across emission lines (see Table II). The following stars also show variable polarization (sources in parentheses) but the observations are too few to include them in this table: EW Lac (3, 5), AX Mon (10), HD 58978 (5, 9, 10), HD 83953 (10), δ Cen (10), HD 100198 (11), HD 113120 (11); for the following stars the variations in polarization are doubtful: κ Dra (5), o Aqr (5, 9, 11), HD 60855 (5, 10), HD 158303 (11). For β Lyrae periodic polarization changes correlated with the light curve are already well understood (Appenzeller and Hiltner, 1967; Coyne, 1970).

[b] Sources for polarimetry: 1. Capps, Coyne and Dyck (1973); 2. Clarke and McLean (1976); 3. Coyne (1975a); 4. Coyne (1975b); 5. Coyne and Kruszewski (1969); 6. Coyne and McLean (1975); 7. Hayes and Illing (1974); 8. Hayes (1975); 9. Serkowski (1968); 10. Serkowski (1970); 11. Serkowski (1975).

TABLE II

Data for Be stars with measured polarization in Hα and Hβ

HD[a] (1)	Star (2)	l^{II} (3)	b^{II} (4)	Sp (5)	Shell (6)	$v \sin i$ (km s^{-1}) (7)	P_B^b (%) (8)	$P_{H\beta}/P_B$ (9)	P_R^b (%) (10)	$P_{H\alpha}/P_R$ (11)	Source[c] polarimetry (12)	Remarks (13)
5 394*	γ Cas	124°	−2°	B0.5IVe$_1$	yes	300	0.97$^+$	0.76	0.61$^+$	0.52	1, 2, 4, 6, 7, 8	
6 811	φ And	126	−16	B8e		71	0.97	1.04	1.01	0.83:	4	sp. bin.
10 516*	φ Per	131	−11	B2Ve$_4$p	yes	493	1.84$^+$	0.72	1.58$^+$	0.55	5, 8	sp. bin.
20 336	HR 985	138	+7	B2.5Ve$_1$n	yes	343	–	–	0.51	0.82:	8	sp. bin.
22 192	ψ Per	149	−6	B5Ve$_2$	yes	390	0.58$^+$	0.64	0.38$^+$	0.48	4, 8	
23 862	28 Tau	167	−23	B8ne	yes	337	0.40$^+$:	1.00:	–	–	4	var. rad. vel.
25 940	48 Per	154	−3	B3Ve$_{1+}$		233	0.92	1.06:	0.89	1.03	2, 6, 7, 8	extreme Be star
28 497*	HR 1423	209	−37	B1.5Ve$_2$	yes	340	–	–	0.36	1.31:	4, 8	
30 076	56 Eri	206	−32	B2Ve$_2$		240	0.47	0.66	0.71	0.17:	4, 8	pole on
32 991	105 Tau	181	−11	B2Ve$_3$		220	–	–	1.50	0.91	8	extreme Be star
37 202*	ζ Tau	186	−6	B4IIIp	yes	310	1.45$^+$	0.65	1.04$^+$	0.27	2, 3, 4, 6, 7, 8	sp. bin.
37 490	ω Ori	201	−14	B2IIIe$_1$		191	–	–	0.38	0.38	8	
41 335	HR 2142	214	−14	B2Ve$_{3+}$n	yes	414	–	–	0.37:	0.19:	8	sp. bin.
44 458*	HR 2284	220	−12	B1Ve$_2$		270	–	–	0.98:	0.57:	8	sp. bin.
45 725*	β Mon A	217	−8	B3Ve$_2$	yes	405	–	–	0.62	0.27	7, 8	var. rad. vel.
45 726–7	β Mon BC	217	−8	B3Ve+B3Ve		360	–	–	0.43:	–	8	
45 995	HR 2370	201	+1	B2.5Ve$_{2+}$		320	–	–	0.43:	0.00	8	
56 014	27 CMa	239	−7	B3IIpe	yes	200	–	–	0.41:	0.57:	8	sp. bin.
58 978*	HR 2855	237	−3	B0.5IVnpe$_1$	yes	260	–	–	1.01:	0.28	8	He shell
60 848	BN Gem	203	+18	O8V:pe		362:	0.56$^+$	0.79	–	–	4	
142 983*	48 Lib	356	+29	B5IIIp	yes	331	0.86$^+$	0.82	0.94$^+$	0.68	4, 7, 8	sp. bin.
148 184*	χ Oph	358	+21	B1.5Ve$_4$		109	0.54$^+$	1.00	0.50$^+$	1.01	4, 7, 8	extreme Be star, sp. bin.
164 284	66 Oph	31	+13	B2Ve$_1$	yes	200	1.06	0.86	0.81	0.77	4, 7, 8	two spectra
178 175	HR 7249	17	−12	B2Ve$_1$		220	–	–	0.54	0.94	8	sp. bin.
183 656	V923 Aq1	40	−7	B6:p	yes	320	–	–	1.12	0.78:	8	

Table II (Continued)

HD[a] (1)	Star (2)	l^{II} (3)	b^{II} (4)	Sp (5)	Shell (6)	$v \sin i$ (km s^{-1}) (7)	P_B^{b} (%) (8)	$P_{H\beta}/P_B$ (9)	P_R^{b} (%) (10)	$P_{H\alpha}/P_R$ (11)	Source[c] polarimetry (12)	Remarks (13)
191 610	28 Cyg	74	+2	B2.5V		310	0.53	0.89	0.66	0.41	4, 8	var. rad. vel.
193 182		77	+2			200	–	–	0.57	0.51	8	
193 911	25 Vul	65	–7	B1IVe	yes	250	–	–	0.31	0.26:	8	
195 325	1 Del	55	–16	B8:	yes	320	–	–	0.33	1.6:	8	
200 120	59 Cyg	88	+1	B1.5Ve$_2$nn	yes	450	–	–	0.83	1.00	8	sp. bin.
200 310	60 Cyg	87	0	B1Vn	yes	320	–	–	0.40	0.53:	8	
203 467	6 Cep	103	+11	B3IVe$_1$		150	–	–	0.65	0.98	8	sp. bin.
205 637*	ε Cap	32	–45	B2.5Vpe	yes	287	1.09	0.73	0.95^{+}	0.55	4, 7, 8	sp. bin.
208 682	HR 8375	106	+9	B2.5Ve$_2$	yes	350	–	–	1.07	0.51	4, 7, 8	
209 409	o Aqr	57	–43	B7IVe$_1$	yes	317	0.42	0.98	0.63	0.66	4, 7, 8	
212 571*	π Aqr	66	–45	B1Ve$_1$	yes	278	1.52	0.62	0.96	0.39	4, 7, 8	sp. bin.
217 050*	EW Lac	104	–10	B4IIIe$_1$p	yes	358	1.97	0.70	1.42^{+}	0.41	4, 7, 8	sp. bin.
217 675	o And	102	–16	B6p	yes	333	–	–	0.46	1.00	4	

[a] An asterisk indicates stars for which the polarization is variable (see Table I). For the following stars the measured continuum polarization is <0.30% with no detected changes across the emission lines: 11 Cam, 25 Ori, ν Gem, ω CMa, κ Dra, HD 174237, HD 189687, HD 192044, V568 Cyg, 16 Peg, 8 Lac.

[b] A cross (+) indicates values for which the interstellar polarization is either negligibly small or has been corrected for.

[c] Sources for polarimetry 1. Clarke and McLean (1974); 2. Clarke and McLean (1976); 3. Coyne (1974); 4. Coyne (1975a); 5. Coyne and McLean (1975); 6. McLean (1976); 7. Poeckert (1975a); 8. Poeckert (1976).

polarization towards the ultraviolet region of the spectrum; (2) a minimum polarization just to the blue side of the Paschen limit and an increase in the polarization to the red side of the Paschen limit. A few observations have been made at near infrared wavelengths (between 1 and 2.2 μm) and these indicate another decrease in the polarization after a maximum is reached in the $\lambda = 1$ μm region.

The multicolor polarimetry done by Serkowski at the Mount Stromlo and Siding Spring Observatories of the Australian National University (e.g. Serkowski, 1968, 1970) and by a group at the Lunar and Planetary Laboratory of the University of Arizona (e.g. Coyne and Gehrels, 1967; Coyne and Kruszewski, 1969; Coyne *et al.*, 1974) have provided the bulk of data on $p(\lambda)$ for the Be stars. Serkowski (1968) first noted the decrease in the ultraviolet polarization and remarked on the similarity to the polarization observed in several supergiant stars, in β Lyrae, and in the recurrent nova T Pyxidis. The discovery by Serkowski and the prediction by Harrington and Collins (1968) of considerable polarization produced in the photospheres of rapidly rotating early-type stars led Coyne and Kruszewski (1969) to an analysis of multicolor polarimetry accumulated over a period of several years on 19 early-type stars with extended atmospheres. While considerable differences were seen from star to star (see Figure 1 of Coyne and Kruszewski 1969) eight of the stars observed showed the typical $p(\lambda)$ curve illustrated at the bottom of Figure 3. Here the mean

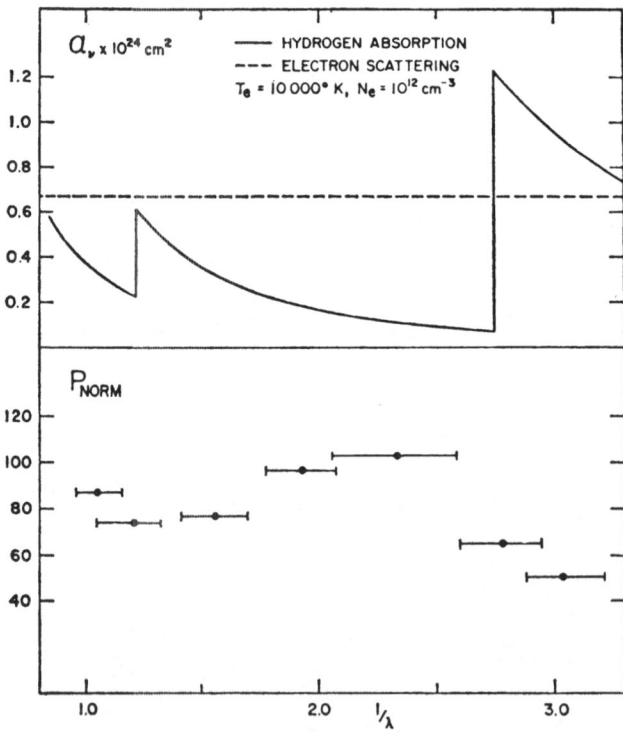

Fig. 3.   At the bottom the mean normalized polarization for eight Be stars is plotted (the horizontal lines indicate the passbands); at the top the opacities for hydrogen absorption and electron scattering are compared. The polarization and hydrogen absorption are inversely related.

polarization over the eight stars is given (after normalizing by equating to 100% the average polarization at the two intermediate filters). The horizontal bars indicate the passband. This curve indicated that hydrogen absorption played a significant role in the observed $p(\lambda)$. The hydrogen absorption coefficient is plotted as the solid line in the upper part of Figure 3 and the anti-correlation with the $p(\lambda)$ curve at the bottom is obvious. This suggested that polarization produced by electron scattering (independent, therefore, of wavelength) but modified by absorption in a hydrogen plasma both before and after scattering was responsible for the observed wavelength dependence. To produce such polarization it was necessary to have a hydrogen plasma characterized by such parameters as to make the electron-scattering opacity dominant at some wavelengths and the hydrogen absorption opacity dominant at others. At the top of Figure 3 a comparison of the two sources of opacity is given and we see that the condition is fulfilled for an electron temperature $T_e = 10\ 000$ K and an electron density $N_e = 10^{12}$ cm^{-3}. It is also necessary, of course, that the scattering optical depth be sufficient to give the observed amount of polarization. We shall see in Section 5 that the path length required in the flattened hydrogen-plasma disk is about 3 to 10 stellar radii, consistent with what is otherwise known about these disks.

The rise of the polarization to the red side of the Paschen limit makes one curious about the polarization in the infrared region of the spectrum. Although such observations are difficult, since Be stars are relatively faint in the infrared, polarimetric observations of Be stars have been extended to 2.2 μm by Dyck (Capps *et al.*, 1973) for ζ Tau and by Vrba (Coyne and McLean, 1975) for ϕ Per. In both cases the polarization decreases in the infrared to a value at 2.2 μm which is about one-half of that at 1 μm where a maximum occurs just longwards of the Paschen limit. In the more quantitative treatment of the electron-scattering disk model (Capps *et al.*, 1973) it is actually free-free emission which mainly governs the wavelength dependence of the polarization in the infrared. We shall discuss this in more detail in Section 5.

In the discussion above we have emphasized the characteristics of the $p(\lambda)$ curve which are common to all the intrinsically polarized Be stars which we have observed. It is necessary to emphasize that there are significant differences from star to star. These differences can be due to the fact that different stars are affected by differing amounts of interstellar polarization and/or to the fact that the intrinsic polarization differs from star to star. The intrinsic differences among various Be stars are illustrated by Figure 4 where the intrinsic polarizations for four Be stars are plotted along with a model calculation (solid curve) to be discussed in Section 5. For two of the stars, γ Cas and ζ Tau, there is none or very little interstellar polarization (see Capps *et al.*, 1973; Clarke and McLean, 1974); for 48 Lib and ψ Per the interstellar polarization has been removed (Coyne, 1975a). The differences among the stars are undoubtedly due to different electron temperatures and densities in the extended atmospheres of these stars.

It is obviously of critical importance, whenever possible, to recover the intrinsic polarization from the observed polarization by removing the interstellar effects. We should, therefore, like to consider the ways in which this is done.

The interstellar polarization for these objects is not directly known. A clue to its presence is the rotation of the plane of the observed polarization with wavelength.

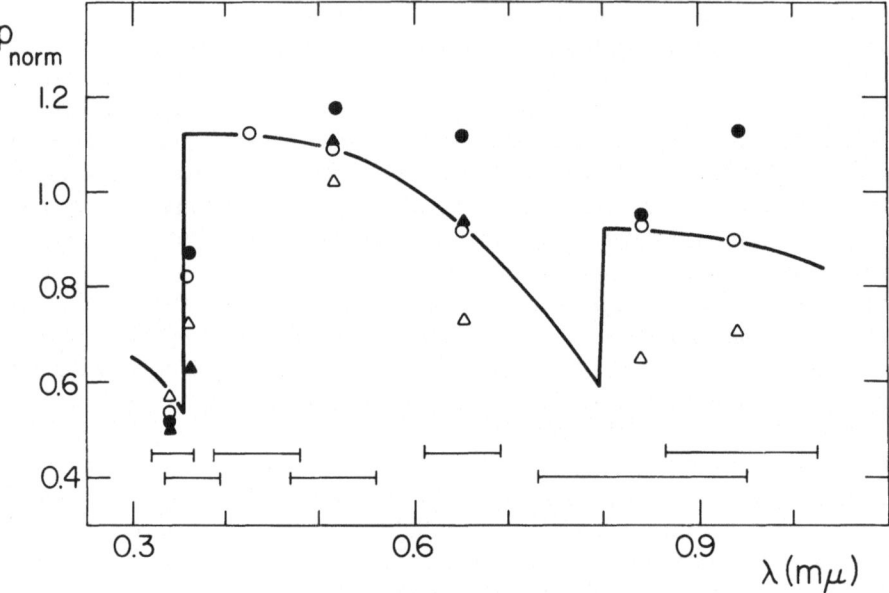

Fig. 4.   The normalized polarization for four Be stars (solid circle: 48 Lib, solid triangle: ψ Per, open circle: ζ Tau, open triangle: γ Cas) are compared to a model calculation (solid line, see Section 5).

One criterion, therefore, for indirectly determining the interstellar polarization is that, when the interstellar effects are removed from the observations, a mean plane of polarization independent of wavelength is found. The amount of interstellar polarization and its orientation may be inferred from observation of stars in the surrounding region. One is, of course, then assuming a certain homogeneity of the interstellar effects over this region, an assumption which is not always verified. The following sources are useful for determining the degree of homogeneity of interstellar polarization effects: Hall (1958); Hiltner (1956); Serkowski *et al.* (1975); Coyne *et al.* (1974). In the latter two sources the wavelength dependence of the polarization characteristic for different regions of the Galaxy is given in terms of P_{max} and λ_{max}, the maximum polarization and the wavelength at which it occurs. The interstellar polarization at various wavelengths is then given by the formula

$$p(\lambda) = p_{max} \exp\left(-1.15 \ln^2 \lambda_{max}/\lambda\right).$$

Serkowski (1970) first showed that the interstellar polarization could be determined graphically if one assumed a certain wavelength dependence for the intrinsic and for the interstellar polarization and also assumed that at any given time the position angle of the interstellar component and the intrinsic component were independent of wavelength. Since we now know that even among the Be stars the intrinsic polarization varies somewhat from star to star, it would be preferable not to assume the $p(\lambda)$ of the intrinsic polarization. For ϕ Per Coyne and McLean (1975) successfully removed the interstellar effects without making any assumptions about the intrinsic polarization except that the plane of the intrinsic polarization be independent of wavelength. The results are shown graphically in Figure 5 where for

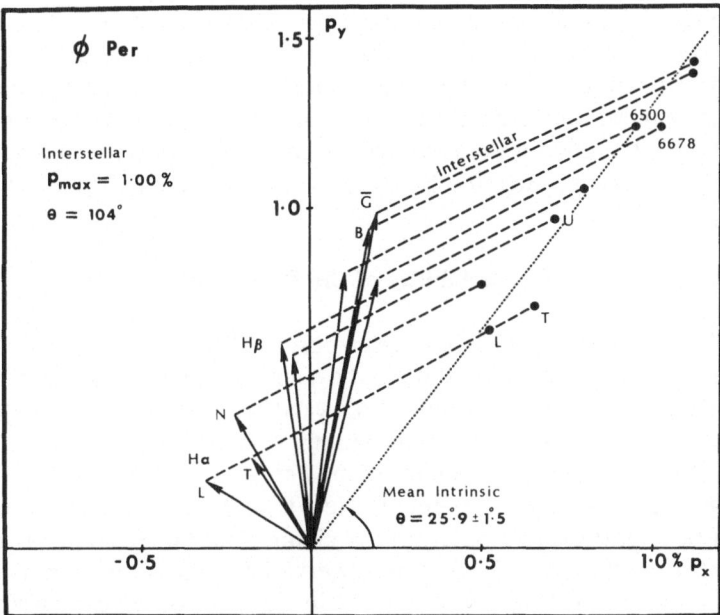

Fig. 5. The observed (arrows), interstellar (dashed line) and intrinsic (solid circles) polarization are plotted in polar co-ordinates for ϕ Per. A well defined direction for the intrinsic polarization is obtained.

ϕ Per the observed, interstellar and resultant intrinsic polarizations at various wavelengths are plotted in polar coordinates. One sees that, except for observational scatter, the resultant plane of the intrinsic polarization is independent of wavelength. The vectors marked Hα and Hβ represent polarizations measured with narrow-band interference filters centered on the emission lines. We shall discuss these in Section 4, but wish now to note the following. If the plane of the intrinsic polarization is the same for the emission line as for the continuum, then independent of any assumptions concerning the interstellar polarization the following must be true. The line which joins the line center polarization in Hα to that for the nearby continuum (λ 6500 Å in Figure 5) must be parallel to the corresponding line for Hβ (continuum measurement is B in Figure 5) and this line defines the direction of the intrinsic polarization. In Figure 5 it is clear that the two lines are indeed parallel and their position angles (26°0 and 26°3 respectively for Hα and Hβ) are exceedingly close to the mean value of 25°9 ± 1°5 derived by removing the interstellar polarization component. This strengthens the conviction that the interstellar component has been correctly estimated. In the same way Coyne (1975a) has been able to extract the intrinsic polarization for several other Be stars plotted in Figure 4. Furthermore, Poeckert (1975 a, b) has shown that if one can assume a certain relationship governing the change in the polarization across the emission line (e.g., that the change is due to unpolarized emission flux) then one can determine the magnitude and direction of the interstellar component from the line and continuum polarization alone. McLean (1976) has shown that both the interstellar and intrinsic components of the polarization can be determined from measurements at two unpolarized emission lines. He also discusses how the intrinsic polarization can be determined if the polarization is variable. It is fortunate that in many cases of Be stars we can successfully remove the interstellar effects or at least make a fair estimate of them. Otherwise an analysis of the observations would be difficult. The observed polarization in Be stars is not greater than 2% and a significant fraction of that may be due to interstellar polarization, since the Be stars are concentrated to the galactic plane.

3. Variability of the Polarization

About 40% of all Be stars which have been observed polarimetrically more than one time with mean errors less than $\pm 0.05\%$ show changes of polarization with time. It is likely that many more Be stars would show variable polarization if the polarimetric observations were made when spectral changes were known to be occurring. No such systematic study has yet been undertaken. Feinstein (1968, 1976) reports small amplitude photometric changes (generally less than $0^{m}.15$) in Be stars but no general correlation of brightness and polarimetric changes in the Be stars has been established. Several of the most frequently observed Be stars are members of binary systems. For instance, ζ Tau, ϕ Per and 48 Lib are spectroscopic binaries, υ Sgr is an eclipsing binary. All show polarimetric changes with time but there is no correlation with either the radial velocity or light curves, as the case may be, despite the fact that in the case of ζ Tau, ϕ Per and υ Sgr observations were made specifically to detect such correlations (Coyne, 1975b). Although in general no systematic study has been made of wide-band polarimetric changes in Be stars in such wise that correlations with other observational parameters, and especially spectral changes, could be sought, it would be useful from existing data to summarize what we do know of the polarimetric changes in these stars. Table I is a summary of polarimetric changes in Be and shell stars based on observations made over the past decade mainly by the Australian and Arizona groups mentioned in the Introduction. There appear to be no serious systematic effects between these two groups of observations (see Coyne *et al.*, 1974; Serkowski *et al.*, 1975). In order to avoid inhomogeneities and because, for our purposes, it was not necessary to have a longer baseline in time we have not included here the polarimetric observations antecedent to the past decade by Behr (1959), Hall (1958), and Hiltner (1956). The sources used are given in Table I. Columns (1) to (6) give the star identification, galactic coordinates, spectral type and projected rotational velocity. The remaining columns summarize the polarimetric data. The spectral types are for the most part from Lesh (1968). The values of $\upsilon \sin i$ are from Bernacca and Perinotto (1971) except for a few values from Boyarchuk and Kopylov (1964) or Uesugi and Fukuda (1970). The average polarization in the visual spectral region, $\bar{P}_{vis}$, the total number of observations in the visual spectral region, n, the maximum observed change in the visual polarization, ΔP_{max}, are given in columns (7) to (9). The visual polarizations are the 'V' observations ($\lambda_0 \cong 0.57~\mu$m) of Serkowski (Sources No. 9, 10 and 11 in Table I), the 'G' observations ($\lambda_0 \cong 0.53~\mu$m) of Coyne (Sources No. 1, 3, 4, 5, 6 in Table I) and for the few observations from other observers in Table I those observations which are closest in effective wavelength to these. For our purposes the difference in the effective wavelengths of the 'V' and 'G' filters is insignificant. The values of Δt in columns (10) to (12) are intervals of time over which respectively the *total* number of observations were made (Δt_{total}), over which the *maximum* change in polarization occurred (Δt_{max}), and over which the polarization changed by more than 0.15% (Δt_0), a value greater in general than three times the standard deviation of a single observation. The values of Δt_0 are rounded off to the nearest five days except for cases where there was clearly a change over a period of one day or less. Because the observations are so non-uniformly scattered in time, no periodicity studies are intended by these intervals of time. They give, however, an idea of the time scale over which significant polarization changes take place.

With the exception of Z CMa and p Car the amplitude of the polarization change, ΔP_{max}, is on the average 40% of the mean value of the polarization. For Z CMa and p Car the amplitude of the polarization change is larger than the mean polarization itself. For a given star the observations have extended over a time ranging from one to eight years during which in general a polarization maximum and minimum have been observed several times. The maximum observed change in polarization has been observed over a time interval significantly shorter in general than the total observing time. Merrill (1956) has shown that the shell typically appears and disappears with a period of about 8 to 10 years. The maximum changes in polarization apparently occur in a much shorter interval of time.

The typical polarization changes for the Be stars can be described as follows. There are apparently longer time scale changes over intervals of some hundreds of days upon which there are superimposed changes on a time scale of days. An example of this is Figure 6 (from Zellner and Serkowski, 1972) where the polarimetric observations in two spectral regions for p Car are plotted. This star is little, if at all, affected by interstellar polarization (Serkowski, 1970).

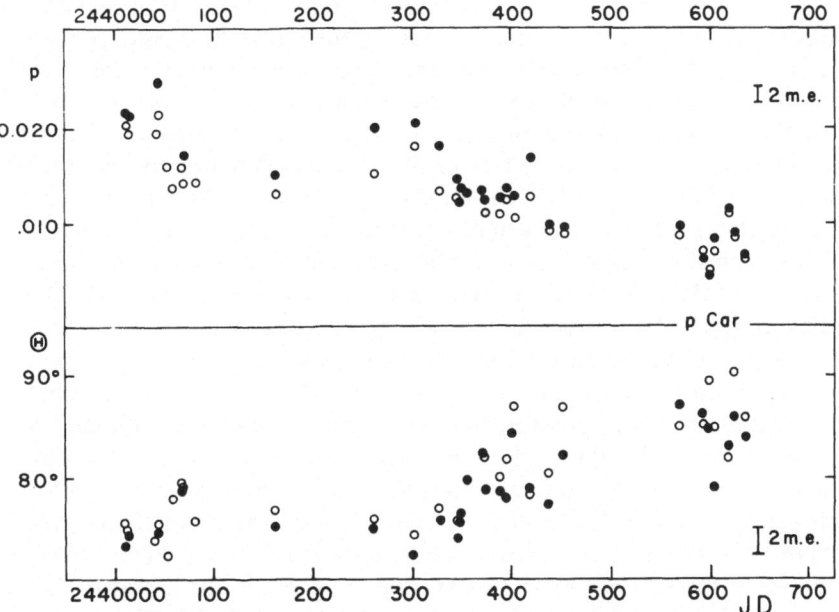

Fig. 6.   Polarization (in magnitudes) for p Car in two spectral regions is plotted vs time.

We see from column (12) of Table I that for all of these stars significant changes in the polarization take place over an interval of days. The value of Δt_0 would undoubtedly be shortened for many of these stars if more frequent and systematic observations were made. The change over a one day interval for 48 Lib is obtained from observations by Serkowski (1970) during the days JD 2440043-45. A change of $\Delta P = 0.3\%$ over an interval of 24 hours has been measured for ζ Tau by Poeckert (1975a). Clarke and McLean (1976) report a change of ~0.2% over an interval of 40 minutes for the polarization in the continuum at λ 4837 in ζ Tau.

For 10 of the stars in Table I (marked with an asterisk in column 1), narrow-band measures of the polarization across the Hα and Hβ emission lines have been made (see Table II, Section 4). All except χ Oph and HD 28497 also show changes in polarization across the emission lines. χ Oph is an extreme Be star (Schild, 1966, 1973). It is a noteworthy case and we shall return to a discussion of it in Section 4. It is difficult to see how these changes in polarization, especially the short interval ones, can be attributed to processes occurring in the stellar photosphere. They are an independent indication that the polarization must be due to processes in an extended circumstellar disk.

4. Polarization Measurements in the Hydrogen Emission Lines

Since the continuum and the emission-line flux originate at different levels in the atmosphere and extended flattened disk about Be stars, one would predict that the polarization produced in the circumstellar regions would vary across the emission lines. In particular, the continuum flux will presumably be scattered through the total path length of the extended atmosphere; the emission flux, originating in the extended atmosphere itself, will be scattered through only a part of that path length. The polarization should, therefore, decrease in the emission lines depending on the relative strength and polarization of the emission to continuum flux. Such changes in polarization across lines are clear proof that at least part of the observed polarization arises in the circumstellar region and an analysis of the changes would assist in determining the physical parameters of the circumstellar region, viz. the electron temperature and density, scattering optical depth and thus the linear dimensions of the circumstellar shell, mean depth at which the emission flux originates, etc.

In recent years there have been a number of studies dedicated to polarimetry in the emission lines of Be and shell stars. In the Introduction, we have reviewed the history of the beginnings of these studies. Most of the observations have been done on the brightest stars at the Hα and Hβ lines. Some measurements have been made at the Hγ line (Hayes and Illing, 1974; Hayes, 1975). Current instrumentation is flux limited and it has not been possible to observe fainter stars or the higher lines in the Balmer series. The general technique is to scan across the emission lines by tilting a narrow-band interference filter to various discrete positions and the spectral resolutions obtained vary between 2 and 12 Å. Since the equivalent width in many Be stars is less than 10 Å, the lines are instrumentally broadened by the wider of these filters. Descriptions of the various polarimeters used are given in the literature (Angel and Landstreet, 1970a, b; Illing, 1973; Clarke and McLean, 1975; Clarke et al., 1975;

Poeckert, 1975a). Coyne (Coyne, 1975a, b; Coyne and McLean, 1975) used wider interference filters (passbands of 8 and 26 Å respectively at Hβ and Hα) and did not tilt them. All investigators simultaneously or quasi-simultaneously measure the continuum at appropriate wavelengths near the emission lines.

The analysis of the polarization measured in the emission line and in the continuum is done in the following way. If both the continuum flux and the emission flux are polarized in the same plane or orthogonal to one another, the following general relation can be written:

$$p_L = \frac{p_A + p_E X}{1 + X},$$ (1)

where p_L is the observed polarization in the line, p_A the polarization at the same wavelength in the underlying absorption, p_E the polarization of the emission flux and X the intensity ratio of the emission flux to that in the underlying absorption. In practice p_A is taken as p_C, the polarization in the adjoining continuum. If the underlying absorption is small with respect to the emission flux then X may be taken as the ratio of the emission flux to the continuum flux. It is customary to take the continuum flux as unity. Then $1 + X = I$, the total intensity of the emission line. With these assumptions, the simplest case is when the emission flux is unpolarized, $p_E = 0$ and $p_L = p_C/I$. If, in other words, the observed $p(\lambda)$ across the emission line varies inversely as $I(\lambda)$ the emission flux is unpolarized. Any deviation from this dependence will indicate either that $p_E \neq 0$ or that Equation (1) is not applicable.

In general the observed values of p_L and p_A or p_C include interstellar polarization effects. Poeckert (1976) has assumed $p_L = p_C/I$ in order to determine the interstellar polarization. Elsewhere in this Symposium McLean (1976) discusses ways in which the observations of polarization in emission lines together with Equation (1) can be used to separate out the interstellar and intrinsic effects and I shall not discuss this more here. McLean (1976) also shows how p_E may be determined from polarimetric and photometric measurements at two different wavelengths within an emission line.

Examples of the results obtained are shown in Figures 7 and 8 taken respectively from Coyne (1975a) and Poeckert (1975a). In Figure 7 is shown for six stars the variation of p and θ with wavelength. The measurements at Hα and Hβ were made with filters having passbands of 26 and 8 Å respectively. The wide and intermediate-band filters used at the other wavelengths are described by Coyne (1975a). o Sco is not a Be star but is a standard highly polarized star and it shows the normal interstellar polarization curve. All other stars show a decrease in the polarization at Hα and Hβ. For 48 Lib and ψ Per there is a dependence of θ on wavelength. Coyne (1975a) has shown that this is due to a combination of the intrinsic and interstellar polarization and has used the effect to remove the interstellar polarization. In Figure 8 we have high resolution (~2 Å) polarimetric and line profile measurements across the Hα line of γ Cas. Compare this to similar measurements by McLean (1976, page 263, Figure 1d). It is clear from both sets of measurements that the polarization in the blue wing of the Hα line is less than obtainable from the simple addition of unpolarized emission flux. Some other mechanism must be causing a decrease in the polarization here. Other stars observed at high resolution also show anomalous polarimetric changes across the emission line, i.e. changes which cannot be explained

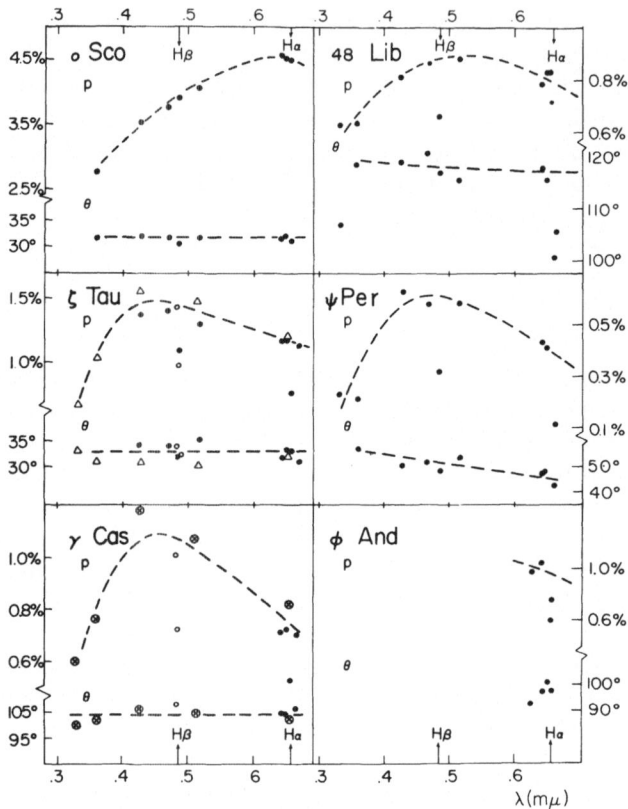

Fig. 7.   A decrease in polarization across the emission lines Hα and Hβ are shown for five Be stars. *o* Sco
is a typical interstellar polarization star.

by the simple addition of unpolarized emission flux. We shall discuss the more important cases shortly but first we wish to summarize the information available for all Be stars for which polarization has been measured across the Hα and/or Hβ emission lines.

Table II presents the essential data obtained from narrow-band polarimetry of Be and shell stars. Columns (1) to (7) give the same fundamental data as in Table I. In column (6) we have attempted to identify all of these stars which have ever shown shell characteristics. Since shell characteristics are transient phenomena and an exhaustive literature search was not possible, this column may be incomplete. The sources for the spectral types and $v \sin i$ are the same as described for Table I. The remaining columns summarize the polarimetric results. Columns (8) and (10) list the mean polarization values for the continuum in the *B*lue region of the spectrum near Hβ and the *R*ed region of the spectrum near Hα respectively. The exact wavelength at which the continuum polarization is sampled differs from author to author but has no effect on these results since $p(\lambda)$ for the continuum near the Hα and Hβ lines does not vary rapidly with λ. The ratios of the polarization in the emission line to that of the continuum are given in columns (9) and (11). The sources of the polarimetry are

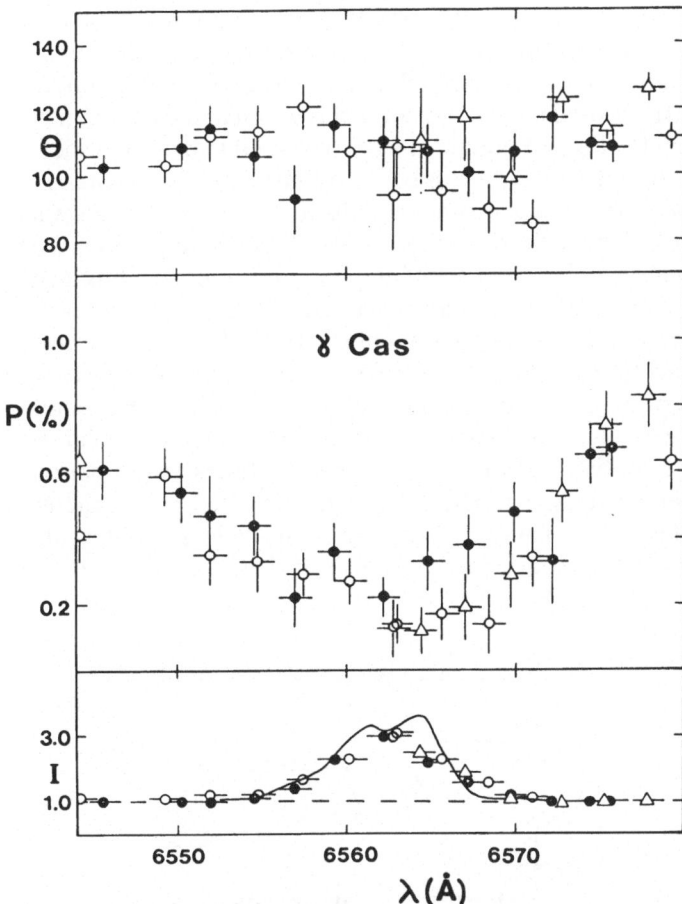

Fig. 8.   The direction and percentage of polarization and the intensity profile across Hα are shown
for γ Cas.

listed in column (12). We have omitted cases where the measured continuum
polarization was less than 0.3% or where the observational errors were too large to
determine the polarization ratios. The average error for a polarization measurement
is ±0.05%. Colons in columns (8) and (10) indicate errors greater than ±0.1%.

Of the 39 stars investigated 21 show significant changes [the polarization ratios are
in italics in columns (9) and (11)] in polarization across Hα and /or Hβ. These have
a mean value of $v \sin i$ of 303 km s^{-1}. For seven stars which do not show changes in
polarization across the emission lines the mean value of $v \sin i$ is 203 km s^{-1}. The star
59 Cyg is an anomaly with $v \sin i = 450$ km s^{-1} and no polarization changes across
the emission line. It is omitted from the means mentioned above. With the possible
exception of 1 Del and o And all stars which are identified as shell stars and which are
measured with sufficient accuracy show polarization changes across the emission
lines. There are five stars which show polarization changes across the emission lines
but are not listed as shell stars (56 Eri, ω Ori, HD 45995, BN Gem and 28 Cyg).

None of the three extreme Be stars (see Remarks column (13) of Table II) of Schild (1966, 1973) shows polarization effects across the emission lines. Three other extreme Be stars with continuum polarization less than 0.3% (not included, therefore, in Table II) showed no decrease of polarization in the emission lines. It appears, therefore, from the limited sample of stars observed to date that shell characteristics are a sufficient but not a necessary condition for the existence of decreased polarization across the emission lines and that extreme Be stars show no such effects.

For a number of stars, the decrease in polarization across the emission lines can be explained simply by the addition of unpolarized emission flux to the continuum polarization. According to Equation (1) this means that, after the removal, if necessary, of interstellar polarization effects, the polarization varies as $1/I$.

Figures 9 and 10, from Poeckert (1975a), give examples of cases where for ζ Tau and 48 Lib respectively the polarization at the center of the Hα emission line varies as $1/I$. In the case of ζ Tau, which has little or no interstellar polarization, although the level of the continuum polarization varies (at times over a period of one day or less), the dependence on I remains the same. In other words the emission flux is, apparently, always unpolarized. The case with 48 Lib is different, since there is significant interstellar polarization for this star. Poeckert (1975a) assumes that the

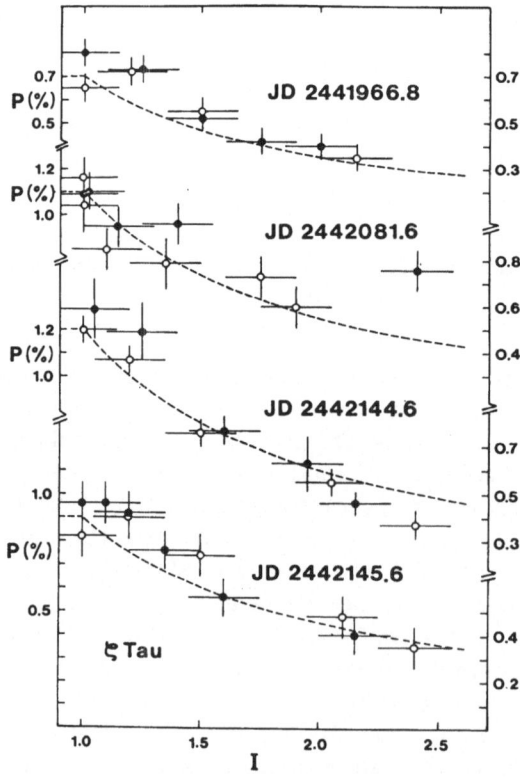

Fig. 9.   The variation of polarization as a function of the intensity across the Hα emission line of ζ Tau.

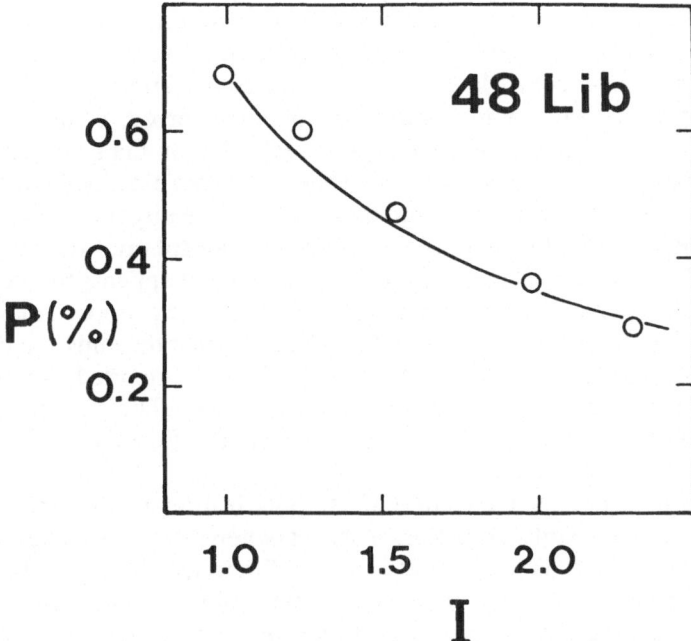

Fig. 10.   The variation of polarization as a function of the intensity across the Hα emission line of 48 Lib.
The interstellar polarization has been removed.

continuum polarization in 48 Lib will vary as $1/I$ and uses this relationship to remove
the interstellar components. Coyne (1975a) has shown that the interstellar polarization derived by Poeckert is essentially correct and results in a well-defined direction
for the intrinsic polarization. Thus in Figure 10 the dependence of the polarization on
$1/I$ represents the intrinsic polarization in 48 Lib and shows that the emission flux is
essentially unpolarized. Other stars for which the interstellar polarization is either
very small or has been corrected for in the manner of 48 Lib are indicated by a cross
(+) in column (8) and/or (10) in Table II.

However, the addition of unpolarized emission flux does not appear to be
adequate explanation for all stars. For some stars there appears to be a residual
polarization after account is taken of the unpolarized emission flux. For instance, for
ϕ Per it appears that the emission flux, at least in Hα, may be polarized about 0.5%
(Coyne and McLean, 1975; McLean, 1976). If the emission flux is polarized as it is in
ϕ Per it probably occurs by scattering if there is sufficient scattering optical depth.
For ϕ Per these conditions are probably fulfilled. It is likely as a matter of fact that we
are viewing the disk edge on; $v \sin i$ for ϕ Per is 493 km s^{-1}. Coyne and McLean
(1975) derive a scattering optical depth of $\tau = 0.04$ corresponding to a path length of
one-half a stellar radius in their simplified model of ϕ Per.

One would like to know from the body of data we now have (see Tables I and II)
whether the values found for the polarization in the emission lines of Be stars can be
typically explained by the simple model of the addition of unpolarized emission flux
or whether other mechanisms must be considered such as suggested for ϕ Per.

Unfortunately simultaneous or quasi-simultaneous photometry and polarimetry are not available for all of the stars of Table II. Furthermore a correct estimate of I from the line profile is difficult. McLean (1976) makes the interesting suggestion that additional shell absorption is responsible for an incorrect estimate of I in the equation $p/p_C = 1/I$ in the sense that I is really larger than measured, because the shell absorption, occurring in a layer further out in the extended atmosphere than the emission layer, affects only the intensity profile. In reality the failure to recognize both this shell absorption and the underlying atmospheric absorption may be responsible for an incorrect estimate of I. Nonetheless it appears that for some stars the emission flux is partially polarized. In the model to be proposed in Section 5 it is not unlikely that this is the case. All that is required is that the emission flux originate at a sufficiently large scattering optical depth in the extended atmosphere. The scattering optical depth for the continuum radiation would be, of course, considerably larger. Alternatively the polarization in the emission flux may be due to resonant scattering.

The differences between the amount of polarization in the emission flux for different stars are probably a function of aspect (when the observer lies in the orbital plane the optical depth is larger) but intrinsic differences between stars cannot be ruled out. Poeckert (1976) has attempted to determine the intrinsic polarization in Be stars by assuming that $p_L = p_C/I$ and then relating the intrinsic polarization to $v \sin i$. He finds only that for $v \sin i \leq 150$ km s^{-1} the polarization is less than 0.2%. For $v \sin i > 150$ km s^{-1} there is a large scatter in the polarization.

Essentially what we have spoken of so far is the central part of the polarimetric profile of the emission lines, i.e. the polarization corresponding to the center of the emission line. High resolution polarimetry across the whole line profile has revealed interesting results for several stars. In the blue wing of the Hα line in γ Cas and extending into the continuum, the polarization is at times lower (McLean, 1976; Poeckert, 1975a) and at times higher (Poeckert, 1975b) than that expected from unpolarized emission flux. Three other stars, β Mon A (Poeckert, 1975a), EW Lac (Poeckert, 1975b) and ζ Tau (McLean, 1976), have shown polarizations larger than the continuum polarization in the blue wing of the Hα line. McLean (1976) suggests that this increase in polarization in the line wings is due either to further electron scattering or to partially polarized flux from the underlying star itself. With respect to the first possibility others have suggested that the broad emission wings in Be stars are due to electron scattering; with respect to the second see the results from the model calculations for rigidly rotating flattened stars by Cassinelli and Haisch (1974) and Haisch and Cassinelli (1976).

In summary a comparison of polarimetric and photometric line profiles for Be stars suggests the following conclusions:

(1) All shell stars show a decrease in the polarization across emission lines.
(2) The emission flux is less polarized than the continuum flux, the ratio depends both on aspect angle and on particular characteristics of individual stars.
(3) Because of possible shell absorption superimposed on the emission and because of uncertainties about the underlying absorption it is difficult to determine the correct value of I, the total intensity at the emission line center normalized to the intensity of the underlying absorption.

(4) The blue wing of the Hα emission line in some stars shows peculiar polarization effects.

There is one anomalous case. χ Oph is a 'pole-on' star (Slettebak and Howard, 1955) and an extreme Be star (Schild, 1973). The polarization is variable (see Table I) with a mean value in the visual of 0.52% and an amplitude of 0.39% and is, therefore, undoubtedly intrinsic. It shows, however, no polarization differences across the Hα and Hβ lines although the respective line intensities relative to the continuum are 18.5 and 4.28 respectively (Gray and Marlborough, 1974). Coyne (1975a) measured the continuum polarization to be 0.53% ± 0.02 with no measurable dependence on λ between 0.36 and 0.66 μm. Thus if unpolarized emission flux were added to polarized continuum radiation then the measured polarization should be 0.03% and 0.12% at Hα and Hβ respectively. None of the other extreme Be stars or 'pole-on' stars observed (48 Per, 11 Cam, 105 Tau, ν Gem, ω CMa, υ Cyg) show either variable polarization (only 48 Per has a large number of observations) or polarization effects across the emission lines. Besides χ Oph only two of these (48 Per and 105 Tau) are polarized more than 0.25%. All other stars which show variable polarization and have been measured in the emission lines show polarization effects across these lines. The observations of χ Oph may be partially explained as follows. χ Oph is a conspicuous example of the stars identified by Schild (1973) as extreme Be stars. It has the strongest hydrogen emission lines among these stars and has shown little intensity variation in the emission lines (contrasted to the classical Be stars) over a period of 50 years. It has, compared to the classical Be stars, strong infrared excess ($E_{V-L} = 0.52$). It differs from other extreme Be stars, including all of those identified in Table II, and those enumerated above in the following two respects: (1) it has a lower $v \sin i$ than is to be expected from the relation of $v \sin i$ to the separation of emission line components for extreme Be stars (see Figure 3 of Schild, 1973); (2) whereas all extreme Be stars are optically thin in the hydrogen lines, χ Oph is optically thick (Burbidge and Burbidge, 1953).

We propose that χ Oph is an intermediate case. It is an extreme Be star but has the classical Be star characteristic of a somewhat flattened extended disk. As contrasted to the other extreme Be stars, χ Oph shows intrinsic polarization because it is optically thick. The polarization is neutral in the wide-band filters and shows no effect across emission lines because we are not viewing equator-on but at a sufficiently large angle of inclination ($i \cong 45°$ would remove the discrepancy in Figure 3 of Schild, 1973) that the path length for electron scattering for the emission and continuum flux through a flattened tilted disk are about the same but the path length is not large enough to introduce the processes responsible for the $p(\lambda)$ curve in shell stars (Coyne and Kruszewski, 1969; Capps et al., 1973). In particular the hydrogen absorption opacity is much less than the free electron opacity (see Figure 2 of Coyne and Kruszewski, 1969). No suggestion is offered to explain the variability of the polarization in χ Oph.

5. Models

A consistent model must explain the major observational results discussed in the previous sections and summarized here: (1) the amount of intrinsic linear polarization in Be and shell stars does not exceed ~1.5%; (2) there is a characteristic wavelength dependence, $p(\lambda)$, for this polarization which differentiates it from that for the interstellar medium. There are differences in $p(\lambda)$ from star to star. The chief characteristics of $p(\lambda)$ are a decrease in the ultraviolet polarization, a polarization maximum at ~0.5 μm, a second maximum or, at least, a plateau longwards of the Paschen limit and a decrease towards 2 μm; (3) the polarization is variable with an amplitude on the average about 40% of its mean value. There are short term variations on a time scale of days superimposed on longer term variations on a time scale of years. Some stars show rapid variations over fractions of an hour; (4) shell stars show a decrease of polarization in the emission lines. In some stars the emission flux may be partially polarized. Some shell stars show peculiar polarization effects in the blue wings of the Hα emission lines.

In general one can think of producing polarization in a star or circumstellar environment by one of the following processes: (1) radiative transfer in a stellar atmosphere which gives unequal polarizations at various regions over a stellar surface so that, when non-sphericity is introduced by flattening, rotational distortion, pole-on-equator radiation differences, eclipses, etc., the integrated flux is polarized; (2) a circumstellar envelope of aligned particles which are either geometrically asymmetric or optically anisotropic; (3) a circumstellar scattering envelope whose configuration is asymmetric.

With respect to number (1) Collins (1970) and Ruciński (1970) have shown that the transfer equations developed for a plane parallel rotationally distorted non-grey atmosphere yield polarizations in the visible region of the spectrum not greater than 0.1%. Cassinelli and Haisch (1974), using transfer equations developed by Cassinelli and Hummer (1971) for Rayleigh scattering in an atmosphere whose density followed an r^{-N} law ($N = 2, 3$), investigated the polarization produced by pure electron scattering atmospheres that are geometrically extended. Because of strong forward peaking of the radiation field polarizations as large as 50% are predicted at the limb of an extended atmosphere (compare to 11% predicted by Chandrasekhar (1946) for a plane parallel model). Polarization from the integrated flux from such an atmosphere is investigated for two geometrical shapes, a disk model and a Roche model, shown in Figure 11. Geometrically the disk model is characterized by 'd', the radius of the core centered in a disk of thickness $2d$. It is assumed that most of the radiation comes from the core and that the material outside the core is optically thin, $\tau \gtrsim 1.0$. Polarizations for the integrated flux for a core at optical depth of one are as large as 6.6%, the exact value depending on the density law used. For the Roche model a maximum polarization of ~2% is obtained for a core at an optical depth of about 2. For this model the core is defined as the volume within a distance of 0.4 times the polar radius where the isodensity surface is distorted by less than 1%. In both of these models the direction of the net polarization is parallel to the polar axis of the star. Haisch and Cassinelli (1976) include absorptive opacities and calculate the $p(\lambda)$ for both the Roche models and the disk model. The Roche model does not yield polarizations greater than 0.3%. While a sufficient amount of polarization can be provided by the disk model the observed wavelength dependence of the polarization cannot be produced.

A model geometrically similar to the disk model of Cassinelli and Haisch (1974) was proposed by Coyne and Kruszewski (1969). In this model, however, the star is simply a point source of flux. The polarization is not produced as a result of the transfer of radiation in the star (in particular by forward peaking of the radiation field) but simply by electron scattering of the radiation from a point source through a flattened disk of hydrogen plasma. The wavelength dependence of the polarization $p(\lambda)$ is obtained by absorption and emission processes in the hydrogen plasma which modify the otherwise neutral polarization produced by electron scattering. In other words polarized radiation is selectively absorbed and unpolarized or partially polarized flux is added.

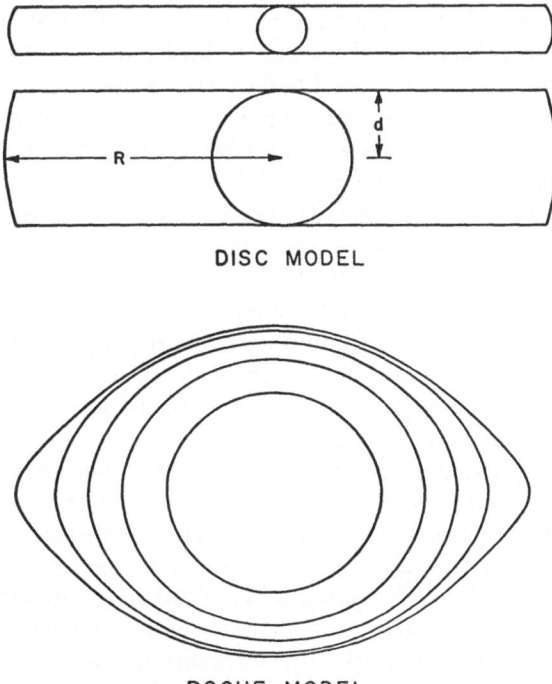

Fig. 11.   Two models used by Cassinelli and Haisch (1974) and Haisch and Cassinelli (1976) to compute the polarization produced in electron scattering atmospheres that are geometrically extended.

In order to have these selective processes change the otherwise neutral polariza-tion, it is necessary that the two sources of opacity, electron scattering and hydrogen absorption, be about equal such that the dominance of one over the other varied with wavelength. Coyne and Kruszewski (1969) showed that this was true for hydrogen plasmas with electron temperatures of $T_e = 10\,000$ K and electron densities of $N_e = 10^{12}$ cm^{-3} (see Figure 3). With such electron densities sufficient optical depths to give the observed amounts of intrinsic polarization (0.5 to 1.5%) were obtained for path lengths of the order of three stellar radii. Polarization produced within this region would then be further modified by absorption and emission processes outside this region to distances of about ten stellar radii.

These parameters of the model compare favorably with what we otherwise know of the extended envelopes about Be stars. From measured line intensities it is estimated that the radius of the emitting region is ten stellar radii and the electron density is of the order of 10^{13} cm^{-3} (Burbidge and Burbidge, 1956). From the width of the stellar absorption lines and the apparent speed of rotation of the emitting gas it is deduced that the main body of gas lies two or three stellar radii from the star (Huang, 1972, 1973). Kitchin (1970) from an analysis of the emission line profiles determines that the radius of the emitting region for ζ Tau is 11 stellar radii. Thus the model of an extended envelope with a more dense region extending to ~3 stellar radii and a less dense region extending to ~10 stellar radii is consistent with both the polarimetric and spectral data. The regions, of course, are not discretely separated but in the main the polarization by electron scattering is produced in the inner envelope region and is selectively modified by absorption and emission processes in the outer region. For instance, the unpolarized (or partially polarized) emission flux which explains the observed decrease in polarization in the emission lines of shell stars is produced in this outer region.

A degree of sophistication was added to this envelope model by Capps *et al.* (1973) who calculated the flux and polarization from 0.3 to 3 μm for a range in T_e from 8 000 to 20 000 K and for N_e from 10^{11} to 10^{12} cm^{-3}. It is shown that the wavelength dependence of the polarization is mainly due to emission in the disk. The variation of P between about 0.9 and 2.2 μm cannot be explained by recombination radiation alone. Free-free emission is required. The shape of the near infrared part of the $p(\lambda)$ curve depends on the relative contribution of these two processes which in turn is quite sensitive to the values of T_e and N_e.

In Figure 12 a comparison of model calculations and polarimetric observations for ζ Tau with a group of high latitude Be stars (Coyne, 1971) is given. For ζ Tau the model fits very well. The departure of the high latitude Be stars in the red and near infrared indicates that free-free emission is a less dominant mechanism for these stars. A slight change in the model parameters would give a fit to the high latitude Be star curve. The same model is plotted in Figure 4 with observations of ζ Tau, γ Cas, 48 Lib and ψ Per and in Figure 13 with observations of ϕ Per. Here again slight changes in T_e and N_e would fit the observations for the various stars by changing the relative contribution of free-free and recombination transitions. While ϕ Per and ζ Tau are comparable, γ Cas has a greater and 48 Lib a lesser contribution from free-free as compared to recombination energy.

Models which would rely upon alignment of geometrically asymmetric or optically anisotropic particles do not seem to be viable. A physical mechanism for the alignment is not available and the appropriate scattering theory has not been developed. Even in those Be stars where, from infrared measurements, one might deduce the presence of grains, the thermal flux is not polarized indicating that the grains are not aligned.

Some combination of the electron scattering models proposed by Cassinelli and Haisch (1974), Haisch and Cassinelli (1976), and Capps *et al.* (1973) appears to be the most fruitful approach to the problem. The variation of the polarization in the wings of emission lines will undoubtedly require selective processes in the radiative transfer calculations of Haisch and Cassinelli (1976).

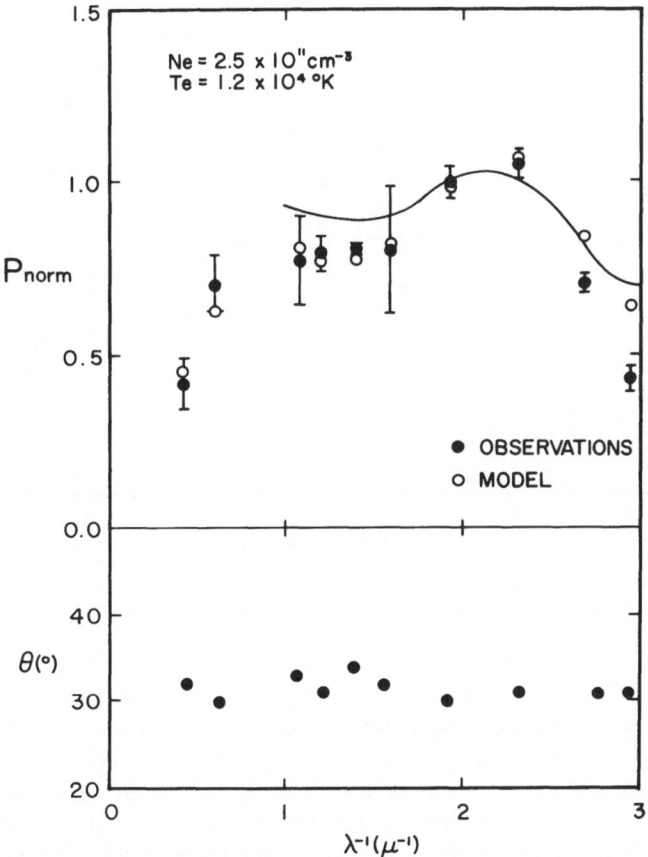

Fig. 12. Model calculations (open circles) for polarization produced in an electron scattering disk about ζ Tau are compared to observations of this star (solid circles) and observations of high galactic latitude Be⁻ stars (solid line).

We have not yet considered the possibility of large grains (i.e. in the Mie scattering range) in the extended envelope of Be stars as polarizing scatterers. Indeed there is some uncertainty as to whether the observed infrared fluxes in Be stars can be explained by free-free emission or whether it is necessary to evoke thermal reemission by grains of absorbed high energy flux (see Stein, 1972). The classical Be stars whose polarization we have considered here give no evidence of the presence of grains. We mention two cases of B-type stars for which there is both polarimetric and infrared evidence for the existence of Mie scattering grains. These are: HD 45677 (Coyne and Vrba, 1976) which shows a $p(\lambda)$ curve which varies on a time scale of about one-half year from a polarization continuously increasing into the near infrared to a curve with a maximum at ~0.6 μm; and HD 44179 (Cohen *et al.*, 1975) which shows a polarization continuously increasing into the near infrared.

It appears that the most urgent need theoretically is to evaluate the relative merits of radiative transfer processes in a stellar atmosphere and electron scattering in an extended stellar envelope. The difficulty with these exclusively photospheric models

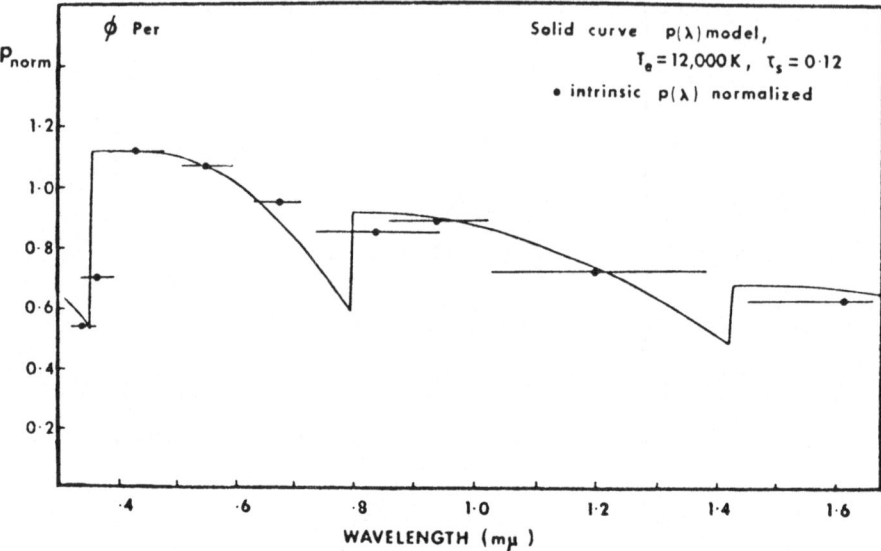

Fig. 13.   Polarization observations of ϕ Per, after removing the effects of the interstellar polarization, are compared to model calculations of polarization produced in a flattened disk with an electron-scattering optical depth of $\tau_s = 0.12$.

is to explain the time variability of the polarization, especially that which occurs on a short time scale of fractions of a day. Such variations could be more reasonably explained by dynamical processes occurring in an extended envelope. On the other hand the assumption in the electron-scattering extended envelope model of an unpolarized point source of radiation at the center of a disk whose height is two stellar radii with the main body of gas lying within three stellar radii and extending to ten stellar radii is only of limited validity. Undoubtedly contributions from processes occurring in both the stellar photosphere and the extended disk are necessary.

Observationally, the most urgent need is for simultaneous high resolution spectrophotometry and polarimetry across emission and absorption lines and across the hydrogen discontinuities. One could then perhaps begin to distinguish at various depths in the extended envelope the various processes of polarization and modification of the polarization. For instance one would like to make a detailed comparison of the polarization in shell lines, in the stellar lines, in emission lines, in the wings of all of these lines and in the continuum and to study the changes in the polarization with changes in the line profiles. Wide- and intermediate-band measurements between 0.6 and 2.2 μm over a variety of types of Be and shell stars would be very valuable. The $p(\lambda)$ in this region of the spectrum is very sensitive to the relative contributions of free-free emission and recombination energy which in turn depend on the relative values of the electron temperature and density in the extended envelope. In fact this may be the most sensitive region of the spectrum for determining these parameters. Finally, it would be valuable to have a polarimetric survey of peculiar Be stars and especially of those which show abnormally large infrared excesses. A concentration on the red and infrared region of the spectrum is suggested since this is where peculiarities in the $p(\lambda)$ curve have been known to occur. The expectation of such a survey would be to detect the presence of grains and eventually to evaluate the relative contribution they make to the analysis of extended envelopes in Be stars.

References

Angel, J. R. P. and Landstreet, J. D.: 1970a, *Astrophys. J. Letters* **160**, L147.
Angel, J. R. P. and Landstreet, J. D.: 1970b, *Astrophys. J. Letters* **162**, L61.
Appenzeller, I. and Hiltner, W. A.: 1967, *Astrophys. J.* **149**, 353.
Behr, A.: 1959, *Nach. Akad. Wiss. Göttingen* **2**; *Math-Phys. Kl.*, No. 7, 185 (= *Veröff. Göttingen*, No. 126).
Bernacca, P. D. and Perinotto, M.: 1971, *Contr. Obs. Astron. Univ. Padova in Asiago*, No. 250.
Boyarchuk, A. A. and Kopylov, I. M.: 1964, *Crimea Izv.* **32**, 44.
Burbidge, G. R. and Burbidge, E. M.: 1953, *Astrophys. J.* **117**, 407.
Burbidge, G. R. and Burbidge, E. M.: 1956, *Vistas in Astronomy* **2**, 1446.
Capps, R. W., Coyne, G. V., and Dyck, H. M.: 1973, *Astrophys. J.* **184**, 173.
Cassinelli, J. P. and Haisch, B. M.: 1974, *Astrophys. J.* **188**, 101.
Cassinelli, J. P. and Hummer, D. C.: 1971, *Monthly Notices Roy. Astron. Soc.* **153**, 9.
Chandrasekhar, S.: 1946, *Astrophys. J.* **103**, 351.
Clarke, D. and Grainger, J. F.: 1966, *Ann. Astrophys.* **29**, 355.
Clarke, D. and McLean, I. S.: 1974, *Monthly Notices Roy. Astron. Soc.* **167**, 27p.
Clarke, D. and McLean, I. S.: 1975, *Monthly Notices Roy. Astron. Soc.* **172**, 545.
Clarke, D. and McLean, I. S.: 1976, *Monthly Notices Roy. Astron. Soc.* **174**, 335.
Clarke, D., McLean, I. S., and Wyllie, T. H. A.: 1975, *Astron. Astrophys.* **43**, 215.
Cohen, M., Anderson, C. M., Cowley, A., Coyne, G. V., Fawley, W., Gull, T. R., Harlan, E. A., Herbig, G. H., Holden, F., Hudson, H. S., Jakoubek, R. O., Johnson, H. M., Merrill, K. M., Schiffer, F. H. III, Soifer, B. T., and Zuckerman, B.: 1975, *Astrophys. J.* **179**, 15.
Collins, G. W.: 1970, *Astrophys. J.* **159**, 583.
Coyne, G. V.: 1970, *Astrophys. J.* **161**, 1011.
Coyne, G. V.: 1971, *Specola Vatic. Ric. Astron.* **8**, 201.
Coyne, G. V.: 1974, *Monthly Notices Roy. Astron. Soc.* **169**, 7p.
Coyne, G. V.: 1975a, *Astron. Astrophys.*, in press.
Coyne, G. V.: 1975b, *Specola Vatic. Ric. Astron.* **8**, 533.
Coyne, G. V. and Gehrels, T.: 1967, *Astron. J.* **72**, 887.
Coyne, G. and Kruszewski, A.: 1969, *Astron. J.* **74**, 528.
Coyne, G. V. and McLean, I. S.: 1975, *Astron. J.* **80**, 702.
Coyne, G. V. and Vrba, F. J.: 1976, *Astrophys. J.*, in press.
Coyne, G. V., Gehrels, T., and Serkowski, K.: 1974, *Astron. J.* **79**, 581.
Derviz, T.: 1970, *Astrophysics (Astrofizika)* **6**, 351.
Eggen, O. J., Mathewson, D. S., and Serkowski, K.: 1967, *Nature* **213**, 1216.
Feinstein, A.: 1968, *Z. Astrophys.* **68**, 29.
Feinstein, A.: 1976, this volume, p. 149.
Gray, D. F. and Marlborough, J. M.: 1974, *Astrophys. J. Suppl.* **27**, 121.
Haisch, B. M. and Cassinelli, J. P.: 1976, this volume, p. 375.
Hall, J. S.: 1949, *Science* **109**, 166.
Hall, J. S.: 1958, *Publ. U.S. Naval Obs.* **17**, Part VI.
Hall, J. S. and Mikesell, A. H.: 1950, *Publ. U.S. Naval Obs.*, Washington, D.C., 2nd Series, **XVII**, Part I.
Harrington, J. P. and Collins, G. W.: 1968, *Astrophys. J.* **151**, 1051.
Hayes, D.: 1975, *Publ. Astron. Soc. Pacific* **87**, 609.
Hayes, D. P. and Illing, R. M. E.: 1974, *Astron. J.* **79**, 1430.
Hiltner, W. A.: 1949, *Science* **109**, 165.
Hiltner, W. A.: 1956, *Astrophys. J. Suppl.* **2**, 389.
Huang, Su-Shu: 1972, *Astrophys. J.* **171**, 549.
Huang, Su-Shu: 1973, *Astrophys. J.* **183**, 541.
Illing, R. M. E.: 1973, Dissertation, Physics Department, Columbia University.
Kitchin, C. R.: 1970, *Monthly Notices Roy. Astron. Soc.* **150**, 455.
Kruszewski, A.: 1974, in T. Gehrels (ed.), *Planets, Stars and Nebulae Studied with Photopolarimetry*, University of Arizona Press, Tucson, p. 845.

Lesh, J. R.: 1968, *Astrophys. J. Suppl.* **17**, 371.
McLean, I. S.: 1974, Dissertation, The University of Glasgow, Scotland.
McLean, I. S. and Clarke, D.: 1976, this volume, p. 261.
Merrill, P. W.: 1956, *Vistas in Astronomy* **2**, 1375.
Nagirner, D. T.: 1962, *Trudy Leningrad Astron. Obs.* **19**, 79.
Öhman, Y.: 1934, *Nature* **134**, 534.
Öhman, Y.: 1946, *Astrophys. J.* **104**, 460.
Poeckert, R.: 1975a, *Astrophys. J.* **196**, 777.
Poeckert, R.: 1975b, private communication.
Poeckert, R. and Marlborough, J. M.: 1976, this volume, p. 277.
Ruciński, S. M.: 1970, *Acta Astron.* **20**, 1.
Schild, R. E.: 1966, *Astrophys. J.* **146**, 142.
Schild, R. E.: 1973, *Astrophys. J.* **179**, 221.
Sen, K. K. and Lee, W. M.: 1961, *Publ. Astron. Soc. Japan* **13**, 263.
Serkowski, K.: 1968, *Astrophys. J.* **154**, 115.
Serkowski, K.: 1970, *Astrophys. J.* **160**, 1083.
Serkowski, K.: 1975, private communication.
Serkowski, K., Mathewson, D. S., and Ford, V. L.: 1975, *Astrophys. J.* **196**, 261.
Shakhovskoi, N. M.: 1962, *Astron. Circ. U.S.S.R.*, No. 228.
Shakhovskoi, N. M.: 1964, *Astron. Zh.* **41**, 1042 (transl. in *Soviet Astron. AJ* **8**, 833).
Slettebak, A. and Howard, R. F.: 1955, *Astrophys. J.* **121**, 102.
Stein, W. A.: 1972, *Publ. Astron. Soc. Pacific* **84**, 627.
Uesugi, A. and Fukuda, I.: 1970, *Contr. Inst. Astrophys. and Kwasan Obs. Univ. of Kyoto*, No. 189.
Vitrichenko, E. A. and Efimov, Y. S.: 1965, *Izv. Krymsk. Astrophys. Obs.* **34**, 114.
Zellner, B. and Serkowski, K.: 1972, *Publ. Astron. Soc. Pacific* **84**, 619.

DISCUSSION

Haisch: I would just like to point out that the emission feature in the Balmer lines is not just polarized because it is subsequently scattered, but the resonance scattering process itself produces polarization, which makes the problem more difficult.

McLean: I would just emphasize that the polarimetric variability observed in ζ Tau over a time interval of some 40 minutes was neither periodic or irregular. What was observed was a steady decrease in the degree of polarization (p) recorded by a monitor channel tuned to the continuum near Hβ. The effect was only apparent 'after the fact', when the data were reduced. A similar observation two weeks later revealed no such changes but the degree of polarization observed was smaller than the lowest value recorded on the night which revealed the variability. Strong night to night changes in p have been reported by other workers.

POLARIZATION MEASUREMENTS ACROSS THE BALMER LINES OF Be AND SHELL STARS

I. S. McLEAN* and D. CLARKE

Department of Astronomy, University of Glasgow, Scotland, United Kingdom

Abstract. We have made linear polarization measurements of several Be and shell stars across the Hα and Hβ lines, all of the stars exhibiting polarization in the continuum, the emphasis here being on measurements made of ζ Tau, 48 Per, φ Per, and γ Cas. Three types of results ensue: some stars show no significant change of polarization across the Balmer features (e.g., 48 Per, X Per); some stars show a reduced polarization across the features (e.g., γ Cas, ζ Tau) indicating the presence of intrinsic polarization; some stars show a change in the degree of polarization but with a marked rotation of the direction of vibration (e.g., φ Per, 48 Lib) which can be attributed to a combination of non-aligned intrinsic (circumstellar) and interstellar polarizations. Interpretations of these results are discussed, and we demonstrate the potential power of line profile polarimetry/photometry as an important new method for separating intrinsic and interstellar polarization effects, thus enabling polarization observations to be used as a constraint on models of Be stars.

1. Introduction

Perhaps the most distinctive feature of the spectra of early-type emission line stars is the appearance of the Balmer series of hydrogen in emission. Although much effort has been devoted to the study of the detailed shape and photometric variations of these lines, their intrinsic polarimetric properties have only recently received attention. Following a preliminary investigation reported to the IAU colloquium on photopolarimetry in 1972 (Clarke and McLean, 1974a), a reduced polarization was discovered at the center of the Hβ emission feature in γ Cas (Clarke and McLean, 1974b) and a similar effect found in ζ Tau. This latter result confirmed unpublished measurements of lower spectral resolution made by Serkowski (quoted in Zellner and Serkowski, 1972). Observations at Hα were first reported by Coyne (1974) for ζ Tau; and, more recently, measurements of this and other stars have been obtained with higher spectral resolution by Poeckert (1975). Polarization changes have also been detected across Hγ in both γ Cas and ζ Tau (Hayes and Illing, 1974; Hayes, 1975, respectively).

Broadband measurements indicate that the wavelength dependence of the intrinsic continuum polarization of shell stars is caused by electron scattering, modified by various wideband absorption and emission processes, deep in the circumstellar envelope (Coyne and Kruszewski, 1969; Capps *et al.*, 1973). The basic observed polarization/wavelength curves for Be and shell stars, however, show a variety of forms, indicating that interstellar effects contaminate the intrinsic polarizations in some cases, putting immediate model fitting out of the question. One of the current hypotheses for explaining the observed polarization changes across the lower Balmer lines, at least to some degree, is that of a dilution of the continuum polarization by addition of unpolarized emission line flux from a higher level of the shell. If this idea

---

* Visiting Astronomer, Lowell Observatory, Flagstaff, Arizona.

A. Slettebak (ed.), Be and Shell Stars, 261–275. All Rights Reserved
Copyright © 1976 by the IAU.

is correct, it has important bearing on separating the obscuring interstellar polarization from the intrinsic effects. No net polarization of the line radiation implies that the region of line formation to the observer is very optically thin for electron scattering. A scattering optical depth $\tau_s \sim 0.12$ (independent of λ) was found applicable to a model of the continuum polarization of ζ Tau (Capps *et al.*, 1973). Therefore, to escape significant polarization, the bulk of the emission must effectively arise in a region with τ_s several times smaller. Whether or not this situation occurs may well depend on the observer's aspect angle. Thus, some intrinsic differences might be expected in line profile effects from star to star and perhaps among different lines in the same star.

In this paper we discuss a variety of new measurements at Hα and Hβ which illustrate most of the observed and intrinsic effects so far encountered in line profile polarimetry of Be stars. We also present briefly methods for separating circumstellar and interstellar effects by line profile observations and by statistical studies of the polarization of non-Be stars in the same region of sky as the Be star.

2. Observations

All of the measurements discussed in this paper were obtained with the dual-channel wavelength scanning polarimeter described by Clarke and McLean (1975a). The observations were made at the Lowell Observatory, Arizona, using either the 79-cm (31-in.) telescope or the 183-cm (72-in.) Perkins telescope (operated jointly by the Ohio State and Ohio Wesleyan Universities and the Lowell Observatory). Line profile scans of Hα and Hβ are obtained by tilt-scanning narrow-band interference filters (Clarke *et al.*, 1975). For most of the measurements given here, passbands (FWHM) of 8.5 Å at Hα and 12 Å at Hβ were employed. A few measurements have been made at Hβ with a passband of 2.3 Å. Since the width of the Hα emission line at the half-maximum intensity point is $\sim$8–9 Å in many Be stars, then the observed Hα profiles are strongly broadened by the instrumental profile, while at Hβ even less spectral detail is obtained. The chosen passbands represent a compromise necessary to retain viable integration times for the polarimetry.

In Figure 1 the observed normalized line profiles, the degree of linear polarization p (in percent), and the position angle (azimuth), θ, of the direction of vibration in equatorial coordinates are plotted for the Hα region of the four major program stars ζ Tau, 48 Per, ϕ Per, and γ Cas. Horizontal error bars, suppressed in the θ diagrams, indicate the uncertainty in the wavelength setting of the filter. A similar diagram for the Hβ region is shown in Figure 2. For ζ Tau, Hβ line profile polarimetry was obtained with a 2.3 Å filter in addition to the 12 Å filter measurements. Therefore, in Figure 2(a), the horizontal error bars represent the filter passband plus wavelength setting error. Different symbols are used for observations taken on separate nights except when those observations are virtually identical both in and out of the lines. It is immediately apparent that the observed changes of polarization across Hα and Hβ differ considerably for these four stars.

With the exception of 48 Per, there is a decrease in the degree of linear polarization p toward the line center, relative to the nearby red and blue continua, for each of

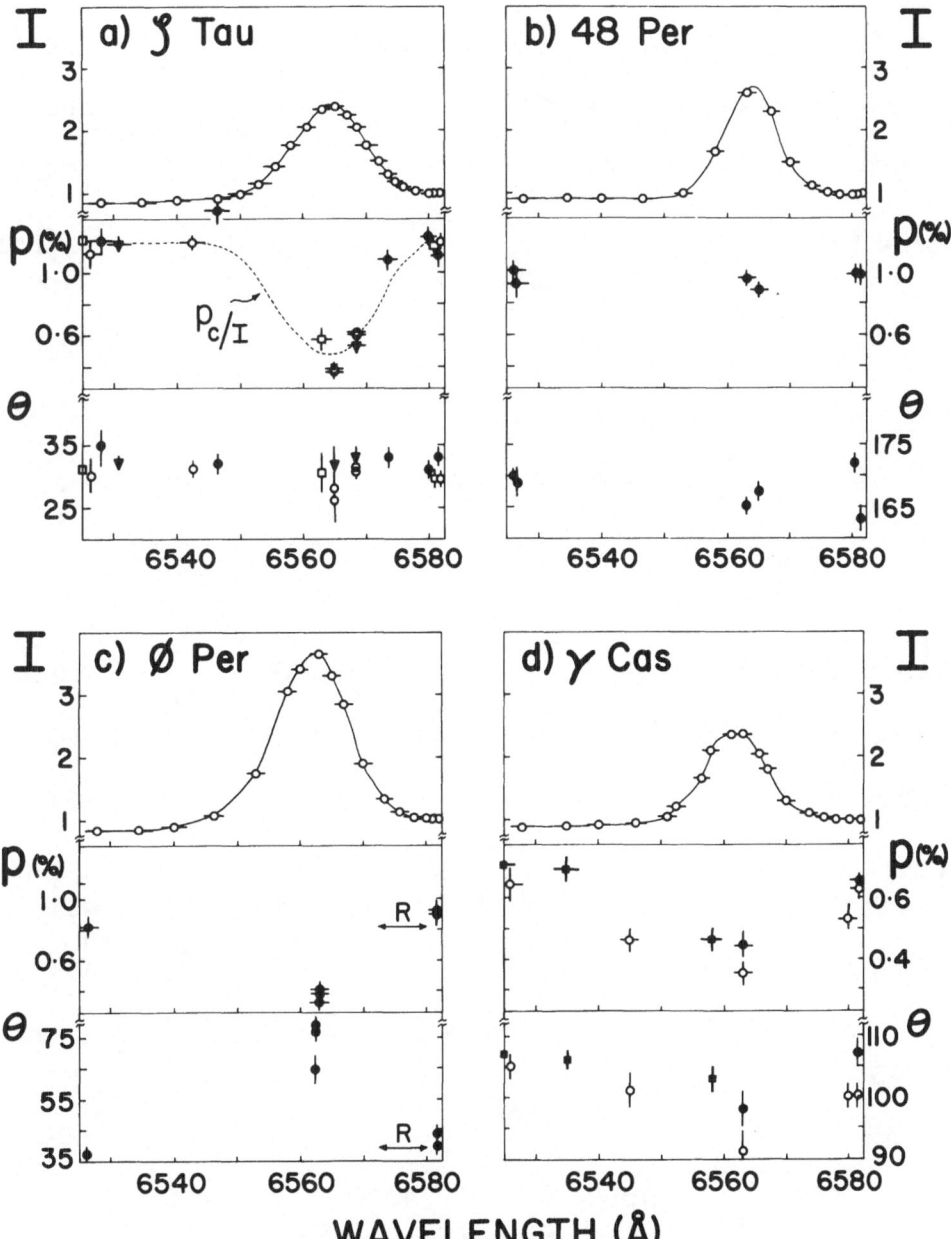

Fig. 1.   The normalized intensity profile obtained by discrete tilt-scanning of an 8.5 Å Hα interference filter is shown, together with the degree of linear polarization p (in percent) and the equatorial position angle, θ, of the direction of vibration for the stars ζ Tau, 48 Per, φ Per, and γ Cas. Horizontal error bars represent the uncertainty of the wavelength setting. For ζ Tau (1(a)), the curve corresponding to dilution of the continuum polarization by unpolarized emission line flux is shown by a dashed line.

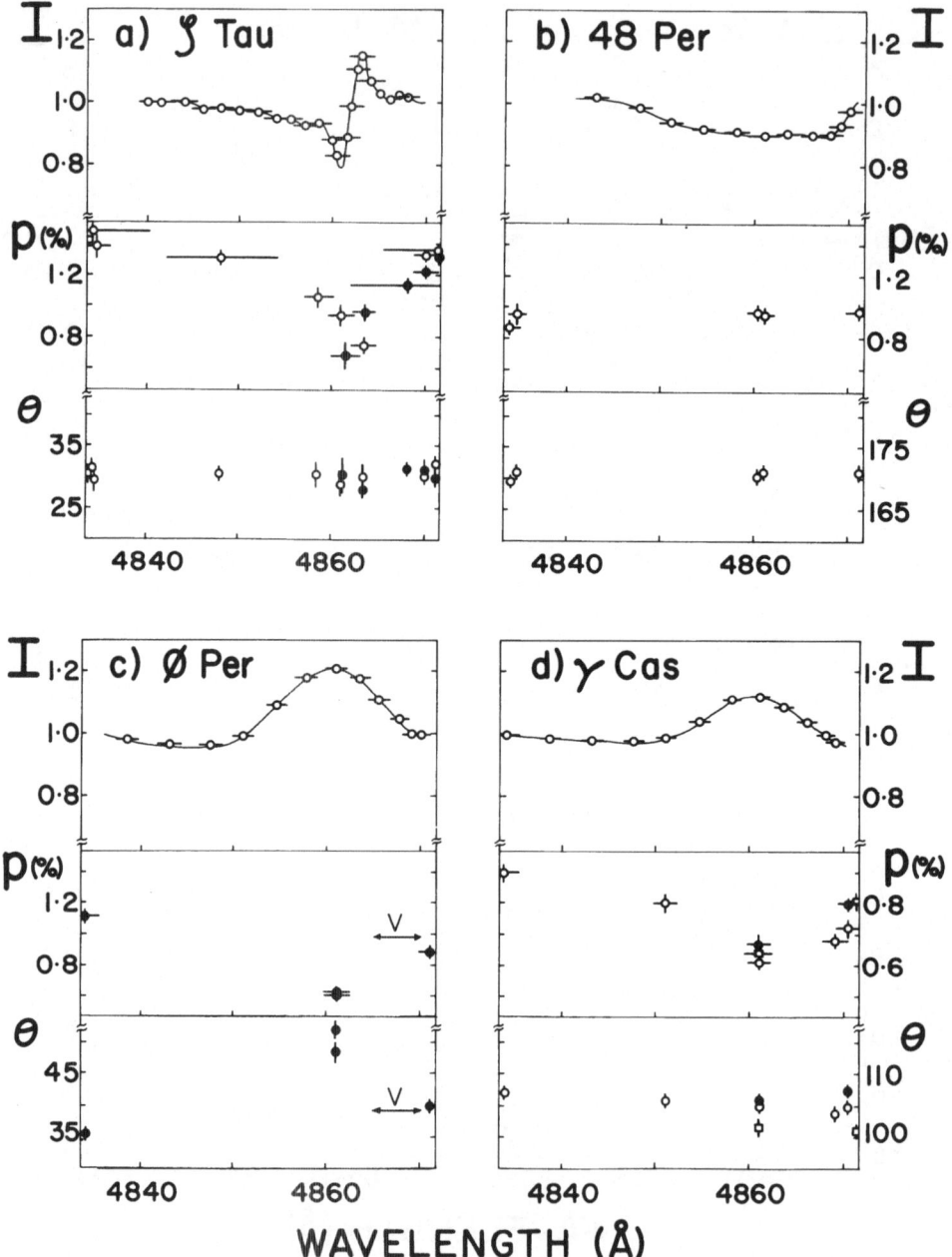

Fig. 2.   As for Figure 1, but for the Hβ region. The tilt scans were made by a 12 Å filter except for ζ Tau, where a 2.3 Å filter was employed to resolve the shell absorption structure. Horizontal bars in Figure 2(a) represent the filter passband. Broad red and yellow filters used in the measurements of φ Per are denoted R and V in Figures 1(c) and 2(c), respectively.

the stars. The change in p is strongest across Hα. However, observations of Hβ in ζ Tau (Figure 2(a)) with a 2.3 Å filter also reveal a strong reduced polarization effect corresponding to a line profile with asymmetric emission wings and a central absorption feature. These measurements on ζ Tau constitute the highest resolution spectropolarimetry so far reported for Hβ. For γ Cas, the changes in p are shallow and commence well out from the line centers (Figures 1(d) and 2(d)); note the change of scale for p. Phi Per exhibits a very marked rotation of the position angle θ which is greatest at the Hα and Hβ line centers (Figures 1(c) and 2(c)), and there is some evidence of slight variability and wavelength dependence in θ for γ Cas. Neither 48 Per nor ζ Tau shows significant variations in θ. It is also clear from both Figures 1 and 2 that, despite the severe instrumental broadening of the line profiles, crude differences of shape are apparent from star to star.

To understand the observations it is necessary to be aware of two factors: (a) the effect of combining different interstellar and intrinsic (circumstellar) polarizations, and (b) the effect of a shell emission line of unknown polarization properties on the circumstellar continuum polarization. A discussion of these two factors is developed in the next section before attempting to interpret the observations of each star individually.

3. Separation of Interstellar and Intrinsic Polarizations

To be able to use polarization measurements as a constraint in modelling the extended envelopes of emission line stars, it is essential to remove from the observations the effect of the interstellar medium. There are several ways to approach this problem and, as stated in the Introduction, the most recently proposed method is that of line profile polarimetry. Alternative methods depend on statistical studies of the polarization of stars in the same region of sky as the program star or on differences between the wavelength dependences of intrinsic and interstellar polarization over a broad spectral region. Where possible, we have considered each method.

3.1. METHODS EMPLOYING THE WAVELENGTH AND TIME DEPENDENCE OF THE INTRINSIC POLARIZATION

The wavelength dependence of the degree of polarization for interstellar grains $p_i(\lambda)$ differs considerably from that expected for the extended atmospheres of early-type emission line stars $p_*(\lambda)$. Serkowski (1973) has shown that $p_i(\lambda)$ follows the relation $p_{max} \exp[-1.15 \ln^2(\lambda_{max}/\lambda)]$, which is a smooth curve, while $p_*(\lambda)$ for Be stars undergoes quite sharp discontinuities at wavelengths corresponding to the Balmer and Paschen limits of hydrogen (Serkowski, 1968; Coyne and Kruszewski, 1969). However, in both cases there is unlikely to be any marked wavelength dependence of the position angles θ_i and θ_*. For most models of the scattering envelope the polarization is expected to be parallel to the star's rotation axis. When interstellar and intrinsic $p(\lambda)$ curves of similar strength are superposed with two different position angles ($\theta_i \neq \theta_*$), then the resultant (observed) position angle will, in general, show a dependence on λ which will depend in turn on p_i, p_* and their relative orientations.

Any time variations in the intrinsic degree of polarization (p_*) alone also lead to variability of both the observed degree of polarization and the observed position angle. This is an important effect since it can immediately yield the intrinsic position angle θ_*. Representing the observed polarization vector by its components, the normalized Stokes parameters $p_x = p \cos 2\theta$ and $p_y = p \sin 2\theta$, yields a point in the $p_x - p_y$ plane. At another epoch a different point will be obtained, but, provided the intrinsic position angle and the magnitude and direction of the interstellar polarization are all constant while p_* alone varies with time, the line joining the two points must be parallel to the line representing the direction of vibration of the intrinsic polarization. This line makes an angle $2\theta_*$ with the p_x axis.

Having obtained θ_* in the above manner, measurements at two different wavelengths can be employed, after some manipulation, to derive λ_{max}, the wavelength at which the interstellar polarization attains its maximum value p_{max}. Assuming a knowledge of the interstellar position angle θ_i ($\theta_i \neq \theta_*$), a series of broadband measurements, covering a wide spectral range and especially including the near ultraviolet, which have distinctly different observed position angles can be used by trial and error to derive p_{max}. Reasonable starting values of p_{max} can be estimated from the observed color excess of the star. The correct value of p_{max} should give values of the interstellar polarization which exactly cancel the wavelength dependence of the observed position angle. Clearly, this method depends on the presence of strong rotation effects in broadbands and on a previous knowledge of θ_i. Over certain small areas of the sky the interstellar material in that direction may be sufficiently uniformly distributed to enable the mean interstellar position angle $\bar{\theta}_i$ to be obtained by averaging the observed position angles for many stars in that area whose light is not expected to be contaminated by intrinsic polarization.

For the present work, a study of the Hall catalogue (Hall, 1958) of polarization measurements was carried out to derive $\bar{\theta}_i$ from non-Be stars contained within a small area ($\pm 4°$ in l^{II} and $\pm 4°$ in b^{II}) centered on the galactic coordinates of the program star in question. The results of this survey are shown in Table I, in which the column headings are as follows. Given in the first two columns, respectively, are the HD numbers and names of the observed stars, and in column three, the MK spectral type taken from the *Catalogue of Bright Stars* (Hoffleit, 1964). The fourth column contains the observed $(B - V)$ color index according to the *Photoelectric Catalogue* of Blanco *et al.* (1968). Using the intrinsic colors given by Johnson (1958), the color excess $E(B - V)$ was calculated and listed in column five. An upper limit to the amount of interstellar polarization corresponding to this reddening, estimated from the approximate relation $p_i(\text{lim}) \lesssim 2.36[3E(B - V)]$ percent (Hiltner, 1956) is given in column six. The last three columns contain the mean observed position angle ($\bar{\theta}_o$) of the program stars, the mean interstellar position angle with its standard error from the catalogue survey, and the number of stars used to form that mean.

Many of the stars in the Hall catalogue are faint and are consequently background stars, but this should not affect the estimate of θ_i if it is independent of distance, that is, if there is essentially a single 'interstellar cloud'. Only for γ Cas is there a very large number of stars within the selected area, and there appears to be strong

TABLE I

Data for the observed stars

HD	Star	SP[a]	$(B-V)$[b]	$E(B-V)$	p_i(lim) (%)	$\bar{\theta}_o$	$\bar{\theta}_i$[d]	N[e]
5 394	γ Cas	B0 IV?e	−0.22	0.08	0.66	105°	95°±1°	117
10 516	φ Per	B1 III–V?pe	−0.04	0.22	1.81	39.5[c]	104±3	16
24 534	X Per	Ope	+0.31[a]	0.62?	5.14	55	88±13	13
25 940	48 Per	B3 Vpe	−0.03	0.17	1.41	171	148±4	13
31 964	ε Aur	F0 Iap	+0.54	0.31	2.57	144	154±4	17
37 202	ζ Tau	B2 IV pe	−0.18	0.05	0.41	31	164±5	19
142 983	48 Lib	B3? pe	−0.09	0.11?	0.91	117	Undefined	
149 757	ζ Oph	O9.5 V	+0.02	0.32	2.65	127	Undefined	

[a] *Catalogue of Bright Stars* (Hoffleit, 1964).
[b] *Photoelectric Catalogue* (Blanco, *et al.*, 1968).
[c] Observed position angle is wavelength-dependent.
[d] Mean position angle for stars, mostly distant, listed by Hall (1958) lying within ±4° in l^{II} and ±4° in b^{II} of the emission line star.
[e] Number of stars within the 8°×8° box. If N was less than 10, the box was increased to ±5° in l^{II} and b^{II} (φ Per only); and if there were less than three stars in this box, the value of $\bar{\theta}_i$ was considered undefined.

evidence that the interstellar position angle in that direction, presumably by coincidence, is not much different from the mean observed azimuth $\bar{\theta}_o$. Of course, since γ Cas is comparatively close to the Sun ($\sim$29 pc) and not strongly reddened, the amount of interstellar polarization present must be small.

So far, the discussion on separating intrinsic and interstellar polarizations has involved only broad-band polarimetry and statistical results which have obvious limitations. Next, we turn our attention to the novel possibility of applying polarimetric measurements across emission line profiles to the same problem.

3.2. METHODS EMPLOYING THE POLARIZATION EFFECTS PRODUCED BY THE SHELL EMISSION LINES

For the idealized case in which the emission line is regarded as arising in a 'layer' of the shell above that in which most of the electron scattering and self-absorption, responsible for the polarization of the stellar continuum flux, occurs then, by adding together the appropriate Stokes vectors, the following simple relations can be derived. If $\chi(\lambda)$ is the ratio of the additional unpolarized emission flux, at some wavelength point (λ) in the line, to the total flux at the same point in the original underlying spectrum, then the line polarization is given by (Clarke and McLean, 1975b)

$$p_*(\lambda) = p_*(\lambda_c)/[1 + \chi(\lambda)], \tag{1}$$

where $p_*(\lambda_c)$ is the degree of polarization in the adjacent continuum. These expressions are, of course, for the *intrinsic* polarization only. When the emission line flux itself is partially linearly polarized with a constant degree of polarization p_e (at the same position angle as the continuum) then the relation becomes

$$p_*(\lambda) = [p_*(\lambda_c) + p_e\chi(\lambda)]/[1 + \chi(\lambda)]. \tag{2}$$

A *negative* value of p_e would correspond to a position angle orthogonal to that in the continuum. Implicit in these formulae is the reasonable assumption that the polarization at wavelength λ in the underlying absorption spectrum is equal to that at λ_c in the adjacent continuum.

At Hα, the shape of the underlying spectrum (the shallow rotationally broadened photospheric absorption feature) can sometimes be neglected in comparison to the emission line strength. In that case, the added flux is essentially relative to the continuum and $1 + \chi$ becomes the observed total intensity (I) with the continuum normalized to unity.

Observationally, we have two values of the observed polarization, $p_o(\lambda_c)$ for the continuum and $p_o(\lambda_1)$ for some wavelength λ_1 in the line (Hα, say) and in addition, $\chi(\lambda_1)$ or at least $I(\lambda_1)$ is known from the line profile scans.

The basic assumption in what follows is that, whatever the emission line flux does to the value of p_* in the adjacent continuum, it does without affecting θ_*, which remains constant and independent of wavelength. This situation can be achieved with unpolarized emission or partially polarized emission either parallel or perpendicular to θ_*.

Plotting once again the observed polarizations in the $p_x - p_y$ plane and joining the two points by a straight line immediately yields, from the length of this line, the intrinsic difference $\Delta p = p_*(\lambda_c) - p_*(\lambda_1)$. If the line through the two points does *not* pass through the origin, then the direction $\lambda_1 \rightarrow \lambda_c$ yields $2\theta_*$. However, if the line *does* pass through the origin, then the intrinsic and interstellar polarization components are either aligned with this direction or are orthogonal to it, and these cases cannot be separated by this method, or there is no significant interstellar effect.

Assuming the unpolarized emission case, $p_*(\lambda_c)$ can be derived from the observed values of Δp and $I(\lambda_1)$ by using Equation (1). If the emission feature is weak, then the underlying line shape must be estimated and $\chi(\lambda_1)$ used instead of $I(\lambda_1)$.

Since $p_o(\lambda_c)$, $\theta_o(\lambda_c)$, $p_*(\lambda_c)$, and θ_* are now known, it is straightforward to complete the vector addition problem and solve for $p_i(\lambda_c)$ and θ_i. Of course, a negligible contribution from the interstellar medium implies that the derived value of $p_*(\lambda_c)$ equals the observed polarization $p_o(\lambda_c)$.

We now have θ_i and θ_*, which are essentially constants, and the value of the interstellar polarization *near* the line (p_i does not change significantly across any given one of the lower Balmer lines). Since $p_i(\lambda)$ depends on two parameters p_{max}, λ_{max}, corresponding to the amount of material in the line of sight and the particle size, respectively, the value of p_i cannot be deduced for a distinctly different wavelength without further observations.

There are two possible courses which can be followed here. Since θ_i and θ_* are assumed known and constant, measurements of the observed polarization $p_o(\lambda)$, $\theta_o(\lambda)$ at any λ will yield simultaneously $p_*(\lambda)$ and $p_i(\lambda)$. Alternatively, repeating the observations at a different emission line (say, Hβ) should provide the same values for θ_i and θ_*, but a new value for p_i. Now, having two sets of results, the equation $p_i(\lambda) = p_{max} \exp[-1.15 \ln^2 (\lambda_{max}/\lambda)]$ can be solved for p_{max} and λ_{max}, hence obtaining the interstellar properties directly.

If the emission line flux is, in fact, uniformly polarized with degree of polarization p_e, as given in Equation (2), it is still possible to obtain some partial results and, in addition, estimate the value of p_e by using two observations within the line. Starting with $p_o(\lambda_c)$, $p_o(\lambda_1)$, $p_o(\lambda_2)$, and the corresponding intensities $I(\lambda_1)$ and $I(\lambda_2)$ and again assuming that θ_* is strictly constant, the length of the lines joining $p_o(\lambda_1)$ to $p_o(\lambda_c)$ and $p_o(\lambda_2)$ to $p_o(\lambda_c)$ in the $p_x - p_y$ plane gives the true change Δp. The three points should also be collinear and define θ_*.

Line profile polarimetry of Be-shell stars appears an attractive method for separating intrinsic and interstellar effects, provided that the simple expressions and assumptions underlying Equations (1) and (2) are at all valid. The basic assumptions are that θ_*, θ_i, and p_e are all constants and that the underlying stellar radiation is polarized near the star. Deviations from the simple formulae presented here should, in principle, indicate the validity of the assumptions. However, very good polarimetry and corresponding line profile photometry is obligatory to differentiate between, for example, unpolarized emission and partially polarized emission with $p_e \sim 0.2\%$. As yet, Hα and Hβ observations are not detailed enough to allow the full potential of the technique to be tested, but results typical of those obtained so far are discussed in the next section.

4. Individual Discussion of the Program Stars

ζ Tau. Referring first to the Hα observations shown in Figure 1(a), it is apparent that the strong change in p across the line is not accompanied by a statistically significant change (rotation) in the direction of vibration. Any interstellar component in the polarization must therefore be either aligned or orthogonal to the intrinsic polarization. Alternatively, for an arbitrary orientation, the interstellar polarization must be very weak. The mean position angle for 19 non-Be stars in the same area of sky as ζ Tau (see Table I) was found to be $164° \pm 5°$ (equivalent to $-16° \pm 5°$). For five stars with tabulated values of p less than half of that for ζ Tau the mean is $10° \pm 6$, while the mean observed position angle for ζ Tau is about $31°$. Alignment of the mean interstellar position angle with the intrinsic angle for ζ Tau is therefore not very convincing, and the small value of $E(B - V)$ implies that the amount of polarization cannot exceed about 0.4%. In fact, p_i must be somewhat smaller, or the above non-alignment would just become observable as a rotation effect at the Hα line center, thus the observed polarization should be nearly all intrinsic, enabling us to test the unpolarized emission hypothesis.

From Figure 1(a) it is clear that the observations of p across Hα do fit quite well the simplified relation (with $1 + \chi$ replaced by I) corresponding to unpolarized emission line flux (see also the measurements by Poeckert (1975), which contain more spectral points). However, the line center measurements are systematically lower than expected, and a single observation on the extremity of the blue wing is higher than expected. These apparently discrepant results cannot be resolved by taking into account the underlying profile or a weak interstellar polarization. If Equation (2) is applied to the measurements, then a suitable agreement can be reached only with a variable emission line polarization such that p_e is zero at the wavelength which fits the p_c/I curve in Figure 1(a), positive in the line wings, and negative (i.e., orthogonally polarized) at the line center. A weak increase in polarization (at the same position angle) in the line wings may not be unreasonable due either to electron scattering (suggested by Marlborough as responsible for the broad emission wings in Be stars) or to partially polarized flux from the underlying rapidly rotating star itself (see, for example, the models of Cassinelli and Haisch, 1974). Orthogonal polarization at the line center is, however, more difficult to understand. One simple explanation worthy of consideration is that the observed value of I at the line center is not the value instrumental in reducing the continuum polarization because the observed profile has been flattened-off by the presence of a shell absorption line. Provided the bulk of the absorption occurs higher in the envelope than the bulk of the emission, then the line center polarization should be unaffected and the absorption only manifests itself in the intensity profile. With this concept in mind, we next refer to the Hβ polarimetry of ζ Tau shown in Figure 2(a).

Evidently the Hβ line profile of ζ Tau is comprised of a sharp absorption line and a displaced emission line. Again, there is no apparent change in $θ$ across Hβ but a remarkably strong reduction in p which is a maximum between the red emission peak and the central reversal. Unlike at Hα, the shape of the underlying absorption cannot be ignored. Since this shape is poorly known, it is difficult to calculate $χ$; however, we can adopt the inverse process. Dividing the observed minimum value of p by the continuum value p_c allows χ to be predicted from Equation 1 for unpolarized emission. The result is $\chi \simeq 1.0$, i.e., the additional flux due to the emission line is equal to the flux at the same point in the original underlying spectrum. If the emission line is superposed on a shallow rotationally broadened photospheric absorption line with a central depth of 0.20 (relative to the continuum as 1.00), then the expected value of I at the line center would be 1.60.

Subtracting from this emission line a shell absorption line with a central depth of 0.80 and with a slight relative displacement results in a situation similar to that which is observed. This simple idea of the effect of the shell absorption enables us to maintain the concept that the shell emission is unpolarized or nearly so at both $H\alpha$ and $H\beta$, and it would also explain the apparently anomalous strong reduction in p observed at $H\gamma$ in ζ Tau by Hayes (1975). However, shell absorption processes, localized to the line center, can be envisaged which will produce the same effect as unpolarized emission (Hayes, 1975).

In reality it may be doubtful that the extended envelope can be broken down into suitable layers of emission and absorption. However, we note that regions of the envelope more distant than 10 stellar radii from the star are thought to be the most likely places for the deep shell lines to be produced on theoretical grounds (Marlborough, 1969) and that Marlborough and Cowley (1974) found that an envelope extending to at least 30 stellar radii was required to reproduce the deep shell line at $H\alpha$ in 1 Delphini.

48 Per. This star shows no significant changes in polarization across $H\alpha$, despite the presence of a strong emission line. A possible slight rotation in θ for the line center values (Figure 1(b)) is confused by the disparity of the two results at λ 6580 Å. No effects are apparent at $H\beta$ (Figure 2(b)), with all the values of θ being closely aligned to 171°, which does seem marginally higher than for the $H\alpha$ values.

Since 48 Per is regarded as an extreme Be star and has a fairly low $v \sin i$, it is reasonable to suppose that it is being viewed nearly pole-on. For an homogeneous scattering medium confined essentially to the equatorial plane, an almost pole-on aspect results in a low net circumstellar polarization because the disk appears symmetric. In contrast, in shell stars such as ζ Tau, the line of sight is nearer to the equatorial plane, and thus the scattering disk appears asymmetric and the polarization is high. The observed polarization, if any, of extreme Be stars should therefore be dominated by interstellar polarization. A comparison with the Hall catalogue yields a mean interstellar position angle of $148° \pm 4°$, while five stars with less observed polarization than 48 Per gave $158° \pm 7°$. The agreement is not as good as might be expected. Taking this fact together with the slight difference in θ between $H\alpha$ and $H\beta$, the possible weak rotation effect at $H\alpha$, and the peculiar wavelength dependence of the observed polarization reported by Coyne and Kruszewski (1969), then the presence of some intrinsic polarization seems likely. For a given value of $\chi(\lambda)$ or $I(\lambda)$, the difference Δp between the line and continuum polarization is proportional to $p_*(\lambda_c)$ which, for an almost pole-on star such as 48 Per, may be only 0.2% and, therefore, for the typical value $I = 3$, Δp would be only 0.13%, compared to 0.80% for ζ Tau. In fact, on the basis of Equation (1) and the limit on Δp set by the observational errors, the intrinsic polarization near $H\alpha$ is most probably less than

0.1%. Very accurate line profile polarimetry is essential to separate intrinsic and interstellar effects for almost pole-on stars.

Even when no line profile effect can be detected by high precision measurements, intrinsic polarization is still possible but with line and continuum flux equally polarized. In such a case, temporal variations are a good indicator of intrinsic polarization.

ϕ *Per.* At Hα and Hβ (Figures 1(c) and 2(c)), there is a very strong change in both p and θ relative to the continuum values, which is most likely attributable to a combination of interstellar and intrinsic polarizations with quite different position angles. Since only line center observations have been obtained so far, it would be premature to apply Equation (1) without additional support.

From Table I, the mean interstellar position angle is $104° \pm 3°$ for stars in the same area of sky as ϕ Per. With this value of θ_i and a value of 0.52 μm for λ_{max}, Coyne and McLean (1975) found that a maximum interstellar polarization of 1.00% was required to remove the strong rotation effects in the observed position angles for a large number of both narrow- and wide-band observations (0.3 μm to 2.2 μm) of ϕ Per. As discussed in the previous section, when the observed line and continuum polarizations are plotted in the $p_x - p_y$ plane, the magnitude and direction of the vector which joins them gives the true difference in intrinsic polarization Δp and the intrinsic position angle θ_*. The value of θ_* derived from the line profile method was in perfect agreement with the result of the broad-band method. However, when the interstellar polarization was removed from the Hα observations, the emission line flux was itself weakly polarized with $p_e \simeq 0.5\%$. Such a result could be consistent with an extensive disk-like envelope seen exactly edge-on since this aspect would provide the maximum optical depth for electron scattering; and, in addition, it should yield a very high intrinsic polarization. The edge-on aspect seems consistent with the very high value of $v \sin i$ for ϕ Per (493 km s^{-1}, Bernacca and Perinotto, 1971). However, intrinsic polarization at Hβ seems more consistent with unpolarized emission if one adopts a maximum depth of 0.15, relative to the continuum, for the underlying rotationally broadened line. If electron scattering is responsible for the polarization of the Hα emission then, since the scattering cross-section is independent of wavelength, the only way Hβ emission flux can escape becoming polarized is for the physical path length through the disk to be lower. This is opposite to what one would expect. Alternatively, some other process such as resonant scattering may be responsible for the Hα line center polarization; and, of course, the observed effect may be confined to the core of the line – the bulk of the emission line flux being unpolarized. Further measurements of ϕ Per at higher spectral resolution are needed to clarify the situation.

Had we assumed Hα to be unpolarized, then somewhat different interstellar values would have been obtained, viz., $\theta_i = 92°$ and $p_{max} \simeq 0.53\%$. These values do still result in a cancellation of the rotation effects observed in broad bands in ϕ Per, but there is about three times more scatter about the mean intrinsic position angle of 26°. When the derived intrinsic polarization is normalized for purposes of comparing it to a model (Coyne and McLean, 1975), then the ultraviolet polarization is found to be very much lower, and the near infrared polarization higher, than obtained with the original interstellar values $\theta_i = 104°$ and $p_{max} = 1.00\%$.

γ Cas. The polarization changes which occur across Hα and Hβ are the most peculiar ones encountered to date. Evidence of variability and slight rotations of θ are just barely apparent for both lines. A similar weak rotation for the Hα line center has also been obtained by Coyne (private communication). If the Hα emission line flux is assumed unpolarized, then method (3.2) and Equation (1) yield the following values when applied to the point just off the line center, viz., $\theta_* \simeq 113°$, $p_*(\lambda_c) = 0.46\%$, $\theta_i = 96°$, and $p_i = 0.27\%$ near the wavelength of Hα. The value obtained for θ_i is in good agreement with that expected from the catalogue survey (see Table I). At Hβ the rotation of θ is barely discernible with the 12 Å filter, but a few preliminary measurements with the 2.3 Å filter gave the value 112° for θ_*. However, the above reasonable estimate of the interstellar polarization seems unable to account for the low values of p obtained at the extremities of the emission lines near the point of crossover with the underlying absorption. These points are well below the unpolarized emission curve and therefore imply an additional intrinsic mechanism which is reducing the continuum polarization in the line wings.

TABLE II

Polarization data on other emission line/peculiar stars

Star	Date	$\lambda/\Delta\lambda$	p (%)	θ (deg)
ε Aur	1974 Dec 11	4861/12	2.17 ± 0.02	144.0 ± 0.3
	1974 Dec 11	5100/51	2.14 ± 0.02	143.3 ± 0.3
	1975 Jan 10	4600/51	1.97 ± 0.03	143.9 ± 0.4
	1975 Jan 10	4872/12	1.98 ± 0.04	143.7 ± 0.6
	1975 Jan 10	4861/12	2.08 ± 0.04	144.3 ± 0.6
48 Lib	1975 May 9	6582/9	0.81 ± 0.11	121 ± 4
	1975 May 9	6563/9	0.79 ± 0.04	101.5 ± 1.5
	1975 Aug 24	4820/12	0.94 ± 0.05	119 ± 1.5
	1975 Aug 24	4861/2.3	0.70 ± 0.08	110.5 ± 3.3
	1975 May 9	4250/1000	0.76 ± 0.03	123.6 ± 1.1
	1975 May 13	5460/1000	0.85 ± 0.05	119.7 ± 1.7
	1975 May 13	3600/500	0.68 ± 0.04	125.0 ± 1.7
	1975 May 13	4600/51	0.83 ± 0.03	122.1 ± 1.0
	1975 May 13	5100/51	0.85 ± 0.06	118.6 ± 2.0
ζ Oph	1975 May 8	6540/9	1.46 ± 0.07	123 ± 1
	*1975 May 8, 14	6563/9	1.39 ± 0.05	126 ± 1
	1975 May 14	6582/9	1.35 ± 0.06	128 ± 1
	1975 Mar 20	5100/51	1.36 ± 0.08	126 ± 2
	1975 Mar 20	4861/25	1.34 ± 0.09	126 ± 2
	1975 May 8	4870/12	1.38 ± 0.04	129 ± 1
	*1975 May 8, 14	4861/12	1.32 ± 0.03	127 ± 1
	1975 May 14	4840/12	1.33 ± 0.05	126 ± 1

* Average of data from the two nights given.

48 Lib. Data for this star are listed in Table II. The intrinsic position angle θ_* derived from the rotation effect at Hα is 155°, while a somewhat more accurate measurement at Hβ gave 139°. The weighted mean is 144°, and the derived value θ_i is 82°. Further measurements will be required, especially at Hβ, before any final conclusions can be drawn regarding the wavelength dependence of the intrinsic polarization of this star.

X Per, ε Aur, and ζ Oph. No line profile polarization effects were found for any of these stars. Only X Per shows appreciable emission (our measurements have been reported elsewhere (Clarke and McLean, 1975c)), but the polarization of all three stars appears to be dominated by the interstellar medium. The observations of ε Aur and ζ Oph are reproduced here to indicate the stability of the polarimeter.

5. Summary and Conclusions

Changes in the degree and position angle of the observed polarization occur across the Balmer emission lines in Be and shell stars. These changes immediately indicate the presence of an intrinsic (circumstellar) polarization component. The variations in the line profile effects are partly due to differing combinations of interstellar and intrinsic polarizations and partly to processes inherent to the extended envelope of the Be star.

Of the stars discussed here, only ζ Tau shows negligible interstellar polarization, and it appears to be possible to attribute the reduction in polarization across both Hα and Hβ to dilution of continuum polarization by unpolarized emission line flux, at least in the line wings. The stars 48 Per, φ Per, γ Cas, and 48 Lib have an intrinsically polarized component superposed on an interstellar component which is usually revealed by a rotation of θ at the line center. For 48 Per the intrinsic polarization is very small (≤0.1%), indicating an almost pole-on aspect; while for φ Per the intrinsic polarization is very high (1.61% near Hα), indicating an almost exactly equator-on aspect. Both φ Per and γ Cas exhibit polarization changes across Hα and Hβ which simply cannot be attributed to unpolarized emission flux alone, but these changes are quite different for the two stars.

We have discussed methods for separating interstellar and intrinsic polarizations and have shown that line profile polarimetry is potentially useful for this purpose. In future, very accurate measurements of the intensity and polarization at several points across, (a) two of the Balmer emission lines or (b) one emission line and in a narrow band at a distinctly different wavelength, should enable one to determine the interstellar properties θ_i, p_{max}, and λ_{max}, and hence the intrinsic (circumstellar) polarization parameters $p_*(\lambda)$ and θ_*. The wavelength dependence, $p_*(\lambda)$, should enable limits to be placed on T_e and N_e, the electron temperature and density of the envelope, while the absolute values of p may give some indication of the geometry of the shell – perhaps yielding sin i and limits on M_*/R_*.

Acknowledgements

It is a pleasure to acknowledge Dr John S. Hall and all the members of his staff for the generous facilities and hospitality provided at the Lowell Observatory. We also wish to thank Mr T. H. A. Wyllie for informative discussions of the observations. This work was supported by a British Science Research Council grant to Dr David Clarke and a post-doctoral fellowship to Dr Ian S. McLean.

References

Bernacca, P. L. and Perinotto, M.: 1971, *Contr. D'Osservat. Astr. D'Univ. Di Padova in Asiago*, No. 249.

Blanco, V. M., Demers, S., Douglass, G. G., and Fitzgerald, M. P.: 1968, *Photoelectric Catalogue*, Publ. U. S. Naval Obs.

Capps, R. W., Coyne, G. V., and Dyck, H. M.: 1973, *Astrophys. J.* **184**, 173.

Cassinelli, J. P. and Haisch, B. M.: 1974, *Astrophys. J.* **188**, 101.

Clarke, D. and McLean, I. S.: 1974a, in T. Gehrels (ed.), *Planets, Stars and Nebulae Studied with Photopolarimetry*, IAU Colloquium No. 23 (1972), Univ. of Arizona Press, Tuscon, p. 752.

Clarke, D. and McLean, I. S.: 1974b, *Monthly Notices Roy. Astron. Soc.* **167**, 27P.

Clarke, D. and McLean, I. S.: 1975a, *Monthly Notices Roy. Astron. Soc.* **172**, 545.

Clarke, D. and McLean, I. S.: 1975b, *Monthly Notices Roy. Astron. Soc.* (1976) **174**, 335.

Clarke, D. and McLean, I. S.: 1975c, *Monthly Notices Roy. Astron. Soc.* **173**, 21P.

Clarke, D., McLean, I. S., and Wyllie, T. H. A.: 1975, *Astron, Astrophys.* **43**, 215.

Coyne, G. V.: 1974, *Monthly Notices Roy. Astron. Soc.* **169**, 7P.

Coyne, G. V. and Kruszewski, A.: 1969, *Astron. J.* **74**, 528.

Coyne, G. V. and McLean, I. S.: 1975, *Astron. J.* **80**, 702.

Hall, J. S.: 1958, *Publ. U.S. Naval Obs.* **17**, Pt. VI.

Hayes, D. P.: 1975, *Publ. Astron. Soc. Pacific* **87**, 609.

Hayes, D. P. and Illing, R. M. E.: 1974, *Astron. J.* **79**, 1430.

Hiltner, W. A.: 1956, *Astrophys. J. Suppl.* **2**, 389, No. 24.

Hoffleit, D.: 1964, *Catalogue of Bright Stars*, Yale Univ. Obs., New Haven, Conn.

Johnson, H. L.: 1958, *Lowell Obs. Bull.* No. 90, 37.

Marlborough, J. M.: 1969, *Astrophys. J.* **156**, 135.

Marlborough, J. M. and Cowley, A. P.: 1974, *Astrophys. J.* **187**, 99.

Poeckert, R.: 1975, *Astrophys. J.* **196**, 777.

Serkowski, K.: 1968, *Astrophys. J.* **154**, 115.

Serkowski, K.: 1973, in J. M. Greenberg and H. C. van de Hulst (eds.), 'Interstellar Dust and Related Topics', *IAU Symp.* **52**, 145.

Zellner, B. H. and Serkowski, K.: 1972, *Publ. Astron. Soc. Pacific* **84**, 619.

DISCUSSION

McLean: I would like to take this opportunity to announce that a program of line profile polarimetry on the eclipsing binary β Lyr was initiated in the spring of 1975. Polarization changes were observed across both Hα and Hβ and these lines were in emission.

Finally, line profile and broad band observations in the early stages of Nova Cyg 1975 did not reveal any evidence of intrinsic polarization.

INTRINSIC LINEAR POLARIZATION OF Be STARS AS A FUNCTION OF $v \sin i$

R. POECKERT and J. M. MARLBOROUGH

University of Western Ontario, London, Ontario, Canada

Abstract. The polarization of 48 Be stars has been measured in two bands near Hα with the aim of determining the relation between intrinsic polarization and $v \sin i$. A technique developed by Poeckert (1975) is used to remove the effect of interstellar polarization. It is found that intrinsic polarization depends strongly on $v \sin i$; stars with low $v \sin i$ having little or no polarization. We have calculated the i dependence of linear polarization for a disk model envelope and find that the polarization is proportional to $\tau_e \sin^2 i$ when the disk is optically thin (τ_e is a characteristic electron scattering optical depth). A comparison of the observed relation between intrinsic polarization and $v \sin i$, and that predicted for the disk model is illustrated. We find that an envelope with an electron density of $\leq 5 \times 10^{11}$ cm^{-3} can account for the degree of intrinsic polarization observed in all the program stars. The fact that stars of low $v \sin i$ have little intrinsic polarization is evidence for the assumption that these stars are seen pole-on and that the envelopes around these stars are axi-symmetric. No apparent difference between pole-on stars and extreme Be stars was obtained.

Reference

Poeckert, R.: 1975, *Astrophys. J.* **196**, 777.

PART IV

LINE FORMATION IN EXPANDING
ATMOSPHERES

LINE FORMATION IN EXPANDING ATMOSPHERES

(*Review Paper*)

D. G. HUMMER*

Joint Institute for Laboratory Astrophysics, National Bureau of Standards and University of Colorado, Boulder, Colo., U.S.A.

Abstract. The current state of understanding of line formation processes in expanding atmospheres is reviewed, and the successes and limitations of current computational techniques are summarized. Some results for differential rotation are also given, although very little work has been done in this area. Special attention is given to the severe difficulties that are encountered in inferring the structure of rapidly expanding or rotating atmospheres from observed line profiles because of the failure under these conditions of the Eddington-Barbier relation in integrated light; the value in this respect of continuum and interferometric observations is emphasized.

1. Introduction

The purpose of this review is to summarize our current understanding of radiative transfer and line formation processes in expanding atmospheres and to indicate the status of theory and computational development in this area. Because very little work has been done on the effects of differential rotation, primary emphasis is given to situations in which the gas flows radially outward; much of the discussion is also relevant to situations in which the gas flows inward. Limitations of space force us to ignore work on optically thin atmospheres despite its historical importance; these essentially geometrical arguments have been instrumental in the determination of the basic structure of the object we are considering, and are included naturally in the most recent computational techniques. It is hoped that this effort will be useful as an introduction to the field for the nonspecialist and as a guide to the literature. Although bibliographic completeness was not the primary consideration, the coverage is quite comprehensive.

It is convenient to consider separately low-velocity flows, in which the dispersion in flow speeds is less than or on the order of the mean thermal speed, and high-velocity flows, which involve speeds much larger than thermal; these cases are discussed in Sections 2 and 3, respectively. Although the important physical processes are the same for both cases, the useful phenomenological descriptions and the computational procedures are very different. The basic ideas of line formation in expanding atmospheres are clearly and completely discussed by Rybicki (1970), who also includes significant original work. Readers unfamiliar with the modern ideas of line formation in static atmospheres may find helpful the brief review by Hummer and Rybicki (1971b).

1.1. NOTATIONAL PRELIMINARIES

Throughout this review $v(r)$ or $v(\tau)$ is used for the radial velocity law, which is here always regarded as known; following the convention of most theoretical work in this

* Staff Member, Laboratory Astrophysics Division, National Bureau of Standards and Department of Physics and Astrophysics, University of Colorado.

A. Slettebak (ed.), Be and Shell Stars, 281–312. All Rights Reserved
Copyright © 1976 by the IAU.

area, v is taken to be positive for motion *towards* an observer situated outside the atmosphere. It is convenient to introduce the mean thermal speed of an atom of mass M at a typical temperature T,

$$v_{\text{th}} = \sqrt{2kT/M}, \tag{1.1}$$

as the unit of velocity, and to define

$$u = v/v_{\text{th}}. \tag{1.2}$$

The frequency displacement from the line center frequency ν_0 is conveniently measured in thermal Doppler units corresponding to v_{th}:

$$x = (\nu - \nu_0)/\Delta_0, \tag{1.3}$$

where

$$\Delta_0 = \nu_0 v_{\text{th}}/c. \tag{1.4}$$

A subtle point arises in the use of the variable x in transforming between inertial frames according to the well-known first-order Doppler formula

$$\nu^c = \nu^s(1 - \mu v/c), \tag{1.5}$$

where ν^s is the frequency of a photon measured in the frame of the stationary observer and ν^c is the frequency in a frame moving with a component of velocity $v\mu$ in the photon's direction of propagation; (below we shall identify the latter frame with the frame co-moving with the gas). If the definition (1.3) is applied consistently in each frame, i.e. $\tilde{x}^c \equiv (\nu^c - \nu_0^c)/\Delta_0^c$, then $\tilde{x}^c = x^s$. It is customary and more useful to define

$$x^c \equiv \frac{\nu^c - \nu_0^c}{\Delta_0^c}, \qquad x^s \equiv \frac{\nu^s - \nu_0^c}{\Delta_0^c} \tag{1.6}$$

with the result that, to first order in v/c,

$$x^c = x^s - u\mu; \tag{1.7}$$

this result is valid as well when Δ_0^c in (1.6) is replaced by Δ_0^s.

The depth in the atmosphere will be measured on the *mean* optical depth scale τ, which is related to the geometrical depth z by

$$d\tau = k\,dz, \tag{1.8}$$

where

$$k = (N_1 B_{12} - N_2 B_{21})h\nu_0/4\pi\,\Delta_0 \tag{1.9}$$

is the integrated line opacity. This definition is unaffected by the velocity field. For finite slabs the optical thickness on the τ-scale is represented by T.

2. Low Velocity Flows

When the maximum flow speed is less than a few times the mean thermal speed and the gradient is small, the most important effect of the velocity field is to distort the monochromatic optical depth scale. The monochromatic optical depth $\tau_{x,\mu}$ of a layer at mean depth τ, measured from the surface of a planar atmosphere for specified values of the frequency x and the direction cosine μ is

$$\tau_{x,\mu}(\tau) = \mu^{-1} \int_0^\tau d\tau' \, \phi[x - \mu u(\tau')], \tag{2.1}$$

which, for a Doppler profile and a linear velocity law

$$u = u_0 + u_1 \tau, \tag{2.2}$$

can be readily evaluated in terms of the error function. The Eddington-Barbier relation states that emergent intensity for specified x and μ is approximately equal to the source function at the mean optical depth $\tau = \tau^*(x, \mu)$ for which the monochromatic optical depth is unity, i.e. where $\tau^*(x, \mu)$ is the solution of

$$\tau_{x,\mu}(\tau) = 1. \tag{2.3}$$

We give, in Table I, values of $\tau^*(x, \mu = 1)$ for $u_0 = 0$ and $u_1 \geq 0$. As changing the sign of u leads simply to a change in sign of x, and giving u_0 a non-zero value is equivalent to replacing x by $x - \mu u_o$, it is sufficient for the present purpose to consider these cases in which the velocity vanishes at the surface (although in reality the flow vanishes below some point in the atmosphere).

We see from Table I that the effect of a positive velocity with a positive gradient (i.e. flow decelerating outward) is to redistribute the line opacity so that the longward side of the line is more transparent than the shortward side (also more transparent than in a static atmosphere). This implies that if the source function increases inwards for a sufficient distance, as in a normal stellar atmosphere, the radiation emerging on the longward side will be more intense than on the shortward side, so that an emission line will appear skewed to the red and an absorption line will be stronger in the blue. However, numerical calculations with $u = 0.01\tau$ and $S(\tau) = \tau$ yield an emission line with the blue component much stronger than the red; the Eddington-Barbier relation holds in the region of the line core and out to about $x = 5$, but fails completely in the red wing. Conditions for the validity of the Eddington-Barbier relations and alternative expressions valid at all frequencies are given by Hummer and Kunasz (1976).

Atmospheres with the flow accelerating outwards have been investigated numerically, using a realistic line-formation theory, by Hummer and Rybicki (1968a, b) for plane slabs of finite thickness and by Magnan (1968) and Mathis (1968) for expanding spherical atmospheres using Monte Carlo techniques and Λ-iteration, respectively. For optical thickness of order 50 and gradients of order $u_1 \simeq -0.1$, the ratio of red to blue peak heights was of order 1.5. In these cases the Eddington-Barbier relation appears to be fairly reliable. Although Magnan (1968) found that radiation from the receding hemisphere strengthened the red component, Hummer and Rybicki (1968b) showed that this skewing to the red was not caused primarily by receding material by considering a model in which the rear half of the slab was at rest, and the front half expanding. The explanation given above for the behavior of

TABLE I

Values of $\tau^*(x)$ for frequency x in the red (r) and blue (b) sides of the lines at which the monochromatic optical depth $\tau_x(\tau) = 1$ for velocity laws $u(\tau) = u_1\tau$. Here $\mu = 1$. Dashes indicate that the equation $\tau_x[\tau^*(x)] = 1$ has no solution, i.e. that any atmosphere is optically thin at these frequencies.

u_1		0	0.25	0.50	0.75	1.0	1.25	1.50	1.75	2.00	2.25	2.50
							$|x|$					
0.0		1.772	1.887	2.276	3.111	4.818	8.456	1.68+1	3.790+1	9.677+1	2.800+2	9.181+2
0.01	r	1.773	1.896	2.303	3.187	5.071	9.530	2.394+1	–	–	–	–
	b	1.773	1.878	2.251	3.041	4.603	7.685	1.369+1	2.452+1	4.115+1	6.239+1	8.604+1
0.03	r	1.774	1.916	2.361	3.363	5.748	1.454+1	–	–	–	–	–
	b	1.774	1.862	2.205	2.919	4.254	6.625	1.052+1	1.612+1	2.312+1	3.092+1	3.907+1
0.10	r	1.791	2.009	2.647	4.532	–	–	–	–	–	–	–
	b	1.791	1.823	2.079	2.606	3.491	4.815	6.583	8.700	1.103+1	1.347+1	1.595+1
0.30	r	1.984	2.795	–	–	–	–	–	–	–	–	–
	b	1.984	1.817	1.902	2.171	2.605	3.185	3.877	4.642	5.446	6.269	7.099

absorption and emission lines in low-speed flows was apparently first given by
Hummer and Rybicki (1968a).

It is interesting to contrast this behavior with that when no gradient exists, as for
example, when a geometrically thin spherical shell of gas expands with a *constant*
velocity. Then both absorption and emission lines in the *flux* as seen by a distant
observer at rest with respect to the center of the star are blue shifted. This simple
consequence of the angular dependence of the Doppler formula (1.5) holds, of
course, for both large and small velocities. Underhill (1947) applied this transformation to discrete-ordinate solutions of the static plane-parallel Schuster problem
visualized as a thin spherical shell, and obtained a blue-shifted absorption profile
with a strong red wing extending nearly to the red edge of the static profile.

Although small velocity fields can have quite profound effects on the emergent
radiation field, the line source function is affected very little. The contrasting
behavior of the emerging radiation and the internal excitation is shown dramatically
in Figures 1 and 2, which represent the line source function and the flux profiles

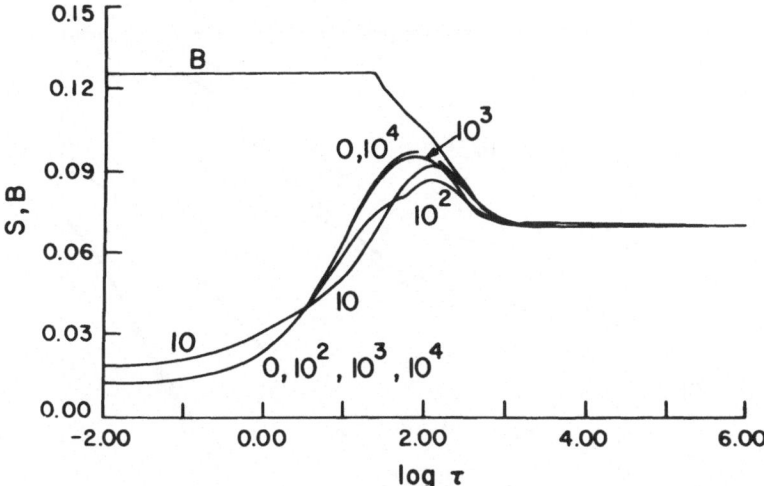

Fig. 1.   Planck functions and line source functions for $\varepsilon = 10^{-2}$ and $u(\tau) = 10/(1 + \tau/T)$, $T = 10, 10^2, 10^3$,
10^4. The static case is labeled by 0 (from Kalkofen, 1970).

calculated by Kalkofen (1970) for a semi-infinite planar atmosphere with the velocity
law $u(\tau) = 10(1 + \tau/T)^{-1}$, where the parameter T has the values 10, 10^2, 10^3, 10^4.
The temperature is taken to be constant below $\tau = 10^3$; above that point it increases
to give the Planck function shown in Figure 1. In many situations the source function
is sufficiently insensitive to the velocity field that it may be calculated in a static
approximation; only in calculating the emergent radiation field must the flow be
taken into account.

This weak dependence of the source function on the flow is readily explained. The
shift of line opacity from the core to the shortward side of the line does not reduce the
core opacity enough to allow significant flux to be carried there, and the effect of
impeding the flow of radiation in one wing is not serious so long as radiation can

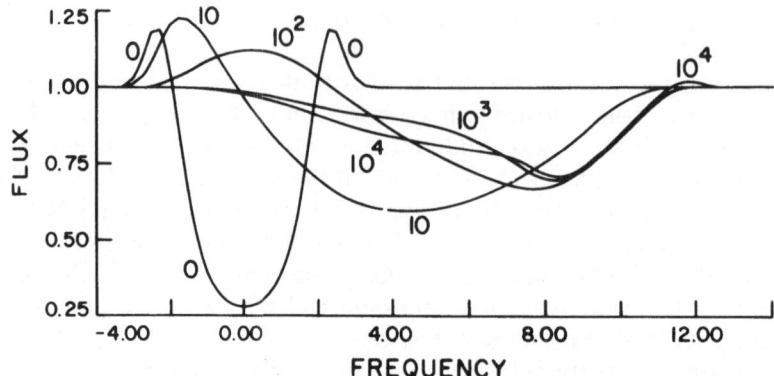

Fig. 2.   The emergent monochromatic flux corresponding to source functions in Figure 2. The frequency is measured from line center in thermal Doppler units (from Kalkofen, 1970).

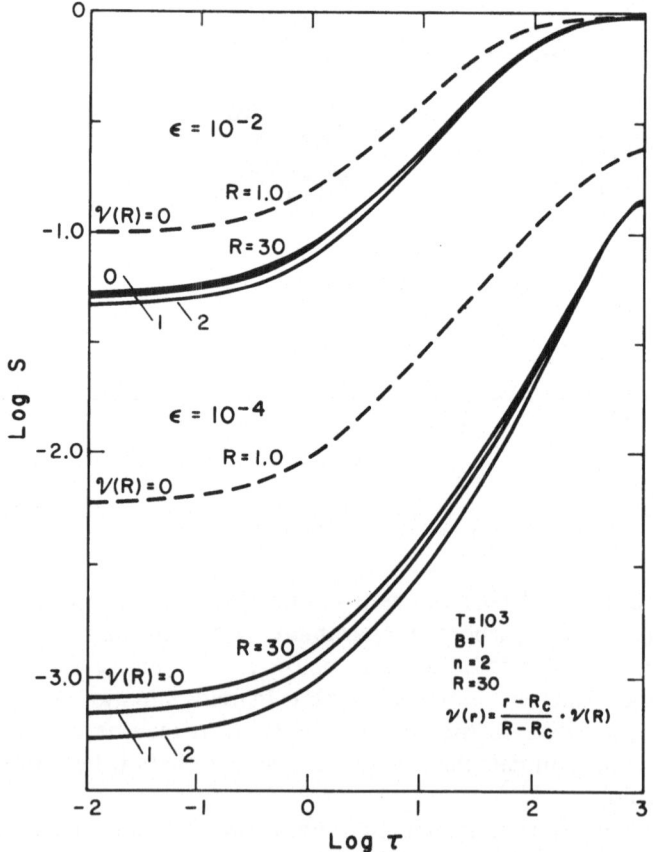

Fig. 3.   Line source functions for $R = 30$, $T = 10^3$, $B = 1$, $k(r) = kr^{-2}$ and a linear velocity law. Curves are labeled by $V(r) = [u(r)$ in text]. Broken lines given corresponding plane-parallel static limits (from Kunasz and Hummer, 1974b).

escape freely in the other wing; consequently the photoexcitation of the line-forming atoms is not substantially altered. Generally, the degree of excitation is slightly reduced for the static values, although in situations in which the velocity field is confined to the surface layers, the excitation is increased because of photoexcitation by the continuum. When the line opacity is spread from the core into both wings, either by thermal gradients, as first discussed by Hummer and Rybicki (1966) and Rybicki and Hummer (1967), or by non-monotonic velocity fields, as considered by Shine and Oster (1973), the source functions at depth can be quite large compared to the isothermal or static values, respectively.

The source function appears to be more sensitive to the geometry of the atmosphere than to its velocity field, at least for velocities less than a few times v_{th}. Figure 3, taken from Kunasz and Hummer (1974b), contrasts the line source functions for an internally excited (by electron collisions, for example) spherical shell with inner and outer radii of 1 and 30, respectively, with the source function for a planar slab of the same optical thickness ($T = 10^3$); in both cases a velocity linear in radius is assumed. The sensitivity to both the geometry and the flow increases as $\varepsilon \approx C_{21}/A_{21}$ decreases, i.e. as the effect of collisions decreases. In spherical geometry a velocity field $v(r)$ also induces a transverse gradient $v(r)/r$ in addition to the radial gradient dv/dr, which facilitates the escape of photons from deep layers. This effect, which has no counterpart in planar atmospheres with flow normal to the surface, is illustrated nicely in Figures 9 and 10 of Kunasz and Hummer (1974), which also show the sensitivity of the flux profile to velocities as small as a few percent of v_{th}.

Rybicki (1970) makes the important distinction between the divergence of the rays of radiation and the divergence of the velocities in a spherical atmosphere. It is quite possible that the rays do not diverge appreciably in an atmosphere in which the velocities do. For example, in a thin spherical shell of radius R and thickness ΔR, where $\Delta R \ll R$, the divergence of the rays is unimportant, but if $v \, \Delta R/R$ is comparable with v_{th}, the lateral motion of two points at the same radius in the shell separated by a distance of order ΔR is clearly significant. In this case it is clearly necessary to allow for the transverse gradient although the usual condition for the plane-parallel treatment of the radiation is satisfied. Rybicki derives a novel form of the planar transfer equation that allows for the inclusion of the transverse gradient.

The effects of rotation, alone and combined with expansion, have been studied (in the low velocity limit) only by Magnan (1970), who used the Monte Carlo method. His results, for a disc with the outer radius four times the inner and an optical thickness of 50, are reproduced in Figure 4. The rotational speed at every radius is $5v_{th}$ and the expansion velocity, in units of v_{th}, labels the curves.

Before going on to discuss numerical techniques, it is worth mentioning a very special transfer problem for which the essential properties of the radiation field can be obtained by approximate analytical arguments. Kahn (1968) has investigated the $L\alpha$ radiation field in an ionized region surrounded by an expanding spherical shell of neutral hydrogen that acts as a partially reflective enclosure. He allows for the reddening of the radiation by scattering from the expanding shell and by scattering within the ionized region; the latter problem is reduced to one of diffusion type in frequency space. By further simple considerations of the leakage of photons through the expanding shell, Kahn shows that the photons escape almost entirely in the

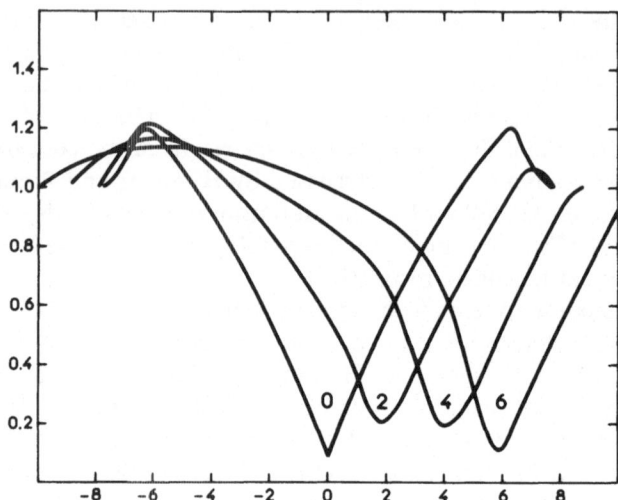

Fig. 4.   Flux profiles relative to continuum vs x for rotating disc with optical radius = 50, outer radius = 4 times the inner radius and rotational velocity of 5 thermal units at all radii. Curves are labelled by expansion velocity in thermal units (from Magnan, 1970).

extreme red wing of the line, with redshifts on the order of one percent of the line center frequency. He also discusses the effect of radiation pressure on the dynamics of the shell. Although these results were obtained in the context of planetary nebulae, the treatment appears to be applicable to the circumstellar region of hot stars.

2.1. NUMERICAL TECHNIQUES

For non-relativistic flow the transfer equation for a differentially moving atmosphere differs from that for the static case only in that the normalized profile of the line absorption coefficient $\phi(x)$ is replaced by $\phi(x - u\mu)$. Here μ is the cosine between the direction of propagation of radiation and the radius vector. Therefore most numerical differential equation techniques can be immediately applied to expanding atmospheres. The practical consequence of expansion is only that in differencing the transfer equation a wider range of frequencies and a large number of angle points must be used; the resulting system of linear algebraic equations can become extremely large, especially in extended atmospheres where the outward peaking of the radiation field necessitates a very fine angular mesh. In the solution of this system, two types of procedures – so-called elimination schemes – are in use, each with its advantages and disadvantages. We digress briefly to describe these schemes.

2.1.1. *Elimination Schemes*

The difference equations are ultimately to be solved for a quantity $y_{\gamma,d} = y(x_i, \mu_j, \tau_d)$, $\gamma = (i, j)$, $\gamma = 1, 2, \ldots, F$, $d = 1, 2, \ldots, N_1$ related to the radiation field with frequency x_i, direction cosine μ_j and optical depth τ_d; F and N are the total number of frequency-angle points and depth points, respectively. In the scheme introduced by Feautrier (1964) the quantities $y_{\gamma,d}$ are organized into vectors labelled by the index d, in which the components are labeled by γ:

$$\mathbf{y}_d \equiv (y_{1d}, y_{2d}, \ldots, y_{Nd}), \qquad d = 1, 2, \ldots, N. \tag{2.4}$$

Rybicki (1971) recognized that in cases for which the number F of frequency-angle points is very large, it is advantageous to use the alternative arrangement

$$\mathbf{y}_\gamma = (y_{\gamma 1}, y_{\gamma 2}, \ldots, y_{\gamma N}) \qquad \gamma = 1, 2, \ldots, F; \tag{2.5}$$

however this scheme can be used only if the line source function is independent of frequency and angle, i.e. depends on y only through a combination of the form

$$J_d = \sum_{\gamma=1}^{F} W_\gamma y_{\gamma d}, \tag{2.6}$$

where the quantities W_j are known constants. Estimates of the number of multiplications and divisions, and of the required storage space in the fast core, are given in Table II; the quantities C, C' and C'' are constants. The superiority of the Rybicki scheme in dealing with large numbers of frequency-angle points is clear.

TABLE II

Comparison of elimination scheme

	Feautrier	Rybicki
Operations	CNF^3	$C'N^2F + C''N^3$
Storage	NF^3	$2N^2$

2.1.2. *Variable Eddington Factors*

Variable Eddington factors, defined as the ratios of certain angular moments of the radiation field, provide the basis for an alternative strategy for cases in which the number F of frequency-angle points is very large, and the dependence of the line source function on the radiation field is more complex than (2.6). Although this procedure does not require angular discretization, so that F is simply the number of frequency points, it does call for a rapidly-convergent iterative process. This procedure was applied to static problems in planar geometry by Auer and Mihalas (1970) and to continuum and static-line problems in spherical atmospheres by Hummer and Rybicki (1971b) and Kunasz and Hummer (1974a), respectively. In this latter work the integrating factor introduced by Auer (1971) is essential.

2.1.3. *Differential Equation Solutions*

The earliest numerical solutions were those of Chandrasekhar (1945a, b), who solved the plane-parallel Schuster problem and the planetary nebula Lyman-α problem in an expanding slab; the techniques used in this work have provided the basis for much of the work on high velocity flow and will be discussed further in Section 3.0. Abhyankar (1964a, b, 1965, 1967) retained Chandrasekhar's assumptions of monochromatic scattering and a two-stream (Eddington approximation), but by using an iterative technique based on transmission and reflection functions for thin layers derived from the transfer equation, he was able to treat more general velocity laws than Chandrasekhar and to use a Doppler profile. Kulander (1967) considered internally-excited atmospheres and used a generalization of the discrete-ordinate method to treat complete frequency redistributions, but was restricted to piece-wise constant velocity laws. In a subsequent paper, Kulander (1968) treated linear expansion by a method based on the superposition of basic solutions of the transfer equations regarded as a one-point boundary problem; the well-known instability of this approach limited him to relatively thin layers. Mathis (1968) used the slowly-convergent Λ-iteration method, but included partial redistribution and assumed spherical geometry. All of these techniques are now primarily of historical interest.

Most of the current work is based on a generalization of the method devised by Feautrier (1964), of which the elimination scheme has just been mentioned. Although this method is ideally suited for use with small velocity flows, and apparently had been so used by L. Auer in unpublished work as early as 1967, the first published results using complete frequency redistribution and internal excitation are those of Hummer and Rybicki (1968), who used the non-linear Riccati method. Kulander (1971) also used this technique to treat a large number of schematic models. The Feautrier method has proved superior to the Riccati method, however, and has been used extensively by many authors in studying moving atmospheres, including Rees (1970), Vardavas (1974), and Cannon and Vardavas (1974). In the latter two papers angle-averaged redistribution functions are used; the results will be discussed in Section 3.5. Cannon and Cram (1974) have generalized the Feautrier procedure further to account for the advection of material in high-speed flows, and Cannon (1974), in considering the propagation of pulses through an atmosphere, has introduced the conservation equations of gas dynamics into the Feautrier procedure. Cannon and Rees (1971) have developed a version of the Feautrier method to treat two-dimensional transfer in the presence of a two-dimensional velocity field; recent developments, including the use of Rybicki-type elimination, allow this problem to be solved much more efficiently (D. Mihalas, private communication). All of the above work has been restricted to planar geometries and has employed Feautrier elimination in solving the linear algebraic system obtained by differencing the transfer equation. Kunasz and Hummer (1974b) obtained solutions to the line transfer problem in expanding (or contracting) spherical atmospheres using the Rybicki elimination scheme.

2.1.4. *Integral Equation Solutions*

Much less work has been done using the integral form of the transfer equation, probably because the generalization of the standard techniques to allow for angle-dependent line opacities is less simple than for the differential equations. The integral equation methods are attractive because they scale like the Rybicki elimination scheme. Kalkofen (1970) has carried out the necessary generalization for expanding planar atmospheres and obtained the results presented above in Figures 2 and 3, for $u_{max} = 10$. This work illustrated clearly the necessity of using a large number of angular points when $v_{th} \gtrsim 1$. For a spherical atmosphere expanding with the law $v(r) = \text{const} \cdot r$, the integral equation takes an especially simple form, which has been exploited by Robbins (1968) to solve the transfer problem including angle-averaged redistribution (cf. Section 3.5) for lines in the helium triplet system.

3. High Velocity Flows

In situations where the flow speeds are much larger than the mean thermal speed, it seems reasonable that the gross features of the radiation field in lines will be established by the macroscopic flow field, while the details are determined by the thermal motion of the atoms. This separation of roles has been exploited in the development of a very useful approximate theory, to be discussed below; of equal importance is the phenomenological picture of the line formation process to which it leads. Moreover, there are important conceptual and computational advantages in describing radiative transfer from the point of view of an observer in the co-moving frame of the gas, rather than from that of an external observer in a stationary frame at rest with respect to the center of the star. The frequency displacement x^s of a photon seen in the stationary frame is related to that in the co-moving frame x^c as given above by Equation (1.7).

As we here suppose that $v \gg v_{th}$, let us assume for the moment that photons are emitted only at line center in the co-moving frame, i.e. with $x^c = 0$. Then all photons seen in the stationary frame with a specified frequency x^s are emitted from surfaces on which

$$u(r)\mu = x^s,$$

i.e. the so-called 'constant velocity surfaces' on which the line-of-sight component of velocity has the value x^s. In Figure 5 appear some constant velocity surfaces for

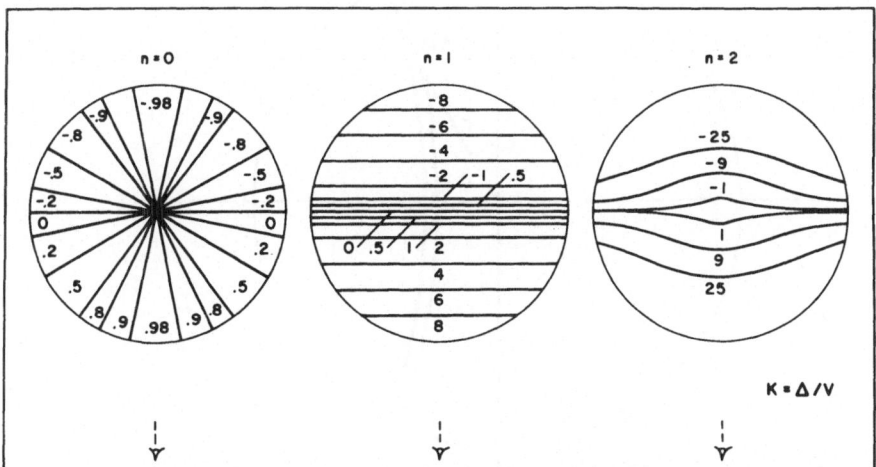

Fig. 5.   Typical constant velocity surfaces for $u = u_0(r/r_0)^n$, labeled by x/u_0. Line of sight is vertical from bottom of figure.

spherical atmospheres in which $u(r) = u_0(r/r_0)^n$, $r > r_0$; $n = 0, 1, 2$; here $r_0 = 1$ and the outer radius is $10\, r_0$. The curves are labelled with the value of (x/u_0).

It is clear at once that a given surface will extend over a non-zero range of radii, Δr, so that there is no way of knowing at what depth radiation with a specific frequency is formed. The familiar Eddington-Barbier relation, which is so ingrained in our thoughts, is now quite inapplicable. Consequently it is *in principle* impossible to infer the depth dependence of quantities such as temperature and density from observed line profiles with a resolution smaller than Δr. Moreover this situation implies that modeling will be severely nonunique. Because the location of the surface is independent of all opacity considerations, no additional information is obtained by using lines of varying strengths. Of course, the state of ionization and excitation will vary with the radius, so that the emission of a particular line *could* be confined to a relatively narrow range of radii, i.e. small compared to r; in other words only those parts of the surface that actually contain ions in the correct state will in fact contribute to the observed line profile.

In order to disentangle the atmospheric structure from the effects of the large-scale velocity field it is most helpful to use continuum observations over a very wide spectral range to infer the gross features of the atmosphere. Unfortunately this approach is at present hampered by the necessity of making large reddening corrections with sufficient accuracy to uncover intrinsic effects on the order of 0.01 mag. The advent of new and increasingly powerful interferometric techniques,

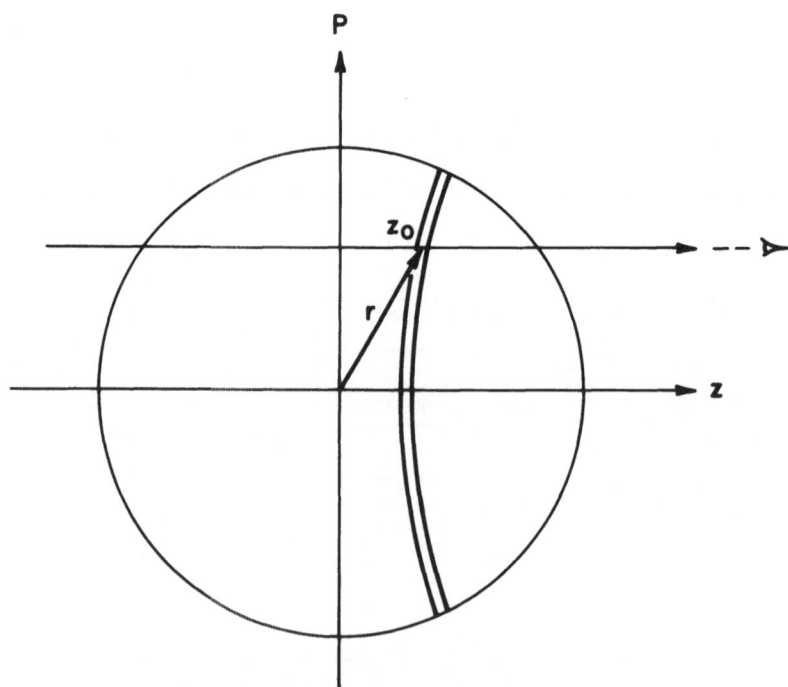

Fig. 6.   Intersection of line of sight with constant velocity surface.

such as speckle and intensity interferometry, promises additional vital information. To capitalize on these innovations, it is essential in all modeling efforts to calculate quantities of interest such as limb-darkening laws and effective disc sizes for a wide range of frequencies.

Unfortunately, not all velocity laws lead to constant velocity surfaces that intersect any line of sight only once, as in Figure 6. Then radiation of a given frequency seen along a given line of sight comes from all of its intersections with the surface and the problem of inferring atmospheric structures from the profile is more difficult than ever. Even worse, from the theoretical point of view, is the fact that the regions around the points of intersection 'see' one another, so that the conditions of excitation at quite different radii are radiatively coupled. It is readily apparent that surfaces intersecting each line of sight uniquely are obtained if the velocity and its radial derivative are both positive, or both negative, throughout the atmosphere. In Figure 7 typical surfaces are shown for two velocity laws that violate these conditions, $u(r) = u_0(r/r_0)^{-1}$ and $u(r) = u_0[(r/r_0) - 2]$.

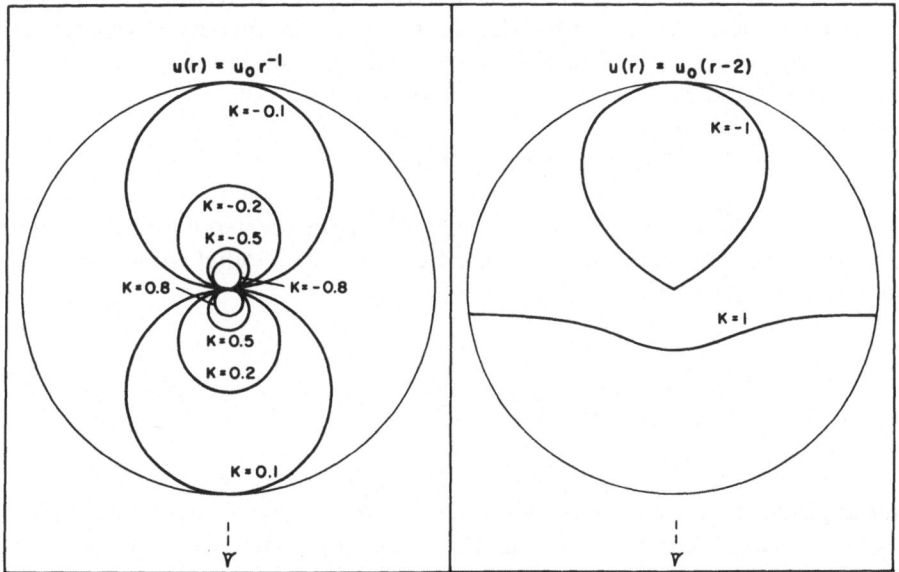

Fig. 7. Typical constant velocity surfaces for $u = u_0(r/r_0)^{-1}$ and $u = u_0(r - 2)$. Line of sight is vertical from bottom of figure.

Because the thermal speeds are not strictly zero, each constant velocity surface is more accurately thought of as a rather fuzzy thin shell with a geometrical thickness along a line of sight given essentially by the ratio of the mean thermal speed to the macroscopic velocity gradient in that direction. The corresponding optical thickness is the geometrical thickness times the total integrated line opacity per Doppler width, i.e., the quantity k defined by Equation (1.9). If we now consider a given line of sight intersecting a shell corresponding to a specific frequency, as illustrated in Figure 6, we see that the emissivity, as well as the opacity, at that frequency is accumulated only within the shell. Mathematically, the optical depth and the emergent intensity are expressed by integrals over a Gaussian function of $x - u\mu$, where $u \gg 1$. To a very good approximation the only contribution to the integral comes from the region where the argument of the Gaussian is zero, i.e., in the shell.

One consequence of this localization of opacity and emissivity is that the transfer problem is confined to the shell. Although the optical depth of the shell is not necessarily small, it is finite, even in an atmosphere extending to infinity in three dimensions. In other words, the velocity field has induced in the atmosphere an *intrinsic escape process* for photons. When the probability of escape arising from the velocity field is sufficiently large, a very great simplification of the line transfer problem results; this fact is the basis of the so-called Escape Probability Method (EPM) which has been so useful in recent years.

3.1. THE ESCAPE PROBABILITY METHOD

Although the escape probability can be calculated in several ways, perhaps the simplest is a direct generalization of the derivation given by Zanstra (1949) for the static case; see Rybicki (1970) for details. For a line of integrated opacity k in a spherical atmosphere expanding with speed $u(r)$, the total optical depth along a line of sight for a photon of frequency x, in the observer's frame, is

$$\tau_x = \int_{-\infty}^{\infty} dz\, k(z)\phi[x - u(z)\mu(z)], \tag{3.1}$$

where z is the distance along the line of sight. Changing the variable of integration to $x' = x - u(z)\mu(z)$ and accounting for the transverse and radial components of the velocity gradient, we have

$$\tau_x = \frac{k}{|f(\mu)|} \int_{-\infty}^{x} \phi(x')\, dx', \tag{3.2}$$

where we have evaluated $k(z)$ and

$$f(\mu) = \mu^2 u'(r) + (1 - \mu^2)u(r)/r \tag{3.3}$$

on the appropriate constant velocity surface, i.e., where ϕ reaches its maximum. The probability that a photon emitted in the frequency interval $x, x + dx$ and in the element of solid angle $d\Omega$ escapes without further scattering is

$$\frac{1}{4\pi} d\Omega\, dx\phi(x) \exp(-\tau_x) \tag{3.4}$$

and summing over all frequencies and directions, we find for the net escape probability

$$\beta(r) = \frac{1}{k} \int_0^1 f(\mu)[1 - \exp(-1/f(\mu))]\, d\mu ; \tag{3.5}$$

the corresponding result for planar geometry is obtained by using $f(\mu) = \mu^2 u'(r)$. Thus the escape probability at radius r is a function only of the velocity and its radial derivative, and of the integrated line opacity. In deriving the expression (3.5) we neglected the probability that the photon would strike the core. This effect is included in another escape probability β_c which is important in treating the interception of photons by the core; β_c is given approximately by $W\beta$, where W is the dilution factor at radius r. By applying the foregoing localization arguments to the exact expression for $\bar{J}$, the mean intensity integrated over the line profile factor $\phi(x)$, which enters the radiation rates in the statistical equilibrium equation, one obtains

$$\bar{J} = (1-\beta)S(r) + \beta_c I_c, \tag{3.6}$$

where I_c is the intensity in the photospheric continuum near the line and $S(r)$ is the line source function. Then $\bar{J}$ can be eliminated using the equation of statistical equilibrium, which for the two-level model atom is

$$S = (1-\varepsilon)\bar{J} + \varepsilon B. \tag{3.7}$$

Here ε is the probability per scattering that a photon is lost from the line by collisional de-excitation and B is the Planck function at ν_0 for the local electron temperatures. The resulting expression for $S(r)$ can be inserted in the usual expressions for the emergent flux profile.

For a semi-infinite isothermal planar atmosphere with a constant gradient (in τ),

$$\gamma \equiv du/d\tau = \text{const}, \tag{3.8}$$

the exact surface value of the source function is readily shown to be

$$S(0) = \varepsilon B/\sqrt{\varepsilon + \beta - \varepsilon\beta}; \tag{3.9}$$

here β is the escape probability (3.5) specialized to the planar case. This result, which contrasts explicitly the two mechanisms of velocity-induced photon escape and photon destruction by collisional de-excitation was implied by Rybicki (1970) and derived explicitly by Magnan (1974a), who also obtained the exact reflection coefficient for the same model. Frisch and Frisch (1975) gave an expression for the mean number of scatterings for a photon escaping from the same model atmosphere; they too derive the value of the surface source function.

3.1.1. *Development and Applications of the EPM*

The EPM was systematically developed by Sobolev (1947, 1952, 1957), who used it to study recombination and other spectra where the emissivity, as obtained by solving the equations of statistical equilibrium, was either independent of depth or varied in a simple known manner. Subsequent applications by Rublev (1961, 1964) and Lyong (1967) involved a variety of *ad hoc* assumptions concerning the distribution of emissivity. The present state of the theory, in which the emissivity at every point is determined by a self-consistent solution of the equations of statistical equilibrium, is due to Castor (1970), Rybicki (1970), and Lucy (1971). Castor and Rybicki both work from the integral forms of the transfer equation and assume complete redistribution in the co-moving frame; Castor gives explicit expressions for all quantities involved for the case of spherical geometry, while Rybicki presents a more discursive treatment, which includes the derivation of the escape probability from several points of view and is applicable to an arbitrary geometry and flow field. Lucy, on the other hand, bases his treatment on the differential form of the transfer equation and carries out his development primarily on the assumption of coherent scattering in the co-moving frame, although he briefly considers complete redistribution in the fluid frame; he quotes numerical results indicating that the difference between the two assumptions is completely negligible. Lucy's transfer equation, which describes the migration of the photon in frequency space at a fixed depth, has been studied in detail by Noerdlinger and Scargle (1972), using the method of addition of layers. Hewitt and Noerdlinger (1974) have discussed the case of a doublet, in which blue absorption features of one component overlap the other component.

Castor and Van Blerkom (1970) have applied the EPM to a 30-level He II ion and Castor and Nussbaumer (1972) have treated a 14-level C III ion, both in the context of a spherical Wolf-Rayet star. Kuan and Kuhi (1975) have used a 12-level hydrogen atom in their work on Balmer line profiles in P Cygni. The EPM has also been used extensively (although not always correctly) in calculating the cooling of interstellar clouds, for which inward motion is usually important.

It is important to note that the EPM, as presently developed, is limited to velocity laws for which the velocity surfaces intersect every line of sight not more than once, i.e. it can be used only for accelerating outflow or decelerating inflow. Although Kuan and Kuhi (1975) account for the contribution of multiple intersections with the line of sight, they ignore, without comment, the radiative coupling between the regions in the vicinity of the points of intersection. When the restriction to unique interactions is heeded and only thermal Doppler broadening is present in the co-moving frame, i.e. when natural and collisional broadening is unimportant, the EPM is accurate for a wide range of conditions; both Castor (1970) and Rybicki (1970) discuss in some detail error estimates and conditions for validity. It follows from Equation (12) of Castor (1970), that if the profile function $\phi(x)$ has Lorentz wings of any non-zero strength, the first omitted term diverges, so that the EPM is invalid in this case. If at the base of the flow the velocity gradient is too small for the EPM to be valid, Lucy (1971) points out that it can still be used with the understanding that the resultant profile within a few thermal Doppler widths of line center will be inaccurate. Finally, because the EPM in its present form takes no account of the static part of the escape probability, i.e. the escape probability in a static atmosphere that is usually negligible with respect to the velocity-induced-escape, it gives source functions that are too large near open boundaries when the static term becomes important.

The EPM can also be applied to more complicated geometries than those with spherical symmetry; Rybicki's (1970) general expressions have already been mentioned. For the study of Be stars, combined differential rotation and expansion in a cylindrically symmetrical configuration is of special interest. One immediate difficulty is that the constant velocity surfaces, at least for pure rotation with reasonable velocity laws, intersect the line of sight more than once. The intersections of typical surfaces with the equatorial plane for $u_{tan} = u_0(r/r_0)^n$, $n = 0, 1, 2$, are very similar to the curves appearing in Figure 6 viewed from the side of the figure. Although this is merely a complicating factor if the source function is assumed or known *a priori*, as in the case considered by Sobolev (1947), it poses a severe problem as discussed above for a self-consistent solution. Fortunately, it appears that differential rotation imposed on rapid expansion leads to tractable surfaces.

Current work on radiation-driven stellar winds (Lucy and Solomon, 1970; Castor *et al.*, 1975) depends crucially on evaluating the force of radiation in a spectral line on a volume of gas in a rapidly expanding atmosphere. Using the EPM, Sobolev (1957) found an expression for the force arising from the diffuse radiation generated in the envelope. In a thorough study relating the line forces to the atmospheric parameters Castor (1974) showed that the force arising from the photospheric continuum in conditions typical of early-type and Wolf-Rayet stars strongly dominated the force from the diffuse field. Castor's expression provided the basis for the Castor *et al.* (1975) theory of stellar winds.

3.2. ACCURATE NUMERICAL SOLUTIONS

We now turn to numerical procedures for the solution of the line transfer equation that can be used for high-velocity flows without making the approximations of the EPM. The transformation to the co-moving frame has already been discussed in connection with the velocity surfaces. This idea, which seems to have been introduced by Milne (1930) appears in the earliest consideration of transfer in moving media (McCrea and Mitra, 1936) and was first exploited in a systematic way by Chandrasekhar (1945a). Rottenberg (1952) also used this transformation in an interesting way. By regarding the extended atmospheres as a small number of geometrically thin shells, although with arbitrary optical thicknesses, he derived two coupled integral equations that could be solved easily by integrating from blue to red. Although he assumed only constant expansion velocities, he did obtain realistic profiles.

There are several reasons for the great utility of transformation to the frequency in the co-moving frame. Consider the scattering integral for an expanding atmosphere as it appears in the stationary frame:

$$\bar{J} = \int_{-\infty}^{\infty} dx \int_{-1}^{1} d\mu \, \phi(x - u\mu) I(x, \mu) . \tag{3.10}$$

In an atmosphere where the maximum velocity in thermal units is u_{max}, the intensity $I(x, \mu)$ contributes substantially to the integral over an interval of length approximately $2u_{max}$; for the static case, or in the co-moving frame the length of the interval is of order five. As u_{max} can be of order 10^3 in stellar winds the use of the co-moving frame offers an enormous advantage. As the angle of the ray changes slightly in the stationary frame, the line opacity profile $\phi(x - u\mu)$ and hence the intensity will change drastically, so that a very fine mesh of μ-points (and consequently of x-points) is required to represent the radiation field and to evaluate the scattering integral. As will be discussed below it is possible to use the so-called angle-averaged redistribution function in the co-moving frame when the presence of strong natural broadening wings prohibits the use of complete frequency redistribution; in the stationary frame this approximation is worthless.

The advantages of working in the co-moving frame are not without cost. The emergent radiation field in the observers frame must be obtained by a separate calculation which is straightforward in principle but troublesome in practice. From the form of the transfer equation in the co-moving frame an even more important fact is apparent. For spherical geometry we have

$$
\mu \frac{\partial I}{\partial r}(\nu, \mu, r) + \frac{1-\mu^2}{r} \frac{\partial}{\partial \mu} I(\nu, \mu, r) - \frac{\nu_0}{c} \left[\mu^2 \frac{d\nu}{dr} + (1-\mu^2)v/r \right] \frac{\partial}{\partial \nu} I(\nu, \mu, r)
$$
$$
= \eta(\nu, r) - \chi(\nu, r) I(\nu, \mu, r),
$$
(3.11)

where χ is the opacity and η the emissivity which in general depends on an angle and frequency integration over $I(\nu, \mu, r)$. In the stationary frame the term in $dI/d\nu$ is absent, so that the equation can be integrated as an *ordinary* differential equation on the characteristics of the differential operator, which are simply the rays parallel to any diameter. When the frequency derivative is present, it is necessary to treat equation (3.11) as a *partial* differential equation.

Before discussing the integration of Equation (3.11), a few comments on its form are appropriate. The term in $\partial I/\partial \nu$, which arises from the transformation to the co-moving frame, contains both radial and tranverse velocity gradients. Although the term is of formal order v/c, relative to the remaining terms, two other terms of the same order that arise from advection and aberration are ignored. Because of the presence of the large number ν_0 in the $\partial I/\partial \nu$ term, it is really of order v/v_{th} relative to the advection and aberration terms, as may be seen by writing the frequency in Doppler units (this was pointed out to us by Dr G. B. Rybicki). A recent study of the effect of these small terms by Mihalas *et al.* (1976b) shows that they lead to effects of order $5(v/c)$ in the source function; fortunately these occur sufficiently close to the surface that their effect on the emergent flux profile is negligible under stellar wind conditions. Although the Equation (3.10) has been known for a long time, problems concerning the correct form of the angle-moment equations derived from it and its use in connection with the gas dynamical equations have been resolved definitively by Castor (1972). Haisch (1976) has derived an alternative expression in tensor form for the transfer equation in the co-moving frame, which leads directly to a stable and conservative differencing scheme in any coordinate system.

As the transfer equation is now a partial differential equation, initial conditions in frequency space must be properly specified for the problem to be well-posed. In addition the differencing schemes in depth and frequency must be chosen, keeping stability criteria in mind. Under the conditions $v \geq 0$, $dv/dr \geq 0$, i.e. the same conditions necessary for unique velocity surfaces, all points recede from one another. Consequently radiation at any point appears redder than at the point where it was emitted. In particular as the extreme shortward edge of the line receives only reddened continuum radiation, the initial condition in frequency space is simply that the line intensity there is continuous with the adjacent continuum intensity at all depths. The frequency and elimination then proceeds from blue to red, as in Rottenberg's (1952) method. Obviously if $v \leq 0$, $dv/dr \leq 0$, all points approach one another and the initial condition is placed in the red wing. If v and dv/dr do not everywhere have the same sign, the frequency integration is more difficult, and will be discussed briefly below.

The first satisfactory solution of the co-moving frame equation was obtained by Noerdlinger and Rybicki (1974) for plane parallel geometry, using Feautrier elimination. Their procedure has proved to be fast, efficient and stable for a wide variety of velocity laws and atmospheric models. A similar procedure, with a slightly different elimination scheme that offers some advantages has been developed by Mihalas *et al.* (1976d), who described modifications necessary to treat non-monotonic velocity fields. An earlier method developed by Simonneau (1973) for plane geometry used co-moving frequencies in the context of an integral equation approach, but is restricted to linear velocity laws with small gradients.

Recently, two different solutions of the co-moving frame transfer equation in spherical geometry have been obtained by Mihalas *et al.* (1975, 1976c). The first method uses Rybicki elimination and is capable of handling very extended atmospheres and arbitrarily large velocity fields, but requires the assumption of complete frequency redistribution in the fluid frame. The second method employs variable Eddington factors (cf. 2.1.2) to eliminate angle quadratures, so that Feautrier elimination is practicable. This scheme allows the use of more accurate descriptions of the scattering in the co-moving frame; unfortunately, both of these solutions require that v and dv/dr have the same sign everywhere.

At present all of the published work on the co-moving frame transfer equation is restricted to two-level atoms. However, Mihalas, Kunasz and Hummer (unpublished) have formulated a multi-level solution using Rybicki elimination and the complete-linearization technique of Auer and Mihalas (1969); coding is now underway.

3.3. TYPICAL RESULTS FOR EXPANDING SPHERICAL ATMOSPHERES

A number of the more important effects of high velocity flow in a self-excited spherical atmosphere have been studied by Mihalas *et al.* (1975). In these models, electron scattering is treated by an assumption of complete redistribution so that the electron scattering opacity and emissivity are approximated by $n_e \sigma_e$ and $n_e \sigma_e J_c$, respectively, where J_c is the mean intensity in the adjacent continuum. The optical depth in the line and continuum are 10^3 and 2, respectively, and the probability per scattering of collisional de-excitation is $\varepsilon = 2 \times 10^{-3}$. The line, continuum and electron-scattering opacity are proportional to r^{-2}, and a velocity law of the form

$$v(r) = v_{\max}[\tan^{-1}(ar+b) - \tan^{-1}(a+b)] \tag{3.12}$$

was chosen to represent a velocity field which increases rapidly at any prechosen radius. The outer radius R has a value of either 3, 30 or 300 times the radius R_c of the core, at which a diffusion boundary condition was applied. Results for some isothermal models are shown in Figure 8. The source functions are labeled by $v_{\max}$ in thermal units; the broken lines represent the mean continuum intensity J_c, obtained by a solution of the continuum transfer equation. In these examples the velocity gradient reaches its maximum at $r \simeq (R+1)/2$. At large depths, the velocity gradient causes the escape probability to increase, so that the source function approaches J_c, which for the cases considered here lies below the source function deep in the

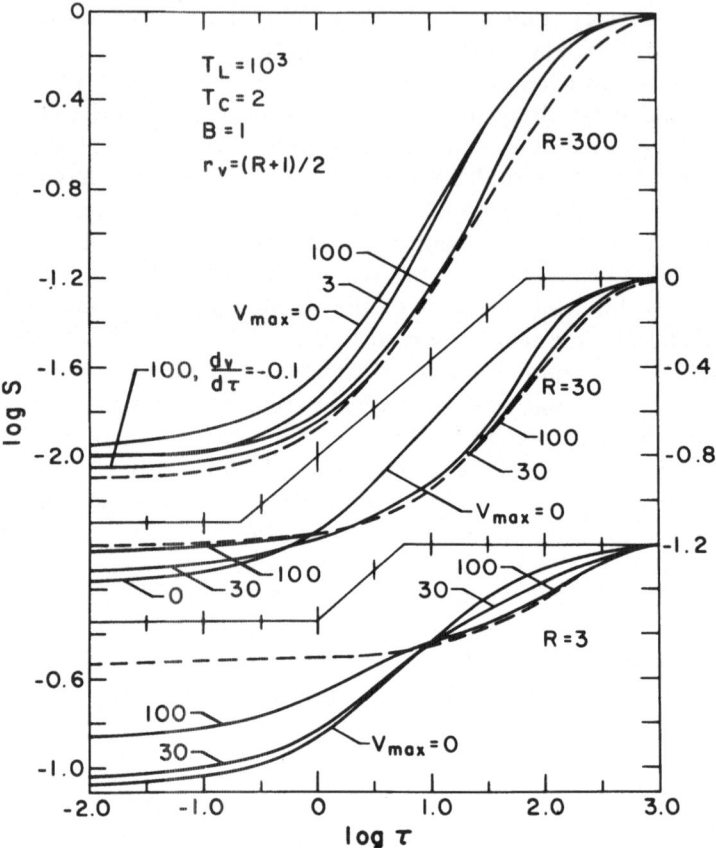

Fig. 8.   Line source function vs line optical depth for isothermal models with various values of R and u_{max}. For each value of R the continuum mean intensity J_c is drawn as a broken line. For $R = 300$ an additional source function is included for $u_{max} = 100$ and $du/d\tau = 0.1$ (from Mihalas *et al.*, 1975).

atmosphere. This effect is largest for small R because the velocity gradient corresponding to a given value of v_{max} is steepest in this case.

In the outer layers the velocity gradient again causes the escape probability to increase, but also causes the line to intercept continuum photons; again the line source function approaches J_c but because, for small R, J_c ($\sim R^{-2}$) is relatively large, the source functions near the surface *increase* with v_{max}. When R is very large, J_c becomes quite small in the outer layers because of electron scattering and dilution, and the source function falls as v_{max} increases. The sensitivity of the source function to the form of the velocity law is illustrated in the top panel of Figure 8 by the case $v_{max} = 100$, $dv/d\tau = -0.1$, where the enhanced escape probability at all depths causes the source functions to be even closer to J_c than for the velocity law (3.12).

In Figure 9, the corresponding flux profiles in the observer's frame are given. The flux is normalized to unity in the continuum and is plotted as a function of the frequency displacement from line center in units of a scale factor x_{max} because of the extreme widths of the profiles. The scale factors x_{max} depend on R and v_{max} and are

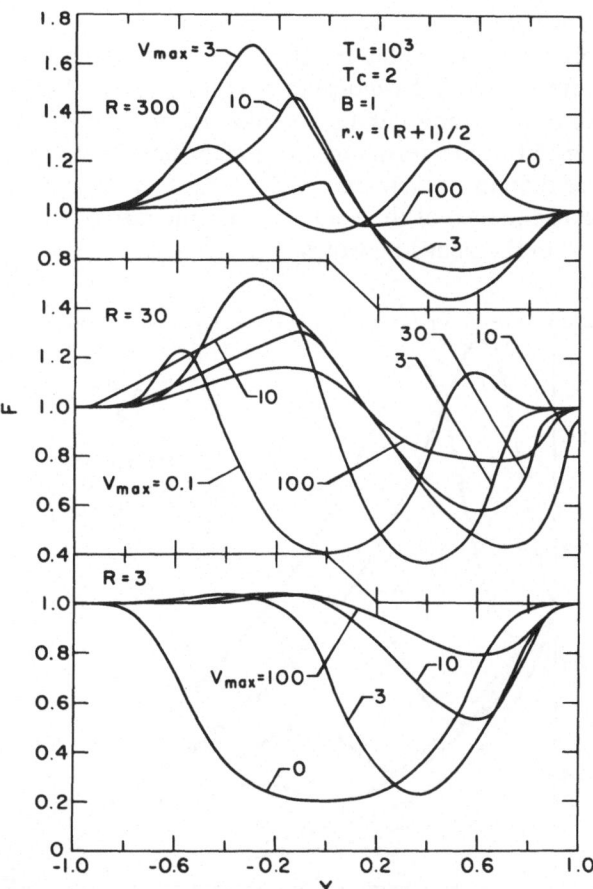

Fig. 9. Emergent fluxes relative to the continuum for some functions of Figure 9 vs displacement $y = x/x_{max}$, where values of x_{max} are listed in Table 2 (from Mihalas *et al.*, 1975).

TABLE III

Scale factors x_{max}

v_{max}	$R = 3$	$R = 30$	$R = 300$
0	3.0	3.5	3.5
3	5.5	6.0	4.5
10	13.0	11.0	12.0
30	–	35.0	–
100	100.0	100.0	100.0

given in Table III. For $R = 3$, there is not enough extension for significant emission, but the shortward shift of the absorption feature is clearly variable. There is also an increase in the absorption equivalent width, which in real spectra with non-zero noise might be confused with 'microturbulence'. In the more extended atmospheres, strong emission lines appear with large blue absorption troughs; as v_{max} increases, the peaks move to shorter wavelengths, while the minima initially move shortward before reversing direction. The peaks assume a distinctive symmetrical triangular shape. For the largest velocities the features become quite indistinct.

For the moderately extended atmospheres with $R = 30$ and $v_{max} = 10$ in the above sequence, the limb-darkening curves are plotted in Figure 10, because of their potential importance for interferometry. It is clear here that limb darkening is very much more severe than in static planar cores because of the geometrical extension; moreover, as a consequence of the velocity field, the disc of the star exhibits bright rings when viewed in the near line wings.

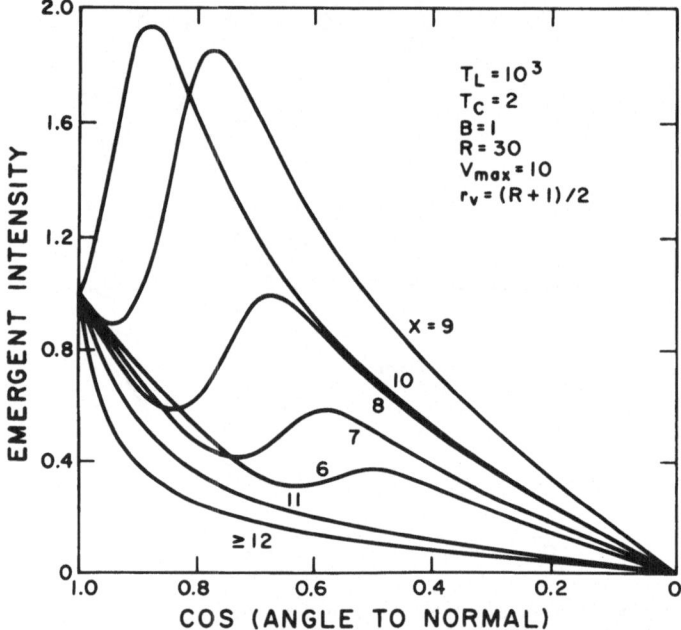

Fig. 10.   Limb-darkening curves normalized to unity at $\mu = 1$ vs μ, for various displacements x from line center, for model with indicated parameters ($V_{max} = u_{max}$ in text and r_v is radius at which velocity gradient is largest) (from Mihalas *et al.*, 1975).

3.4. THE TEMPERATURE DISTRIBUTION IN AN EXPANDING ATMOSPHERE

To examine in a highly idealized way the effects of a large velocity field on the temperature structure of the atmosphere, Mihalas *et al.* (1976a) have calculated radiative-equilibrium picket-fence models of expanding, spherical atmospheres, in which the run of density and velocity are specified *a priori*. Of particular interest were the heating and cooling arising from the shift of line opacity into the continuum, and from the enhanced photon escape rate, respectively. As is well known, the effect in a static atmosphere of introducing a line into the continuum is to lower the temperature at the surface, while raising it at depth ('backwarming'); as deviations from LTE increase, the surface effect becomes quite small. In expanding atmospheres, it is useful to think in terms of three mechanisms that influence the temperature structure: (a) the *irradiation effect*, in which surface layers are heated by intercepted continuum radiation; (b) *band-width constriction*, in which the line opacity shifted to the wings impedes photon escape and causes heating in the deeper layers; and (c) *escape enhancement*, which can lead to cooling deep in the atmospheres if large velocity gradients are present there. An example of this effect is seen in the temperature drop at a velocity discontinuity, as obtained, for example, in the schematic models of Mihalas (1969). The first two effects are clearly seen in Figure 11, where the Planck function is plotted against the continuum optical depth for

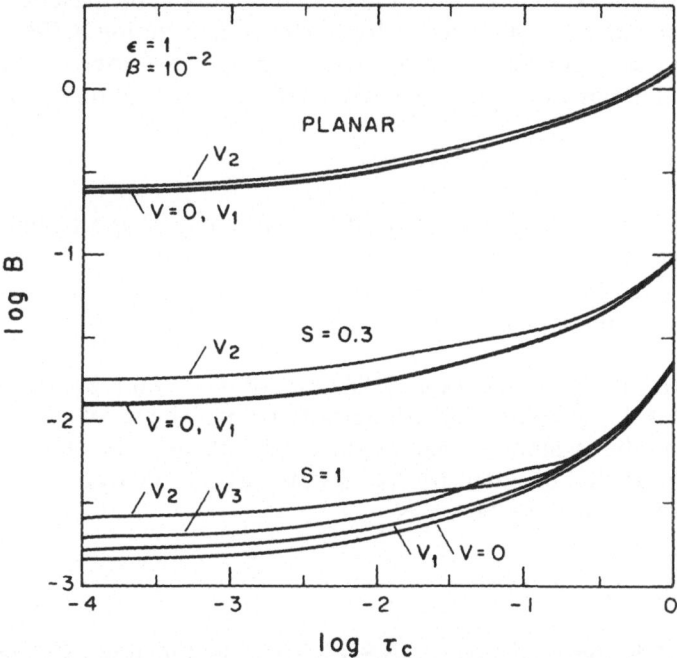

Fig. 11.   Picket fence models for static and expanding atmospheres with $\varepsilon = 1$, $\beta = 10^{-2}$ and various velocity laws with $u_{max} = 12$; $V = 0$ is static case; V_1 designates constant velocity gradient; V_2 and V_3 designate case with sharp rise at line optical depth of 1 and 10, respectively.

moderately strong lines that cover approximately 0.2 of the spectrum. These models range from planar to moderately extended, depending on the value of the parameter S, the ratio of the mean free path in the continuum at the outer boundary to the radius of the atmosphere. In all cases except $v = 0$, the maximum flow speed at the surface was 12 v_{th}. Curves labeled V_1 are for models with a constant value of dv/dr, while those marked V_2 and V_3 correspond to a sharp rise in flow speed at line optical depths of 1 and 10, respectively. As the lines depart further from LTE the effects are qualitatively the same, but become smaller in magnitude. When the speed is small at large depths and rises suddenly to v_{max} at line optical depth unity (case V_2) both the irradiation and bandwidth constriction effects are apparent; the boundary temperature is raised by 16% in the most favorable case. If the maximum gradient occurs near line optical depth of ten, local heating occurs but the distinction between the two effects is blurred.

Although these results are based on very simple models, they do show that velocity fields can lead to substantial modifications of the temperature distributions by purely radiative means, quite apart from any dissipation of mechanical energy. Such temperature modifications are large enough to influence significantly the dynamical processes occurring in the atmosphere.

3.5. THE QUESTION OF FREQUENCY REDISTRIBUTION

Nearly all of the work on line formation in static media in the past decade has been based on the assumption of complete frequency redistribution (CFR), in which the frequency and direction of a photon before and after scattering are completely uncorrelated. The corresponding emissivity of scattered radiation is proportional to

$$\phi(\nu) \int d\upsilon' \int d\mathbf{n}' \, \phi(\nu')I(\nu', \mathbf{n}') . \tag{3.13}$$

However, this approximation is the third in a hierarchy of approximations derived from the exact scattering term

$$\int d\nu' \int d\mathbf{n}' \, R(\nu, \mathbf{n}, \nu', \mathbf{n}')I(\nu', \mathbf{n}') , \tag{3.14}$$

where ν' and $\mathbf{n}'$ are the frequency and direction of the photon before scattering and $R(\nu, \mathbf{n}, \nu', \mathbf{n}')$ is the so-called redistribution function. The second level of approximation involves the assumption that the variation with direction of the radiation field is unimportant, so that the intensity $I(\nu', \mathbf{n}')$ can be replaced by its mean value $J(\nu')$, in which case the emissivity becomes

$$\int d\nu' \, R(\nu, \nu')J(\nu') , \tag{3.15}$$

where $R(\nu, \nu')$ is the angle-averaged redistribution function. The redistribution functions have been discussed in detail by Hummer (1962). The assumption of CFR can be justified when the radiation field is approximately isotropic and white near the line center, as was apparently first shown by Bieberman (1947) and Holstein (1947). Extensive numerical calculations by Hearn (1964), Hummer (1969), and Milkey and Mihalas (1973) have confirmed the validity of CFR for static media.

On the other hand, because the presence of a velocity field reduces the degree of both whiteness and isotropy of radiation in the lines, the validity of CFR in moving media must be reconsidered. Magnan (1968), on the basis of Monte Carlo calculations for a spherical shell expanding with a constant velocity of one thermal unit found very little difference between the exact scattering expression and CFR, applied in the co-moving frame. On the other hand Vardavas (1974) and Cannon and Vardavas (1974), working in the observer's frame and using an angle-averaged redistribution function due to Hummer (1968) that included the effect of the velocity field, found very large effects from CFR. Magnan (1974b) criticized the use of angle-averaged functions in the observer's frame for moving atmospheres. Mihalas *et al.* (1976d) repeated the calculations of Cannon and Vardavas, using both CFR and an angle-averaged redistribution function in the *co-moving* frame, and obtained nearly identical line profiles. It appears that in the derivation of the angle-averaged redistribution function for moving atmospheres in the observer's frame, the flow velocity is treated as a random velocity that is averaged together with the thermal velocity; consequently the photons are distributed over much too wide a spectral region, at least for $v \gtrsim v_{th}$, and the resulting function is also independent of the direction of the flow.

Caroff *et al.* (1972) have made an extensive investigation of different redistribution mechanisms in a rapidly expanding spherical shell with large velocity gradients, using the Monte Carlo techniques in which they follow the history of individual photons emitted with a known frequency at the inner boundary. They considered three cases: (1) coherence in the atom's frame; (2) coherence in the co-moving frame; (3) complete redistribution in the co-moving frame. The first of these is identical to the use of the full redistribution function in the co-moving frame; the second was used extensively by Lucy (1971). Caroff *et al.* pointed out that the correlation between frequency and direction that is explicitly enforced when coherence in the atom's frame is assumed gives rise to a skewing of the profile of the emergent radiation to longer wavelengths. This effect, which is different from others so far discussed, can be understood with the aid of Figure 12. A photon of frequency ν_e emitted in the core

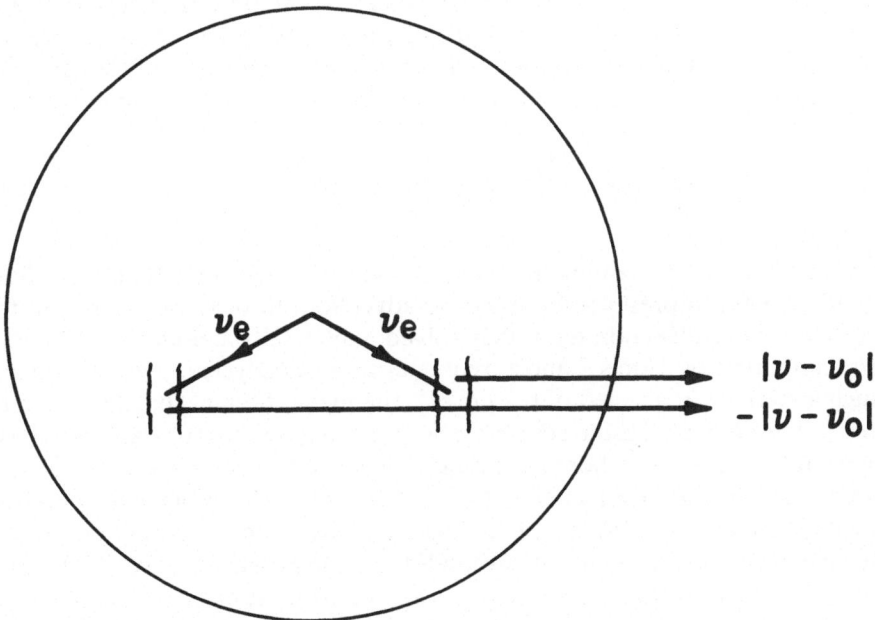

Fig. 12.   Photons emitted with frequency ν_l are seen with frequency displacements $\pm |\nu - \nu_0|$.

and scattered from constant velocity surfaces symmetrically located in front and behind the core will be seen by the stationary observer with equal displacements from the line center ν_0. However, the photon with the shorter wavelength is seen slightly reddened by the atom in its constant velocity surface and has therefore a higher probability of being absorbed than has the longer wave photon, which has an *inward* component of velocity as it travels through its shell. Occulation produces the opposite effect, however, and it is likely that in most cases of interest this is not a major effect. For high velocity flows it appears that the precise assumption about frequency redistribution in the co-moving frame is not of great consequence, so long as the thermal Doppler effect is the only important broadening mechanism. If natural broadening is significant, as in H I or He II $L\alpha$, and if the optical thickness is large enough for most of the transfer to occur in the wings, it is essential to use the appropriate angle-averaged redistribution function in the co-moving frame, just as for static problems. The numerical solutions of Mihalas *et al.* (1975) for expanding extended spherical atmospheres, using CFR, coherent scattering and angle-averaged redistribution functions for both pure Doppler and combined Doppler and natural broadening verify explicitly the above conclusions. This work shows also that the assumption of complete redistribution for *electron* scattering is valid for a wide range of conditions.

3.6. MATERIAL ADVECTION

For rapid flow, the equation of statistical equilibrium must contain terms accounting for the material carried into and out of the volume element under consideration, i.e. for material advection. Rybicki (1970) showed by essentially dimensional arguments that advection was negligible for resonance lines in Wolf-Rayet stars, although it could become significant for forbidden lines. Cannon and Cram (1974) derived an expression for the source function accounting for material advection of the form

$$S = (1-\varepsilon)\bar{J} + \varepsilon B - \frac{(1-\varepsilon)}{A_{21}}\frac{\mathrm{d}S}{\mathrm{d}t}, \tag{3.16}$$

where the last term, containing the time derivative of the source function following the fluid element, represents the effect of advection. In order to demonstrate the importance of the advection term, they solved numerically a model in which a wave with an amplitude of 2 km s^{-1} and a frequency in the kilocycle range travels upward through a medium at a speed of 0.2 km s^{-1}; the mean thermal speed is taken to be 1 km s^{-1}. The numerical solutions for the source function show deviations as large as 30 percent from the case where advection is neglected. Although one would expect the source function to be affected by this velocity field, it is unfortunately difficult to reconcile these results with simple estimates of the effect of advection. For their model one unit of optical depth corresponds to a geometrical distance of 10^{-4} km; on the other hand during the lifetime of 10^{-8} s assumed for the excited state of the atom, a typical distance it could move at the maximum velocity of particles in the wave is 10^{-8} s $\times$ 2 km s^{-1} = 2 $\times$ 10^{-8} km. Because the scale for advection is nearly four orders of magnitude smaller than the optical scale, it is difficult to understand how advection could have any appreciable effect in this case. It appears likely that their numerical integration, in which they used 1 second time steps to integrate a system with a relaxation time of order 10^{-8} s, may be at fault.

It is straightforward to derive from Cannon and Cram's expression explicit conditions that indicate when material advection can be neglected. For a macroscopic *steady* radial flow of speed $v(\tau)$, we can rewrite the Lagrangian time derivative of the source function as

$$\frac{\mathrm{d}S}{\mathrm{d}t} = \frac{\mathrm{d}S}{\mathrm{d}\tau}\frac{\mathrm{d}\tau}{\mathrm{d}t} = k\frac{\mathrm{d}S}{\mathrm{d}\tau}\frac{\mathrm{d}z}{\mathrm{d}t} = kv\frac{\mathrm{d}S}{\mathrm{d}\tau}, \tag{3.17}$$

where k is given by Equation (1.9). Using the well known relation among the Einstein coefficients, we find

$$\frac{(1-\varepsilon)}{A_{21}}\frac{\mathrm{d}S}{\mathrm{d}t} = \frac{(1-\varepsilon)}{8\pi} \; N_1\lambda_0^3(\omega_2/\omega_1)(v/v_{\mathrm{th}})(\mathrm{d}S/\mathrm{d}\tau), \tag{3.18}$$

where we have neglected the stimulated emission term in k. Here ω_1 and ω_2 are the statistical weights of the lower and upper levels respectively, N_1 is the density of the line-forming ion and λ_0 is the line-center wavelength (in centimeters). It is interesting that Equation (3.18) contains no reference to the radiative transition probability.

When $\varepsilon \ll 1$, as is the case for strong lines, the production of excited atoms near the outside of the atmospheres is controlled by radiation flowing up from great depths in the line wing rather than by thermal excitation, i.e. the $(1-\varepsilon)\bar{J}$ term is the controlling one and not εB. Although a precise statement of the proper magnitude for comparison involves rather elaborate arguments in terms of the thermalization length, it is sufficient for our purpose to compare the advection term with $(1-\varepsilon)\bar{J}$, or equivalently, with S. Thus the condition that advection is unimportant near the surface can be written as

$$\frac{v}{v_{\mathrm{th}}} \ll \left(\frac{8\pi}{(1-\varepsilon)N_1\lambda_0^3}\right)\left(\frac{\omega_1}{\omega_2}\right)\left(\frac{S}{\mathrm{d}S/\mathrm{d}\tau}\right). \tag{3.19}$$

Now $S/(\mathrm{d}S/\mathrm{d}\tau)$ is very roughly approximated by τ; a better estimate can be obtained from Ivanov's (1973) approximate expression for the source function in a semi-infinite isothermal atmosphere:

$$S \simeq \sqrt{\varepsilon}B[\varepsilon + (1-\varepsilon)K_2(\tau)]^{-1/2}, \tag{3.20}$$

where

$$K_2(\tau) = \int_{-\infty}^{\infty} \mathrm{d}x\,\phi(x)E_2[\tau\phi(x)] \tag{3.21}$$

is a function that decreases monotonically from $E_2(0) = 1$ *as* τ increases from 0. The desired result is

$$\frac{S}{\mathrm{d}S/\mathrm{d}\tau} \simeq \frac{\varepsilon + (1-\varepsilon)K_2(\tau)}{(1-\varepsilon)K_1(\tau)} \xrightarrow[K_2 \gg \varepsilon]{} K_2(\tau)/K_1(\tau), \tag{3.22}$$

where

$$K_1(\tau) = (-1/2)\,\mathrm{d}K_2/\mathrm{d}\tau; \tag{3.23}$$

the properties of K_1 and K_2 are discussed in detail by Avrett and Hummer (1965). For Doppler broadening $K_2/K_1 = 1.7, 3.2$ and 15 at $\tau = 0.1, 1.0$ and 10 respectively, and for Lorentz broadening these values should be increased by a factor of approximately two. Thus at $\tau = 1$, the product of the last two factors in Equation (3.19) is roughly unity and for $\lambda_0 = 5000$ Å and $N_1 = 10^{10}$, which is typical of an abundant ion density in a stellar wind, the remaining factor is of order 2000.

Deep in the atmosphere, the advection term must be compared with the thermal excitation term εB. The last factor in (3.19) must therefore be replaced by $\varepsilon B/(dS/d\tau)$, which can be estimated in the same way. For $\varepsilon = 10^{-4}$ and $\tau = 10^3$, 10^4 and 10^5, this factor is approximately 0.5, 15 and 1300, respectively; for the values of N_1 and λ_0 used above and setting $\omega_1/\omega_2 = \frac{1}{3}$, the corresponding values for the right side of the condition (3.19) are roughly 300, 10^4 and 10^6.

From these arguments it appears that material advection in stellar winds is of at most marginal importance as far as the lines are concerned. However, as this conclusion is based on the model of a two-level atom, it is possible that the interplay of slow and fast rates could cause an observable effect.

3.7. ELECTRON SCATTERING

In regions where rapid flow occurs, the densities can become quite low and the material will be largely ionized. Consequently a significant density of electrons can be expected in the flow, which suggests that electron scattering could play a role in the line formation process. This problem was apparently considered by Edmonds (1950), but his treatment is based on a highly artificial separation of the electrons into a reversing layer.

The treatment of electron scattering on the same basis as scattering by line atoms is difficult, even for static situations, because the thermal Doppler width for the electrons is so much greater than that for the atoms. This problem has been attacked by Auer and Mihalas (1968a, b) and in a simplified way, by Weymann (1970) and Mathis (1970).

Auer and Van Blerkom (1972) have used the Monte Carlo technique to examine the scattering of photons by electrons in a spherical shell expanding with either a constant or a linear velocity law. They suppose that the photons are initially emitted as the result of recombination, but that the atoms do no scattering. They obtain results for electron optical thickness of 0 and 1 and for cases in which the thermal velocity of the electrons is either negligible compared to the flow velocity, or equal to its value at the inner boundary of the shell. For unit optical thickness they obtain emission lines with core widths in agreement with the flow velocities, but with very strong, extended red wings. The red wings are a consequence of the spherical distribution of the electrons, as can be seen in Figure 13. When a photon is scattered at radius r, the value of $\mu = \cos \theta$ is symmetrically distributed between about $\mu = 0$, so that $\langle \mu \rangle = 0$. If the next scattering occurs at a radius r_s that is larger than r, it makes an angle θ_s with the radius vectors such that $\mu_s = \cos \theta_s \geq \mu$. As the distribution of direction cosines in the case is biased towards 1, we have $\langle \mu_s \rangle > 0$. The direction cosine μ'_s after scattering is again symmetrically distributed and $\langle \mu'_s \rangle = 0$. If the thermal speed is small compared to the flow speed, the frequencies involved in the scattering at r_s are related by

$$x' - x = u(r_s)(\mu'_s - \mu_s) . \tag{3.24}$$

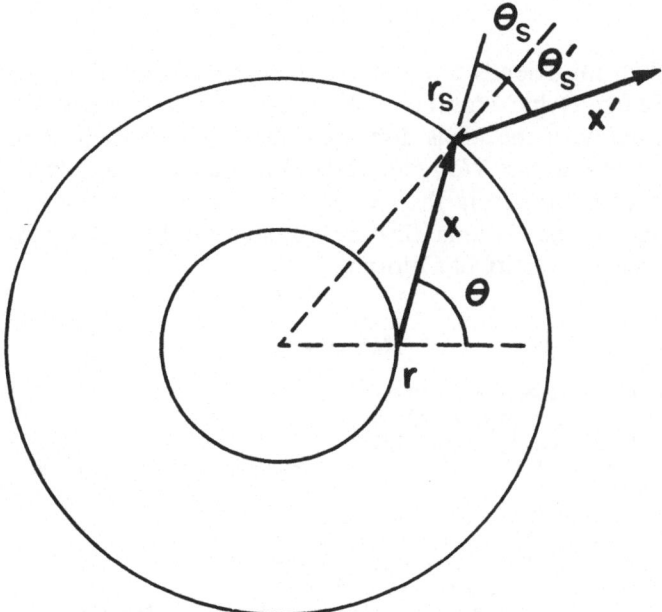

Fig. 13. Photons scattered at $r_s > r$ have $\langle \mu_s \rangle \geq 0$ and $\langle x' - x \rangle \leq 0$.

Averaging over many events, we have

$$\langle x' - x \rangle = -u(r_s)\langle \mu_s \rangle , \tag{3.25}$$

so that the frequency shift is redward. Since the photons are flowing from smaller to larger radii, this reddening effect is preponderant. It appears that this effect is not restricted to optically thin lines; the fact that emission lines are not usually observed to have extended red wings may imply certain limits on the densities and dimensions of the expanding atmospheres in question.

4. Conclusion

At present an understanding of line formation and our computational ability to solve line transfer problems for accelerating radial flow in spherically symmetrical configurations are both well developed. As the computations are unfortunately extremely heavy and require the largest available computers, work in this area is now restricted to a comparatively small number of people. The treatment of line formation in cylindrically symmetrical flows by either the escape probability method or by direct numerical integration of the transfer equation is very much more difficult than for the spherical case, and little progress has been made. Even when accurate numerical solutions are available meaningful comparisons of model results with observed line profiles will be difficult because of the smearing of structural details by the line formation process in high-speed flows.

Acknowledgements

I am indebted to my friends and colleagues John Castor, Paul Kunasz, Dimitri Mihalas and George Rybicki for their patient assistance over the years in helping me to understand many of the ideas discussed here and for their comments on this review. I am grateful to Drs Wolfgang Kalkofen and Christian Magnan for permission to reproduce figures from their papers. The writing of this review has been supported in part by the National Science Foundation through Grant No. MPS72-05026 A02 to the University of Colorado.

References

Abhyankar, K. D.: 1964a, *Astrophys. J.* **140**, 1353.
Abhyankar, K. D.: 1964b, *Astrophys. J.* **140**, 1368.
Abhyankar, K. D.: 1965, *Astrophys. J.* **141**, 1056.
Abhyankar, K. D.: 1967, in M. Hack (ed.), *Modern Astrophysics, A Memorial to Otto Struve*, Gauthier-Villars, Paris, p. 199.
Auer, L. H.: 1971, *J. Quant. Spectrosc. Radiat. Transfer* **11**, 573.
Auer, L. H. and Mihalas, D.: 1968a, *Astrophys. J.* **153**, 245.
Auer, L. H. and Mihalas, D.: 1968b, *Astrophys. J.* **153**, 923.
Auer, L. H. and Mihalas, D.: 1969, *Astrophys. J.* **158**, 641.
Auer, L. H. and Mihalas, D.: 1970, *Monthly Notices Roy. Astron. Soc.* **149**, 65.
Auer, L. H. and Van Blerkom, D.: 1972, *Astrophys. J.* **178**, 175.
Avrett, E. H. and Hummer, D. G.: 1965, *Monthly Notices Roy. Astron. Soc.* **130**, 295.
Bieberman, L. M.: 1947, *J. Exp. Theor. Phys.* **17**, 416.
Cannon, C. J.: 1974, *J. Quant. Spectrosc. Radiat. Transfer* **14**, 745.
Cannon, C. J. and Cram, L. E.: 1974, *J. Quant. Spectrosc. Radiat. Transfer* **14**, 93.
Cannon, C. J. and Rees, D. E.: 1971, *Astrophys. J.* **169**, 157.
Cannon, C. J. and Vardavas, I. M.: 1974, *Astron Astrophys.* **32**, 85.
Caroff, L. J., Noerdlinger, P. D., and Scargle, J. D.: 1972, *Astrophys. J.* **176**, 439.
Castor, J. I.: 1970, *Monthly Notices Roy. Astron. Soc.* **149**, 111.
Castor, J. I.: 1972, *Astrophys. J.* **178**, 779.
Castor, J. I.: 1974, *Monthly Notices Roy. Astron. Soc.* **169**, 279.
Castor, J. I., Abbott, D. C., and Klein, R. I.: 1975, *Astrophys. J.* **195**, 157.
Castor, J. I. and Nussbaumer, H.: 1972, *Monthly Notices Roy. Astron. Soc.* **155**, 293.
Castor, J. I. and Van Blerkom, D.: 1970, *Astrophys. J.* **161**, 485.
Chandrasekhar, S.: 1945a, *Rev. Mod. Phys.* **17**, 138.
Chandrasekhar, S.: 1945b, *Astrophys. J.* **102**, 402.
Edmonds, F. N.: 1950, *Astrophys. J.* **112**, 324.
Feautrier, P.: 1964, *Compt. Rend.* **258**, 3189.
Frisch, U. and Frisch, H.: 1975, *Monthly Notices Roy. Astron. Soc.* **173**, 167.
Haisch, B. M.: 1976, *Astrophys. J.*, in press.
Hearn, A. H.: 1964, *Proc. Phys. Soc.* **84**, 11.
Hewitt, T. G. and Noerdlinger, P. D.: 1974, *Astrophys. J.* **188**, 315.
Holstein, T.: 1947, *Phys. Rev.* **72**, 1212.
Hummer, D. G.: 1962, *Monthly Notices Roy. Astron. Soc.* **125**, 21.
Hummer, D. G.: 1968, *Monthly Notices Roy. Astron. Soc.* **141**, 479.
Hummer, D. G.: 1969, *Monthly Notices Roy. Astron. Soc.* **145**, 95.
Hummer, D. G. and Kunasz, P. B.: 1976, to be published.
Hummer, D. G. and Rybicki, G. B.: 1966, *J. Quant. Spectrosc. Radiat. Transfer* **6**, 661.
Hummer, D. G. and Rybicki, G. B.: 1968a, *Astrophys J.* **153**, L107.
Hummer, D. G. and Rybicki, G. B.: 1968b, in R. G. Athay, J. Mathis, and A. Skumanich (eds.), *Resonance Lines in Astrophysics*, National Center for Atmospheric Research, Boulder, Colorado, p. 213.
Hummer, D. G. and Rybicki, G. B.: 1971a, *Ann. Rev. Astron. Astrophys* **9**, 237.
Hummer, D. G. and Rybicki, G. B.: 1971b, *Monthly Notices Roy. Astron. Soc.* **152**, 1.
Ivanov, V. V.: 1973, in D. G. Hummer (ed.), *Transfer of Radiation in Spectral Lines*, NBS Spec. Publ. No. 385 (Washington: Govt. Printing Office), p. 441.

Kahn, F. D.: 1968, in D. E. Osterbrock and C. R. O'Dell (eds.), 'Planetary Nebulae', *IAU Symp.* **34**, 236.

Kalkofen, W.: 1970, in H. Groth and P. Wellmann (eds.), *Spectrum Formation in Stars with Steady-State Extended Atmospheres*, NBS Spec. Publ. No. 332 (Washington: Govt. Printing Office), p. 120.

Kuan, P. and Kuhi, L. V.: 1975, *Astrophys. J.* **199**, 148.

Kulander, J. L.: 1967, *Astrophys. J.* **147**, 1063.

Kulander, J. L.: 1968, *J. Quant. Spectrosc. Radiat. Transfer* **8**, 273.

Kulander, J. L.: 1971, *Astrophys. J.* **165**, 543.

Kunasz, P. B. and Hummer, D. G.: 1974a, *Monthly Notices Roy. Astron. Soc.* **166**, 19.

Kunasz, P. B. and Hummer, D. G.: 1974b, *Monthly Notices Roy. Astron. Soc.* **166**, 57.

Lucy, L. B.: 1971, *Astrophys. J.* **163**, 95.

Lucy, L. and Solomon, P. M.: 1970, *Astrophys. J.* **159**, 879.

Lyong, L. V.: 1967, *Soviet Astron.* **11**, 224.

Magnan, C.: 1968, *Astrophys. Letters* **2**, 213.

Magnan, C.: 1970, *J. Quant. Spectrosc. Radiat. Transfer* **10**, 1.

Magnan, C.: 1974a, *J. Quant. Spectrosc. Radiat. Transfer* **14**, 123.

Magnan, C.: 1974b, *Astron. Astrophys.* **35**, 233.

Mathis, J. S.: 1968, in R. G. Athay, J. Mathis, and A. Skumanich (eds.), *Resonance Lines in Astrophysics*, National Center for Atmospheric Research, Boulder, Colorado, p. 299.

Mathis, J. S.: 1970, *Astrophys. J.* **162**, 761.

McCrea, W. and Mitra, K.: 1936, *Z. Astrophys.* **11**, 359.

Mihalas, D.: 1969, *Astrophys. J.* **157**, 1363.

Mihalas, D., Kunasz, P. B., and Hummer, D. G.: 1975, *Astrophys. J.* **202**, 465.

Mihalas, D., Kunasz, P. B., and Hummer, D. G.: 1976a, *Astrophys. J.* **203**, 647.

Mihalas, D., Kunasz, P. B., and Hummer, D. G.: 1976b, *Astrophys. J.* in press.

Mihalas, D., Kunasz, P. B., and Hummer, D. G.: 1976c, *Astrophys. J.* in press.

Mihalas, D., Shine, R. A., Kunasz, P. B., and Hummer, D. G.: 1976d, *Astrophys. J.*, in press.

Milkey, R. W. and Mihalas, D.: 1973, *Astrophys. J.* **185**, 709.

Milne, E. A.: 1930, *Z. Astrophys.* **1**, 98.

Noerdlinger, P. D. and Rybicki, G. B.: 1974, *Astrophys. J.* **193**, 651.

Noerdlinger, P. D. and Scargle, J. D.: 1972, *Astrophys. J.* **176**, 463.

Rees, D. E.: 1970, *Proc. Astron. Soc. Australia* **1**, 384.

Robbins, R. R.: 1968, *Astrophys. J.* **151**, 511.

Rottenberg, J. A.: 1952, *Monthly Notices Roy. Astron. Soc.* **112**, 10.

Rublev, S. V.: 1961, *Soviet Astron.* **4**, 780.

Rublev, S. V.: 1964, *Soviet Astron.* **7**, 492.

Rybicki, G. B.: 1970, in H. Groth and P. Wellmann (eds.), *Spectrum Formation in Stars with Steady-State Extended Atmospheres*, NBS Spec. Publ. No. 332 (Washington: Govt. Printing Office), p. 87.

Rybicki, G. B.: 1971, *J. Quant. Spectrosc. Radiat. Transfer* **11**, 589.

Rybicki, G. B. and Hummer, D. G.: 1967, *Astrophys. J.* **150**, 607.

Shine, R. A. and Oster, L.: 1973, *Astron. Astrophys.* **29**, 7.

Simonneau, E.: 1973, *Astron. Astrophys.* **29**, 357.

Sobolev, V. V.: 1947, *Moving Envelopes of Stars*, Leningrad State University (in Russian), trans. S. Gaposchkin, Harvard University Press, Cambridge, Massachusetts (1960).

Sobolev, V. V.: 1952, in V. A. Ambartsumyan (ed.), *Theoretical Astrophysics* (in Russian), trans. J. B. Sykes, Pergamon Press, London (1958).

Sobolev, V. V.: 1957, *Soviet Astron.* **1**, 678.

Underhill, A. B.: 1947, *Astrophys. J.* **106**, 128.

Vardavas, I. M.: 1974, *J. Quant. Spectrosc. Radiat. Transfer* **14**, 909.

Weymann, R. J.: 1970, *Astrophys. J.* **160**, 31.

Zanstra, H.: 1949, *Bull. Astron. Inst. Neth.* **11**, 1.

DISCUSSION

Snijders: I agree with your comment that it is very difficult to get anything useful out of one or two lines, but what if you take a number of lines which have different excitation potentials and different ionization potentials, for example, Si II and Si IV, and combine this with N V, say?

Hummer: Clearly the more information you have the better off you are. However, a constant velocity surface encompasses a range of radius. If you have a lot of different stages of excitation you have emission only in a relatively small radius band, and therefore that will help to isolate where on the surface you are. So if you have conditions where the degree of ionization changes rapidly, the depth resolution will be increased and lines of different ions will then give more information than just one line.

Kalkofen: One of your slides in which you compare the effects of velocity fields and geometry shows dramatic effects due to geometry. What was the ratio of the outer radius to the mean free path of photons?

Hummer: The radius is small relative to the mean free path of photons.

RADIATIVE TRANSFER IN DYNAMIC STELLAR ATMOSPHERES

RICHARD I. KLEIN

National Center for Atmospheric Research, Boulder, Colo., U.S.A.

Abstract. In order to understand the Be star phenomenon, and move in directions away from ad hoc models, detailed calculations have been made that treat nonlinear hydrodynamics *self-consistently* with non-L.T.E. radiative transfer.

We investigate the radiating shock wave produced by a constant-velocity piston moving into an atmosphere in radiative, hydrostatic and statistical equilibrium. Self-consistent numerical solutions are obtained to the equations of hydrodynamics, radiative transfer and level population. Only Lyman continuum radiation is considered, although calculations including the transfer of radiation in higher continua are underway. The results are interpreted in terms of the relaxation lengths for the collisional and radiative processes, and by comparing the 'radiative' case both with one in which only collisional transitions occur and with the adiabatic case.

The shock quickly develops a temperature spike. At large optical depth, the structure of the spike is controlled by collisional ionization and three-body recombination. When the shock approaches $\tau_{912} \approx 209$, precursor radiation drives an ionization front rapidly through the remaining atmosphere. Preliminary calculations for multi-continuum radiative transfer indicate a radiation precursor effect from the shock front when the shock is much deeper in the atmosphere at earlier times. This is a result of the gas becoming optically thin to photons of frequencies very different from those in the Lyman continuum. At small optical depth, escape of recombination radiation narrows the temperature spike producing a radiation dominated shock. Long collisional recombination times compared to gas travel times in the upper atmosphere produce a large plateau of nearly constant ionization behind the shock.

DISCUSSION

Sonneborn: In a star rotating very rapidly near breakup velocity, if you hit it hard enough with a shock, could you get mass ejection?

Klein: This is quite possible but our models do not include the effects of rotation.

Cardona: Is it possible to include the effects of rotation in your formulation?

Klein: The inclusion of rotation is difficult on two levels: in the hydrodynamics and in the radiative transfer. In the latter case it is not clear that the problem can be done, given the finite constraints of the 7600 computer. The hope for a model which includes both the hydrodynamics and the radiative transfer is therefore a long way off.

Noerdlinger: To study ejection of matter you would probably want to include radiation force in your first equation. The shock could well thin out the material in the upper layers to the point where radiation pressure, especially in the lines, is important.

Klein: Absolutely.

Haisch: In your calculations what takes the place of the piston in a real star?

Klein: In the Be stars radiation pressure might act as a self-starting mechanism.

A. Slettebak (ed.), Be and Shell Stars, 313. *All Rights Reserved*

Si II LINES IN THE SHELL OF ζ TAU

SARA R. HEAP

Laboratory for Optical Astronomy, Goddard Space Flight Center, Greenbelt, Md., U.S.A.

Abstract. Non-LTE strengths of Si II lines were computed in an effort to reproduce the observed strengths of Si II lines in the shell spectrum of ζ Tau. This effort was unsuccessful, probably because it assumed that the shell could be represented by a finite, plane-parallel slab. The main conclusion of this study, then, is that in order to understand the shell spectra of Be stars, a more realistic geometry must be assumed, in which the shell is represented by a thick disk.

The availability of simultaneous ultraviolet and visual spectra of B-shell stars has made it possible for the first time to observe and measure shell lines resulting from all the important transitions of an ion. An example of these fortunate circumstances is the ion, Si II. Figure 1 shows all the important transitions of Si II. They include the

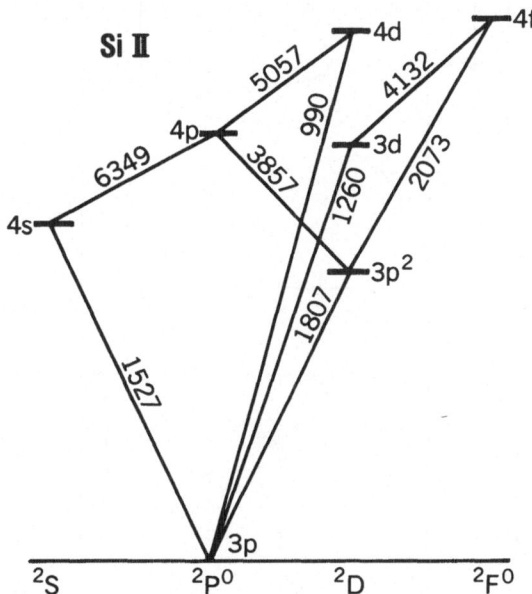

Fig. 1.   Important transitions of the ion, Si II. The wavelength of each transition is indicated.

resonance lines, which all lie in the ultraviolet, and lines from excited states, most of which lie in the visual region of the spectrum. All these lines, with two exceptions, were observed in the shell spectrum of ζ Tau. The two exceptions are the λ 990 Å line and the λ 2073 Å line, both of which are beyond the range of our spectrograms. The ultraviolet spectrograms of ζ Tau were obtained in November 1972 by a rocket experiment (cf. Heap, 1975), while the visual spectrograms were obtained by Helmut Abt at the coudé focus of the 84-in. telescope at Kitt Peak National Observatory, just one night before the rocket launch. Figure 2 shows the profiles of some of these Si II

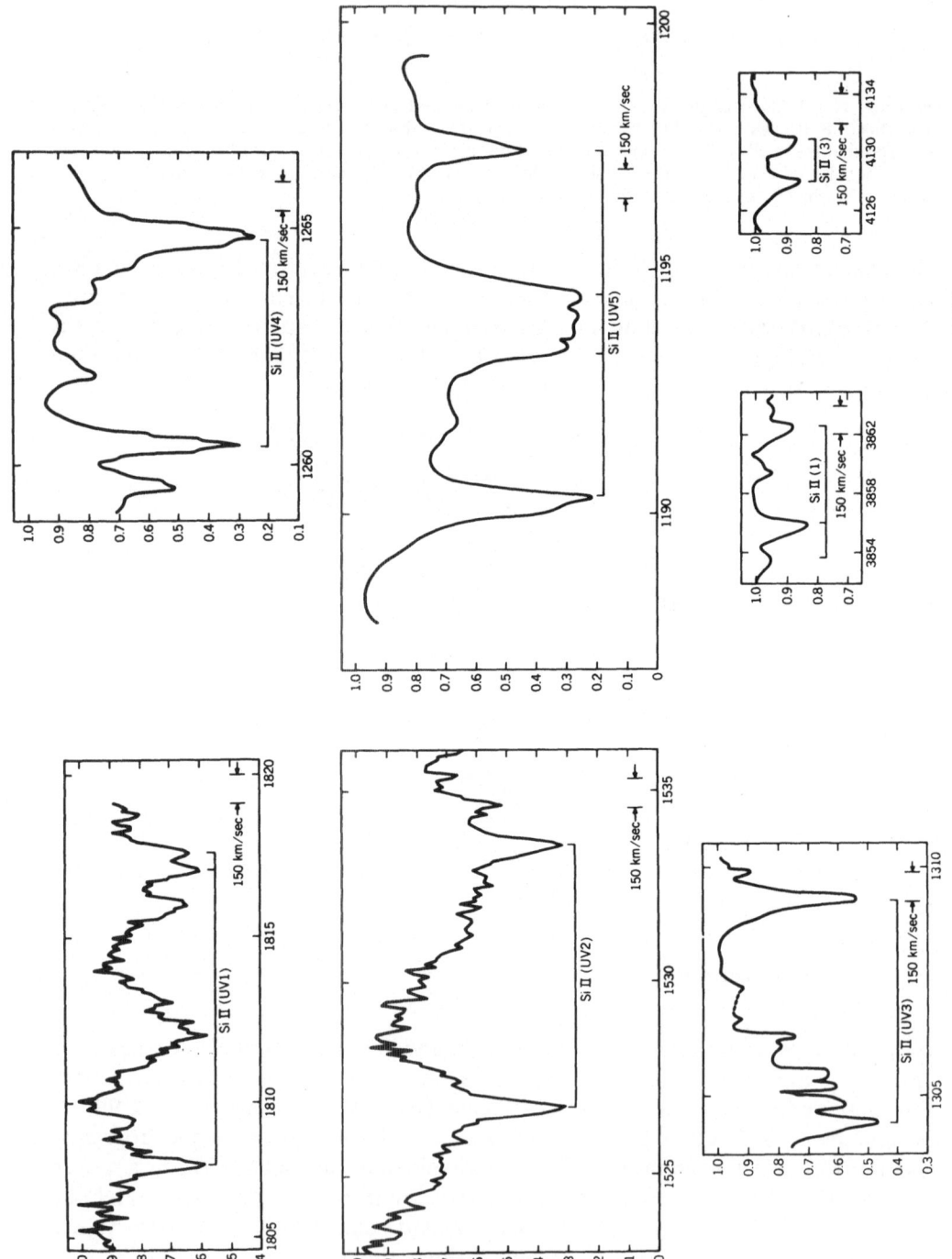

Fig. 2.   Profiles of some ultraviolet and visual Si II shell lines in the spectrum of ζ Tau.

shell lines. The most important feature to note is that all the shell lines are in absorption.

I was curious to find out if these shell lines could be reproduced theoretically on the assumption that dilute radiation from the exciting star is incident on the shell, which is represented by a finite, plane-parallel slab. In order to estimate the non-LTE strengths of the Si II shell lines, I used a modified version of Kamp's (1975) non-LTE code, DRIVER. Table I gives the assumed parameters for the shell and exciting star.

TABLE I

SHELL PROPERTIES

- N_e=1.23 x 10^{11} cm^{-3}
- T_e=9457 °K
- COLUMN DENSITY=0.67 gm cm^{-2}
- HYDROGEN AND HELIUM ARE IN LTE
- STATIC SHELL

PROPERTIES OF INCIDENT RADIATION

- STELLAR PARAMETERS: T_{eff}=27,500 °K, LOG g=4.0
- DILUTION FACTOR=25X
- SPECTRUM: PURELY CONTINUOUS (NO LINES)

The adopted value of the electron density is compatible with that obtained from the Inglis-Teller relationship as applied to the Balmer shell lines. An electron temperature of about 10 000 K is assumed by most studies of the shell around ζ Tau, but the temperature is of little importance because at an electron density of 10^{11} cm^{-3}, collisional transitions are very rare compared to radiative transitions. The LTE population of the $n = 2$ level of hydrogen along with the total column density given in the table yields a number column density of the $n = 2$ level of hydrogen of 4.9×10^{15} cm^{-2}, a value which is compatible with that derived from the observed strengths of the higher members of the Balmer series. The shell was assumed to be static because the radial velocity of the H Balmer shell lines was constant with increasing upper quantum number, and hence with decreasing gf-value, and therefore, increasing depth of formation in the shell. The shell was assumed to lie five stellar radii away from the exciting star, which emits purely continuous radiation, characterized by the parameters, $T_{eff} = 27\,500°$, and log $g = 4.0$. This assumption of continuous radiation, even at the wavelengths of the Si II lines, is probably a good approximation (with one exception) since the stellar Si II lines must be very weak. The one exception is the Si II λ 990 Å line. Very probably, the N III line is blue-shifted so that the stellar radiation at Si II λ 990 Å is smaller than estimated here. The boundary conditions for the equation of transfer of radiation are: (1) radiation from the star impinges normally on the inner side of the shell, and (2) no radiation from outside the shell impinges on the outer edge of the shell. The transfer of radiation, subject to these two boundary conditions, was solved via Feautrier's method (cf. Mihalas, 1970). The model silicon atom consisted of 13 levels of Si II and the lowest 3 levels of Si III. A bound-free transition(s) from each Si II level and 25 line transitions were taken into account. The equivalent two-level-atom approach (cf. Mihalas, 1970) was used to estimate the line source function.

The results of the computation were that the resonance lines and a few lines from excited levels were in absorption, as observed, but other lines from excited levels were in emission, contrary to observation. By looking at the *net* rates of the various transitions, we can see why some of the visual lines were computed to be in emission. Figure 3 shows the net rates between various levels for the mass point, $m = 0.05$, a

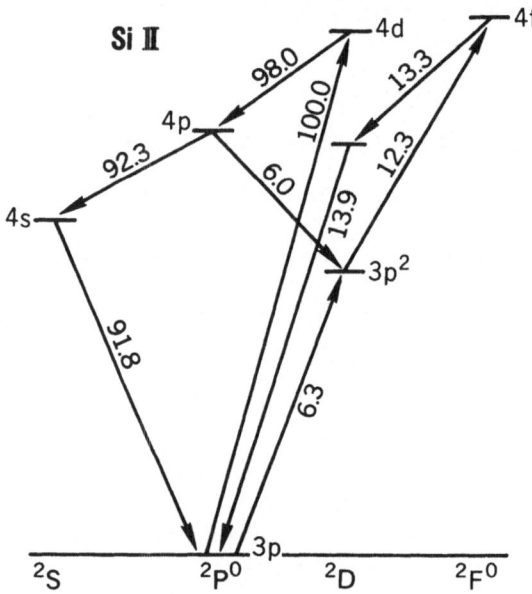

Fig. 3.   Net rates between various excitation states of Si II.

layer at which the optical depth of some of the weak visual lines is about unity. The net rates in Figure 3 are normalized arbitrarily so that the strongest net rate is equal to 100. The figure shows a pattern familiar in non-LTE physics: the existence of loops or 'Rosseland cycles'. Two are present here: one powered by the λ 990 Å transition, and the other powered by the λ 1808 Å transition. In order to stop the excited lines from going into emission, we have to stop these cycles from forming, which in effect means that we have to stop the strong net upward rates of the resonance lines. How can this be done?

An examination of the model shows two internal contradictions. The first is that the geometrical thickness of the shell is about 10^{13} cm, so spherical symmetry should be taken into account. The second contradiction, which was not apparent until after the calculations were made, is that the mean distance before destruction of a Si II line photon is greater than the probable height of the shell, which is estimated to be about

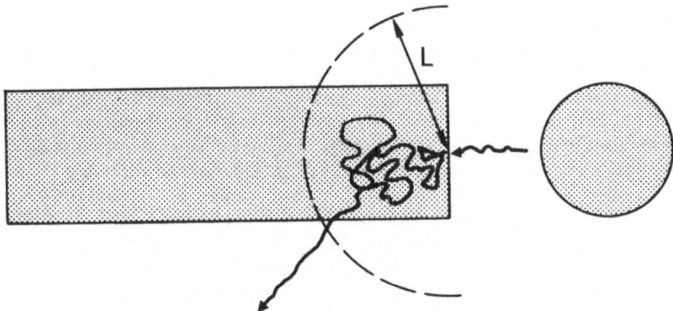

Fig. 4. Scattering of a Si II photon by the shell. The two shaded regions represent the cross-sections of the star (right) and the shell (left). See text for details.

one stellar diameter, or about 5×10^{11} cm. Figure 4 illustrates the situation. The computations show that an incoming photon from the star will be scattered many times before it is destroyed. In the process of scattering, the photon will travel a distance, L, given by the product of the mean-free path per scattering and the square root of the number of scatterings per destruction. For the Si II resonance lines, this distance is near 10^{12} cm. The net result is that a resonance-line photon will be scattered to either the upper or lower edge of the shell and escape the shell, and therefore, not be available to drive the Rosseland cycle.

This means that the reason why all the Si II lines in the shell spectrum of ζ Tau appear in absorption is probably that they escaped out the upper and lower boundaries of the shell. The distinct possibility of escape from the upper and lower boundaries of the shell violates the plane-parallel assumption, which implies, first, that the thickness of the shell is small with respect to its distance from the star, but more importantly, that the shell is infinite in the directions perpendicular to incident radiation from the star. Thus there is an internal contradiction in our approach. The main conclusion, then, is that in order to understand the shell spectrum of ζ Tau, the assumption of a plane-parallel shell must be discarded, not in favor of the assumption of spherical symmetry, but rather that of cylindrical symmetry, in which the shell is represented by a thick disk.

References

Heap, S. R.: 1975, *Phil. Trans. Roy. Soc. London A.* **279**, 371.
Mihalas, D. M.: 1970, *Stellar Atmospheres*, W. H. Freeman, San Francisco.

DISCUSSION

Plavec: I think this is an example of a problem which might be solved very well by the Monte Carlo method.

Poeckert: In the case of pole-on Be stars, would you expect that these lines would be in emission? Since you are allowing photons to escape below and above the plane, is that something to look for?

Heap: I don't know.

Peters: No, these lines are not observed to be in emission in pole-on stars.

Conti: I didn't quite understand why the shell had to be more or less a donut instead of having spherical symmetry.

Heap: A thick disk provides a large surface from which the photons can escape.

Conti: I am still worried about the effect of the observed wind on the shell. Surely the wind cannot begin outside the shell unless the radius is five stellar radii and the dilution factor $W = 25$. The wind must be inside the shell. How can the shell remain stable in the presence of this wind? Or is the shell maintained by the wind?

Heap: I can't tell you in physical terms; I can only tell you in observational terms: I see no evidence for any effect (e.g. heating) of the stellar wind on the shell.

Hummer: Since the asymmetric profiles of the absorption lines indicate that there is radial motion (at least) in the shell, it may be that the missing ingredient in your model is the velocity-induced photon escape.

Limber: Were the centers of the shell lines displaced?

Heap: At the time of the rocket observations, the radial velocity of the Balmer shell lines was about 20 km s^{-1} *greater* than that of the Balmer photospheric lines.

THE COMPLEX STRUCTURE OF THE Ca II H AND K LINES IN THE SPECTRUM OF THE A0ep STAR WITH INFRARED EXCESS HD 190073

J. SURDEJ and J. P. SWINGS

Institut d'Astrophysique, B-4200 Cointe-Ougrée, Belgium

Abstract. Radial velocities and profiles of the components of the H and K lines in the spectrum of HD 190073 are analyzed in a 24-spectrum sample covering the period 1943–74. A 2-to-1 ratio in the radial velocities of some components is shown not to be significant on a spectrum-to-spectrum basis, contrary to suggestions by Merrill (1951) and Scargle (1973). The details of the H and K complex structure are correlated with the profiles of the Balmer lines. Radiative forces acting selectively via a resonance scattering mechanism of Ca$^+$ atoms are capable of producing the main features of the complex profiles of H and K in HD 190073 and in other stars exhibiting similarities to HD 190073 (such as P Cygni Balmer lines and infrared excess).

DISCUSSION

Bidelman: Why is this such a rare phenomenon?

Swings: It is not a rare phenomenon. I have a list of about 10 objects which show several components in the H and K lines.

Bidelman: Isn't it probable that the H1 and K1 line components are interstellar?

Swings: The velocity of H1 and K1 corresponds exactly to the radial velocity of the star, to within ± 1 km s^{-1}. The lines also show emission wings around both sides which, it seems to me, come from the chromosphere of the star.

Harmanec: Can you comment about the behavior of the sodium lines in this same object?

Swings: The sodium lines are in emission. There might be a small absorption component.

Doazan: Can you comment on the infrared spectrum?

Swings: The Ca II triplet is in emission but at 230 Å mm^{-1} dispersion you do not see any structure. But Polidan may comment on that, since he has better spectra.

Polidan: We have looked at the star and it shows no structure whatsoever in the calcium triplet. The lines are simple, very strong emission lines, much stronger than one would expect from the K-line strength. They also appear to be optically thick; that is, all three lines have the same intensity.

Snijders: We have heard now about the 4s-4p H and K resonance lines and the 4p-3d infrared triplet. Yesterday emission in the forbidden 3d-4s transitions in some Be stars was mentioned. Were these forbidden lines observed in this star?

Swings: The forbidden lines of Ca II at λ 7291 and λ 7323 were not observed in this star.

Goldberg: Calcium is one part in a million of the total amount of gas in stellar atmospheres and I don't see how you can consider the motion of the calcium independently of the drag expected on it by the surrounding gas.

Swings: In the second of our two papers (submitted for publication in Astronomy and Astrophysics), we try to answer the question "Why is the ejection of atoms from HD 190073 limited to Ca II?" The probability for an atom to be ejected will actually be important if five requirements are fulfilled: Ca II meets all of them. Next, I believe it was shown a long time ago that Ca II emission was observed in the solar chromosphere up to heights greater than for, say, hydrogen. We therefore do not consider a 'pure calcium' star but we assume that there are layers in the extended atmosphere of HD 190073 having a high calcium abundance. We also consider collisions to be negligible in the regions where the resonance scattering mechanism is supposed to be efficient.

A. Slettebak (ed.), Be and Shell Stars, 321. All Rights Reserved

THEORETICAL EMISSION-LINE PROFILES COMPUTED AT ONDŘEJOV

SVATOPLUK KŘÍŽ

Astronomical Institute of the Czechoslovak Academy of Sciences, Ondřejov, Czechoslovakia

(Paper read by P. Harmanec)

Abstract. A rotationally symmetrical disk-shaped envelope was considered and the steady-state equations for level populations solved using the escape probability approximation. The line profiles were then calculated and compared with observations.

The recent theoretical work by Marlborough (1969, 1970) represents a considerable progress in the quantitative interpretation of emission lines of Be stars. We feel that such an approach is necessary if one wants to know the real physical conditions in gaseous envelopes, their origin, etc. Therefore we decided to construct a similar program for computations of theoretical profiles of emission lines from envelopes of Be stars.

We started with a very simplified model. Our original program assumed the envelope to be a flat hydrogen disk, axially symmetrical about the axis of rotation of the star. The conditions in the disk are assumed to be independent of the distance from the equatorial plane, i.e. all the properties of the matter are functions of the distance from the axis of rotation only. Also it is assumed that the support of the envelope is centrifugal and thus the velocity of an arbitrary point is the Keplerian velocity of circular motion. Finally, the continuum radiation of the central star is assumed to be Planckian.

The input data of the program are as follows: the mass, radius and effective temperature of the central star, the profiles of the absorption lines formed in the atmosphere of the central star, the density of atoms, the electron temperature, and the thickness of the disk as functions of distance from the axis of rotation.

First, the steady-state equation for level populations is solved for a given grid of points in the envelope. The departure coefficients from LTE populations, b_n, are calculated for the levels of principal quantum number $n \leq 20$. The levels of $n > 20$ are assumed to have the populations appropriate to thermodynamic equilibrium with free electrons. The radiative bound-bound transitions are treated by a modified Sobolev's escape probability method (see Kříž, 1973), assuming the absorption, emission and induced emission profiles to be identical and having the Gaussian thermally broadened shape. Only the diluted radiation of the central star is considered in calculating the bound-free radiative transitions. The rate coefficients for collisional processes are calculated by Johnson's (1972) formulae.

The intensity of the line radiation emitted at each point of the disk toward the observer is calculated (for the required lines and required wavelengths) by means of the precomputed level populations, assuming the angle of inclination $i = 90°$. Finally, the flux of energy emitted by the entire envelope is integrated. For these calculations we used the simplified solution of the equation of line-radiation transfer in media

A. Slettebak (ed.), Be and Shell Stars, 323–325. All Rights Reserved
Copyright © 1976 by the IAU.

with large velocity gradients given by Kříž (1973) – see Equation (45) of that paper. Consequently we cannot compute the central parts of emission lines because of the small corresponding velocity gradient.

As a result we have obtained a series of theoretical line profiles for different models of gaseous disks. We have tried to match these profiles to the observed profiles of Be stars. It is very easy to find a model which gives excellent agreement with observations of one line (we have used Hα). Unfortunately, we are able to construct a large number of models giving the same profile of Hα. Therefore a comparison of the profile of one line has no meaning and we must compare as many lines as possible.

It is very difficult to construct a model which matches more observed lines. Figures 1 and 2 show observed profiles of Hα and Hβ in EW Lac (Gray and Marlborough, 1974), and the corresponding theoretical profiles. The differences are clearly larger than the observational errors.

We therefore arrived at the conclusion that our models are oversimplified. We are now working on a new, more sophisticated program. It will enable us to construct models of elliptical rings (see Huang, 1973), use arbitrary angles of inclination, compute centres of the lines, etc.

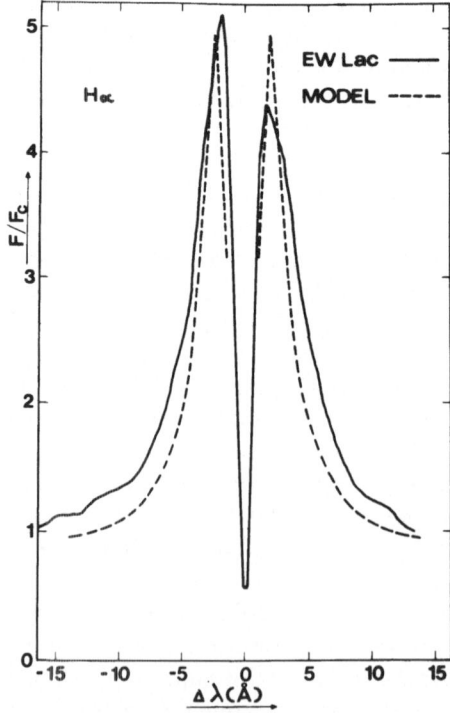

Fig. 1.   The profile of Hα in EW Lac observed by Gray and Marlborough (1974) compared with a theoretical model characterized by the following parameters: mass, radius and effective temperature of the central star are $10\,M_\odot$, $4.66\,R_\odot$, $20\,000\,K$, respectively; the disk has radius $172.4\,R_\odot$, thickness $13.6\,R_\odot$; number of hydrogen atoms per cm^3 is 10^{11}; and the electron temperature equals $20\,000\,K$ (everywhere in the disk).

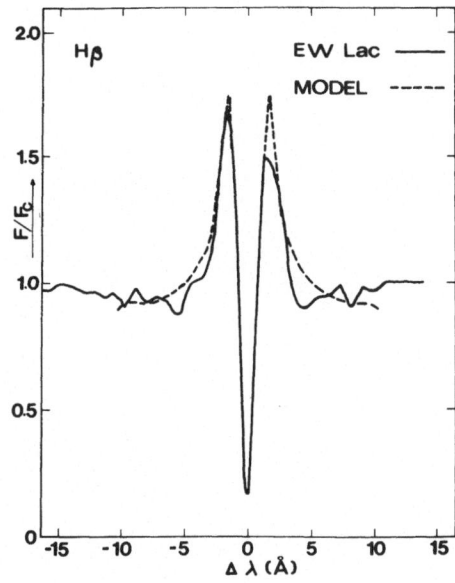

Fig. 2. Comparison of observed profile of Hβ with theoretical model. For details, see caption for Figure 1.

Acknowledgements

I am very indebted to Mr P. Koubský who performed the comparison of the theoretical and observed profiles.

References

Gray, F. G. and Marlborough, J. M.: 1974, *Astrophys. J. Suppl.* **27**, 121.
Huang, S.-S.: 1973, *Astrophys. J.* **183**, 541.
Johnson, L. C.: 1972, *Astrophys. J.* **174**, 227.
Kříž, S.: 1973, *Bull. Astron. Inst. Czech.* **25**, 143.
Marlborough, J. M.: 1969, *Astrophys. J.* **156**, 135.
Marlborough, J. M.: 1970, *Astrophys. J.* **159**, 575.

MOTIONS IN THE SHELLS AND ATMOSPHERES OF V923 AQL AND EW LAC AND THEIR MANIFESTATION IN THE SPECTRUM

N. F. VOJKHANSKAYA

Special Astrophysical Observatory, Academy of Sciences, U.S.S.R.

The stars V923 Aql (HD 183656) and EW Lac (HD 217050) are marked in the General Catalog of Variable Stars (Kukarkin *et al.*, 1969) as unique, similar objects with shells. The two stars have quasiperiodic variations in brightness (Lynds, 1960; Walker, 1953). It is assumed here that the variation of brightness is caused by an inhomogeneous distribution of brightness over the surface of the star. A consideration of the literature data on the radial velocities V_r which we have carried out has shown that the variation of V_r is different (Figure 1).

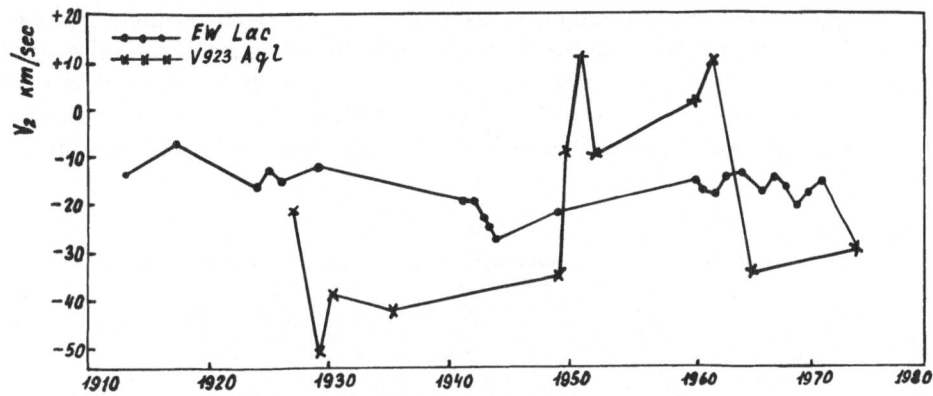

Fig. 1.   Radial velocities of EW Lac and V 923 Aql.

For the spectrophotometric analysis, spectrograms with dispersion 29 Å mm^{-1} in the region λ 6700–3600 Å have been used. Twelve spectrograms for EW Lac and ten for V923 Aql were obtained. In addition, ten spectrograms of V923 Aql obtained at the Crimean Astrophysical Observatory in 1965 with a dispersion of 15 Å mm^{-1} in the region λ 4700–3700 Å were loaned to us by T. M. Rachkovskaya.

As a result of our analysis it has been determined that the shell spectrum of EW Lac varies markedly during the period of the order of 24 hours, which corresponds approximately to the photometric period. The lines arising in the atmosphere of the star do not vary. The behaviour of the He I lines is peculiar. In spite of the fact that He I lines most likely arise in the star's atmosphere, their intensities vary greatly, and the character of variation is similar to the intensity variation in the shell spectrum. Figure 2 shows the record of the portion of the spectrum containing the He I line λ 3926 Å in terms of relative intensities. Variation of the hydrogen lines with brightness phase is observed in the spectrum of V923 Aql. Line intensity

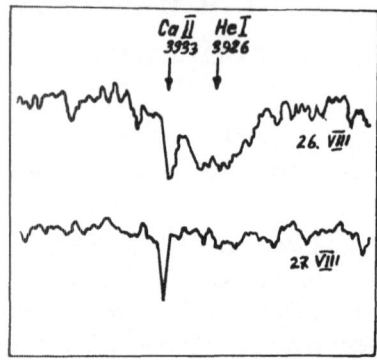

Fig. 2.　Variations in He I λ 3926 Å in the spectrum of EW Lac.

variation of other elements, including He I, with light phase does not exceed measurement errors.

For radial velocity measurement we used a device in which setting on a line is performed by scanning the contour back and forth, thereby permitting splitting the observed contour into the shell line and that of the star. We have assumed here that the line core is formed in the shell and the outer parts of the contour in the star's atmosphere. The measured radial velocities are presented in Tables I and II.

TABLE I

Results of radial velocity measurements in the spectrum of V923 Aql

Phase	0		0.5		Error $(km\ s^{-1})$
Line	1965	1974	1965	1974	
H (shell absorption)	−32	−33	−37	−21 ⎫	
Hα (shell emission)	−	0	−	0 ⎬	±10
H (star)	−59	−14	−85	−44	±20

TABLE II

Results of radial velocity measurements in the spectrum of EW Lac

Element	26 VIII, 1974	27 VIII, 1974	28 VIII, 1974
Fe I	+92±25	+124±5	
Fe II, Cr II, Ti II, Sc II, V II	+28±13 ⎫		
Mg II	−34.5		
Ca II	−39.2	−20±8	
Si II	−21.9±0.3		
H (shell, absorption)	−30±10 ⎭		
Hα (shell, emission)	−114.7	−48.9	−64.9
H (star)	−92±20	−80±19	
Ca I	−171	−224	
Na I	−45±5	−50±1	
O I	−59.4	−	
O II, C II, N II, N III	−8±8	+64±10	
He I	−21±10	−82±22	

Velocities for EW Lac have been measured from lines of other elements (see Table II). If V_r is determined from a single line, the error is not given in Table II.

The difference between the hydrogen line contours in 1974 is greater than in 1965 (Figure 3). It can be seen from Table I that the velocities measured from shell lines during these years are the same, within the errors of measurement, while the

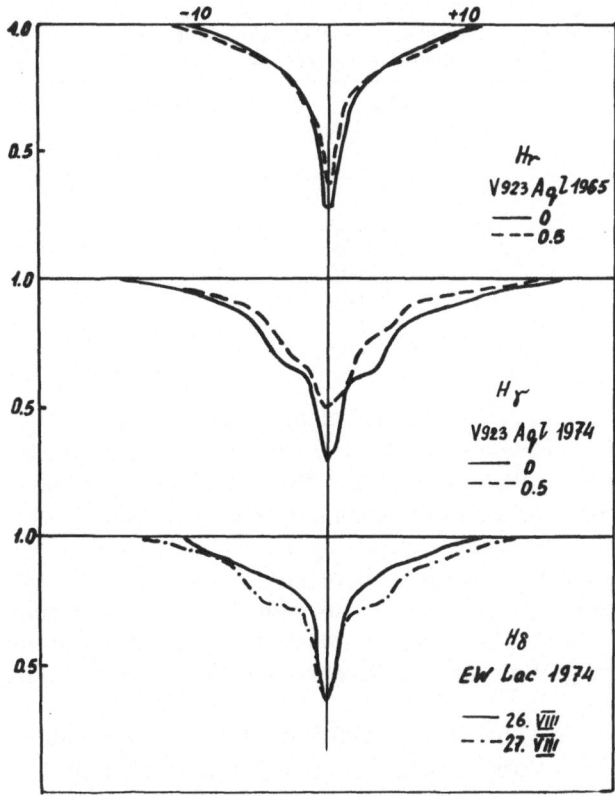

Fig. 3.   Balmer line profiles in V 923 Aql and EW Lac.

velocities V_* measured from lines of the star are different. Comparison of the contours with the Table I data has shown that the larger the ratio of velocities V_* (in absolute value) in phases 0.5 and 0, the greater the difference between the contours in these phases. However the value of the velocity V_* is not the major factor determining the shape of the contour. A more important factor is, apparently, the dynamics of motions in the atmosphere and in the shell of the star. If can be seen in Figure 1 that after 1962 the velocity of expansion of the shell of V923 Aql had been rapidly increasing. After 1965 it ceased to increase. As a result of this, considerable inhomogeneities must have developed which would lead to broadening of the contours.

Hydrogen line contours in the spectrum of EW Lac vary in a different way. The radial velocity V_* defined from these lines turned out to be constant. In deeper layers of the star's atmosphere, where the lines of high excitation potential (C II, O II, N II, N III, He I) arise, the velocity V_* varies.

Taking EW Lac as an example we shall trace in detail the relation between the motion in the shell and the physical parameters determined from the spectrum. In Table III we present the dilution coefficients and the distances in units of the star's

TABLE III

Heights of spectrum formation of different elements and dilution coefficients for EW Lac

Element	26 VIII, 1974		27 VIII, 1974	
	r/R_*	W	r/R_*	W
Fe I	8.9	0.003	–	–
Fe II	6.6	0.006	–	–
Mg II	2.2	0.05	1.1	0.2
Si II	4.7	0.01	2.7	0.03
Ca II	4.2	0.01	4.5	0.01
O I	3.9	0.02	–	–
He I	0.8	0.4	0.7	0.5

radius at which spectra of the different elements arise. The velocity found from C II, O II, N II, N III lines is assumed to be the velocity of rotation of the star. It follows from Tables II and III that on August 26 the surface layers of the star which border the inner layers of the shell were expanding at a mean velocity of about -8 km s^{-1}, and on August 27 they were contracting at a velocity of $+64$ km s^{-1}. The deeper layers of the star's atmosphere, where He I lines arise were first expanding at -21 km s^{-1} and then at -82 km s^{-1}. Thus, radial velocity measurements have shown that the surface layers of the star do not remain stationary, but move at a variable velocity in both magnitude and the direction, which is certain to affect the state of the shell.

It can be seen from Tables II and III that on August 26 the outermost layers of the shell, where Fe I lines arise, were contracting at a velocity of $+92$ km s^{-1}, and on August 27 the velocity of contraction reached a value of $+124$ km s^{-1}. The deeper layers, in which the lines of singly ionized metals Fe II, Cr II, Ti II and Sc II are formed, were contracting at $+28$ km s^{-1} on August 26, while still deeper layers situated closer to the surface of the star, where H_{shell}, Mg II, Ca II, and Si II lines originate, were expanding at a mean velocity of -30 km s^{-1}. On August 27 all these elements had a velocity of about -20 km s^{-1}. The turbulent velocity defined from the curve of growth was 4 km s^{-1} on August 26, and on August 27 it had increased by a factor of 2. This fact can be accounted for if one refers to the analysis of the measured radial velocities. On August 26 the inner and outer layers of the shell were approaching each other. At some moment they collided, after which the magnitude and the direction of the velocity was determined by the motion of the more massive hydrogen envelope, and the radial velocities became equal. However, as a result of the collision of two fluxes turbulence developed and the turbulent velocity increased. This led to clearing of the shell: the optical depth in the Balmer lines was reduced by a factor of two. For example, in the Hδ line $\tau = 3.2 \times 10^2$ on August 26, and on August

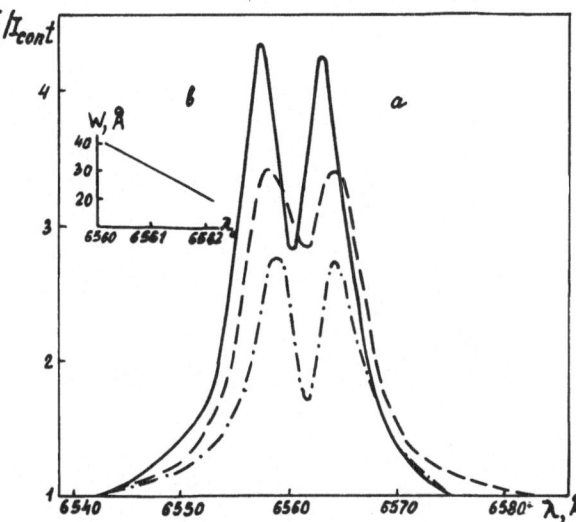

Fig. 4.   Hα emission line profiles in EW Lac.

27 $\tau = 1.6 \times 10^2$. A still more obvious relation between the motion of matter in the shell and the intensity of the spectrum follows from Figure 4, where it can be seen that with increasing shortward shift of the Hα emission line, its intensity increases.

Balmer absorption lines formed at distances of 5–6 R_* have smaller velocities than those corresponding to the shift of the Hα emission lines. On the basis of the results presented in Table III it may be suggested that emission in Hα arises at a distance of $\geq 10\ R_*$. If this is so, the velocity of the hydrogen atoms would then increase toward the outer border of the shell. It is difficult to say whether the observed acceleration will lead to a gradual spread and loss of the envelope. If the mass of the star is considered to be 12–13 $M_\odot$ and $R_* = 5\ R_\odot$ (Strand, 1969), the escape velocity at a distance 10 R_* would be about 300 km s^{-1}. The observed velocities are considerably smaller.

References

Kukarkin, B. V., Kholopov, P. N., Efremov, Yu. N., Kukarkina, N. P., Kurochkin, N. E., Medvedeva, G. I., Perova, N. B., Fedorovich, V. P., and Frolov, M. S.: 1969, *General Catalog of Variable Stars* (3d ed.) Moscow.

Lynds, C. R.: 1960, *Astrophys. J.* **131**, 390.

Strand, K. Aa.: 1969, *Stars and Stellar Systems* **3**, 273.

Walker, M. F.: 1953, *Astrophys. J.* **118**, 481.

PART V

MODELS

MODELS FOR THE CIRCUMSTELLAR ENVELOPES
OF Be STARS
(Review Paper)

J. M. MARLBOROUGH

Dept. of Astronomy, University of Western Ontario, London, Ontario, Canada

Abstract. A survey is presented of the theoretical attempts to determine the structure of the circumstellar matter around Be stars. The general equations describing the structure and dynamics of Be star envelopes are given. The complications introduced by various physical phenomena are briefly discussed and initial attempts to solve restricted problems are considered. The various *ad hoc* models proposed for Be stars are discussed and comparisons of the observations with predictions of these models are illustrated. The strengths and weaknesses of these models are evaluated and areas where progress is being or should be made are considered.

1. Introduction

Be stars were first discovered by Secchi (1867) when, from visual inspection of spectra, he noted the Hβ line in emission in γ Cas. Since that time many astronomers have contributed to the vast array of observational data pertaining to Be stars and some astronomers, notably D. B. McLaughlin and to a lesser extent P. W. Merrill, have devoted a significant fraction of their professional careers to the study of Be stars.

The first suggestion to account for the origin of the emission lines in Be stars was due to Struve (1931; also Struve and Swings, 1932). Struve proposed that the emission lines arose from matter ejected from the equatorial region of a rapidly rotating star. This ejected material was assumed to form a nebulous ring around the star. If the emitting matter had an orbital speed less than that of the star, this hypothesis could easily account for the observed fact that the width of the emission lines was less than that of the underlying photospheric features. In many respects this idea qualitatively resembled the rings of the planet Saturn, as Struve himself noted. Furthermore, by restricting the ejected matter to regions near the equatorial plane, Struve was able to account for Merrill *et al.*'s (1925) observations that all stars with double emission at Hβ had 'nebulous' absorption lines (large $v \sin i$) whereas those with single emission at Hβ had 'sharp' absorption lines (low $v \sin i$). Finally, if the orbits of individual emitters in the ring were elliptical rather than circular, the two components of a double emission line would not necessarily have the same strength. A rotation of the line of apsides of the ring might then lead to a periodic variation of the strength of the two emission components thus providing an explanation for the V/R variation, V and R being measures of the strength of the violet and red components of the emission line, respectively.

McLaughlin (1933, 1938) extended the original idea of Struve and included those of Rosseland (1926) concerning the transformation of high energy radiation to radiation of lower energy to explain the origin of emission lines in stellar spectra. McLaughlin realized that the emitting region around a Be star is large compared to

A. Slettebak (ed.), *Be and Shell Stars*, 335–370. *All Rights Reserved*

the dimensions of the star, in fact similar to the dimensions of a small planetary nebula, but of lower excitation. Photoionization of hydrogen atoms by Lyman continuum radiation and the subsequent recapture and cascade of electrons through various energy levels lead to the production of emission lines. In the inner part of the emitting region hydrogen would be predominantly ionized while, further out, geometrical and physical dilution will significantly deplete the stellar Lyman continuum so that hydrogen becomes predominantly neutral.

To account for some of the observed spectral variations McLaughlin proposed a rotating-pulsating model for Be stars. This suggestion was a combination of the rotational model of Struve and the idea of an expanding atmosphere introduced by Beals (1930) to explain the observations of stars of the Wolf-Rayet and P Cygni types. Assumed temperature variations of the star yield a variable radiation pressure force on the envelope material thus producing expansion and contraction. Although this idea and a similar intermittently expanding-rotating model are no longer considered relevant to the Be star phenomenon and were rejected by McLaughlin (1961) on observational considerations and by Huang (1973) on theoretical considerations, they are important for two reasons. First, McLaughlin suggested the idea that radiation pressure might be important for Be stars, i.e., some other force apart from gravitation and those arising from the pressure gradient and centrifugal effects must be considered. Secondly, McLaughlin emphasized the importance of the velocity gradient for the line optical depth. In addition, McLaughlin (1961) further extended the suggestion of Struve concerning the possibility of an elliptical ring to account for the V/R variation.

Gerasimovic (1934) objected to Struve's rotational model for a variety of reasons; Struve (1942) later showed that most of Gerasimovic's objections were no longer significant. Gerasimovic suggested that the material responsible for the emission lines is ejected from the equatorial region of a rapidly rotating star by radiation pressure, both line and continuum, aided by the rapid rotation of the star. For this type of gas flow the streamlines are spirals and the circumstellar envelope is confined between two coaxial cones with vertices near the center of the star and which extend along the rotation axis of the star in the positive and negative directions. The initial velocity of the outflowing matter is assumed to be small in the radial direction since no large line asymmetries are observed in Be star spectra. To explain some of the time variations, Gerasimovic further proposed that matter is ejected spasmodically, regulated by radiation pressure due to an inward flux of radiation scattered back toward the star from the envelope. Dissipation of the outer envelope reduces this inward flux and allows matter to escape from the star again.

Many of the above ideas can be combined to provide the following simple *schematic* picture of a Be star. Consider a rapidly rotating B star with an extended moderately dense atmosphere confined chiefly to the equatorial plane as illustrated in Figure 1. For definiteness suppose the origin of the coordinate system is at the center of the star, the z axis parallel to the rotation axis of the star and the $x - y$ plane lying in the star's equatorial plane and consider standard spherical polar coordinates r, θ, and ϕ.

In Figure 1a is shown a schematic picture of the star and circumstellar envelope. For an observer in the direction indicated and not necessarily lying in the equatorial

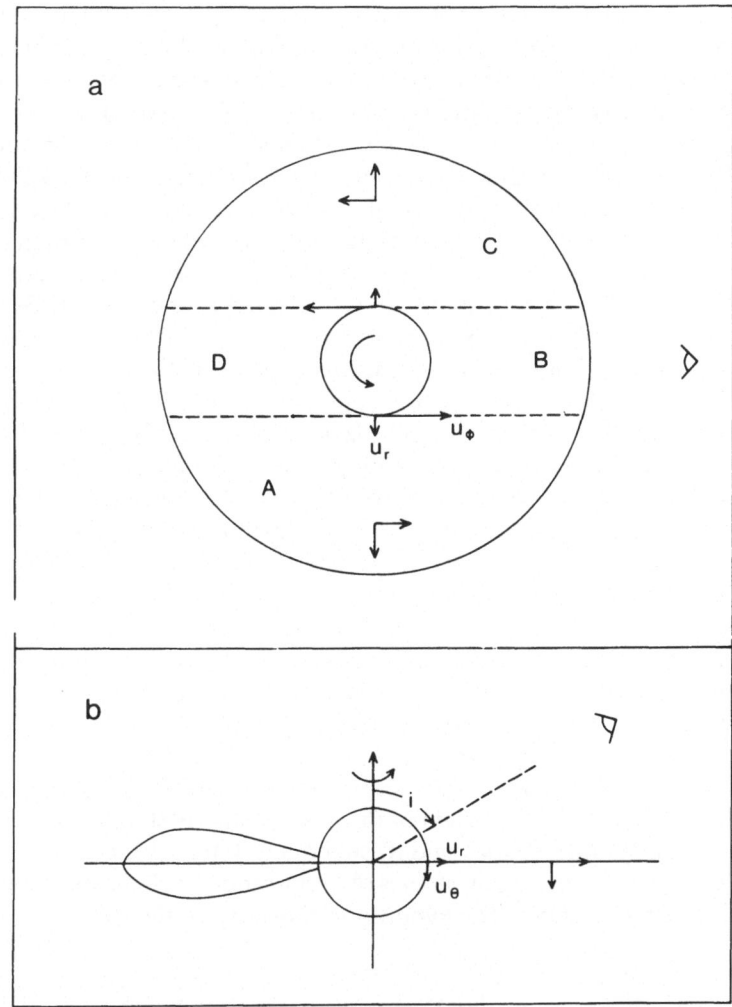

Fig. 1. A schematic representation of a Be star envelope. Part a is an equatorial plane view. The observer is to the right but not necessarily in the equatorial plane. The magnitude, direction and variation with r of the velocity components u_r and u_ϕ are illustrated schematically. Part b is a meridional plane view. The curve shown is a possible equidensity contour and the envelope is presumably symmetric about the rotation axis. The star has been drawn as spherical for simplicity. The magnitude, direction and possible variation with r of u_r and u_θ are illustrated.

plane, regions A, B, C and the part of D not occulted by the star would contribute to the emission in a given line; the part of region B projected against the star would produce the absorption component of the observed line. The observational data indicate that in the region of the envelope in which lines seen in the visible part of the spectrum are formed, $u_r \ll u_\phi$ where u_r and u_ϕ are the r and ϕ components of velocity, respectively, and this is illustrated schematically in Figure 1a. Recent evidence by Snow and Marlborough (this volume, p. 179) indicates that for some early Be stars of large $v \sin i$, u_r is an increasing function of r (see Marlborough and

Zamir, 1975, for a discussion of the inferences on $u_r(r)$ from theoretical models). This increase in $u_r(r)$ with increasing r is also illustrated schematically in Figure 1a.

In Figure 1b is shown a meridional plane view of a schematic Be star envelope. The appearance of the envelope in this case is uncertain. The curve shown may be thought of as representing an equidensity contour. Whether or not any significant amount of matter is located above the poles of the star is not known. If the model illustrated in Figure 1 is to represent both Be and shell stars, then the matter in the envelope must extend to at least one stellar radius from the equatorial plane to account for the deep shell absorption lines observed in the spectra of shell stars. Observational evidence on the magnitude of u_θ, the θ component of velocity, is uncertain. If Be stars of low $v \sin i$ are actually rapid rotators seen at small angles of inclination i, then, since the relatively sharp emission lines present in the spectra of Be stars of low $v \sin i$ are observed at their expected wavelength, the sum of the components of u_r and u_θ projected on to the line of sight cannot be large. Consequently, it is realistic to expect $u_\theta \lesssim u_r$, at least in the region of the envelope where the emission lines are formed.

A qualitative understanding of many of the observations of Be and shell stars can be obtained in the context of the schematic model illustrated in Figure 1 by allowing the density $\rho(\mathbf{r})$, the velocity $\mathbf{u}(\mathbf{r})$, the temperature $T(\mathbf{r})$ and the angle i to vary. Perhaps the strongest observational support for this model are the observations that $\Delta\lambda/\lambda$ is a constant for the emission lines in most Be stars and that many Be stars have a non-zero intrinsic linear polarization – presumed to arise from electron scattering – implying that for some values of angle i the envelope projected onto the plane of the sky, as seen by the observer, is not circularly symmetric.

Ideally one would like to predict $\rho(\mathbf{r}, t)$, $\mathbf{u}(\mathbf{r}, t)$ and $T(\mathbf{r}, t)$ as functions of position $\mathbf{r}$ and time t from a consideration of the relevant physical phenomena. In addition, one would hope to understand the origin of the cirumstellar material – does it arise from the star or a nearby companion – and its subequent fate. Furthermore, a knowledge of the mass loss rate (or mass gain rate) is of importance for determining what role the Be phenomenon may play in the evolutionary history of the star.

2. Structure and Dynamics of Be Star Envelopes

Let us now consider the general problem of determining $\rho(\mathbf{r}, t)$, $\mathbf{u}(\mathbf{r}, t)$ and $T(\mathbf{r}, t)$ for the circumstellar material around a Be star. Some initial attempts to solve for these variables have been investigated and will be discussed below.

2.1. GENERAL THEORY

We will assume that the circumstellar matter in a Be star envelope can be treated as a continuum and will use a single fluid description of this continuum. Morgan (1975b) has recently investigated both of these assumptions and concluded they are justified for Be star envelopes.

A complete hydrodynamic treatment of the structure and dynamics of Be star envelopes requires the simultaneous solution of the following equations

$$\frac{\partial\rho}{\partial t} + \nabla \cdot \rho\mathbf{u} = 0 , \tag{1}$$

$$\frac{\partial \mathbf{u}}{\partial t} + \mathbf{u} \cdot \nabla \mathbf{u} = -\frac{1}{\rho}\nabla P_g + \mathbf{F} + \mathbf{f}, \tag{2}$$

$$\frac{\partial E_g}{\partial t} + \mathbf{u} \cdot \nabla E_g + P_g \frac{\partial}{\partial t}\left(\frac{1}{\rho}\right) + P_g \mathbf{u} \cdot \nabla \left(\frac{1}{\rho}\right) = G - L, \tag{3}$$

where Equations (1), (2), and (3) represent the equation of continuity, the equation of motion, and the equation for the conservation of energy, respectively. In these equations ρ is the density, $\mathbf{u}$ is the velocity, P_g is the gas pressure, $\mathbf{F}$ is the total body force per unit mass on a fluid element, $\mathbf{f}$ is the total boundary force per unit mass on a fluid element, E_g is the internal energy of the gas per unit mass, G is the rate of energy gain per unit mass of the gas, L is the rate of energy loss per unit mass of the gas, and all these variables are in general functions of position $\mathbf{r}$ and time t. The basic Equations (1), (2) and (3) are discussed in any book on fluid dynamics, e.g. Zel'dovich and Raizer (1966). If the gas is assumed to be a simple perfect gas of constant mean molecular weight, then $E_g = c_v T$ where T is the temperature and c_v, the specific heat at constant volume, is generally assumed to be constant if the gas is non-degenerate and non-relativistic (see Cox and Giuli, 1968, for details). For such a situation, Equations (1), (2), and (3), together with the equation of state, constitute six equations for the six dependent variables ρ, $\mathbf{u}$, P_g, and T if $\mathbf{F}$, $\mathbf{f}$, G and L do not introduce any new dependent variables. In general, however, this is not the situation.

The total body force $\mathbf{F}$ includes gravitation due to the central star and others it the star is a member of a multiple star system, self-gravitation of the envelope matter, the radiation pressure force, and the Lorentz force if a magnetic field is present. Normally, self-gravitation of the circumstellar material is neglected for Be star envelopes and this neglect seems justified based on estimates of the total mass of circumstellar matter (see Weidelt, 1970; and Biermann and Kippenhahn, 1973, for complications when self-gravitation is included).

If the radiation pressure force makes an important contribution to $\mathbf{F}$ and/or radiative energy gains and losses occur in G and L, then the basic equations will contain terms involving the specific intensity $I_\nu(\mathbf{r}, t)$ at frequency ν, the source function $S_\nu(\mathbf{r}, t)$ and various moments of $I_\nu(\mathbf{r}, t)$, together with the mass absorption coefficient κ_ν. Since both S_ν and κ_ν can be expressed in terms of the populations of atomic energy levels contributing to the radiation at frequency ν, a set of equations determining these atomic energy level populations together with the equation of transfer must be solved simultaneously with Equations (1), (2), (3) and the equation of state. Some idea of the complexity involved even for the case of a steady-state, spherical symmetry and a very simplified treatment of the radiative transfer, can be obtained from the investigations of Cassinelli and Castor (1973) and Castor *et al.* (1975).

If a magnetic field plays an important role in the dynamics of Be star envelopes, as has been suggested by Crampin and Hoyle (1960) and Limber and Marlborough (1968) among others, then in general Equations (1) to (3) together with the equation of state must be solved simultaneously with Maxwell's equations describing the electromagnetic field. Of course, if the effects of both radiation and a magnetic field are included, the relevant equations become horrendously complex.

Whether or not viscous forces are important – i.e. whether or not $\mathbf{f} = 0$ – is presently unclear. Morgan (1975b) compared the ratio of viscous to inertial terms in Equation (2) and concluded viscous effects were completely negligible. However Limber and Marlborough (1968) concluded that outward transport of angular momentum by some viscous agent – either magnetic or turbulent viscosity – was important for the dynamics of Be star envelopes. Recent investigations (see Section 2.2) seem to support both these results in that additional forces such as those arising from a magnetic field or the absorption of radiation are important in order to account for the envelope dynamics while the inclusion in Equation (2) of terms arising from the viscous stress tensor may not be necessary.

Due to differential motions in Be star envelopes, a further complication arises when the radiation field plays an important role. Equations (1) to (3) are valid in an inertial reference frame. However the equation of transfer and the equations for the population of atomic energy levels are normally expressed in a form valid only in a frame at rest with respect to the local macroscopic velocity of the matter, i.e. the fluid frame. For example, it is only in this latter frame that κ_ν is independent of the direction of the radiation being absorbed. This problem is discussed in detail in various books (see Sampson, 1965; and Pomraning, 1973). Castor (1972) has discussed the problem in detail for a spherically symmetric situation.

It should also be pointed out that it may be necessary to include non-radiative energy sources in G and L. Thomas (1970) has constantly emphasized the possible importance for stellar atmospheres – considered to be the transition zone between the stellar interior and the interstellar medium – of a supply of non-thermal kinetic energy. The recent suggestion of the existence of a chromosphere in Vega (A0V) by Praderie *et al.* (1975) may be direct evidence in support of Thomas' suggestion for relatively early type stars. Smith (1970) has shown that in rotating stellar atmospheres meridional circulation currents become unstable near the surface and generate a turbulent layer with turbulent velocities which in general are only slightly subsonic. Although the energy dissipated in the turbulent eddies is predicted to be small compared to the stellar luminosity, a non-thermal energy source of this kind may be important in the Be star envelopes.

On account of the complexity of the basic equations describing the dynamics of Be star envelopes, little progress has been made in solving the general problem for the structure and dynamics as a function of both position and time. In general, the basic equations are at least a function of two independent variables, i.e. r and θ, and thus consist of a set of non-linear partial differential equations. Such a set of equations is notoriously difficult to solve. The various attempts to date have employed a number of simplifying assumptions to reduce the general problem to one for which some solution can be found in restricted cases.

2.2. Specific solutions

In essentially all considerations so far, a steady state has been assumed. It is important to point out, however, that in many, if not all Be stars, spectroscopic and photometric features normally attributed to the circumstellar material vary on some time scale which can be anywhere from the order of minutes or less to years. Limber

(1970) has discussed the theoretical implications of these time scales. All attempts to solve the basic equations for some restrictive set of assumptions are concerned with stellar wind solutions, i.e. solutions for which $\rho(r, \theta = \pi/2)$ is non-zero for $R_* \le r < \infty$ and for which $u_r(r, \theta = \pi/2)$ is generally an increasing function of r for $R_* \le r < \infty$ (R_* is the stellar radius).

For the case of continuing rotationally forced ejection with gravitation and gas pressure gradient being the only forces acting, Limber (1964, 1967) has constructed steady-state solutions which are symmetric about the equatorial plane and which are limited to regions near the equatorial plane. The solutions also require the specification of the functional dependence of $u_\phi(r, \theta = \pi/2)$ on r. In these solutions, the effect of the stellar radiation field is not included explicitly but can be included by reducing the effective gravitational mass of the star. Comparisons of predictions based on these solutions with observational data are given in Section 3.

Cassinelli and Castor (1973) and Castor *et al.* (1975) have included the effects of the stellar radiation field explicitly in Equations (2) and (3) and investigated the solutions for the case of spherical symmetry in order to account for the violet displaced absorption lines observed in the far ultraviolet spectra of early type, luminous stars. Marlborough and Zamir (1975) extended the solutions of Cassinelli and Castor (1973) to the equatorial plane of rapidly rotating stars. Under the assumption that $u_\phi(r, \theta = \pi/2)$ is a known function of r and using other assumptions described in that paper Marlborough and Zamir obtained the solution for $u_r(r, \theta = \pi/2)$. If the radiation field of the star is sufficiently strong and/or the rotation rate sufficiently large, then a solution for $u_r(r, \theta = \pi/2)$ deviating considerably from hydrostatic behavior is obtained in which u_r is subsonic near the star and supersonic at larger distances. Such a solution is qualitatively similar to ones obtained by Limber (1967). Marlborough and Zamir further suggested that the solution for u_r for $\theta = 0, \pi$ would probably deviate negligibly from hydrostatic behavior. Heap (1975) discovered that all stellar lines in the ultraviolet spectrum of ζ Tau (B3IV), obtained on a rocket flight, are shifted to the violet corresponding to velocities up to 120 km s^{-1}. Analysis of *Copernicus* data by Snow and Marlborough (this volume, p. 179) indicates that several early Be stars of large $v \sin i$ show asymmetric resonance line profiles, particularly the Si IV doublet $\lambda\lambda$ 1393, 1402, indicating mass loss with typical velocities of order several hundred km s^{-1}. Other Be stars of low $v \sin i$, observed to date by *Copernicus* do not appear to show this line asymmetry. These observations are consistent with the predictions of Marlborough and Zamir. Therefore, at least for early type Be stars, the Be star phenomenon may be the result of a radiatively driven stellar wind. Massa (1975) has also concluded that the Be phenomenon is to be understood in terms of a radiation driven stellar wind. However, it should be noted that the asymmetric line profiles are also at least qualitatively consistent with any stellar wind model in which u_r is an increasing function of r in the region where the absorption lines are formed.

Two recent investigations have included the effects of a magnetic field on the dynamics of the envelope. Limber (1974) has modified the investigations of Weber and Davis (1967) concerning the angular momentum of the solar wind and extended their analysis to the region of Be star envelopes in and near the equatorial plane. Stellar wind solutions were found in which $u_r(r, \theta = \pi/2)$ is such that the radial

Alfvènic Mach number $M_A \equiv (4\pi\rho u_r^2/B_r^2)^{1/2}$ varies from <1 near the star to >1 at larger distances (B_r is the radial component of the magnetic field). This model possesses several interesting features. For essentially all values of r except very close to the star, $u_\phi(r, \theta = \pi/2) \propto r^{-1}$, indicating that centrifugal support is not important except near the star. For physically acceptable steady state solutions corresponding to realistic values for the mass of the star, its radius, the envelope temperature and the density and angular velocity at the stellar surface, B_r at the stellar surface is of order 10 to 100 gauss; the total field strength would be in the range 100 to 1000 gauss. As B_r increases from much smaller values than those considered above, the mass loss rate increases, the magnetic field transporting energy from the star to the envelope. For B_r much larger than the above values, the mass loss will most likely be suppressed due to the inability of the matter to modify the geometry of the field lines. Finally, there exists the possibility that many of the steady-state solutions are unstable leading to time dependent solutions which may account for some of the observed time variations in Be stars.

Saito (1974) has also constructed a model for a Be star envelope including the effects of a magnetic field. His analysis is restricted to the equatorial plane. In general, the solutions obtained are similar to those of Limber and the required magnetic field strengths at the stellar surface for steady state solutions are also in the range 100–1000 gauss.

It is of interest to consider briefly whether such field strengths are presently detectable. According to Landstreet (1975), there are three methods which can be used to determine magnetic field strengths. Photographic techniques are restricted to sharp line stars, those for which $v \sin i < 30$ km s^{-1}, and for a sixth magnitude B star would yield a typical standard error of a few hundred gauss. The photoelectric line scanner approach is also limited to sharp line stars. For eight B stars brighter than apparent magnitude 4.5 with $v \sin i < 30$ km s^{-1}, recent measurements of the magnetic field strength by Landstreet yield standard errors in the range 30–120 gauss. Finally, interference filters can be used to observe hydrogen lines and other strong photospheric features and should be capable of detecting magnetic fields with standard errors of 50–100 gauss in bright B stars with $v \sin i < 200$–300 km s^{-1}. It therefore seems possible that in the near future, one will be able to test these magnetohydrodynamic models by directly measuring the stellar magnetic field. Landstreet and Marlborough have recently initiated such a program for Be stars.

Morgan (1975a) has subjected Limber's isothermal, hydrostatic model (1964) to a linear stability analysis. He found that some oscillating temporal and angular variations could occur on a time scale of the order of a few minutes. Even though no consideration of the phenomena initiating these variations was investigated, the results are interesting because they provide a possible explanation for the rapid variations observed by Delplace *et al.* (1969), Kupo (1971), and Bahng (1971), among others (see also Hutchings' article in this volume, p. 13). To explain these rapid variations Limber (1970) has suggested that changes in the energy of the magnetic field due to reconnection of field lines can convert magnetic energy to kinetic energy thus producing local heating of small regions, while Hutchings (1970b) proposed short lived condensations in the envelope near the stellar surface as a possible explanation. On the other hand, Lester (1975) has suggested that photometric variability observed in EW Lac (B2pe) on a time scale of $0^{\text{d}}7$ is a result of temperature changes in the stellar photosphere and not phenomena occurring in the extended envelope. Fernie (1975) reported photometric variability of a few hundredths of a magnitude in π Aqr (B1Ve) over a time scale of hours and attributed this variation to a pulsation of the star. However North and Olofsson (1974) reported no variations greater than $0\overset{m}{.}01$ in lines or continua in the same star based on their observations in 1972. Thus a plethora of suggestions exists to explain the observed short term variation. Whether any of these is correct and what relation, if any, they have to the variations predicted by Morgan (1975a) is presently unknown.

3. *Ad Hoc* Models

Up to the present time, the solutions discussed in Section 2 have not been used to make specific predictions concerning observational quantities such as line profiles, linear polarization, etc., which can then be compared directly with observational data. The predictions concerning observable features which do exist are based on simplifications and extensions of the models discussed in Section 2 and other ideas in order to obtain the complete three dimensional structure of the envelope. In this way, many approximate models have been constructed to represent the envelopes of Be stars. Essentially all models of this type are *ad hoc* to a greater or lesser extent in that either they do not satisfy the basic Equations (1), (2), and (3) or that they have not been tested to see if they satisfy these equations. In addition, many of these models may not be self-consistent. Consequently, some caution must be exercised in using these *ad hoc* models and in evaluating the comparison of model predictions with observational data.

Some purists may argue that reliance on any *ad hoc* model is futile at best and in general may be highly misleading. However, this is not necessarily so. For example, Rybicki (1970) has emphasized that much of our present conception of extended stellar atmospheres is based to a significant extent on the formal solution of the equation of transfer employing *ad hoc* assumptions for the structure of the extended atmosphere and the source function. In the preface to his book Stellar Evolution, Struve (1950), while noting the reluctance of most scientists to rely on *ad hoc* assumptions or uncertain hypotheses, nevertheless asserts that such hypotheses or assumptions can be of value in providing guidance for future studies. Finally, to quote Chandrasekhar (1934) in the context of Wolf-Rayet stars and novae, but also applicable to Be stars at our present level of understanding: "A dynamical theory of the ejection process itself is necessary only insofar as we require the emission per unit volume and the radial velocity as functions of r, and consequently even without any underlying dynamical theory we can formally examine the type of band contours that could be expected from assumed hypothetical laws of velocity variation, and from a comparison with the observed contours infer something about the actual conditions obtaining in stellar atmospheres where such emission bands originate."

3.1. STEADY-STATE, STELLAR WIND MODELS

A variety of authors have constructed steady-state, stellar wind models to represent the envelopes of Be stars. Hutchings (1970a, 1971) has compared observed line profiles for the Be stars γ Cas, κ Dra and HD 142926 with theoretical ones based on his model (Hutchings, 1968) originally applied to OB supergiants. The particulars of the model are described in detail in the latter paper; only a brief summary is given here. In the radiative transfer problem electron scattering is assumed to be the dominant opacity source. Doppler broadening is taken as the dominant broadening mechanism and lines are assumed to be formed by absorption and re-emission of photospheric radiation. The star is assumed to radiate like a black body near each line of interest. Each stellar photon is assumed to undergo at most one pure scattering and one electron scattering interaction. Allowing one more scattering per photon made only a small change according to Hutchings. The atomic energy level populations were taken from Baker and Menzel (1938) modified according to a procedure described by Hutchings (1968). Each of $\rho(\mathbf{r})$, $\mathbf{u}(\mathbf{r})$, and $T(\mathbf{r})$ was written as a simple function of $\mathbf{r}$ containing some adjustable parameters and the envelope model derived empirically by modifying these adjustable parameters to reproduce

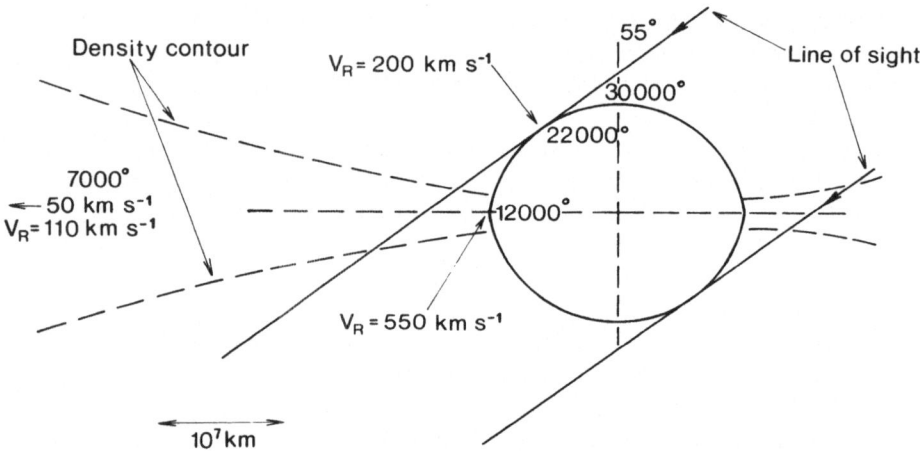

Fig. 2.   Schematic model of γ Cas (Hutchings, 1970a).

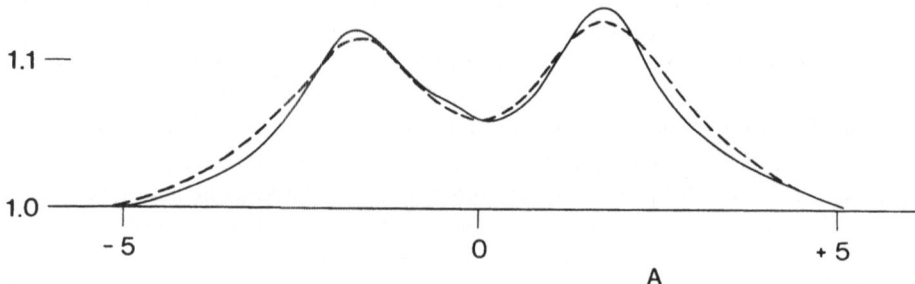

Fig. 3.   Theoretical and observed Hγ line profiles for γ Cas: – – – observed; ——— theoretical. Ordinate is flux relative to the continuum (Hutchings, 1970a).

the peak separation of the two components of a double emission line, the V/R ratio, the central depth, the wing extent and the total peak height. The model thus derived for γ Cas (B0 IVe) is illustrated schematically in Figure 2 and the comparison between the observed and theoretical $H\gamma$ line profile for γ Cas is given in Figure 3. A similar comparison between the observed and computed $H\gamma$ profiles for κ Dra (B7p) is shown in Figure 4 and for HD 142926 (B9e) in Figure 5. The final envelope model for the latter two stars is qualitatively similar to that for γ Cas (Figure 2). The mass loss rates for each star, as given by the models, are: γ Cas, $10^{-7} M_\odot$ yr^{-1}; κ Dra, $3.3 \times 10^{-8} M_\odot$ yr^{-1}; HD 142926, $2.5 \times 10^{-8} M_\odot$ yr^{-1}.

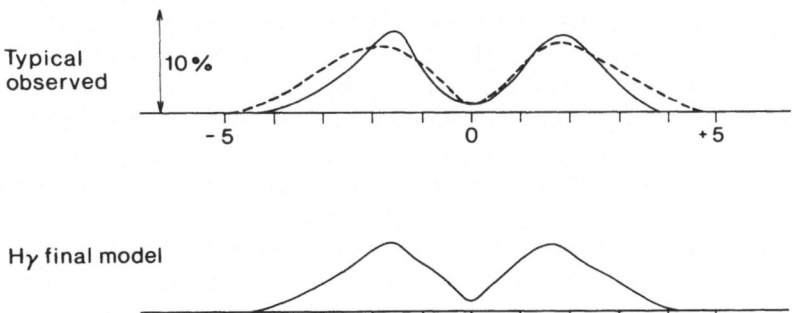

Fig. 4.   Theoretical and observed $H\gamma$ line profiles for κ Dra. Ordinate same as Figure 3 (Hutchings, 1971).

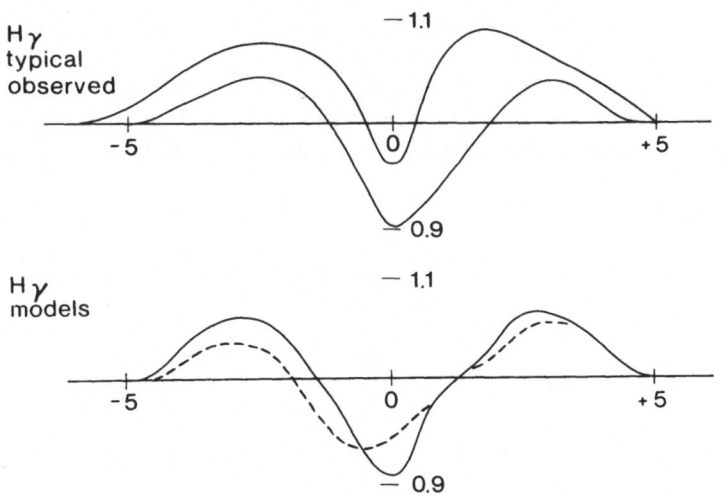

Fig. 5.   Theoretical and observed $H\gamma$ line profiles for HD 142926. Ordinate same as Figure 3 (Hutchings, 1971).

Marlborough (1969, 1970) and Marlborough and Roy (1971) have discussed a steady-state stellar wind model to represent Be star envelopes. The details of the model are described in the 1969 and 1970 papers. The state of excitation and ionization of the envelope was determined in a manner similar to that employed in nebular studies. The statistical equilibrium equations included photoionization from all energy levels and collisional transitions between levels of principal quantum number differing by ± 1. Terms in the statistical equilibrium equations pertaining to the radiation field contain contributions from the star and the envelope. The continuous radiation at any point in the envelope was assumed to be due solely to the star reduced by geometrical dilution and bound-free absorption between the star and the point considered; continuous radiation produced by the envelope was completely neglected. The radiation emitted by the star was taken from an appropriate model atmosphere. The contribution of the star to line radiation was neglected. The envelope contribution to line radiation in the statistical equilibrium equations was taken into account in the following approximate manner. The statistical equilibrium equations were solved for each of three cases: case I (classical nebular case A), case II (classical nebular case B) and case III for which the envelope was assumed to be optically thick in all line radiation. The final atomic energy level populations at a given point were obtained by taking weighted averages of either cases I and II or II and III, the pair to be chosen depending on the optical depths in $L\alpha$ and $H\alpha$ between the point considered and the edge of the envelope in a direction parallel to the rotation axis. This direction was chosen because in general it is the one for which line photons have the greatest chance of escaping from the envelope. The details of this weighting and averaging procedure are given by Marlborough (1969). The envelope was assumed to be isothermal and $u_\phi(r, \theta = \pi/2)$ taken to vary as $r^{-1/2}$. If the meridional projections of streamlines are assumed to be straight lines originating at some arbitrary point below the star's surface, then $\rho(r, \theta = \pi/2)$ is obtained from the equation of continuity once $u_r(r, \theta = \pi/2)$ and the arbitrary point at which the streamlines originate are specified. Finally $\rho(\mathbf{r})$ is obtained from $\rho(r, \theta = \pi/2)$ by using the straight streamline assumption and by assuming Limber's hydrostatic solution (1964) is applicable at some arbitrary point $(r', \theta = \pi/2)$. Again, the details and relevant formulae are given by Marlborough (1969), as is the method used to calculate the theoretical line profiles. In Figure 6 is shown a comparison of the

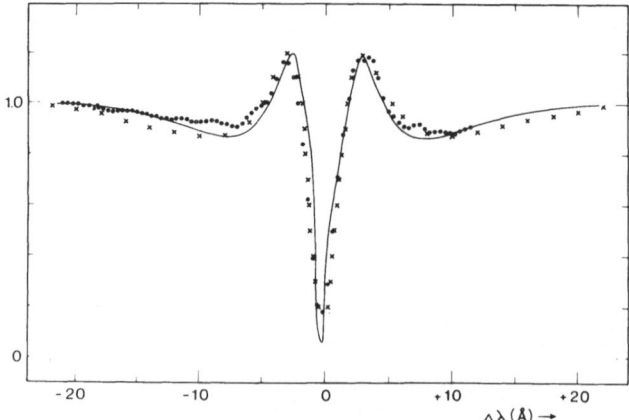

Fig. 6.   Theoretical and observed $H\alpha$ line profiles for 1 Del: ——— theoretical profile; ... photoelectric observations (Gray and Marlborough, 1974); $\times \times \times \times$ mean photographic profile (Marlborough and Cowley, 1974). Ordinate same as Figure 3.

theoretical and observed Hα line profiles for the shell star 1 Del (A0pe). The model envelope is similar to that used by Marlborough and Cowley (1974) except that the number density $N(r, \theta = \pi/2) = 1.35 \times 10^{12}$ cm^{-3} at $r = 4\,R_*$, $\theta = \pi/2$ where R_* is the radius of the star and the resulting mass loss rate is $8.8 \times 10^{-9}\,M_\odot$ yr^{-1}. In Figure 7 is shown the results of a preliminary attempt to reproduce the observed Hα line

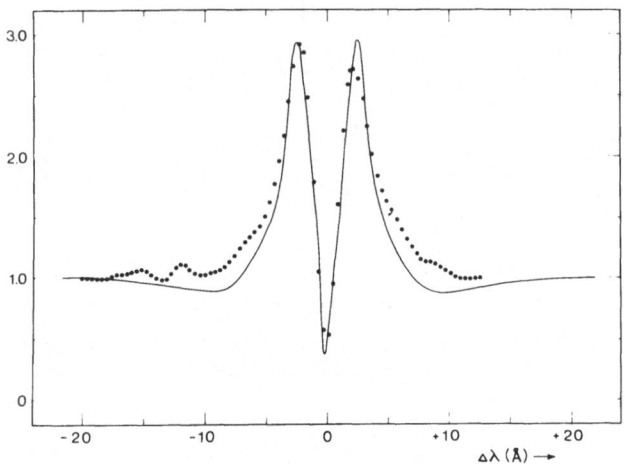

Fig. 7.   Thereotical and observed Hα line profiles for HD 193182: ———— preliminary theoretical profile; . . . photoelectric observations (Gray and Marlborough, 1974). Ordinate same as Figure 3.

profile in the shell star HD 193182 (B7IV–Ve). The envelope model is the same as that used for 1 Del except that $N(r, \theta = \pi/2) = 2.3 \times 10^{12}$ cm^{-3} at $r = 4\,R_*$, $\theta = \pi/2$ and the corresponding mass loss rate is $1.5 \times 10^{-8}\,M_\odot$ yr^{-1}.

Doazan (1965) has constructed a stellar wind model for the Be star HD 50138 (B8e). From the analysis Doazan concluded that this star is probably intermediate in type between Be and P Cygni stars. Despite this however it is useful to consider the results of her model because in that model the radiative transfer problem is treated by Sobolev's method (1960; see also Rybicki, 1970; and Hummer's article in this volume, p. 281). For this model, calculations were confined to the equatorial plane. It was assumed that $N_e(r, \theta = \pi/2) \propto r^{-2}$, $u_\phi(r, \theta = \pi/2) \propto r^{-1}$, $u_r(r, \theta = \pi/2) \propto r^s$ and $u_r(R_*, \theta = \pi/2)/u_\phi(R_*, \theta = \pi/2) = k$ where s and k are adjustable parameters and N_e is the electron density. The observed and calculated Hβ profiles are illustrated in Figure 8 for $s = -1.4$ and $k = 2$. In Figure 9 is shown $u_r(r, \theta = \pi/2)$ for the model illustrated in Figure 8 and also a modified form in which $u_r(r, \theta = \pi/2)$ initially increases with r. The Hβ profile for this second model is shown in Figure 10. The resulting mass loss rate is approximately $5 \times 10^{-7}\,M_\odot$ yr^{-1}.

For the stellar wind models considered above, the mass loss rates inferred from the comparison of observed and theoretical line profiles lie in the range 10^{-7} to $10^{-9}\,M_\odot$ yr^{-1}. The main sequence lifetimes for B stars vary from approximately 10^7 yr for B0V to 5×10^8 yr for A0V. If the Be phase occupies a fraction $f(\leq 1)$ of the main sequence lifetime where f includes the transient nature of the presence of circumstellar matter as deduced from the appearance and disppearance of emission

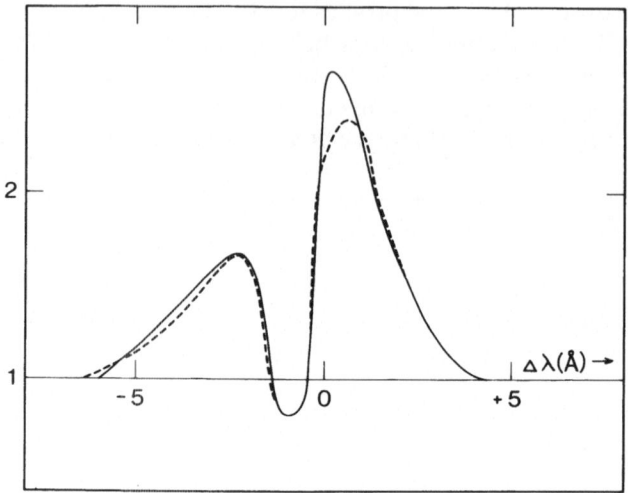

Fig. 8.   Theoretical and observed Hβ line profiles for HD 50138: ———— theoretical profile for $s = -1.4$, $k = 2$; – – – observed profile from plate W 1783. Ordinate same as Figure 3 (Doazan, 1965).

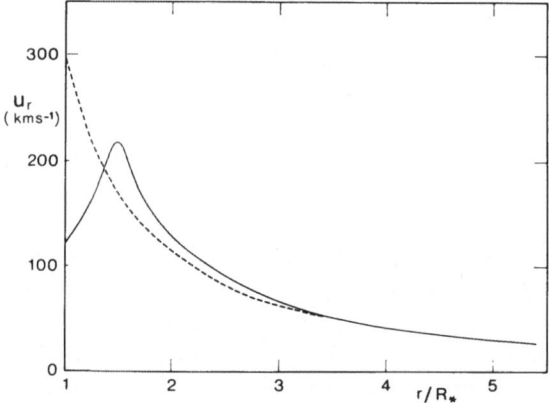

Fig. 9. $u_r(r, \theta = \pi/2)$ for theoretical profiles in Figures 8 and 10: – – – u_r for Figure 8; ———— u_r for Figure 10 (Doazan, 1965).

and/or shell lines and if the above estimates of mass loss are realistic, it would seem that the mass loss during the Be phase will not have any significant effect on the star's future evolution, at least as far as the total mass is concerned. Whether the corresponding angular momentum loss will significantly reduce the initial total angular momentum of the star is presently unclear for these stellar wind models, although for Limber's hydromagnetic model (1974 and Section 2.2), the angular momentum loss over a period of 10^7 yr is predicted to be about one tenth of the total initial angular momentum.

In all of the stellar wind models the circumstellar envelope extends to large distances from the star. If the envelope is sufficiently extensive, some Be stars may be

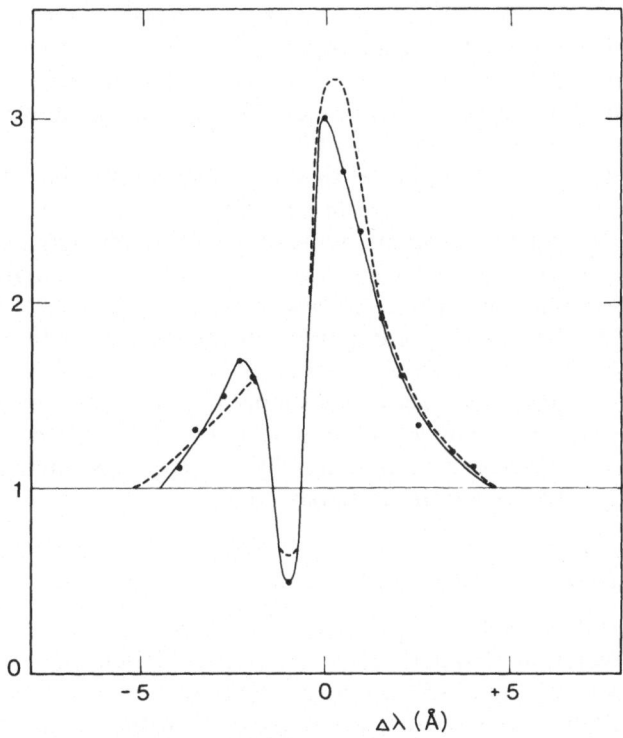

Fig. 10. Theoretical and observed Hβ line profiles for HD 50138: ——— theoretical profile; ‑‑‑
observed profile from plate W 576. Ordinate same as Figure 3 (Doazan, 1965).

detectable as thermal radio sources. Trasco *et al.* (1970) have reported the detection
of κ Dra (B7p) at λ = 6 cm at a flux level of 0.020 Jy (see the article by Purton, this
volume, p. 157, for a discussion of the observations). No predictions however have as
yet been made of the expected radio continuum emission for any of the models
discussed in Section 3.

It would be useful to be able to decide which of the above models best represents
the envelopes of Be stars. Unfortunately, this does not appear possible at the present
time and is partly due to the fact that the comparison of observational data with
theoretical line profiles is made for different stars. In the absence of a comparison of
predicted line profiles from different models for the same star one might perhaps
decide the best model based on a criterion such as the smallest number of *ad hoc*
assumptions. Even this is not completely satisfactory, however, because the arbitrar-
iness of a particular assumption is at least partly, if not greatly, a subjective decision.
The fact that each of the above models uses a different approach to determine the
excitation and ionization state of the envelope and to solve the radiative transfer
problem and at the same time can satisfactorily reproduce the observed line profiles
may simply be an indication that each model possesses a sufficient number of
adjustable parameters. More likely, however, the agreement is an indication that to
lowest order, the gross structure of line profiles is dominated by the geometry and
velocity field and less by the radiative transfer effects. When the effect of Doppler

broadening is large, as it is in Be stars – at least for moderate to large values of angle i – the intrinsic shape of the line will not have great significance in determining the observed profile. Nevertheless, better treatments of the line transfer problem are necessary if only for the fact that the methods used in the above three models, particularly the first two, are primitive at best.

A glance at Figures 3 to 10 indicates that the stellar wind models discussed above can satisfactorily reproduce observed line profiles for which $V = R$ and for which $R > V$. It is by no means apparent, however, whether models in which $u_r(\mathbf{r})$ is everywhere positive can generate a line profile for which $V > R$. At present this is the severest limitation of the stellar wind models because they appear incapable, without the addition of further *ad hoc* assumptions, of accounting for the observed V/R variation, particularly those cases in which V/R becomes significantly greater than unity. Alternately, this discrepancy may simply be a reflection of the elementary radiative transfer solutions employed. Magnan (1970, 1972) has demonstrated that complex line profiles can arise in differentially rotating slabs when a more realistic treatment of the radiative transfer is employed.

3.2. Time dependent, stellar wind models

Limber (1969) introduced a simple time dependent, stellar wind model for the structure and evolution of the envelope to explain Pleione's (B8p) shell episode between the years 1938 and 1954. In this model Limber predicted the profile of the Balmer line H25 by assuming that the population of the atomic energy level of principal quantum number $n = 2$, at any point r in the envelope, was proportional to some power of the number density of hydrogen atoms there and by neglecting re-emission in the envelope. The model is one in which $u_\phi(r, \theta = \pi/2) \propto r^{-1/2}$, the hydrostatic solutions (Limber, 1964) are used for the envelope structure perpendicular to the equatorial plane and $u_r(r, \theta = \pi/2)$ increases from small values for $r \simeq R_*$ to larger values at greater r and is independent of time. The time dependence in this model is introduced by the variation with time of the mass loss rate as illustrated in Figure 11. A comparison of the predictions of the model and the observations is also seen in Figure 11. As is evident, Limber's model accounts very well for Merrill's observations (1952) of the approximate strength of the shell lines and for the radial velocity of H25. No agreement could be obtained for situations in which $u_r(r, \theta = \pi/2)$ was a decreasing function of r.

Marlborough (1971), for the same model, determined the time variation of the line profiles of Hα, H15, H25 and H35, the latter three being included to possibly account for the observed Balmer progression (Merrill, 1952). Re-emission effects were neglected for H15, H25 and H35. The predicted line profile changes are shown in Figures 12, 13, 14, 15, 16 and 17, and in Figure 18 the upper Balmer lines are superimposed to show the shift to shorter wavelength with time after 1945. Limber's model reproduces the observed equality of R and V for Hα for most of the shell phase, the change to $R > V$ during the later stages, and a qualitative similarity of the predicted Hα profile in 1945 to a tracing of the observed Hα profile given by Merrill (1952). However there are also some important disagreements. Specifically, according to the model the emission at Hα does not begin early enough – emission at Hα

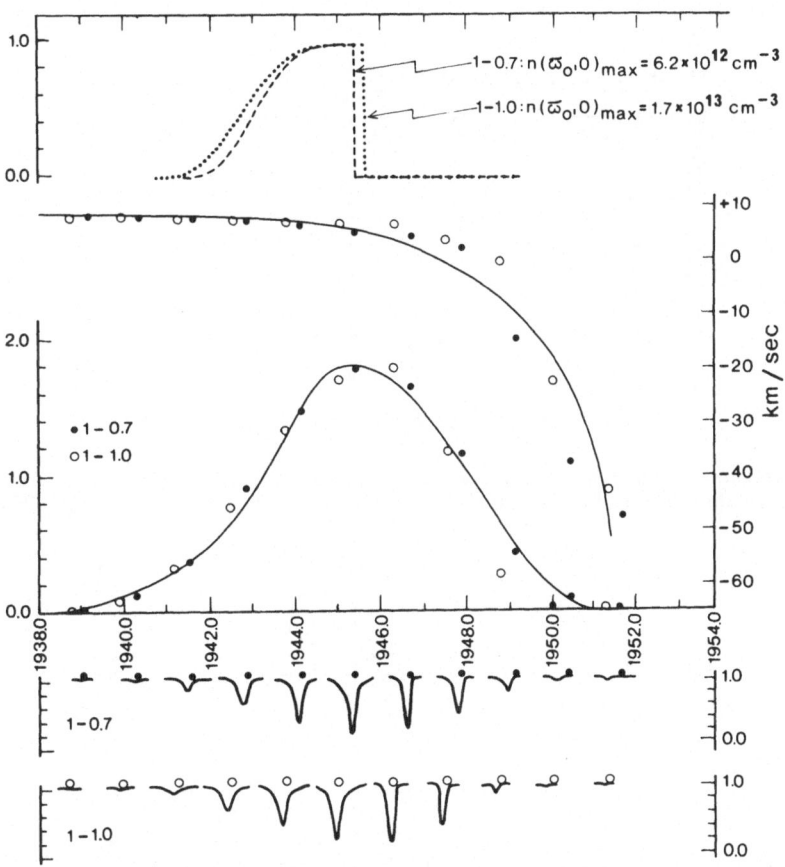

Fig. 11. Comparison of Limber's model with observations of Pleione. Dashed and dotted curves in the upper part give the rate of mass loss with time for two models. The upper solid curve is the smoothed version of the observed radial velocity of H25. The lower solid curve represents Merrill's observations of the time variation of the strength of the shell lines. The filled and open circles are the predicted variation for each of the two models. At the bottom are the line profiles for the two models and the filled and open circles represent the position of the line center with respect to the center of the star. Both models are for the case that $u_r(r, \theta = \pi/2)$ generally is an increasing function of r (Limber, 1969).

being reported in October, 1938, by McLaughlin (1938) and Mohler (1938) – and the emission at Hα disappears too rapidly near the end of the shell phase – Slettebak (1954) reported emission at Hβ in 1954 (see also Burd, 1954, for a description of spectral changes after 1947). In addition, no apparent Balmer progression is evident in Figure 18.

Marlborough and Gredley (1972) included reemission from the envelope in the upper Balmer lines to see if this would produce better agreement between the observed and theoretical Balmer progression. No significant change was produced by the inclusion of envelope reemission in these lines. In summary, Limber's model accounts very well for a number of observational results pertaining to Pleione's shell phase, but it is unsatisfactory in accounting for the early appearance of emission at

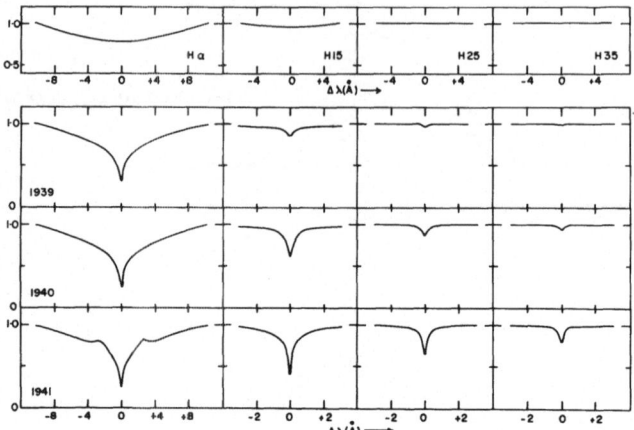

Fig. 12.   Theoretical Balmer line profiles for Limber's model for Pleione for 1939–1941. The first row is
the rotationally broadened stellar lines. Ordinate same as Figure 3 (Marlborough, 1971).

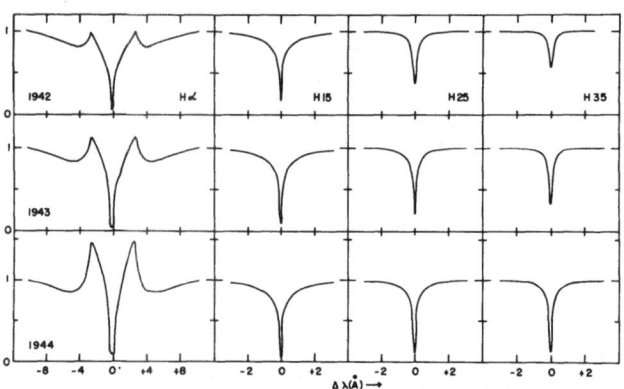

Fig. 13.   Theoretical Balmer line profiles for 1942–1944 (Marlborough, 1971).

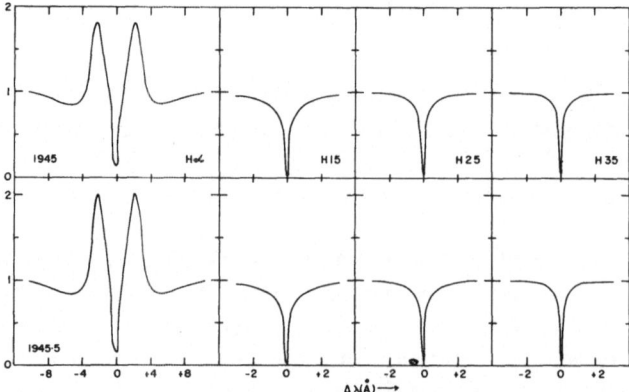

Fig. 14.   Theoretical Balmer line profiles for 1945 and 1945.5 (Marlborough, 1971).

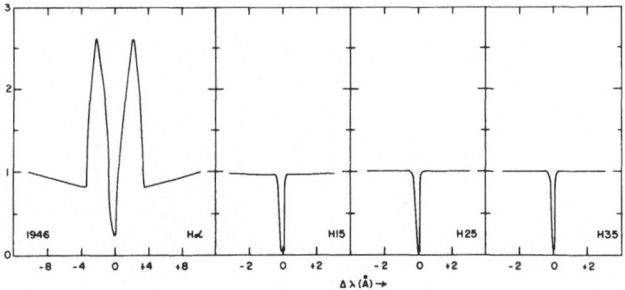

Fig. 15. Theoretical Balmer line profiles for 1946 (Marlborough, 1971).

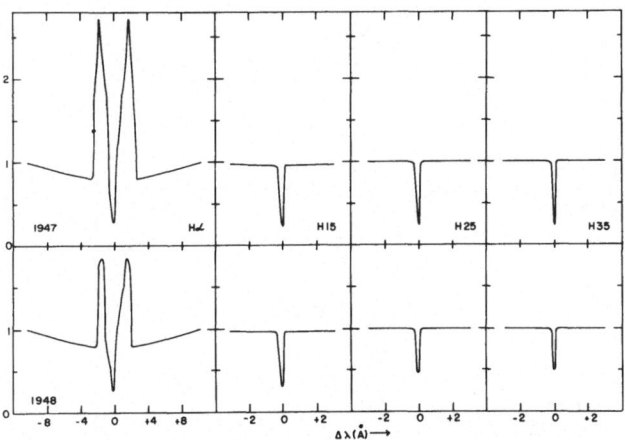

Fig. 16. Theoretical Balmer line profiles for 1947 and 1948 (Marlborough, 1971).

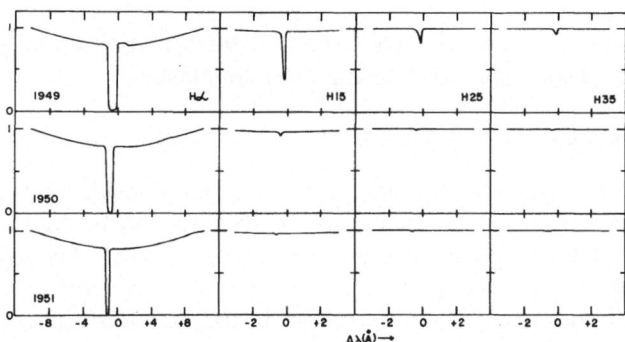

Fig. 17. Theoretical Balmer line profiles for 1949–1951 (Marlborough, 1971).

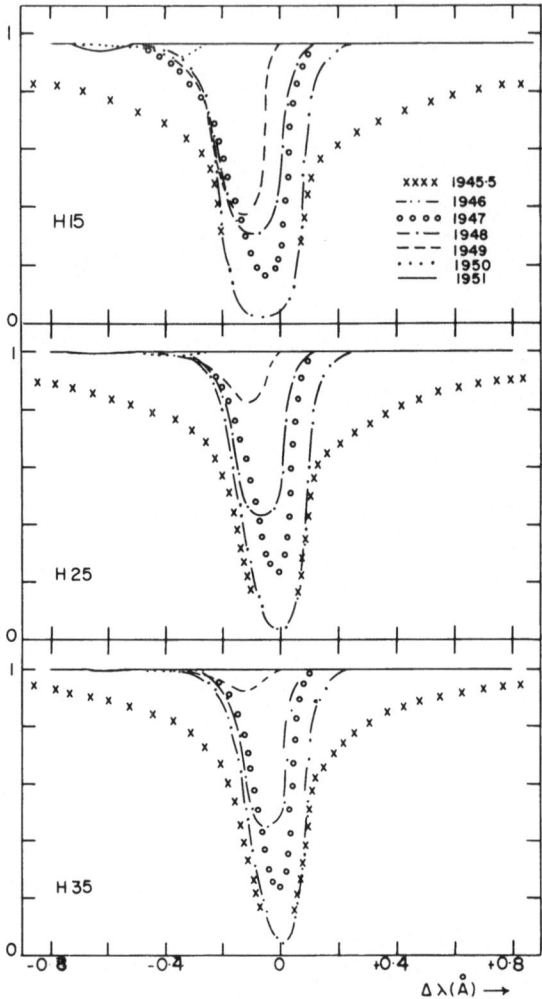

Fig. 18. Theoretical profiles for upper Balmer lines for 1945.5–1951. Ordinate same as
Figure 3 (Marlborough, 1971).

Hα, the persistence of emission at Hα when the shell lines have essentially vanished
and the Balmer progression especially during the later stages.

3.3. ELLIPTICAL RING MODEL

Huang (1972, 1973) and Albert and Huang (1974) have reconsidered the elliptical
ring model originally suggested by Struve and elaborated on by McLaughlin. A
qualitative summary of the major ideas is given by Huang (1975). According to this
picture the emitting and absorbing atoms in a Be star envelope are strongly
concentrated to the equatorial plane and move on individual orbits according to
Keplerian motion in the same way that the particles comprising the ring of Saturn
revolve around the planet. In each of the papers detailed line profiles are not

computed because the radiative transfer problem in the ring is not considered. Instead what is done is to calculate the broadening function $B(t)$ which relates the observed line profile $F_{obs}(\Delta\lambda)$ to $F(\Delta\lambda)$, the line profile in an individual radiating element of the system seen in a frame at rest with respect to the local macroscopic velocity of the matter. Apart from possible normalization factors, the relationship between $F_{obs}(\Delta\lambda)$ and $F(\Delta\lambda)$ is given by

$$F_{obs}(\Delta\lambda) = \int_{-\infty}^{+\infty} B(t)F(\Delta\lambda - t)\,dt. \tag{4}$$

In the situation that the intrinsic line profile is extremely narrow and/or the Doppler broadening large, the observed line profile is then approximately given by the broadening function. In practice, the observed profile will be wider than the broadening function due to other broadening mechanisms such as thermal motions.

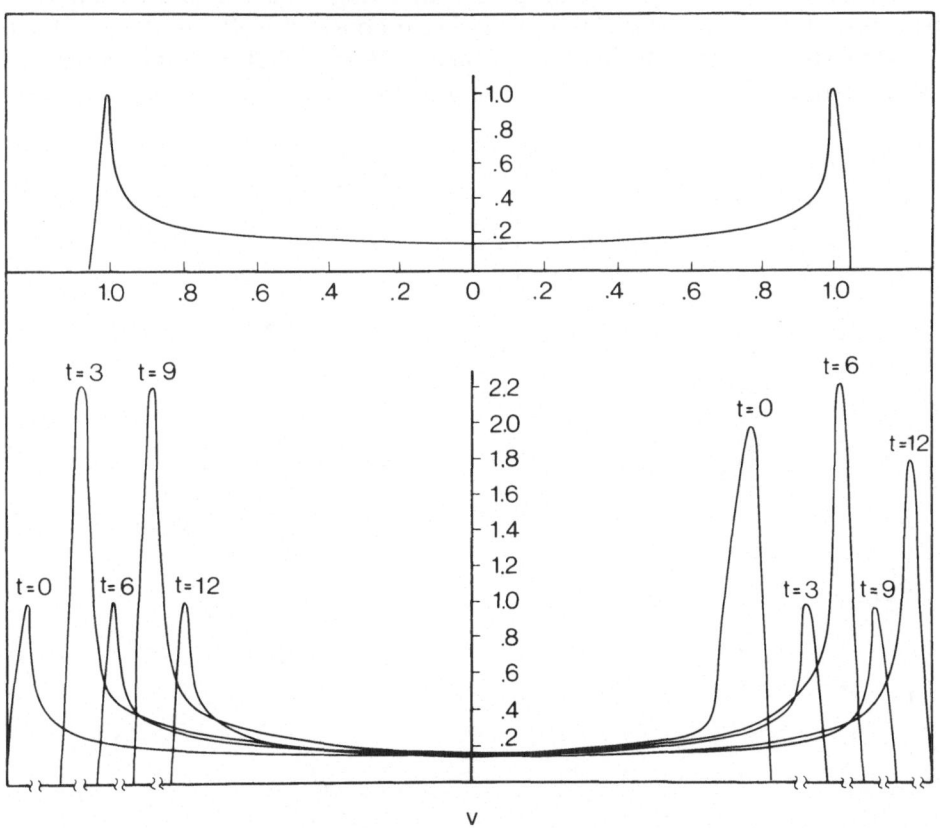

Fig. 19. The profile of the broadening function normalized so that its maximum value is unity for a circular ring for an axisymmetric and a non-axisymmetric distribution of matter. Abscissa scale is the radial velocity expressed in units of the Keplerian orbital speed of a particle on a reference circle taken to be the outer boundary of the circular ring and is the same for both profiles. The top profile is for the axisymmetric distribution; the bottom profiles are for a non-axisymmetric distribution given by Equation (5) for $\rho_1 = 2\rho_0$ normalized in such a way that the peak emission is unity when $\Delta\rho = 0$ (Huang, 1972).

In the first paper Huang (1972) computed the broadening function for a circular ring for both a symmetric and an asymmetric distribution of atoms in the ring. Due to differential revolution the asymmetric distribution predicts profiles which change continuously with time, the time scale being of the order of the revolution period of an atom in the ring, i.e. hours or days. A typical result for the simple case

$$\rho(r, \phi, t) = \rho_0(r) + \Delta\rho(r, \phi, t), \tag{5}$$

with

$$\Delta\rho(r, \phi, 0) = \rho_1 = \text{constant}, \qquad r_1 \leq r \leq r_2, \quad 0 \leq \phi \leq \alpha$$
$$= 0 \text{ elsewhere},$$

is shown in Figure 19.

In the second paper Huang (1973) considers an elliptical ring. The essence of this model is illustrated schematically in Figure 20. The major result of the model is that the revolution of the major axis of the elliptical ring, assumed to arise due to the oblateness of the central star (Johnson, 1958), produces a V/R variation. In Figure 21 is shown the change of the broadening function for an elliptical ring of eccentricity 0.4 and width 0.093 in units of the semi-major axis of the outer edge. Note

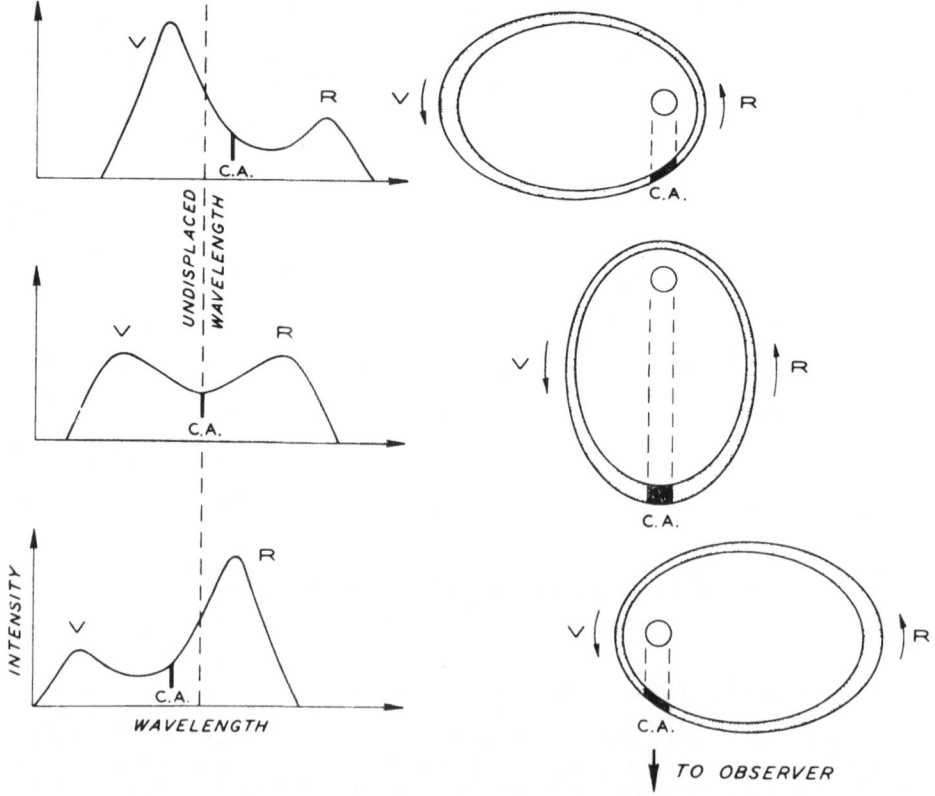

Fig. 20.   Schematic representation of the elliptical ring model showing V/R variation (Huang, 1975).

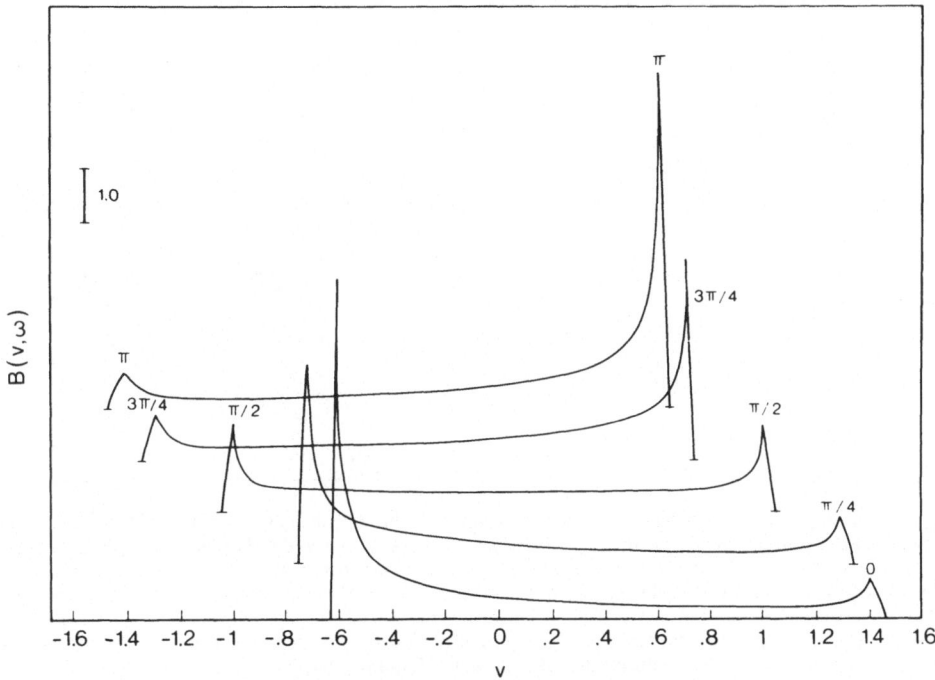

Fig. 21.   The profile of the broadening function for an elliptical ring. Abscissa scale is the radial velocity in units of the Keplerian orbital velocity of a particle in an elliptical orbit at the outer boundary of the ring. The profiles are for a ring of eccentricity 0.4 and $a_1 = 1$, $a_2 = 0.907$ where a_1 and a_2 are the semi-major axes of the outer and inner edges, respectively. Each profile is labelled by ω, the longitude of periastron. Only the cycle for $\omega = 0$ to $\omega = \pi$ is shown, the other half being given by the condition $B(v, \omega) = B(-v, \pi \pm \omega)$. If the elliptical ring is undergoing apsidal motion, ω measures time. Note that the profile changes shape as well as position with time (Huang, 1973).

specifically that during the cycle, the shape of the line, the position of the line and the V/R ratio all change. A comparison of McLaughlin's observations (1966) of the V/R ratio and the radial velocities of the red and violet emission edges and the central absorption for the Be star 105 Tau (B2Vp) and the predicted variations for a narrow elliptical ring with eccentricity 0.2 and a revolution period for the major axis of 11.5 yr is illustrated in Figure 22. Albert and Huang (1974) have been able to account for similar observational data for the Be stars HD 20336 (B2Ve), 25 Ori (B1Vpe) and β^1 Mon (B3Ve). Satisfactory agreement with the predictions of the elliptical ring model is obtained. In each case the semi-major axis of the outer edge of the ring is about 3–4 R_* and $0.2 \gtrsim e \gtrsim 0.3$, where e is the eccentricity.

Despite this satisfactory agreement between predictions of the elliptical ring model and the observational data for the stars studied, the model has some drawbacks. The ring must be narrow, according to Huang (1973), i.e. $(a_2 - a_1) \ll a_2$, where a_2 and a_1 are the semi-major axes of the outer and inner edges, respectively, in order to preserve the elliptical shape for long times to explain the V/R variation. However, it is difficult to see how a narrow ring confined chiefly to the equatorial plane of the star can produce the central absorption features seen in Be stars. For example, for $i \approx 90°$, which is thought to correspond to the case of shell stars, it seems

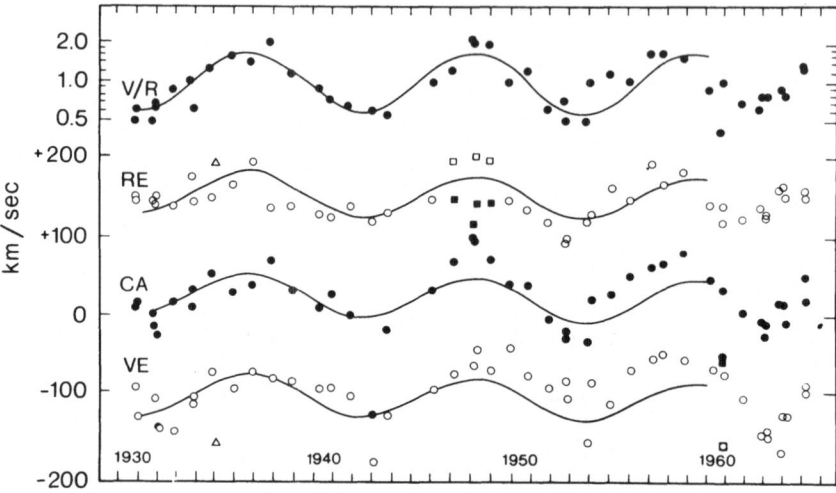

Fig. 22.   Comparison of observations and predictions of the elliptical ring model for 105 Tau. Upper curve is V/R variation on a log scale. Lower curves are radial velocity of red emission edge (RE), central absorption (CA), and violet emission edge (VE). For the lower three curves: open circles are emission edges (mean of $H\gamma$ and $H\beta$); filled circles are central absorption; open squares are $H\gamma$ emission edges; filled squares are $H\beta$ emission edges; triangles are $H\alpha$ emission edges. The solid curves are derived from a narrow elliptical ring with eccentricity 0.2, assumed period of V/R variation of 11.5 yr, and velocity amplitude of 132 km s^{-1} (Huang, 1973).

unlikely that the elliptical ring can produce the deep metallic shell lines. The well known shell star 48 Lib (B3p) has shown conspicuous V/R variations and the self reversals have approached zero intensity (see Underhill, 1960, p. 145 and Figure 2). It would therefore appear that in order for the elliptical ring model to account for the observations of 48 Lib a large absorbing region, presumably lying outside the elliptical ring and extending to at least one stellar radius from the equatorial plane, would have to be added. Where this matter comes from and how it is related to Huang's picture of the formation and evolution of the ring is unclear. Perhaps it represents the remains from the dissipation of an earlier ring or rings.

For smaller values of the angle i, other difficulties arise. Let us assume that the central absorption in the hydrogen lines occurs due to the fact that the ring is seen projected against the stellar disk in the observer's line of sight. For a spherical star of radius R_* and an elliptical ring of eccentricity e, semi-major axis of the inner edge a_1 and situated in the equatorial plane, the minimum angle, i_{min}, for which the observer's line of sight is tangent to both the inner edge of the ring at periastron and the star's surface is given by

$$i_{min} = \cos^{-1}\left[\frac{R_*}{a_1(1-e)}\right]. \tag{6}$$

If the ring extends to a height h perpendicular to the equatorial plane, then i_{min} for which the observer's line of sight is tangent to both the inner edge of the ring at height h at periastron and the star's surface is related to h, a, and e by

$$(h/R_*)\sin i_{min} = [a_1(1-e)/R_*]\cos(i_{min}) - 1. \tag{7}$$

In Table I are the data pertaining to the elliptical ring model for the stars discussed by Albert and Huang (1974). Columns 2 and 3 are the eccentricity e and semi-major axis a_1 for the inner edge of the ring, in units of R_*, obtained from fitting the data. Column 4 is the inclination i deduced by Huang from $v \sin i$ and the equatorial rotation speed of the star. Column 5 is the angle i_{min} from Equation (6) determined

TABLE I

Elliptical ring model

Star	e	a_1/R_*	i	i_{min}	h/R_*
105 Tau	0.20	3.4	26°	68°	3.3
HD 20336	0.25	3.2	40°	65°	1.3
25 Ori	0.26	3.9	35°	70°	2.4
β^1 Mon	0.32	3.8	43°	67°	1.3

for each star using the values of e and a_1/R_* in columns 2 and 3, respectively, for a ring in the equatorial plane. It is only for angles $i > i_{min}$ that the specific ring considered would be projected against the stellar disk in the observer's line of sight at periastron. In column 6 are the values of the extent of the ring above the equatorial plane, in units of R_*, if the inner edge of the ring is to be projected against the stellar disk at periastron for an observer for whom the angle of inclination i is given by the values in column 4. Again, the numbers in column 6 represent the minimum extent of the ring from the equatorial plane for given a_1/R_*, e, and i if the central absorption is due to the ring seen projected against the stellar disk.

A comparison of columns 4 and 5 leads to the immediate conclusion that it is geometrically impossible for the rings considered to be projected against the stars for the values of i deduced by Huang if the ring lies predominantly in the equatorial plane. The star 105 Tau possesses a conspicuous absorption in Hγ and presumably Hβ as well (see Burbidge and Burbidge, 1953, Figure 12). Of course, as Huang (1973) noted, considerable uncertainty exists in the values of i in column 4. Nevertheless the discrepancy between columns 4 and 5 is disturbing. If the ring is not confined to the equatorial plane, then it must extend to $\sim$1–3 R_* for the stars considered if Huang's values of i are approximately correct. It is not immediately apparent how such a structure could survive for a period of time long enough to produce several cycles of the observed V/R variation. If the central absorption does not arise from direct absorption of stellar radiation then the above arguments are not applicable. The central absorption in the elliptical ring model would then arise from self absorption along the line of sight through the ring. The limited geometrical extent assumed for the ring by Huang makes this possibility rather unlikely. In either case, it would be useful in assessing the relevance of the elliptical ring model if some line profile calculations based on this general model were available in order to determine whether or not the above discussion represents a major source of difficulty.

3.4. Disk and disk-like models

A variety of disk and/or disk-like models have been considered. In most of these models the circumstellar matter is assumed to lie within two coaxial cylinders whose axis coincides with the star's rotation axis, the extent of the cylinder along the rotation axis and the inner and outer radii being free parameters. One of the earliest disk-like or lenticular models was employed by Burbidge and Burbidge (1953) in their analysis of the outer atmospheres of some Be stars.

Kogure (1969) has discussed a disk model to interpret the observations of several pole-on Be stars. The model is illustrated schematically in Figure 23. The envelope was assumed to be isothermal, V_2 and V_3 were assumed to be zero where V_2 is

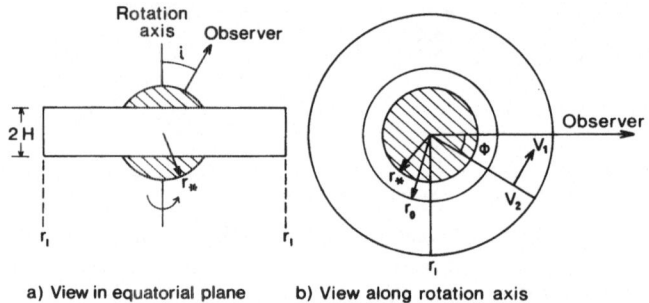

a) View in equatorial plane   b) View along rotation axis

Fig. 23.   Schematic representation for a Be star envelope (Kogure, 1971).

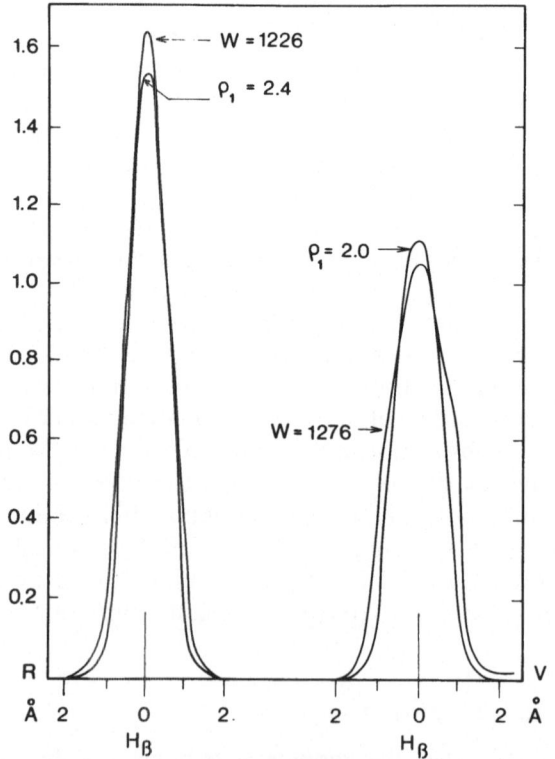

Fig. 24.   Observed and theoretical $H\beta$ profiles for HD 58343. ρ_1 is the ratio of the outer radius of the envelope to the stellar radius. Ordinate same as Figure 3 (Kogure, 1971).

$u_r(r, \theta = \pi/2)$ in the above notation, V_3 represents the velocity component perpendicular to the equatorial plane and $V_1 \propto r^{-1}$ where V_1 is $u_\phi(r, \theta = \pi/2)$. The radiative transfer solution used is that based on the work of Miyamoto and Kogure for a static plane parallel atmosphere for which the optical depth in the Lyman continuum is taken to be approximately unity and for which a variety of assumptions are employed for optical depths in other continua and discrete lines (see Kogure, 1967, and references contained therein for details). The Eddington approximation is used in the solution for the radiation field. Caution should be used in applying radiative transfer solutions for static plane parallel models to physical situations where the extent of the atmosphere is large so that curvature terms in the transfer equation are important and where velocity gradients exist (see Böhm, 1973; and Hummer's article in this volume, p. 281). In the case being considered where the model is applied to pole-on stars, the differences between the static and dynamic atmosphere predictions may not be too great because the velocity gradient for small angles i may actually be small and the effective emitting and absorbing part of the extended atmosphere may be strongly concentrated to the equatorial plane. For all cases the total envelope extent parallel to the rotation axis was assumed to be one stellar radius. The predicted and observed Hβ profiles for the stars HD 58343 (B3Ve) and HD 212076 (B2Ve) are shown in Figures 24 and 25, respectively. The model satisfactorily reproduces the observed Hβ profiles.

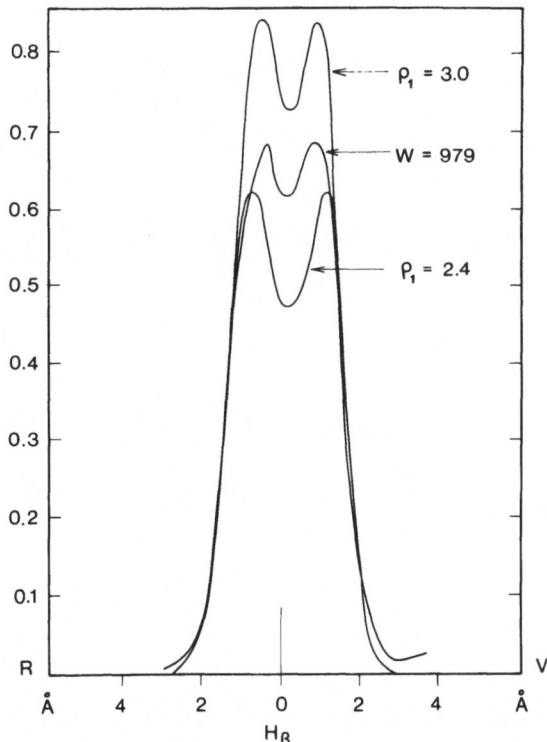

Fig. 25.   Observed and theoretical Hβ profile for HD 212076. ρ_1 is same as Figure 24 and ordinate same as Figure 3 (Kogure, 1971).

Observations of the continuous infrared emission for Be stars have been published by various authors (see Gehrz *et al.*, 1974, and Swings' article in this volume, p. 219). For none of the models discussed in Sections 3.1, 3.2 and 3.3 have predictions been made concerning the expected infrared continuum radiation. Predicted infrared emission has been made for thin disk models for which all physical parameters are treated as constants for simplicity. The infrared observations of ϕ Per (B1III–Vpe) and βCMi (B7Ve) are compared with predictions for a free-free continuum in Figures 26 and 27, respectively. There does not appear to be any serious discrepancy in accounting for the infrared flux as arising from free-free emission in a disk-like circumstellar envelope.

Disk or disk-like models have also been employed in connection with the observations of linear polarization in Be stars. For a discussion of the observations see the article by Coyne in this volume, p. 233. To explain the polarization observations of ζ Tau (B2IVep), Capps *et al.* (1973) considered a homogeneous isothermal thin disk of completely ionized hydrogen illuminated by a point source star and included both free-bound and free-free radiation from the disk. The

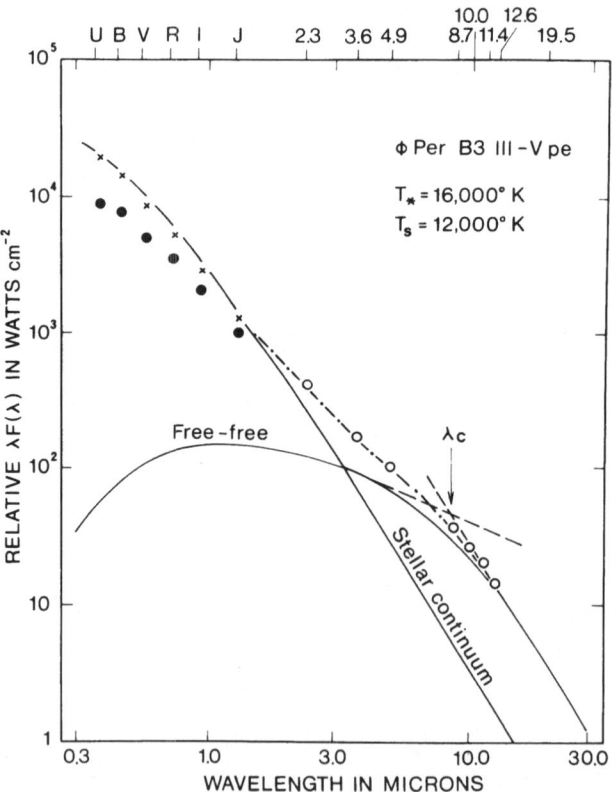

Fig. 26.   Observed and theoretical infrared emission for ϕ Per for a free-free shell which is optically thick at large λ. Filled circles are *UBVRIJ* data and crosses represent reddening corrections. Open circles are infrared measures of Gehrz *et al.* λ_c is wavelength at which $\tau(\lambda_c) \equiv 1$ for intersection of optically thin and optically thick segments of shell spectrum (Gehrz *et al.*, 1974).

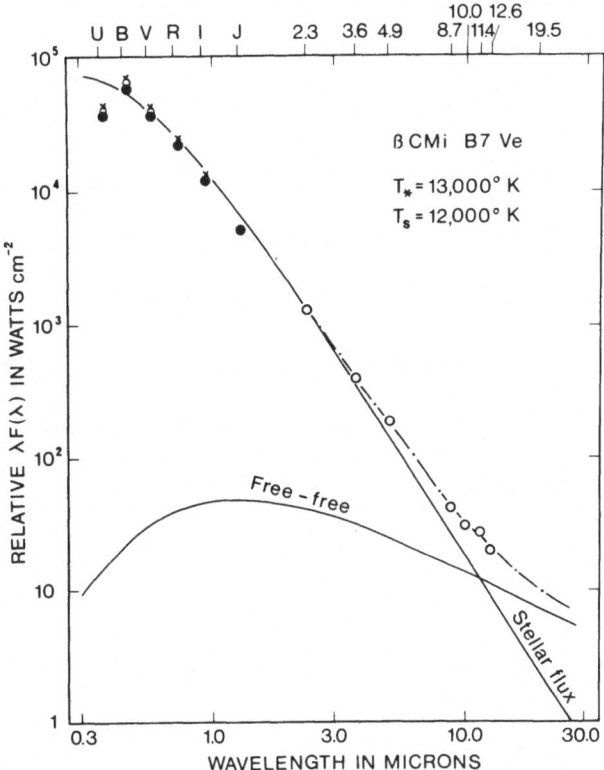

Fig. 27.   Same as Figure 26 for β CMi for an optically thin free-free shell (Gehrz *et al.*, 1974).

observations and predicted polarization are shown in Figure 28. More elaborate calculations have recently been performed by Haisch (1975, and the paper in this volume, p. 375). These involve a more rigorous treatment of the radiative transfer problem for more realistic models for the extended atmosphere than those provided by isothermal homogeneous disks. R. Poeckert is presently considering the predictions of linear polarization for the stellar wind model considered by Marlborough (Section 3.1) in connection with detailed linear polarization observations as a function of wavelength for the Be Stars EW Lac (B2pe) and γ Cas (B0IVe).

3.5. OTHER MODELS

Numerous other models have been considered. In general, however, they have either not been investigated quantitatively or have not been subjected to as detailed a comparison with observations as have those considered above.

Nariai (1970) has proposed a qualitative model which combines circulatory motion of gas in the extended envelope together with precession of the primary star and applied this model to the shell star ζ Tau. In a meridional plane matter in the circumstellar envelope moves outward in and near the equatorial plane and returns to the vicinity of the stellar surface from above and below the equatorial plane.

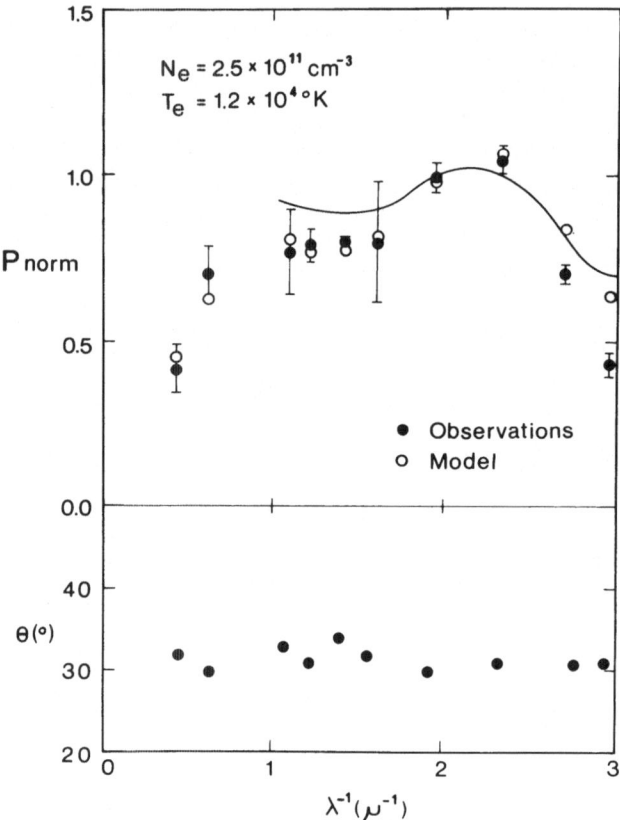

Fig. 28. Comparison of observed polarization for ζ Tau normalized so that $P = 1$ at $\lambda^{-1} = 2\mu^{-1}$ and theoretical predictions for a homogeneous isothermal completely ionized disk. Solid curve is mean relation for high latitude Be stars (Capps *et al.*, 1973).

Because of precession, the observer sometimes sees the outward moving part of the circulation pattern while at other times he sees the inward motion. This model does not appear to have been investigated further.

Hazlehurst (1967) suggested a simple hydromagnetic model to account for the loss of matter from the circumstellar envelopes of Be stars and applied it to Pleione's shell phase from 1938 to 1954. In this model the most important effect of the magnetic field is to transfer angular momentum outward from the star to the shell, ultimately allowing the shell material to escape. The model specifically predicts that $u_r(r, \theta = \pi/2)$ should be a decreasing function of r. However, Limber's study of Pleione (Section 3.2) has conclusively demonstrated that no agreement with observations could be obtained for cases in which $u_r(r, \theta = \pi/2)$ decreases with increasing r. It would therefore appear that Hazlehurst's model is not relevant to the Be star problem.

Henriksen (1969) has attempted to explain Pleione's behavior with a hydromagnetic model initially suggested by Crampin and Hoyle (1960). In this picture the radial component of the magnetic field plays a dominant role. Differential rotation of the matter in Keplerian orbits near the star leads to the production of an azimuthal component of the magnetic field. The resultant amplification of this component ultimately leads to large radial forces acting and subsequent explosive dissipation of the circumstellar ring. This model predicts an increase of $u_r(r, \theta = \pi/2)$ with r and is qualitatively similar with Merrill's observations (1952) for Pleione. Beginning with the outer part of the envelope successive rings of material are dissipated by this process until the entire shell has disappeared. This model has some interesting features in that it does contain an explanation for the dissipation of the shell. For this reason it would be useful to compute line profiles for several upper Balmer lines to compare with observations. One must keep in mind, however, that this model is not complete. As mentioned in Section 3.2, the entire envelope could not have disappeared about 1951 since emission lines were still visible at that time and also later on.

3.6. BINARY STAR MODELS

All the models discussed to this point have explicitly or implicitly assumed that the Be star phenomenon involves a single star. If a specific Be star happens to be a member of a binary system, the companion may exert considerable influence upon the dynamics and state of ionization and excitation of the circumstellar envelope, but it is not assumed to be the origin of the circumstellar envelope around the Be star.

On the other hand, various authors have advanced the suggestion that some Be stars are members of binary systems in which the circumstellar envelope arises due to mass transfer from a companion filling its inner Lagrangian surface. In particular, Kriz and Harmanec (1975) have tentatively suggested that all Be stars are members of mass-exchanging binary systems. A schematic picture of this process is shown in Figure 29.

Over the past few years binary stars have again become a subject of considerable interest. This renewed interest arose primarily because of the discovery of x-ray sources in binary systems and also because of the possible discovery of black holes. A variety of disk or disk-like models are available for the disk surrounding the accreting star. Most of these disks, however, do not necessarily represent rigorous solutions of the basic equations. The hydrodynamic problem for the transfer of matter in binary systems is complicated. Prendergast and Taam (1974) have investigated a numerical solution of the gas flow in a semi-detached binary system to account for the observations of the binary system U Cep. Lubow and Shu (1975) have recently concluded a detailed study of the gas dynamics in semi-detached binaries. Neither of these studies however presents detailed models for the disk-like region around the accreting star so that a detailed comparison of observations and theoretical predictions can be made. An initial attempt to study line profiles in a binary system undergoing mass exchange is due to Sima (1973).

Qualitative models involving disks and gas streams have been advanced to account for the observations of several Be stars. Many of these are discussed by Plavec (1973) and by Plavec and Harmanec in this volume and for this reason will not be repeated here.

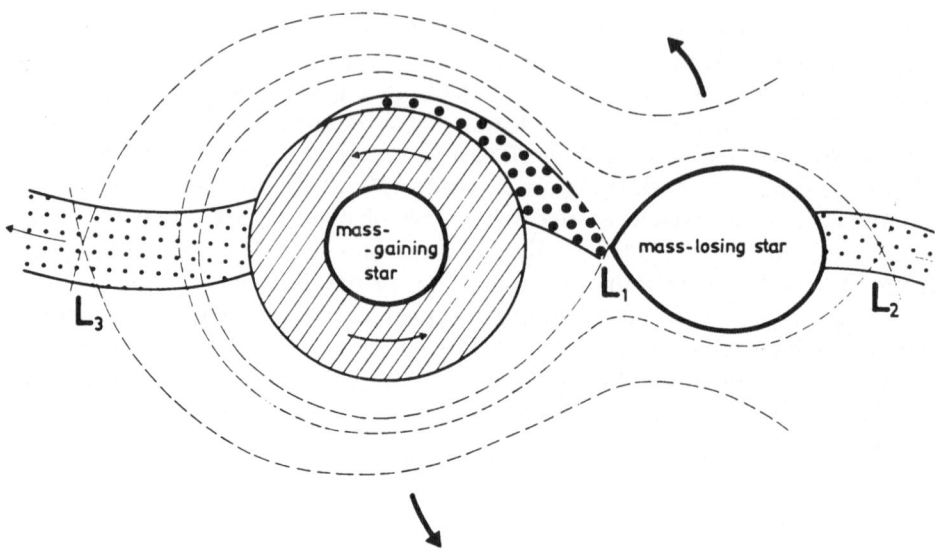

Fig. 29.   A schematic picture of the distribution of circumstellar matter in a close binary. The dashed lines represent Roche equipotentials. The region with large dots represents matter flowing from L_1 to the disk (hatched area). The regions with small dots near L_2 and L_3 represent possible escape of matter from the system (Kriz and Harmanec, 1975).

4.   Conclusions

In the above discussion we have considered the general problem for the solution of the basic equations pertaining to the structure of Be star envelopes and described some of the attempts to solve restricted problems as rigorously as possible. The observations obtained in the far ultraviolet from rockets and satellites provide some support for stellar wind solutions but do not provide a preference for the radiation driven solutions compared to those involving a magnetic field. The required magnetic field strengths are not beyond experimental detection especially if they lie toward the upper end of the expected range and future measurements can be expected to provide direct evidence for or against the hydromagnetic models.

The various *ad hoc* models account satisfactorily for some observations. The far ultraviolet observations appear to provide some support for the *ad hoc* stellar wind models, particularly those for which $u_r(r, \theta = \pi/2)$ is an increasing function of r. However it is difficult to see how the stellar wind models in their present form can account for the V/R variation, particularly when $V > R$. On the other hand the elliptical ring model can satisfactorily account for the V/R variation, but it may encounter difficulties in matching line profiles, especially the absorption components.

Clearly what is needed is to subject the predictions of each of the models to more observational tests. For example the stellar wind models should be used to predict the infrared continua, linear polarization as a function of wavelength, etc. Likewise the elliptical ring model should be employed to predict line profiles, linear polarization, etc., in order to see whether or not one or both approaches is completely inadequate.

At the same time, many aspects of these *ad hoc* models can be improved. One immediate step is a more realistic treatment of the radiative transfer problem. An important step in this direction has been initiated by Kriz (1974), and the paper in this volume. It would also be useful to include better representations of $\rho(\mathbf{r})$, $\mathbf{u}(\mathbf{r})$ and $T(\mathbf{r})$, perhaps using those obtained from solutions described in Section 2. In this way the number of adjustable parameters characterizing these models might be reduced.

Finally the reader is wise if he does not take any of these models too seriously. Nature is undoubtedly much more subtle that we can imagine but also, perhaps, simpler than we think.

Acknowledgements

I wish to thank R. L. Poeckert for his comments on the manuscript and J. D. Landstreet for the discussion relating to the measurement of the magnetic field in B stars. The work on this paper was supported by the National Research Council of Canada.

References

Albert, E. and Huang, S. S.: 1974, *Astrophys. J.* **189**, 479.
Bahng, J. D. R.: 1971, *Astrophys. J. Letters* **168**, L75.
Baker, J. B. and Menzel, D. H.: 1938, *Astrophys. J.* **88**, 52.
Beals, C. S.: 1930, *Publ. Dominion Astrophys. Obs.* **4**, 294 and 297.
Biermann, P. and Kippenhahn, R.: 1973, *Astron. Astrophys.* **25**, 63.
Böhm, K. H.: 1973, in A. H. Batten (ed.), 'Extended Atmospheres and Circumstellar Matter in Spectroscopic Binary Systems', *IAU Symp.* **51**, 148.
Burbidge, G. R. and Burbidge, E. M.: 1953, *Astrophys. J.* **117**, 407.
Burd, S.: 1954, *Publ. Astron. Soc. Pacific* **66**, 208.
Capps, R. W., Coyne, G. V., and Dyck, H. M.: 1973, *Astrophys. J.* **184**, 173.
Cassinelli, J. P. and Castor, J. I.: 1973, *Astrophys. J.* **179**, 189.
Castor, J. I.: 1972, *Astrophys. J.* **178**, 779.
Castor, J. I., Abbott, D. C., and Klein, R. I.: 1975, *Astrophys. J.* **195**, 157.
Chandrasekhar, S.: 1934, *Monthly Notices Roy. Astron. Soc.* **94**, 522.
Cox, J. P. and Giuli, R. T.: 1968, *Principles of Stellar Structure*, Vol. 1, Gordon and Breach, New York, Ch. 9.
Crampin, J. and Hoyle, F.: 1960, *Monthly Notices Roy. Astron. Soc.* **120**, 33.
Delplace, A. M., Herman, R., and Peton, A.: 1969, in L. Detre (ed.), *Non-Periodic Phenomena in Variable Stars*, D. Reidel, Dordrecht-Holland, p. 223.
Doazan, V.: 1965, *Ann. Astrophys.* **28**, 1.
Fernie, J. D.: 1975, *Astrophys. J.* **201**, 179.
Gehrz, R. D., Hackwell, J. A., and Jones, T. W.: 1974, *Astrophys. J.* **191**, 675.
Gerasimovic, B. P.: 1934, *Monthly Notices Roy. Astron. Soc.* **94**, 737.
Gray, D. F. and Marlborough, J. M.: 1974, *Astrophys. J. Suppl.* **27**, 121.
Haisch, B. M.: 1975, Ph.D. Thesis, Univ. Of Wisconsin, Madison (unpublished).
Hazlehurst, J.: 1967, *Z. Astrophys.* **65**, 311.
Heap, S. R.: 1975, *Phil. Trans. Roy. Soc. London* **279**, 371.
Henriksen, R. N.: 1969, *Astron. Astrophys.* **1**, 457.
Huang, S. S.: 1972, *Astrophys. J.* **171**, 549.
Huang, S. S.: 1973, *Astrophys. J.* **183**, 541.
Huang, S. S.: 1973, in A. H. Batten (ed.), 'Extended Atmospheres and Circumstellar Matter in Spectroscopic Binary Systems', *IAU Symp.* **51**, 22.

Huang, S. S.: 1975, *Sky Telesc.* **49**, 359.
Hutchings, J. B.: 1968, *Monthly Notices Roy. Astron. Soc.* **141**, 329.
Hutchings, J. B.: 1970a, *Monthly Notices Roy Astron. Soc.* **150**, 55.
Hutchings, J. B.: 1970b, in A. Slettebak (ed.), *Stellar Rotation*, Gordon and Breach, New York, p. 283.
Hutchings, J. B.: 1971, *Monthly Notices Roy. Astron. Soc.* **152**, 109.
Johnson, M.: 1958, in *Etoiles a raies d'emission*, Institute d'Astrophysique, Liège, p. 219.
Kogure, T.: 1967, *Publ. Astron. Soc. Japan* **19**, 30.
Kogure, T.: 1969, *Astron. Astrophys.* **1**, 253.
Kriz, S.: 1974, *Bull. Astron. Inst. Czech.* **25**, 143.
Kriz, S. and Harmanec, P.: 1975, *Bull. Astron. Inst. Czech,* **26**, 65.
Kupo, I. D.: 1971, *Alma Ata Publications* **16**, 165.
Landstreet, J. D.: 1975, private communication.
Lester, D. F.: 1975, *Publ. Astron. Soc. Pacific* **87**, 177.
Limber, D. N.: 1964, *Astrophys. J.* **140**, 1391.
Limber, D. N.: 1967, *Astrophys. J.* **148**, 141.
Limber, D. N.: 1969, *Astrophys. J.* **157**, 785.
Limber, D. N.: 1970, in A. Slettebak (ed.), *Stellar Rotation*, Gordon and Breach, New York, p. 274.
Limber, D. N.: 1974, *Astrophys. J.* **192**, 429.
Limber, D. N. and Marlborough, J. M.: 1968, *Astrophys. J.* **152**, 181.
Lubow, S. H. and Shu, F. H.: 1975, *Astrophys. J.* **198**, 383.
McLaughlin, D. B.: 1933, *Proc. Nat. Ac. Sci.* **19**, 44.
McLaughlin, D. B.: 1938, *Pop. Astron.* **46**, 361.
McLaughlin, D. B.: 1938, *Astrophys. J.* **88**, 622.
McLaughlin, D. B.: 1961, *J. Roy. Astron. Soc. Can.* **55**, 13 and 73.
McLaughlin, D. B.: 1966, *Astrophys. J.* **143**, 285.
Magnan, C.: 1970, *J. Quant. Spectrosc. Radiat. Transfer* **10**, 1.
Magnan, C.: 1972, *Astron. Astrophys.* **21**, 361.
Marlborough, J. M.: 1969, *Astrophys. J.* **156**, 135.
Marlborough, J. M.: 1970, *Astrophys. J.* **159**, 575.
Marlborough, J. M.: 1971, *Astrophys. J.* **163**, 525.
Marlborough, J. M. and Roy, J. R.: 1971, *Astrophys. J.* **169**, 327.
Marlborough, J. M. and Gredley, P. R.: 1972, *Astrophys. J.* **178**, 477.
Marlborough, J. M. and Cowley, A. P.: 1974, *Astrophys. J.* **187**, 99.
Marlborough, J. M. and Zamir, M.: 1975, *Astrophys. J.* **195**, 145.
Massa, D.: 1975, *Publ. Astron. Soc. Pacific* **87**, 777.
Merrill, P. W.: 1952, *Astrophys. J.* **115**, 145.
Merrill, P. W., Humason, M. and Burwell, C. G.: 1925, *Astrophys. J.* **61**, 389.
Mohler, O. C.: 1938, *Astrophys. J.* **88**, 623.
Morgan, T. H.: 1975a, *Astrophys. J.* **195**, 391.
Morgan, T. H.: 1975b, *Astrophys. Space Sci.* **33**, 99.
Nariai, K.: 1970, *Publ. Astron. Soc. Japan* **22**, 313.
Nordh, H. L. and Olofsson, S. G.: 1974, *Astron. Astrophys.* **31**, 343.
Plavec, M.: 1973, in A. H. Batten (ed.). 'Extended Atmospheres and Circumstellar Matter in Spectroscopic Binary Systems', *IAU Symp.* **51**, 216.
Pomraning, G. C.: 1973, *The Equations of Radiation Hydrodynamics*, Pergamon, Oxford.
Praderie, F., Simonneau, E. and Snow, T. P.: 1975, *Bull. Am. Astron. Soc.* **7**, 359.
Prendergast, K. H. and Taam, R. E.: 1974, *Astrophys. J.* **189**, 125.
Rosseland, S.: 1926, *Astrophys. J.* **63**, 218.
Rybicki, G. B.: 1970, in H. G. Groth and P. Wellmann (eds.), *Spectrum Formation in Stars with Steady-State Extended Atmospheres*, NBS Spec. Publ. 332, p. 87.
Saito, M.: 1974, *Publ. Astron. Soc. Japan* **26**, 103.
Sampson, D. H.: 1965, *Radiative Contributions to Energy and Momentum Transport in a Gas*, Interscience, New York, Ch. 3 and App. B.
Secchi, A.: 1867, *Astron. Nachr.* **68**, 63.
Sima, Z.: 1973, *Astrophys. Space Sci.* **24**, 421.
Slettebak, A.: 1954, *Astrophys. J.* **119**, 460.
Smith, R. C.: 1970, *Monthly Notices Roy. Astron. Soc.* **148**, 275.
Sobolev, V. V.: 1960, *Moving Envelopes of Stars* (Engl. Transl.), Harvard Univ. Press, Cambridge.
Struve, O.: 1931, *Astrophys. J.* **73**, 94.
Struve, O.: 1942, *Astrophys. J.* **95**, 134.
Struve, O.: 1950, *Stellar Evolution*, Princeton Univ. Press, Princeton, p. ix.

Struve, O. and Swings, P.: 1932, *Astrophys. J.* **75**, 161.
Thomas, R. N.: 1970, in H. G. Groth and P. Wellman (eds.), *Spectrum Formation in Stars with Steady-State Extended Atmospheres*, NBS Spec. Publ. 332, p. 259.
Trasco, J. D., Wood, H. J., and Roberts, M. S.: 1970, *Astrophys. J. Letters* **161**, L129.
Underhill, A. B.: 1960, in J. L. Greenstein (ed.), *Stellar Atmospheres*, Univ. of Chicago Press, Chicago, Ch. 10.
Weber, E. J. and Davis, L. D.: 1967, *Astrophys. J.* **148**, 217.
Weidelt, R. D.: 1970, *Astrophys. Space Sci.* **6**, 205.
Zel'dovich, Ya. B. and Raizer, Yu. P.: 1966, *Physics of Shock Waves and High Temperature Hydrodynamic Phenomena*, Academic Press, New York, Ch. 1.

DISCUSSION

Slettebak: You discussed radiation driven stellar wind models (for example, Hutchings' calculations for γ Cas) in which gravity darkening was taken into account and an equatorial temperature of 12 000° was predicted. But what about a late B-type shell star like Pleione: a B8 star? When you include gravity darkening surely the equator must be very cool. Do you have enough radiation then to drive material into the shell?

Marlborough: I don't really know whether one could account for the phenomenon in Pleione or not, in terms of a radiation driven stellar wind. Perhaps radiation from the hotter, polar regions might be sufficient to generate a stellar wind from the outer regions of a disk in the equatorial plane. However, the problem remains as to how to form the disk or extended atmosphere.

Heap: What observations lead you to conclude that the shell surrounding a Be star is concentrated toward the equatorial plane?

Marlborough: I would argue that the distribution of electrons (linear polarization is assumed to arise from electron scattering) could not be spherically symmetric and produce a non-zero intrinsic linear polarization, so there has to be flattening in some sense. But I think that is the only strong evidence we have concerning the overall structure of the circumstellar envelope. In the case of shell stars the absorbing matter must extend to at least one stellar radius from the equatorial plane to account for the very low flux in the cores of strong shell lines.

Hutchings: I would like to make a comment which essentially agrees with what you have just said. I have computed a number of models of the type you have described, *ad hoc* models with quite a range in geometries, in order to see if one could find a dependence on i or on the geometries, or any other rather obvious parameters. It turned out, as far as I could make out, that nothing was very sensitive to anything. You could have very nearly spherically symmetrical envelopes and still reproduce profiles fairly satisfactorily.

Marlborough: If you consider only one piece of observational data like one line profile, for example, then generally these *ad hoc* models can account for the observational evidence considered because the models contain many free parameters. You can always adjust the parameters within realistic bounds to reproduce one line profile. If one considers several line profiles, one may be able to restrict the models a little more, but this has not yet been done.

Doazan: How do you interpret the change in the observed Balmer progression as a function of the velocity law?

Marlborough: I do not know how at the present time.

Delplace: How can you explain the variation of the period of the radical velocities which is observed in some shell stars like ζ Tau and 48 Lib?

Marlborough: In this type of picture, i.e. a stellar wind model, I cannot. This is one difficulty with any of these *ad hoc* models: they explain some observations but there are many things they cannot explain. And as all the observations which have been presented over the past few days indicate, the amount of data that the models can explain relative to what they cannot explain seems to be smaller and smaller. Maybe we should be very pessimistic, throw out all these models, and suggest that there be no new theories as far as Be stars are concerned for 10 years. Or alternatively, no new observations for 10 years.

Peters: There are a number of 'equator-on' Be stars which show featureless or only slightly structured Hα profiles. These objects typically show Fe II emission and O I λ 7774 emission. In your opinion, what physical parameters and/or geometries characterize the envelopes of these objects?

Marlborough: In one of his review articles McLaughlin comments that Be stars with strong hydrogen emission generally have Fe II in emission also. In the context of stellar wind models, this fact suggests a higher density envelope. Perhaps O I λ 7774 emission can be explained in the same way. Qualitatively one might account for a featureless Hα by having a region of moderate density and small extent in the equatorial plane just above the surface of the star to explain the Hα, Fe II and O I emission. If the outer edge expanded very rapidly, the density in the wind might be low enough so no strong absorption component to Hα occurred.

Hummer: I would like to stress the necessity of testing models for a number of lines, covering all parts of the model. In order to do this, it is essential that observers obtain, as near to one time as possible, and publish profiles for many lines, including photospheric lines, for a few typical objects. Concentration on only one or two 'interesting' features in an object gives very little assistance in inferring its structure in any reasonably unique way.

Marlborough: I strongly support these comments.

ON THE POSSIBLE ROLE OF MAGNETIC FIELDS
IN THE DYNAMICS OF THE Be PHENOMENON

D. NELSON LIMBER

Dept. of Astronomy, University of Virginia, Charlottesville, Va., U.S.A.

A problem that has intrigued me for some years is that of the dynamics of the Be phenomenon. For, although the Be phenomenon has been extensively studied for decades, the problems associated with the dynamics of the circumstellar envelopes about Be stars – the problems associated with their formation, with their maintenance, and with their dissolution – have, until very recently, received little more than qualitative consideration.

In what follows, I shall direct my remarks to those Be stars for which close binary companions – if any – play no major role in the Be phenomenon. (It is my belief – perhaps, hope – that the set so restricted is not an empty set.) For such Be stars, there appears to be no question but that the rapid rotation of the star involved plays a decisive role in the phenomenon. However, it is also clear that rotation, alone, is not enough to either move centrifugally supported matter outward from the star's equator to form a circumstellar envelope or to move the matter within such an envelope, once formed, off to infinity.

This raises the question as to just what physics it is that, in cooperation with these stars' rapid rotations, accounts for the envelope dynamics. Two such possibilities have been at least semi-quantitatively explored within the last few years: (1) radiation pressure and (2) magnetic field lines coupling the rapidly rotating star to the matter in the circumstellar envelope. Some aspects of the first of these possibilities are being presented in the contributed papers by others. It is to certain aspects of the second of these that I should like to direct your attention.

The study of steady-state situations – situations in which the partial derivatives of all macroscopic quantities vanish – is the obvious problem with which to begin such a discussion. Not because most or all of the Be stars are observed to be in situations that appear to be steady-state (or nearly steady-state) ones (because they are not), but because at least certain basic aspects of this situation are mathematically tractable, because at least some Be stars do show little if anything in the way of time variations over decades, and because steady-state solutions provide a natural point of departure for the consideration of all other situations, that is, for the consideration of the non steady-state situations that clearly play such an important part in the phenomenon.

The search for steady-state solutions having the required mathematical properties for the magneto-rotational model has been found to constitute an eigenvalue problem (Limber, 1974). When these mathematical requirements are supplemented by the physical requirement that the magnetic energy per unit volume not significantly exceed that in the form of the mass motions at any point within the circumstellar envelope, the acceptable solutions are still further restricted. Nevertheless, families of steady-state solutions have been found that satisfy these requirements

A. Slettebak (ed.), Be and Shell Stars, 371–373. *All Rights Reserved*
Copyright © 1976 by the IAU.

and that appear to be generally consistent with the observations for values for the relevant parameters that appear to be reasonable – reasonable, that is, if one considers pre-outburst, photospheric magnetic field strengths in the range from a few gauss to several hundred gauss to be reasonable and if one considers values for the equatorial rotational velocities not much more than 5 or 10% smaller than the critical rotational velocities to be reasonable.

At least as interesting, however, is the finding that for many combinations of the relevant parameters which, at least at first sight, appear not unreasonable, no acceptable steady-state solutions are found to exist. Thus, for example, for one somewhat oversimplified formulation of the problem: given a star and given the rate at which angular momentum is supplied to the equatorial region (presumably through evolutionary effects in the deep interior), the star can achieve acceptable steady-state solutions by adjusting its equatorial angular velocity to an appropriate value if the value of the pre-outburst photospheric magnetic field strength lies within a certain range. For all other values for this field strength, no possible adjustment of the star's equatorial angular velocity will lead to acceptable steady-state solutions. Thus, if at any time the rate at which angular momentum is being supplied to the star's equatorial region is inconsistent with steady-state solutions for the star's given photospheric magnetic field strength, then its subsequent photospheric and envelope evolution will, of necessity, be time-dependent.

Further, if after an interval during which a steady-state solution is possible and has been set up, there is a change in the rate at which angular momentum is supplied to the equatorial regions from the interior, then either the circumstellar envelope will be found to undergo changes in its structure from its initial steady-state situation that will bring it to another steady-state situation appropriate to the new input parameters or, if no such steady-state solution exists for the new parameters, the initial steady-state situation will be followed by a time-dependent situation for which there is no steady-state relief in sight. Since such changes result from changes in the rate at which angular momentum is supplied to the star's equatorial region, such time-variation might be expected to be characterized by time-scales of months, years, decades, or longer, but not to minutes or hours. In this connection it may be worth noting that, although the analogy with the Be phenomenon is not a very close one in a number of ways, interactive effects of differential rotation and magnetic fields appear to be capable of accounting for the observed properties of the sun-spot cycle in a very reasonable, semi-quantitative way (Leighton, 1969); and in the Sun's case, the characteristic time-scale for the phenomenon is one or two decades. It is thus not altogether unreasonable to consider the possibility that similar interactive effects of differential rotation and magnetic fields might give rise to the periods of this same order that are so often observed in the Be phenomenon.

It is of interest in this discussion of the observed time variations in the Be phenomenon to point out that for the steady-state solutions that exist for the cases studied, the matter moving outward through the circumstellar envelopes is not, except very close to the photospheres, even approximately centrifugally supported; that is, the effects of the magnetic fields are not in any major way to increase the angular momentum of the matter as it moves outward from the stars' equators. Consequently, in breakdowns of steady-state solutions, either by reason of decreases in the rate of angular momentum input into the stars' equatorial regions or by reason of hydromagnetic instabilities, sizable portions of these circumstellar envelopes could fall back toward the stars' surfaces.

This brings us to the last point that I should like to make concerning the possible role of magnetic fields in the Be phenomenon: that of the possibility of hydromagnetic instabilities within the envelopes. Whether the envelopes of Be stars are formed through effects of the kind here described, involving magnetic fields, through the effects of radiation pressure, or through still other mechanisms, if there are pre-outburst photospheric magnetic field strengths of the order of tens of gauss or more and if the outward velocities within the inner part of the envelopes are quite small with respect to the circular components of velocity there, then the mass motions within the inner portions of the envelopes, by reason of the combination of differential circular motion and outward motion, will lead to a severe stretching and winding around of the magnetic field lines until conditions would appear to become favorable for the snapping and reconnection of field lines. Although the theory for the reconnection of field lines is not yet very quantitative, solar flares bear solid testimony to the fact that such events not only can take place but actually do so – and under conditions that may be no more extreme than these encountered in Be envelopes. It thus appears quite possible that such snappings and reconnections of field lines will take place from time to time within the envelopes of Be stars, giving rise to flare-like outbursts. And it is very tempting for me to think in terms of such an explanation for the short-term time-variations in the range of minutes to hours that are observed in the profiles and integrated strengths of emission and shell lines. I, personally, feel that it is very important to continue the observational study of the short-term time variations that take place within Be envelopes both for the purpose of learning more about the physical and geometrical properties of the individual 'flares' and also for the purpose of learning more of possible temporal relationships between flare activity and other developments in the history of an envelope. We should learn more of the short-term time-variations in this way; we might learn something of the role of magnetic fields in the Be phenomenon, as well.

References

Leighton, R. B.: 1969, *Astrophys. J.* **156**, 1.
Limber, D. N.: 1974, *Astrophys. J.* **192**, 429.
Marlborough, J. M. and Zamir, M.: 1975, *Astrophys. J.* **195**, 145.

THEORETICAL WAVELENGTH DEPENDENCE OF
POLARIZATION IN EARLY-TYPE STARS

BERNHARD M. HAISCH and JOSEPH P. CASSINELLI

Washburn Observatory, University of Wisconsin, Madison, Wis., U.S.A.

Abstract. We have examined the theoretical wavelength dependence of linear polarization produced by electron scattering and modified by an absorptive opacity in the extended, distorted atmospheres of Be and Wolf-Rayet stars. A model atmosphere representing a Wolf-Rayet star recently calculated by Cassinelli and Hartmann was used. This model takes into consideration the subsonic portions of a radial, steady-state flow together with the requirement of radiative equilibrium. The model was then scaled down in temperature to approximate the cooler Be stars, again assuming radiative equilibrium. There were four models altogether, one WR star, and three Be models. All the atmospheres are quite extended by radiation pressure.

We then wrote the equation of transfer in spherical coordinates for I_L and I_R, the intensities in the radial and tangential planes of polarization, took the zeroth and first moments of this equation, and combined them into a second order equation to be solved numerically by iterating on a variable Eddington factor, K/J. The results showed polarization as high as 70% at the limbs of the atmospheres. However, in order to produce polarization in the net flux, an asymmetry must exist in the shape of the star. We therefore constructed disc models by simply truncating a spherical atmosphere, and Roche models, in which equipotential surfaces are assumed also to be surfaces of equal source function with a modification to account for the forward peaking of the radiation field. The polarization in the net flux was then calculated for all four model atmospheres for each of these two distorted shapes.

For the disc models we find that the polarization increases from about $\frac{1}{3}$% at 1 μ^{-1} to about 1% near 2.5 μ^{-1}, with drops across the absorption edges of H and He. The reduction in polarization across these edges is due to the absorption of polarized light and subsequent thermal re-emission of unpolarized light. The overall rise of the polarization toward shorter wavelength is due to the steeper gradient of the Planck functions at higher frequencies. The strength of the polarization decrease across the absorption edges, as well as the overall strength of the polarization, changes moderately from the hottest to the coolest models.

For the Roche models, the behavior is qualitatively similar, but the overall strength of the polarization is markedly reduced (0.2 to 0.3%). In all the models, the net polarization is perpendicular to the polar axis. None of the models predicts an upturn in the polarization longward of the Paschen edge as is cited in some wide-band polarimetric observations. The drop in the polarization across the Balmer limit is severe only for the coolest models. The slope between the Balmer and Paschen edges agrees well with observations. However, it is found that only a rather flattened disc having a ratio of thickness to radial depth of about $\frac{1}{3}$ can provide polarization as high as is observed.

A. Slettebak (ed.), Be and Shell Stars, 375–376. All Rights Reserved
Copyright © 1976 by the IAU.

The energy distributions of the Roche models show Balmer continuous emission, except for the very coolest model which shows a Balmer discontinuity of zero. This is due to the extreme geometrical extensions of the atmospheres.

DISCUSSION

Hummer: Are your models based on the LTE assumption? If so, I would doubt your Balmer discontinuities in emission. In a large number of extended spherical, non-LTE model atmospheres computed by Kunasz, Mihalas and myself (*Astrophys. J.*, in press), the Balmer jumps are *always* in absorption by roughly 0.05 mag.

THE POSSIBLE ROLE OF RADIATIVE ACCELERATION IN SUPPORTING EXTENDED ATMOSPHERES IN Be STARS

R. L. KURUCZ and R. E. SCHILD

Center for Astrophysics, Harvard College Observatory, and Smithsonian Astrophysical Observatory Cambridge, Mass., U.S.A.

Abstract. A detailed calculation of the radiative acceleration in B-type stars shows it to be a double-peaked function of effective temperature at small optical depths. The two peaks are shown to coincide approximately with peaks in the distribution of mean Hα emission strength as a function of $B - V$ color in Be stars. These facts suggest that radiation may play an important role in the support of the Be star extended atmosphere.

1. Introduction

Whereas the earliest papers on the Be stars stressed the role of rotation (Struve, 1945; Slettebak, 1949), more recent work has tended to stress the role of radiation processes. The most recent paper by Marlborough and Zamir (1975) suggests that stellar winds may be important in maintaining a mass of gas at distances of several stellar radii above the rotationally distorted stars. Thus our conception of Be stars as rapidly rotating stars in which centrifugal force maintains a shell of gas several stellar radii above the photosphere, is slowly giving way to a picture wherein rotation effectively reduces the surface gravity to the point where the radiation field generates a stellar wind, which constantly replenishes the shell which is in turn constantly expanding away from the star.

We wish to further substantiate this view with results of model atmosphere calculations and a comparison with data for normal Be stars. Schild and Romanishin (1975) have recently analysed the data for 41 Be stars in clusters and shown that the strength of Hα emission is a double peaked function of intrinsic $(B - V)$ color, with peaks near $(B - V) = -0.28$ and $(B - V) = -0.12$ corresponding to spectral types B1 and B8. We are not aware of any explanation of these peaks in terms of strictly rotational effects in the post-main-sequence evolution of stars. In this paper we present results of model atomosphere calculations which may explain the double peak in terms of an effect of radiative acceleration.

2. Calculations

Kurucz (1975) has recently completed a new model atmosphere grid covering the range 5500 K to 50 000 K. This grid is a revision and extension of the work by Kurucz *et al.* (1975) on early type stars. The new models have an improved treatment of line opacity and essentially encompass the range of temperatures for models that can be computed without including molecular lines. In the new calculations the line

A. Slettebak (ed.), Be and Shell Stars, 377–382. All Rights Reserved
Copyright © 1976 by the IAU.

list computed by Kurucz and Peytremann (1975) was edited down to fewer than 1 000 000 lines and the statistical distribution function representation of the line opacity was tabulated at many additional temperature values. Wavelength bandpasses were reduced from 100 Å to 25 Å in the visible to allow calculation of intermediate band colors.

These new models seem to be good representations of real stars. The model for Vega is in excellent agreement with the new calibration of Hayes and Latham (1975) and reproduces the Balmer profiles of Peterson (1968). The *UBV* and *uvby* colors predicted from these models are normalized so that the model for Vega has the colors of Vega. Kurucz is currently checking these models further by comparing detailed spectral calculations with satellite and ground-based observations.

The limitations on these models that affect the discussion here are several. First, these are plane-parallel models for non-rotating stars. Rotating the models in the manner of Collins (1966) or Maeder and Peytremann (1972) would change the colors somewhat. Second, the models are not numerically reliable above $\tau_{\rm Ross} = 10^{-4}$ because the surface was set at 3×10^{-5} or 10^{-5}. The calculation should extend another decade or two to investigate the Hα core. Third, at these shallow optical depths, non-LTE becomes important. Nonetheless we feel that these models are adequate to make the qualitative argument presented here and are likely to be an improvement over the models used by Lucy and Solomon (1970) in their discussion of radiative effects.

Radiative acceleration is determined by the integral

$$a = \frac{4\pi}{c} \int \kappa_\nu H_\nu \, d\nu,$$

where κ_ν is the total monochromatic mass absorption coefficient and H_ν is the flux. Since the total flux $\int H_\nu \, d\nu$ is proportional to $T_{\rm eff}^4$, a is roughly proportional to $T_{\rm eff}^4$. However since κ_ν varies through the atmosphere, and with gravity and effective temperature, and since the radiation field is highly non-gray, we must work with a grid of models to predict where the observable effects might appear.

Since the Be effect appears first in the core of Hα let us look at a shallow layer $\tau_{\rm Ross} = 10^{-4}$ which would influence the Hα core. In Figure 1 we plot log radiative acceleration against log effective temperature for the new models of effective temperature 10 000 K and hotter. The gravities range from 2 to 5 in steps of 0.5. A line with a slope of 4 is indicated to show a proportional to $T_{\rm eff}^4$. Note that there are two peaks above this slope at $T_{\rm eff} \sim 12\,000$ to 13 000 K and $T_{\rm eff} \sim 40\,000$ to 50 000 K with a minimum in between. Note that the higher gravity models generally have higher accelerations because increased pressures lead to decreased ionization and increased opacity. At greater optical depths we find that the minimum disappears and that increasing opacity shifts the curves upward.

In Figure 2 we plot $\kappa_\nu H_\nu$ for three models to show the source of the minimum and maxima. The first maximum, at effective temperature 13 000 K arises from absorption by singly ionized iron group lines in the region around 2600 Å. The second maximum, at 40 000 K, arises from the absorption lines of multiply ionized light elements spread throughout the far ultraviolet. The minimum reflects the lack of

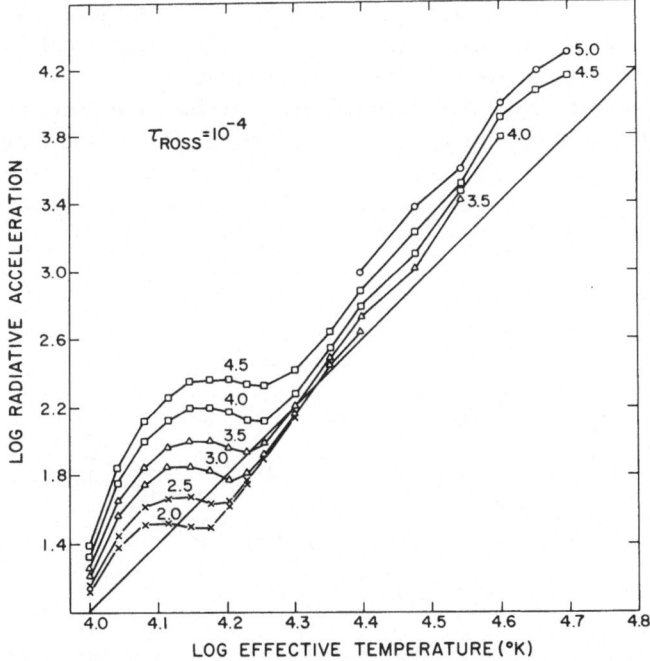

Fig. 1. Radiative acceleration is shown as a function of effective temperature for several choices of log g as shown for each curve. The straight line represents the factor proportional to T_{eff}^4.

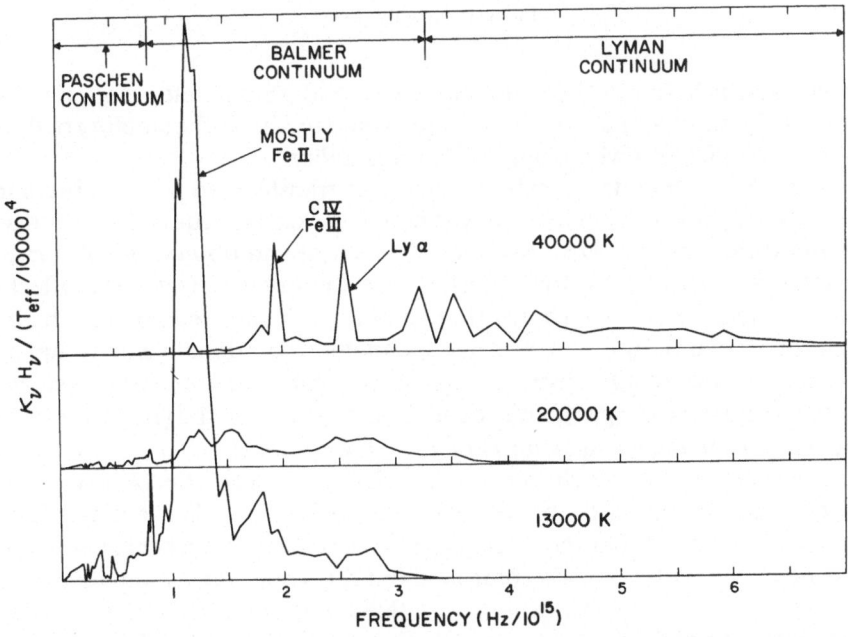

Fig. 2. The monochromatic term $\kappa_\nu H_\nu$ plotted as a function of frequency, to identify the contributors to the radiative acceleration. Note especially the large peak due to singly ionized iron group elements in the near ultraviolet for late B-type stars.

suitable absorption lines at intermediate temperatures. A more detailed investigation could easily be performed (if computer time were available) by directly
synthesizing the spectra and identifying individual lines.

In Figure 3 we replot the data from Figure 1 to show the maxima and minimum
more clearly by removing the T_{eff}^4 dependence so the slope is nominally zero. Since

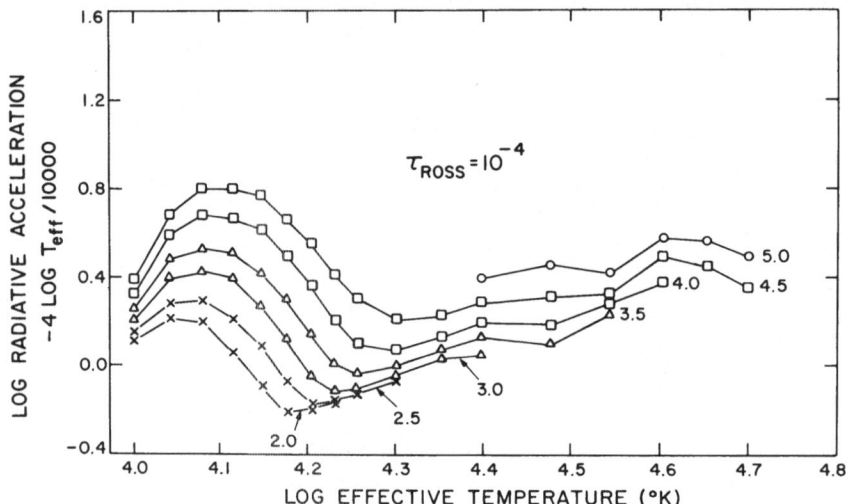

Fig. 3.   Same results as in Figure 1, but the factor of T_{eff}^4 has been divided out to
emphasize opacity effects.

the gravity at which a model becomes unstable to radiative acceleration, $g = a$ at any
depth, roughly varies as T_{eff}^4, this figure approximately shows the stability of the outer
layers as the model as a whole approaches instability.

In Figure 4 we show the results of Figure 2 transformed to the observational
variable $(B - V)$ for a comparison with the measured Hα equivalent widths from
Schild and Romanishin (1975). Both the theoretical and observational curves may
have systematic errors. As noted earlier, the theoretical curves are affected by
rotation; Collins (1966) has shown that rotating stars are redder than their non-
rotating counterparts, by amounts that depend upon rotational velocity and aspect.
Also, because the energy distributions of Be stars are generally distorted by
continuum emission from gas above the stellar photosphere, the $(B - V)$ colors given
by Schild and Romanishin are estimated from the colors of normal stars in the color
magnitude diagrams of the clusters studied. Both of these effects may amount to
errors of a few hundredths of a magnitude in $(B - V)$. We believe from the
comparison in Figure 4 that the strength of Hα emission in post-main-sequence Be
stars is correlated with $(B - V)_0$ color in about the same way as radiative acceleration. This suggests, but by no means proves, that the strength of the gas shell depends
upon the strength of the radiative acceleration near the surface of the star.

Finally, we note a comparison with the work of Lucy and Solomon (1970). Their
work on radiative acceleration took into account only the resonance lines of the light

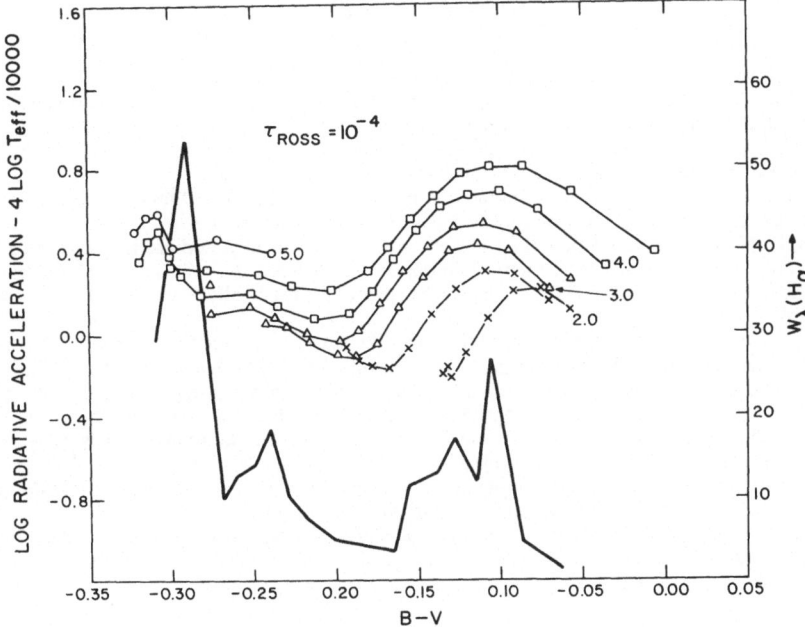

Fig. 4.   Radiative acceleration as function of $(B - V)$ compared with data on Hα emission line equivalent width for Be stars in clusters.

metals, which are in the ultraviolet spectral region near the peak of the Planck function for the Be stars. As noted by Schild and Romanishin (1975), their results predict a secondary peak near the minimum of the Hα equivalent width curve. We attribute our 13 000 K secondary peak to the inclusion of weaker lines in our calculations. It would be particularly desirable to compute a self-consistent Be star model which includes effects of radiation transfer and a detailed calculation of envelope support or a stellar wind, analogous to the self-consistent Of star model of Castor, Abbott, and Klein (1975).

References

Castor, J., Abbott, D., and Klein, R.: 1975, *Astrophys. J.* **195**, 157.
Collins, G. W., II.: 1966, *Astrophys. J.* **146**, 914.
Hayes, D. and Latham, D.: 1975, *Astrophys. J.* **197**, 593.
Kurucz, R.: 1975, in preparation.
Kurucz, R. and Peytremann, E.: 1975, *Smithsonian Astrophysical Observatory Special Report* No. 362.
Kurucz, R., Peytremann, E., and Avrett, E.: 1975, *Blanketed Model Atmospheres for Early-Type Stars* (Washington: Smithsonian Institution Press), 189 pp.
Lucy, L. and Solomon, P.: 1970, *Astrophys. J.* **159**, 879.
Maeder, A. and Peytremann, E.: 1972, *Astron. Astrophys.* **21**, 279.
Marlborough, J. M. and Zamir, M.: 1975, *Astrophys. J.* **195**, 145.
Peterson, D.: 1968, *Smithsonian Astrophysical Observatory Special Report* No. 293.
Schild, R. and Romanishin, W.: 1976, *Astrophys. J.* **204**, 493.
Slettebak, A.: 1949, *Astrophys. J.* **110**, 498.
Struve, O.: 1945, *Pop. Astron.* **53**, 201, 259.

DISCUSSION

Slettebak: Do you take gravity darkening into account in your computations?

Schild: No. In fact there has not been a calculation of a stellar model here. All this has been is Kurucz's calculations of the line opacity for normal stars.

Slettebak: If you believe the von Zeipel theorem or a variation thereof, a darkened equator would play a smaller role in supporting an extended Be star atmosphere.

Schild: We do expect the temperature gradient in Be stars to differ from that of a normal star, but we will not get to these questions until we have a self-consistent Be star model.

Hutchings: These Fe II lines below the Balmer limit: where are they, and are they observed?

Schild: I'm not sure if you expect them to be observed. As I recall, spectra at wavelengths near the atmospheric cutoff begin to show large numbers of iron lines, especially if you go to somewhat lower temperatures.

Snijders: In answer to Hutchings' question, I should like to point out that the results by various authors (mostly in press now) obtained with both the S-59 experiment and the *Copernicus* satellite show that the Fe II lines dominate the spectrum between λ 2000 and λ 3000 Å in late B-type stars, in good agreement with the Schild and Kurucz predictions. Further, Schild warned that his results are based on the LTE assumption and that this could be a serious source of errors. I should like to mention as yet unpublished non-LTE results for late B-type stars obtained by me which suggest that at $\tau_{\text{Rosseland}} \lesssim 10^{-4}$ the excited levels (when not metastable) are systematically underpopulated. This could considerably diminish the importance of radiation pressure from weak lines in late B-type stars.

Goldberg: Circumstellar shells were originally postulated because lines from metastable levels were found to be abnormally strong. Therefore, it's hard to understand how the excited levels can be populated in LTE.

PART VI

SINGLE VERSUS BINARY STARS

DUPLICITY OF Be STARS AS SEEN FROM ONDŘEJOV

(Review Paper)

PETR HARMANEC and SVATOPLUK KŘÍŽ

Astronomical Institute of the Czechoslovak Academy of Sciences, Ondřejov, Czechoslovakia

Abstract. Observed Be stars are considered to be interacting binaries undergoing mostly case B mass exchange. Their observed properties can be explained satisfactorily as the consequences of different modes of mass transfer. Suggestions for further observational and theoretical studies to verify or to disprove the hypothesis presented here are briefly outlined.

1. History

Although the similarity of Be envelopes to the disks observed in some close binaries with hydrogen emission lines had been pointed out by Struve and other astronomers as early as the 1930's, the concept of the possible binary nature of Be stars appeared only after numerical studies of mass exchange in close binaries performed by several teams (for review see, e.g., Plavec, 1970a; or Paczyński, 1971) became available.

Cowley (1964) in her detailed spectroscopic study of AX Mon interpeted this shell star as an interacting binary. In 1968, at the symposium on 'Mass Loss from Stars', Kříž (1969) suggested that the shell star ζ Tau may be a result of case B mass exchange. At the same symposium, Plavec and Horn (1969) offered a similar interpretation of V367 Cyg. Plavec (1970b) pointed out that the excess angular momentum, brought to the mass-gaining star by the infalling matter, is high particularly for case B mass exchange and may lead to the formation of an envelope around this star. He remarked that some shell stars could be formed by this process. In 1971, Harmanec, Koubský and Krpata in Ondřejov started systematic observations of selected shell stars in an effort to detect possible duplicity of these objects. They succeeded in disclosing periodic radial-velocity changes of 88 Her and 4 Her, which were confirmed later by American, Canadian and French observations (see Harmanec *et al.*, 1972a, b, 1973, 1974, 1976; Doazan, 1973; Heard *et al.*, 1975). Also Peters (1973), Hutchings and Redman (1973) and Kříž (1973) interpreted (on the basis of new spectroscopic observations) the Be stars HR 2142, HD 187399, and β Lyr, respectively, as case B mass-exchanging binaries. Harmanec *et al.* (1972b) suggested therefore that the possibility be considered tentatively that *all* shell stars are interacting binaries. Plavec (1973) objected against such a general hypothesis and pointed out some problems of binary interpretation for o And, ζ Tau, φ Per and several other shell stars. Nevertheless, the team of Plavec, Peters and Polidan at Los Angeles has also been studying very thoroughly the possible evolutionary connections between Be stars and binaries (cf. Plavec *et al.*, 1973a; Polidan *et al.*, 1973). In recent years, their observational effort has been particularly directed to infrared spectroscopy. Very recently Kříž and Harmanec (1975) published a completely general hypothesis about the binary nature of all Be stars. The rest of this paper is devoted to the outline and further development of this hypothesis.

A. Slettebak (ed.), Be and Shell Stars, 385–400. All Rights Reserved

2. Duplicity of Be Stars

Let us consider the appearance of case B mass-exchanging systems. It is necessary to mention that most of the available computations deal with conservative cases of mass exchange (with no mass and angular momentum lost from the system). Concerning the loss of angular momentum, Wilson and Stothers (1975) have estimated that the part of this loss transformed into the increase of the rotation of the mass-gaining component cannot change the sign of the expected period change dP/dt. No reliable estimate of mass loss is available, however. The peak rates of mass exchange in case B are very high in some cases and a substantial mass loss from the system cannot be excluded. If it really occurs, such mass loss can be self-accelerating (because the size of the binary system shrinks) and can lead to the rapid formation of a contact system (Van den Heuvel and De Loore, 1973). Plavec *et al.* (1973b) analysed this problem and concluded that some modes of mass loss can *decrease* the actual rate of mass transfer very substantially and thus serve as a regulating mechanism for the whole process. The actual course of mass exchange may then be qualitatively similar to the corresponding conservative case. Thus, the situation is not very clear and we must admit that the lack of our knowledge of actual mass and angular momentum loss from interacting binaries is probably the weakest point of all the considerations regarding the consequences of large-scale mass transfer in binaries. At the moment, we can only accept the observed existence of long-period systems with clear signs of continuing mass transfer as an empirical proof of the fact that mass and angular momentum loss from the system is not in most cases so substantial as to change the sign of dP/dt. If so, available computations provide us with a qualitatively good picture of mass-exchanging systems. Since the first, rapid part of the mass transfer is at least several times shorter in duration than the rest of the mass exchange, we can expect that the later stages of mass exchange, when the role of the components is already interchanged, are statistically the most significant. We therefore should observe mostly long-period systems (10^1–10^2 days) with less massive mass-losing components. These components should be at their limit of stability (usually at the Roche limit). Mass-gaining components of such systems are expected to be essentially unevolved B-type (or A-type) stars with very small relative radii. Computations indicate typical mass ratios of 0.1–0.2; Roche lobes around mass-gaining stars are therefore very large and allow the formation of very extended envelopes within them. Dynamical considerations show that the formation of such envelopes by mass transfer is indeed very probable (cf. Lubow and Shu 1975). Hydrogen emission and shell lines may originate in such envelopes.

As a consequence of relatively large masses and long periods, the semi-amplitudes of the radial-velocity curves of mass-gaining components must be very small, in many cases even on (or below) the limit of detectability.

A wide range of spectral types of mass-losing stars must be expected depending on initial parameters such as masses of the components, mass-ratio, evolutionary age, etc. In neither case, however, should the spectral type of the mass-losing component be earlier than that of the mass-gaining star. In some cases, the spectrum of the mass-losing component may be dominant, but more frequently, the mass-gaining star will be the brighter of the two in the visual part of the spectrum.

When observed spectroscopically, similar systems may appear as single Be stars rather than binaries.

Case B mass exchange proceeds on the Kelvin time scale (10^3–10^6 years) and the existing computations do not describe possible changes taking place on the dynamical time scale which, however, is observationally most interesting. Thus, the possible dynamical effects must be considered separately. We can start with the analogy with other types of interacting binaries. Biermann and Hall (1973) discussed a possible mechanism of cyclic changes in the rate of mass transfer between binary components connected with the dynamical instability of the surface convective layers of the mass-losing star. Their theory was successfully applied by Hall (1975) to explain alternate period changes of U Cep.

Available observational data clearly indicate the existence of several modes of mass transfer:

(i) Continuous steady-state transfer (β Lyr, SV Cen)

(ii) Non-continuous transfer proceeding in bursts (novae)

(iii) Variable transfer (U Cep, SW Cyg).

A typical time-scale of non-steady events is 10^0–10^1 years, i.e. similar to the time scale of the long-term variations of Be envelopes. This led us to the idea of considering tentatively three analogous modes of mass transfer in Be binaries.

Recent progress in the hydrodynamical treatment of gas dynamics in close binaries enables us to predict roughly the structure of circumstellar matter in binary stars. Let us consider a stream of gas which begins to flow from the conic surface of the mass-losing component in the neighbourhood of the Lagrangian point L_1 towards the other star. As already mentioned, the relative radii of mass-gaining stars in case B mass exchange are invariably very small. Consequently, the stream does not strike the surface of the mass-gaining component, but proceeds to orbit asymmetrically around the star and interacts with itself. This interaction produces a deflection of the primary stream, followed by a change in the returning stream and so on. According to Lubow and Shu (1975), the subsequent transient behaviour may last for hundreds to thousands of orbital periods. During this phase, a density maximum can be formed above the advancing hemisphere of the mass-gaining star, such as indicated by the computations of Prendergast and Taam (1974). If the gas outflow from the contact component continues, the envelope around the mass-gaining star eventually becomes so dense as to represent a 'solid wall' to the incident stream. This steady-state situation results in the formation of a very nearly symmetrical envelope, the inner parts of which spiral slowly in toward the underlying star.

On the other hand, if the mass flow between components were to vary substantially within several years (being very weak or absent in some intervals), the transient behaviour of the envelope and its variable asymmetry might be rather typical of such systems (for a binary with $P = 50$ days, for example, 100 orbital periods represent 13.7 years). Keeping the above considerations in mind, we can now discuss the expected properties of the three idealized types of Be binaries.

2.1. TYPE 1: STEADY-STATE MASS TRANSFER BINARIES

The density of gaseous streams in case B mass exchange may be high enough for these streams to produce observable effects. Observationally, we can expect (provided $i \neq 0°$):

(i) Double emission lines arising in the envelope around the mass-gaining star.
(ii) Emission from the stream which can manifest itself by some asymmetry of the observed profiles or by periodic V/R variations (with a period equal to the orbital period).
(iii) Hydrogen absorption lines, which are frequently used to derive the radial-velocity curves, will contain components which can hardly be separated from the underlying star, the envelope, and (at least in some orbital phases) also the stream. It may result in some of the following distortions of the velocity curve:

(a) The well-known Struve (Barr) effect of observing a curve with spurious eccentricity and $\omega \sim 0°$ due to an artificial increase of the observed velocity by the contribution from the (receding) stream as observed shortly before conjunction of the two stars with the mass-losing component in front. Velocity curves of SX Cas or 4 Her may serve as examples; see Figure 1.

In cases when the mass-losing star dominates in the spectrum, an analogous effect with $\omega \sim 210°$ may occur. There exists indeed a local maximum in Batten's (1968) ω's histogram and we suggest that HD 72754, HD 187399, W Cru, and possibly even 17 Lep, may be examples of such systems, the evidence being particularly strong for HD 72754 and W Cru (see Figure 2). Although no pronounced effect of a similar kind is known for β Lyr, it is interesting to note that also its velocity curve accepted for Batten's (1967) 'Sixth Catalogue' has $\omega = 217°$.

(b) The absolute dimensions of the systems in question are so large that different spectral lines from the stream may arise effectively in somewhat different parts of the stream. Provided the binary is not eclipsing, the effect of inclination can contribute and radial-velocity curves of individual lines (affected by the Struve effect) may differ one from another in the semi-amplitude and in the time of maximum velocity. This is because when a line arises from a more extended region of the stream, its contribution (in projection against the stellar disk) will be maximum closer to the conjunction of the stars (provided, of course, $i \neq 90°$) when the projected velocity of the stream is higher, and consequently, the maximum of the velocity curve occurs later. If mean velocities of several lines are used in such cases, the above effect contributes again to the scatter of the velocity curve, thereby decreasing its reliability.

(c) As already mentioned, it is quite possible that (especially in systems with a large mass exchange) a fraction of the gas escapes from the system. Although the detailed mechanism is not known as yet, it seems that this material can form an envelope around the whole system. In favourable cases, parts of this envelope projected against the disk of the brighter component may produce shell-type absorptions observable in the spectrum. Using a simplified assumption that the outer envelope has the form of a circular ring revolving around the centre of gravity of the two stars, we found that the radial velocity derived from the shell lines of this envelope follows the velocity curve of the brighter component of the binary, but with a substantially smaller amplitude. The following relation holds at any phase

$$\frac{RV_e}{RV} = \sqrt{\frac{A^3}{d^3}},$$

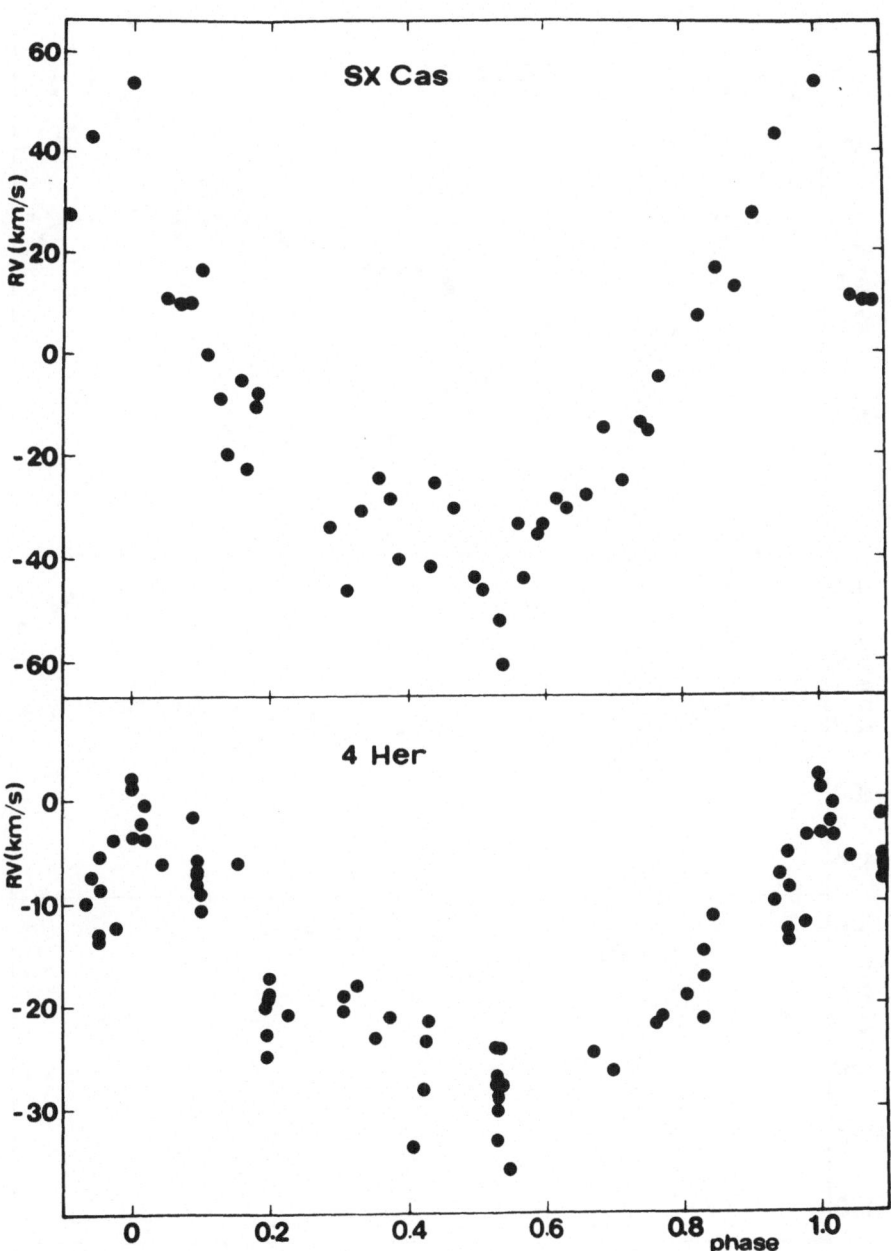

Fig. 1. Examples of the Struve effect on the velocity curves of SX Cas (Struve, 1944a) and 4 Her (Heard *et al.*, 1975).

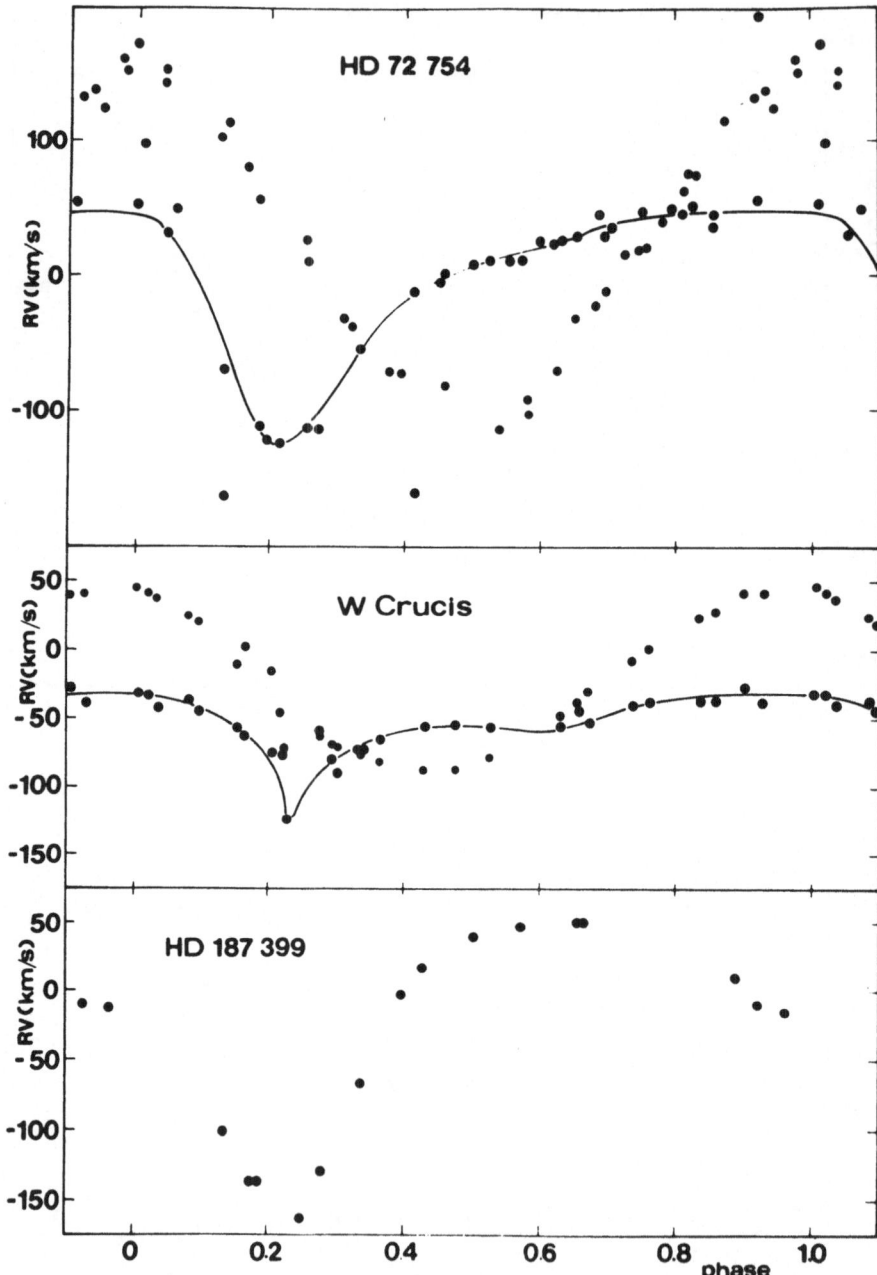

Fig. 2. Examples of the distortion of the hydrogen velocity curves (filled circles) of mass-losing components by absorption in gas streams: HD 72754 (Thackeray, 1971), W Cru (Woolf, 1962) and HD 187399 (Hutchings and Redman, 1973). In the first two cases, the metallic velocity curves of mass-losing components, free of the distortion, are also shown (by open circles).

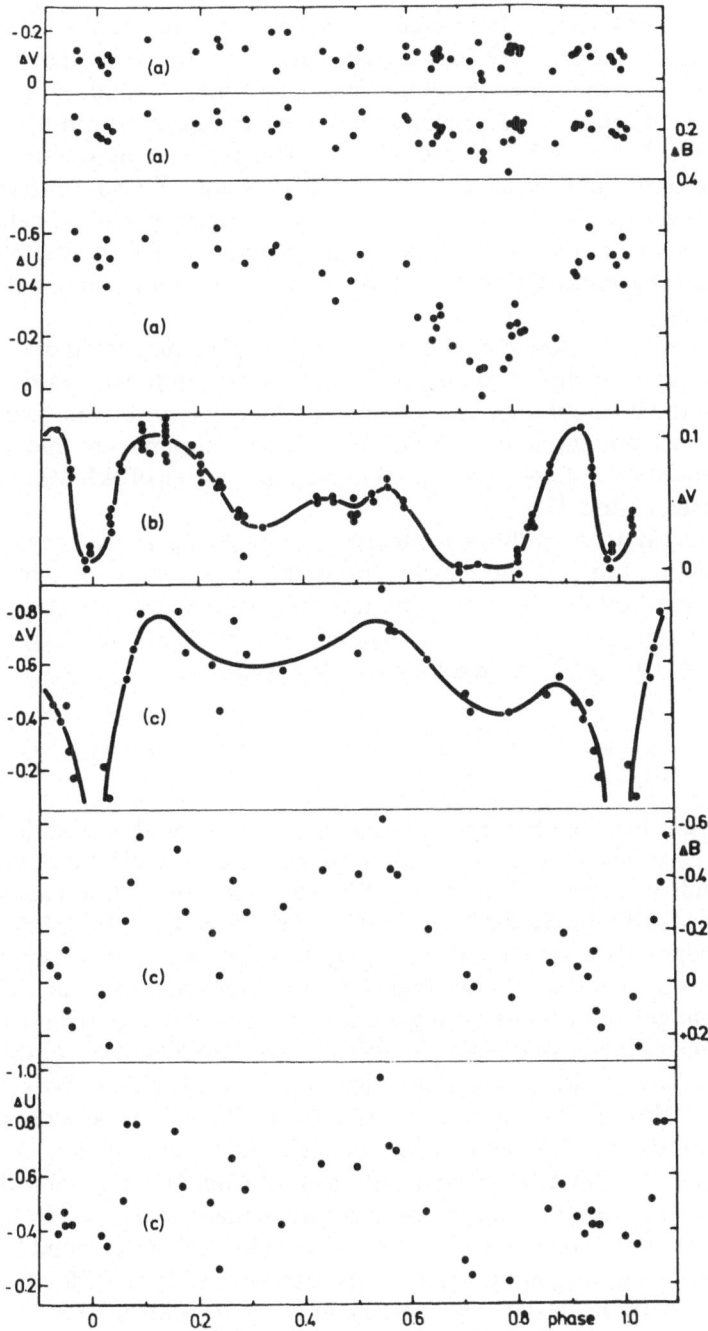

Fig. 3. Photoelectric light curves of AX Mon (a) (Magalashvili and Kumsishvili, 1969), HD 187399 (b)
(Hutchings and Redman, 1973) and W Ser (c) (Lynds, 1957).

where RV_e and RV denote the radial velocity of the envelope lines and of the star, respectively, A is the distance between the centres of the components, and d is the effective radius of the outer envelope (we have found a small error in a similar formula by Thackeray, 1971). Lines of the outer envelope with this behaviour are known for HD 72754, V 367 Cyg and W Ser. The presence of an outer envelope is detectable also in the systems β Lyr, HD 187399 and W Cru. In these cases, the velocity changes of the envelope lines are not so clearly visible; rather, a slight expansion prevails. Notably, all of the systems just mentioned show different signs of a very strong mass transfer (for most of them, the *mass-losing* component dominates in the spectrum).

Let us now consider possible photometric effects of circumstellar matter. If the gas stream is dense enough to produce continuous absorption, we can expect a broad minimum in the light curve of the system in or shortly after the elongation with the mass-losing component approaching. Such broad minima are indeed observed (around photometric phases 0.75–0.85) in the light curves of AX Mon, HD 187399 and W Ser (see Figure 3).

For eclipsing systems, additional effects can be expected, such as asymmetry of the primary minimum due to the gas stream, or an ultraviolet excess in the primary minimum caused by the light from the (partially uneclipsed) envelope around the mass-gaining star. Such effects were demonstrated by Hall and Garrison (1972) in the UBV light curves of the Algol binary SW Cyg.

2.2. TYPE 2: BINARIES WITH NON-CONTINUOUS MASS TRANSFER GOING IN RAPID BURSTS

If mass transfer between binary components occurs in rapid bursts, the duration of which is at least one order of magnitude shorter than the orbital period, then the ejected cloud of matter will form an elliptical ring around the mass-gaining star during several orbital periods. The behaviour of this ring can be followed approximately by means of the restricted three-body problem. We followed a particular case in a system with the mass-ratio 0.1 (which is rather typical for the final stages of case B mass exchange) in the non-rotating frame of reference with the origin in the centre of the mass-gaining star (see Figure 4). The gradual apsidal motion of the ring due to the attractive force of the mass-losing component is clearly visible. We estimated that the period of this apsidal motion corresponds to 24 binary periods. Considering binaries with $P = 50$, 100 and 1000 days, we arrive at 3.3, 6.6 and 66 years, respectively, for the period of apsidal motion of the ring. This fits well with the typical 'periods' of cyclic long-term variations observed in many Be stars. The hypothesis of duplicity of Be stars provides us therefore with a physical background for the model of an elliptical ring (originally suggested by Struve, 1931; and McLaughlin, 1961) which has recently been developed by Huang (1973) and successfully applied by Huang (1973) and Albert and Huang (1974) to observations of the Be stars 105 Tau, HD 20336, 25 Ori and β^1 Mon.

We think that probably the best known example of the presence of an elliptical ring is in the Be binary ζ Tau. The orbital period of the system is 133 days and the semi-amplitude of the velocity curve of the B2e component $K = 9$ km s^{-1}. Delplace

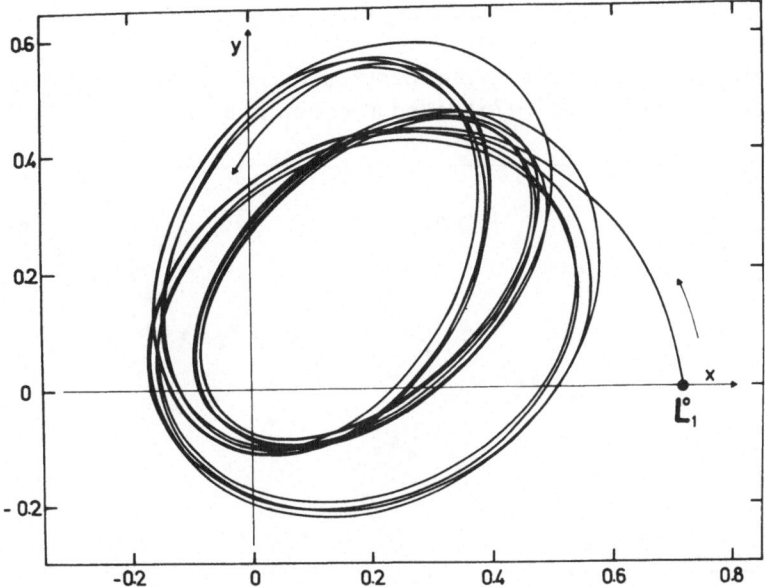

Fig. 4. Evolution of the orbit of a gaseous cloud formed by rapid ejection of matter from the vicinity of the inner Lagrangian point L_1. A non-rotating frame of reference is used. L_1^0 denotes the position of the Lagrangian point at the moment of ejection.

(1970) showed that (after some event in 1958) the lines of the envelope had followed a sinusoidal velocity curve with $K = 60$ km s^{-1} and $P^* = 6.8$ years during the period from 1960 to 1967, with the binary motion with $K = 9$ km s^{-1} still detectable along this long-period curve. Considering the long change as a consequence of the formation of an elliptical envelope, we obtain a reasonably good agreement between the observed values

$$P^*/P = (365 \times 6.8)/133 = 19, \qquad \text{and } K^* = 60 \text{ km s}^{-1}$$

and their theoretical predictions

$$P^*/P = 24, \qquad \text{and } K^* = 63 \text{ km s}^{-1}.$$

Very probably, the tidal forces will tend to circularize the ring so that the long-term variations do not repeat with exactly the same period and may gradually cease after several cycles, provided no new outburst occurs in the meantime.

2.3. TYPE 3: BINARIES WITH VARIABLE MASS TRANSFER

This type represents an intermediate case between the two previous cases. The periods in which mass transfer is strong and weak (or completely absent) are comparable and alternate on a time scale of years. Stars alternating between Be and B phases may belong to this type, e.g., 4 Her, Pleione, 88 Her, o And and some others. Observationally, these objects can exhibit some strictly periodic variations in intervals when mass transfer is acting similarly to those of Type 1.

Some additional effects can be observed however:

(a) As already mentioned, the envelope around the mass-gaining component can be notably asymmetric in non-stationary cases and a density maximum can exist above the advancing hemisphere of the star. As indicated by computations of Prendergast and Taam (1974), this dense material moves slowly towards the underlying star. Harmanec (1976) called attention to a consequence of this effect:

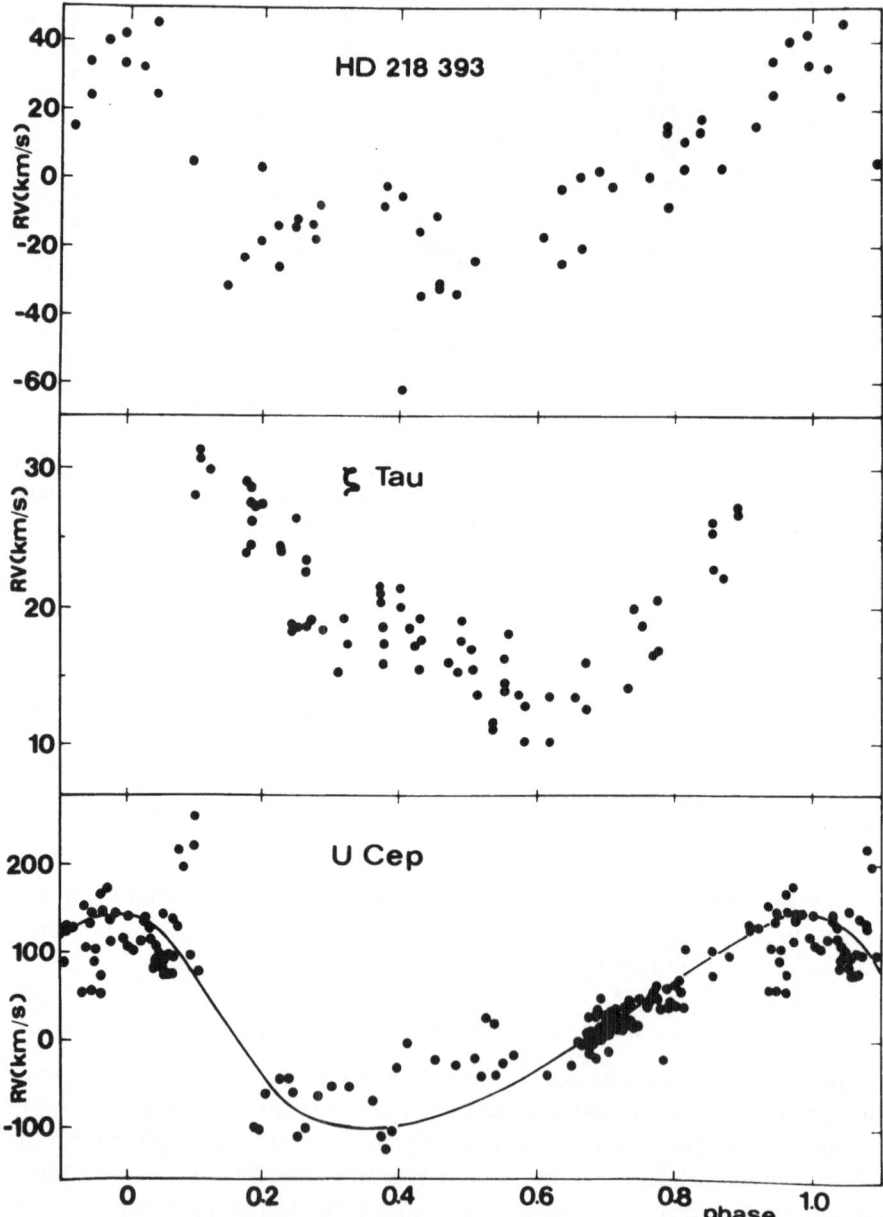

Fig. 5. Examples of the hump on the velocity curves of Be stars HD 218393 (Struve, 1944b) and ζ Tau (Hynek and Struve, 1942) and of the well-known eclipsing binary U Cep (Batten, 1974). Different symbols are used to denote different observed cycles.

line absorption in this part of the envelope, as seen in appropriate orbital phases, may cause a weaker effect similar to the Struve effect: a hump on the velocity curve in the neighbourhood of the minimum velocity. He demonstrated that the effect is indeed observed in many velocity curves of very different kinds of objects. HD 218393 or ζ Tau can serve as examples of the effect among Be stars (see Figure 5). It is important to realize that, under non-stationary conditions, the strength and the position of the density maximum can vary very sensitively, thus causing irregular variations of the shape of the velocity curves. This may even result in the failure of an attempt to establish a period of velocity changes in particular cases (this probably happened in the case of HD 218393).

(b) The density maximum may in some cases even produce photometric effects: a broad shallow minimum around photometric phases 0.2–0.4. Light curves of HD 187399 and W Ser seem to display this effect (see Figure 3).

In addition to the changes correlated with the orbital period, the systems with variable transfer will certainly exhibit (often even more pronounced) long term spectral, light, and colour variations.

When the contact component of such a system becomes stable against mass loss and the mass exchange ceases for some time, the further behaviour of the gaseous envelope around the mass-gaining component probably depends on the duration of the previous mass-exchange period. If this period was relatively short and the envelope still has the form of an elliptical ring, cyclic V/R and velocity variations similar to those of Type 2 binaries can be expected. In the case of a long duration of the previous mass-exchange period, however, when the envelope is almost cir-cularized already, no such variations should occur. Only some effects of a gradual dissipating of the envelope may be observed.

It must be stressed that the three types just described were introduced only to explain different possible effects of mass transfer. We feel that Type 3 of variable mass transfer is probably the most frequent among real Be binaries and that the two other types are idealized limiting cases only which, however, will illustrate possible essential differences in the behaviour of circumstellar matter under different condi-tions.

In real systems, even more complicated situations may be possible.

It cannot be excluded, for example, that rapid bursts occur even in systems with continuous mass transfer so that an elliptical ring is formed 'in addition' to the existing smaller envelope. In fact, we have to assume such a situation in the case of ζ Tau for which some effects of continuous transfer were detected in the past.

3. Observed Variations of Be Stars and their Possible Causes

It may be useful now to summarize the observed types of variations of Be stars and to comment on their possible causes. They are as follows:

3.1. SPECTROSCOPY

3.1.1. *Long-term Spectral Variations* (time-scale 10^0–10^1 years)
(a) Alternate, mostly irregular, changes in the intensity of emission and shell lines,

sometimes a complete disappearance of these lines. Such changes may reflect variations in the intensity of mass transfer between components caused possibly by the dynamical instability of the outer convective zone of the mass-losing star. Though a complete theory of this phenomenon is still lacking, the variability of mass transfer between binary components is well established in the case of Algol binaries (variable Struve effect, $O - C$ diagram of alternate period changes and recent appearance of emission lines in the spectrum of U Cep may serve as a good example).

(b) Cyclic V/R and velocity variations with transient periods of several years. These agree well with the model of an elliptical rotating envelope which we have already discussed.

3.1.2. *Periodic Velocity, V/R and Profile Variation* (time-scale 10^1–10^2 days)

In our opinion, these changes reflect directly or indirectly the orbital motion of the Be binary. The V/R variations of this type are probably caused by the asymmetry of the Be envelope connected with the presence of a gas stream. They are, similarly to the long-termed V/R variations, in phase with the velocity curves. Concerning the velocity changes, it is important to realize that in many cases they can reflect mainly the contribution of the stream (the Struve effect), the amplitude of the velocity changes of the star itself being very small. Even such a curve (or the observed V/R variations) can, in principle, serve to derive the orbital period. It is necessary to realize, however, that in intervals when mass transfer is absent one can observe a *constant* radial velocity which is more negative than the 'systematic' velocity of the previously observed curve. Quite possibly, this is also the case for 4 Her.

3.1.3. *Rapid and Ultrarapid Spectral Variations* (time-scale: minutes–several days)

An increasing effort has been made in recent years to detect rapid spectral variations of Be stars. For several reasons, this problem is very delicate. We personally are not convinced that every rapid variation announced so far is indeed well established. The unclear situation may be illustrated by the following examples:

(a) Ringuelet and Machado (1974) reported rapid periodic velocity variations for π Aqr with a period of 0.087 days. During one week in October 1973 Haefner *et al.* (1975) performed simultaneous spectroscopic, photometric and polarimetric observations of this star and detected no rapid variations from minutes to days. In particular, a periodogram of their velocities and the velocities of Ringuelet and Machado did not reveal any significant period.

(b) Hutchings *et al.* (1971) reported ultrarapid variations of the Hα profile of κ Dra, detected by TV techniques. Nordh and Olofsson (1974) followed the star for three nights by means of narrow-band Hα photometry (one narrow and two broad-band filters) with 10–15^m time-resolution. No changes were detected. Accordingly, we leave the problem of rapid spectral variations for future investigation.

3.2. PHOTOMETRY

Curiously enough, systematic photoelectric observations of Be stars are very rare so far, although they are urgently needed. Consequently, it is hard to classify the light changes of Be stars. Available observations indicate the presence of long-term light and colour variations in many cases but nothing definite is known so far. To the best of our knowledge, no clear correlation between photometric and spectral changes has been found as yet.

Probably the best documented are the periodic changes reflecting the binary motion: light curves of eclipsing binaries and eclipses of stars by gaseous streams (see, e.g. AX Mon or HD 187399, Figure 3). Rapid photometric variations (shorter than 1 day) with transient periods and variable amplitude are well documented in several cases – see, e.g. EW Lac (Walker, 1953; Lester, 1975) or V932 Aql (Lynds, 1960). It would be tempting to speculate about the possible similarity of these variations to the flickering of novae but, clearly, more data must be accumulated before a more serious analysis can be made.

Lester (1975) explains the photometric variations of EW Lac as temperature changes in the atmosphere connected with some sort of pulsation and refuses the model of a dark spot originally suggested by Walker (1953).

4. Suggestions for Further Investigation

Finally, we want to mention briefly which further studies would be desirable in our opinion to confirm or to disprove the outlined hypothesis and to improve our understanding of Be stars in general.

Probably the weakest point of the present reductions of observed data is the semi-qualitative and partly intuitive character of the present-day theories. No complete physical theory of processes taking place in Be envelopes and gaseous streams between binary components is available.

Several very formidable tasks must be solved:

(1) The complete hydrodynamical solution of the gas dynamics in binary stars. Recent progress in this field is very promising (Prendergast and Taam, 1974; Lubow and Shu, 1975) but the available results are still not sufficient. For example, Prendergast and Taam published their computations for only one set of binary parameters corresponding to U Cep (i.e. the mass-ratio $m = 0.49$ and the relative radius of the mass-gaining star $r_1 = 0.19$). In their case, most of the material strikes directly upon the mass-gaining star and only a small fraction of it contributes to the formation of an envelope. At least one other computation with the parameters relevant for a typical Be binary, say $m = 0.1$ and $r_1 = 0.05$, would be highly desirable. Moreover, non-stationary solutions are necessary to study the long-term variations, formation of an elliptical ring and so on.

(2) A complete theory of line formation in circumstellar matter (including non-LTE populations of atomic levels, heating of the envelope, etc.) based on the hydrodynamical solutions of gas dynamics.

(3) An improved analysis of the light curves of eclipsing binaries, taking into account the effects of circumstellar matter.

Only after the above-mentioned problems have been satisfactorily solved, is a reliable quantitative analysis of line profiles, radial velocities, and light changes, and derivation of absolute physical data for individual systems possible. At the moment, such data are, in fact, completely lacking. For any known Be binary (perhaps only with the exception of β Lyr) it is not known with a reasonable certainty, for example, whether or not the mass-losing component fills the corresponding Roche lobe, and so on.

Before the complicated theoretical problems are solved, a lot of work can be done observationally. The following items are what we consider most desirable:

(1) Systematic spectroscopic observations of selected Be stars in a wide spectral range with the highest dispersion possible.

(2) Publishing and analysing radial velocities of individual lines, not just mean values.

(3) Systematic photoelectric photometry of all the brighter Be stars, aimed at discovering possible relations between spectral and light changes and, possibly, disclosing new eclipsing systems among Be stars.

(4) Checking the suspected duplicity of Be stars, for which the (supposedly orbital) period and therefore approximate dimensions of the system are known, by means of new observational techniques such as lunar occultations or speckle interferometry.

(5) Accumulating the most complete and best possible data for the known eclipsing Be binaries β Lyr, V367 Cyg, W Ser, W Cru, RX Cas, SX Cas, UX Mon, HD 72754 and HD 187399, for which a future comparison of observations and theory may be most promising.

(6) Continuing in the effort to detect rapid variations of Be stars. To detect and to follow systematically possible rapid variations of a known Be binary would be especially interesting as it could help to clarify the question of the possible relation of these variations to the phase of the binary period.

(7) An improved comparative statistics of Be stars and Be binaries among B-type stars, taking into account the transient character of the Be phenomenon, evolutionary effects and other factors, would also be interesting.

References

Albert, E. and Huang, S.-S.: 1974, *Astrophys. J.* **189**, 479.
Batten, A. H.: 1967, *Publ. Dominion Astrophys. Obs.* **13**, 119.
Batten, A. H.: 1968, *J. Roy. Astron. Soc. Can.* **62**, 344.
Batten, A. H.: 1974, *Publ. Dominion Astrophys. Obs.* **14**, 191.
Biermann, P. and Hall, D. S.: 1973, *Astron. Astrophys.* **27**, 249.
Cowley, A. P.: 1964, *Astrophys. J.* **139**, 817.
Delplace, A. M.: 1970, *Astron. Astrophys.* **7**, 68.
Doazan, V.: 1973, *Astron. Astrophys.* **27**, 395.
Haefner, R., Metz, K., and Schoembs, R.: 1975, *Astron. Astrophys.* **38**, 203.
Hall, D. S.: 1975, *Acta Astron.* **25**, 1.
Hall, D. S. and Garrison, L. M.: 1972, *Publ. Astron. Soc. Pacific* **84**, 552.
Harmanec, P.: 1976, to be published.
Harmanec, P., Koubský, P., and Krpata, J.: 1972a, *Bull. Astron. Inst. Czech.* **23**, 218.
Harmanec, P., Koubský, P., and Krpata, J.: 1972b, *Astrophys. Letters* **11**, 119.
Harmanec, P., Koubský, P., and Krpata, J.: 1973, *Astron. Astrophys.* **22**, 337.
Harmanec, P., Koubský, P., and Krpata, J.: 1974, *Astron. Astrophys.* **33**, 117.
Harmanec, P., Koubský, P., Krpata, J., and Žďárský, F.: 1976, *Bull. Astron. Inst. Czech.* **27**, 47.
Heard, J. F., Hurkens, R., Harmanec, P., Koubský, P., and Krpata, J.: 1975, *Astron. Astrophys.* **42**, 47.

Huang, S.-S.: 1973, *Astrophys. J.* **183**, 541.
Hutchings, J. B., Auman, J. R., Gower, A. C., and Walker, G. A. H.: 1971, *Astrophys. J.* **170**, L73.
Hutchings, J. B. and Redman, R. O.: 1973, *Monthly Notices Roy. Astron. Soc.* **163**, 209.
Hynek, J. A. and Struve, O.: 1942, *Astrophys. J.* **96**, 425.
Kříž, S.: 1969, in M. Hack (ed.), *Mass Loss from Stars*, D. Reidel Publ. Co., Dordrecht-Holland, p. 257.
Kříž, S.: 1973, *Nature* **245**, 36.
Kříž, S. and Harmanec, P.: 1975, *Bull. Astron. Inst. Czech.* **26**, 65.
Lester, D. F.: 1975, *Publ. Astron. Soc. Pacific* **87**, 177.
Lubow, S. H. and Shu, F. H.: 1975, *Astrophys. J.* **198**, 383.
Lynds, C. R.: 1957, *Astrophys. J.* **126**, 81.
Lynds, C. R.: 1960, *Astrophys. J.* **131**, 390.
Magalashvili, N. L. and Kumsishvili, J. I.: 1969, *Bull. Abastumani*, No. 37.
McLaughlin, D. B.: 1961, *J. Roy. Astron. Soc. Can.* **55**, 13 and 73.
Nordh, L. and Olofsson, G.: 1974, *Stockholm Obs. Report*, No. 7.
Paczyński, B.: 1971, *Ann. Rev. Astron Astrophys.* **9**, 183.
Peters, G. J.: 1973, in A. H. Batten (ed.), 'Extended Atmospheres and Circumstellar Matter in Spectroscopic Binary Systems', *IAU Symp.* **51**, 174.
Plavec, M.: 1970a, *Publ. Astron. Soc. Pacific* **82**, 957.
Plavec, M.: 1970b, in A. Slettebak (ed.), *Stellar Rotation*, D. Reidel Publ. Co., Dordrecht-Holland, p. 133.
Plavec, M.: 1973, in A. H. Batten (ed.), 'Extended Atmospheres and Circumstellar Matter in Spectroscopic Binary Systems', *IAU Symp.* **51**, 216.
Plavec, M. and Horn, J.: 1969, in M. Hack (ed.), *Mass Loss from Stars*, D. Reidel Publ. Co., Dordrecht-Holland, p. 242.
Plavec, M., Polidan, R. S., and Peters, G. J.: 1973a, A paper presented at the 141st meeting of AAS.
Plavec, M., Ulrich, R. K., and Polidan, R. S.: 1973b, *Publ. Astron. Soc. Pacific* **85**, 769.
Polidan, R. S., Peters, G. J., and Plavec, M.: 1973, A paper presented at the 141st meeting of AAS.
Prendergast, K. H. and Taam, R. E.: 1974, *Astrophys. J.* **189**, 125.
Ringuelet, A. E. and Machado, M. E.: 1974, *Astrophys. J.* **189**, 285.
Struve, O.: 1931, *Astrophys. J.* **73**, 94.
Struve, O.: 1944a, *Astrophys. J.* **99**, 89.
Struve, O.: 1944b, *Astrophys. J.* **99**, 75.
Thackeray, A. D.: 1971, *Monthly Notices Roy. Astron. Soc.* **154**, 103.
Van den Heuvel, E. P. J. and De Loore, C.: 1973, *Astron. Astrophys.* **25**, 387.
Walker, M. F.: 1953, *Astrophys. J.* **118**, 481.
Wilson, R. E. and Stothers, R.: 1975, *Monthly Notices Roy. Astron. Soc.* **170**, 497.
Woolf, N. J.: 1962, *Monthly Notices Roy. Astron. Soc.* **123**, 399.

DISCUSSION

Cowley: You present a model which very nicely fits observed characteristics of some stars, but what is the evolutionary history of these systems? In particular, what are the counterparts of these binaries in earlier evolutionary stages, what are their masses and periods, and do we observe them?

Harmanec: Present Be binaries originate, according to our suggestion, from normal main-sequence binaries with relatively short periods. A system consisting of $5 M_\odot + 2 M_\odot$ with an orbital period $P = 2$ days may serve as an illustrative (but surely not the only possible) example. For a great variety of initial masses, mass ratios, and orbital periods, case B mass exchange leads to the formation of a wide system with a mass-gaining star of spectral type B or A enclosed by an extended envelope, i.e., to a Be binary as we understand it. Many 'detached' binaries from Kopal's 1956 catalogue are probable progenitors of future Be binaries. During subsequent evolution, more massive components of these systems expand to their Roche limits and case B mass exchange begins. The first part of the process is very rapid and we have little chance to observe it. Yet, it seems that the recurrent nova T CrB and the eclipsing binary UX Mon may be examples of such systems; their mass-losing components are the more massive components in both cases, which indicates that they are at a very early stage of mass exchange. Continuing mass transfer then leads to the formation of typical semi-detached binaries (U Cep may probably serve as a good example) which represent an evolutionary phase immediately preceding the Be stage. It thus seems that we do observe the counterparts of present Be binaries in all earlier evolutionary stages.

Hutchings: If the ring circularizes over some decades one should see the V/R variations damping in amplitude. What we in fact seem to see is a constant amplitude which suddenly begins and suddenly ends.

Harmanec: I think that observations of long-term V/R and velocity variations do indicate damping in amplitude before the variations cease; see, for example, the cases of HD 20336 or β^1 Mon.

With regard to Plavec's and Peters' doubts regarding the reality of long term V/R variations, I agree completely that with scattered observations you can fail to detect a shorter-period variability. On the other hand, I am convinced that in at least some cases (25 Ori, β^1 Mon, ζ Tau, and 48 Lib) the long-term variations are well documented and real. In my opinion, they differ in their origin from shorter and periodic V/R variations, related to binary motion.

Heap: How did you get your radial velocity curve for the star, ζ Tau? What lines did you use?

Harmanec: The hydrogen shell lines and other shell lines (i.e. silicon).

Heap: What is the inclination of the orbital plane of ζ Tau with respect to the line of sight?

Harmanec: I don't know. The star is not known to be an eclipsing binary; however, no systematic photometry of the object has been performed.

Conti: I agree that you have shown that in at least some cases, a Be star can be the result of a 'Case B' mass exchange. However, there seem to be two difficulties with concluding that all or many Be stars are of this type.

First: the fraction of Be stars is $\simeq 20\%$ of all B stars. The fraction of Be-type binaries is about $\frac{1}{3}$ to $\frac{1}{2}$. So the hypothesized post mass-exchange time scale is $\simeq 60$–40% of the lifetime of the pre mass-exchange time scale. This seems much too long for a disk to remain about a mass-gaining star, since it is a significant fraction of its remaining lifetime.

Second: There seem to be more early B-type Be stars than other types, whereas the mass-exchange hypothesis would seem to produce no dependence of frequency with spectral type. In particular, there should be A-type emission-line main sequence stars. But we do not observe such late stars.

Harmanec: I must confess I do not believe that the present data on the relative number of Be stars and of spectroscopic binaries among B stars are very meaningful. But even if we accept them, the reasoning concerning lifetimes is not as simple as you suggest. It is necessary to take into account the large differences in evolutionary lifetimes along the main sequence. Because during any case B mass exchange the original mass-ratio is more than reversed, the relative fraction of mass-exchanging binaries with primaries of a given spectral type among all Be stars of this spectral type is enhanced. We have published (Kříž and Harmanec, 1975) simple statistical computations which indicate that the observed percentage of Be stars can be explained in this way. I cannot answer your second objection precisely but it seems again that the different evolutionary lifetimes should be considered.

My present position is the following: we have good observational and theoretical arguments to believe that many Be stars are in fact interacting binaries. We therefore suggest serious consideration of the possibility that the Be phenomenon is always caused by mass exchange, along with other existing hypotheses. This assumption may serve as a good working hypothesis to stimulate further investigation of Be stars.

Doazan: What are the final criteria used to determine whether a particular Be star is a binary or not?

Harmanec: To detect a secondary component, either directly in the spectrum, or indirectly, from photometry in the case of eclipsing binaries, or by means of lunar occultations or possibly by speckle interferometry in other cases. Otherwise the evidence is necessarily indirect.

Goldberg: Can you give an estimate of the angular separation of the binaries?

Harmanec: I have not computed this, but it can be estimated from the fact that the typical separation of the components is several hundred solar radii.

Goldberg: I am just wondering whether they are large enough so that there is even a possibility of observing them by speckle interferometry. You cannot do any better than the theoretical resolving power of the aperture. For the Kitt Peak 4-m telescope using visible light this is about $0''.03$.

Cowley: If, on the main sequence these stars were of approximate masses 5 $M_\odot$ and 2 $M_\odot$, and the Be phenomenon is a result of Case B mass transfer from the 5 $M_\odot$ star, then one should observe no Be stars in clusters in which there are stars with $M > 5\ M_\odot$ still on the main sequence.

Harmanec: I used 5 $M_\odot + 2\ M_\odot$ as an illustrative example; case B mass exchange in a system with, say, 12 $M_\odot + 7\ M_\odot$, leads to a similar result. Moreover, P. Hintzen and J. Scott (preprint) consider seriously the possibility that the evolutionary ages of individual stars in a particular cluster need not be the same.

Thompson: The variations of Hα reported in κ Dra by Hutchings *et al.* (*Astrophys. J.* **170**, L73, 1971) are probably partly instrumental effects, principally the instability of the scanning raster of the Isocon camera, used in those observations. Hutchings reports as yet unpublished observations of Hα in κ Dra obtained with a Reticon detector (which is not susceptible to these instabilities) which confirm these rapid variations. Further observations of this type are obviously very important.

Hutchings: As I mentioned in my talk, I am aware of the pitfalls in detecting these effects. However, there are now sufficient independent and different experiments to convince me that they are real.

ON THE DETECTION OF BINARY Be STARS

R. S. POLIDAN

Dept. of Astronomy, University of California, Los Angeles, Calif., U.S.A.

Abstract. Possible methods for detecting the presence of a cool companion to a Be star are discussed. Photometric observations are shown to be incapable of detecting companions in all but the most extreme cases. Spectroscopic investigation is also unlikely to yield many new discoveries. It, however, remains the most promising method for the detection of binary Be stars. The four known binary Be stars are also discussed.

Infrared calcium triplet emission in Be stars is discussed in detail. The lines are shown to originate in a region of large optical thickness and low temperature ($T \sim 5000$ K). The possible connection between the presence of calcium triplet emission and binary nature is briefly discussed.

1. Introduction

If one wishes to discuss the possibility that some Be stars are interacting binary stars it becomes necessary to define in as exact terms as possible the physical properties of the system. As pointed out by Dr Harmanec, mass transfer calculations have not yielded a unique model for Be binary stars. It is possible to obtain a wide variety of systems by simply changing the starting conditions or the stage of mass transfer in which the binary is viewed. There is, however, a specific type of system which we at Los Angeles feel is representative of many of the Be binary stars. This is a system that bears a close resemblance to the semi-detached eclipsing systems collectively known as Algols. In the case of Be stars the brighter component (primary) is an early-type (B0 to A0) main-sequence band object surrounded by an extensive ring of gas. The cooler companion (secondary) is a star of spectral type G or K (possibly M) and of luminosity class III or IV (in some cases II). The physical size of the system must be larger than that of a typical Algol system in order to accommodate the large disk of gas. Periods greater than 10 to 20 days will satisfy this requirement. An upper limit to the period will be set by the condition that mass transfer must take place; this in principle can be 1000 days or more. However, periods in excess of a few hundred days are expected to be rare. Regarding masses, the secondary is expected to show evidence of mass loss: a very low mass for its position in the HR diagram. The primary would be expected to have a mass typical for its location in the HR diagram. It also may have a slight overluminosity due to the accretion of mass from the disk; this is expected to be significant only in the early stages of mass transfer. Typically, mass ratios in the range 5 : 1 to 15 : 1 in favor of the primary are expected.

This definition of a Be binary star still has a considerable range of physical parameters. It does, however, allow discussion of some of the observable properties of these systems. In particular it allows the question of detectability to be discussed.

2. Detectability

The question of the detectability of a companion to a Be star must be answered on many fronts. A companion can be detected through photometric anomalies in the Be

A. Slettebak (ed.), Be and Shell Stars, 401–415. All Rights Reserved.

star (eclipses or a composite flux distribution), radial velocity variations indicative of orbital motion, periodic activity in the spectrum of the star, and perhaps a few others. Since we are considering a hot star combined with a cooler larger star photometric detection would seem to be the most promising way to search for a secondary. In particular the recent advances in far infrared ($\lambda > 2\ \mu$) photometry should aid in the detection of the red component. The presence of infrared free-free emission in Be stars should reduce the probability of detection, in some cases making impossible the detection of the companion by photometric means. If we look at the Be stars with known companions to resolve this question we find a surprising and somewhat unsettling answer. Allen (1973) has obtained infrared (HKL) observations of the Be binaries for which we detect the presence of a companion in the near infrared. If we look at the stars with M-type companions, 17 Lep (Slettebak, 1951; Cowley, 1968) and XX Oph (Lockwood et al., 1975), we find that the companion is photometrically visible. Both are classified by Allen as having infrared colors similar to those of late type (i.e. M) stars. Allen has also observed the star BD + 14°3887. An infrared spectrum obtained of this object this summer shows a spectrum very similar to that of XX Oph, possibly even to the faint M-star absorption spectrum that is seen in XX Oph. Allen also photometrically classifies this star as having infrared colors similar to that of an M-type star. Thus, it would seem that detection of M-type companions is possible from infrared photometric observations. However, the model for the binary Be star suggested above states that a G- or K-type star is a much more likely secondary component than an M-type star. Fortunately, Allen has obtained HKL colors of the three Be stars known to contain G or K-type companions: AX Mon, HD 218393, and HR 894 (HD 18552). All three were found to show no evidence of a companion; the infrared colors could be represented by an early-type continuum plus emission from an optically thin ionized hydrogen envelope. This is in spite of the fact that at λ 8500 the companions contribute respectively 65%, 40%, and 25% of the total light of the system. This pronounced invisibility of the K-type companion becomes even more apparent if one looks not at a classical Be star (though the star does show very weak emission at Hα) but at the eclipsing binary star β Per (Algol). This may appear to be an unfair comparison, but using the results of Hill and Hutchings (1970) for the three components of Algol it is found that the B-type star contributes only 60% of the total flux at any point beyond approximately 2 μ. Of the remaining flux, 10% is contributed by the A-type component and 30%, almost $\frac{1}{3}$, is contributed by the K-type subgiant. This binary is then quite representative of what is expected for a Be binary star. Algol also has high quality photometric observations available from 0.36 μ (U) through 2.2 μ (K) (Johnson et al., 1966) and from 2.3 μ to 11.4 μ (Gerhz et al., 1975). Gerhz et al. used Algol as a representative normal B8V star in their 2.3 μ to 19.5 μ photometric study of Be stars. They also observed the truly normal unreddened B8V star β Lib as did Johnson et al. A comparison of these two B8V stars gives a surprising and as yet not completely understood result. NO significant difference is found between the flux distributions of β Per and β Lib that cannot be accounted for by reddening ($E(B-V)$ for β Per is 0.04). The 2.3 μ to 11.5 μ distributions are significantly different only at the 5 μ point. What is most surprising is that the 8-color observations of Johnson et al. reveal no infrared excess, even through the K-type secondary contributes fully 30% of the total light at the K (2.2 μ) magnitude.

The results of Gerhz *et al.* should not be entirely unexpected. In the region beyond 2 microns all three components are on the Rayleigh-Jeans portion of their energy distributions. Hence, the distribution depends only on the ratio of their temperatures and is, therefore, independent of wavelength. This is the primary reason for the failure of far infrared photometry to detect the companions in AX Mon, HD 218393, HR 894, and in β Per. Far infrared photometric observations cannot distinguish differences in the distributions of flux of stars hotter than approximately 3600 K.

The results of the observations of Johnson *et al.* are much more difficult to understand. From the parameters obtained by the solution of the light curve of Algol (Hill and Hutchings, 1970) a $V - K$ excess in excess of 0.4 mag. is expected. The observed $V - K$ excess is only 0.13 mag., very nearly what is expected from reddening and the observed $B - V$ excess of Algol. The reason for this lack of excess is unknown. It is unlikely that the cause is uncertainty in the physical parameters, radii and temperatures, of the components of Algol. More likely is our lack of knowledge of the effects of mass loss or the proximity of the critical Roche surface on a stellar atmosphere.

The conclusion that one is forced to accept based on the above observed results is that M-type companions to Be stars should be easily detected photometrically, but the expected most common type of secondary, a G or K-type star, is photometrically invisible.

One other photometric method of detection should at this point be mentioned: eclipses. If we have binary Be stars, then we should observe some of them to undergo eclipses. The question is one of what is the probability of an eclipse occurring in such a wide system and, with only scattered observations, what is the probability that the eclipse will be detected. It must be mentioned that many eclipsing 'Be' stars do exist: they are usually called Algol-type eclipsing binary stars. As Plavec pointed out in his introductory address to this Symposium many Algol systems have emission spectra quite similar to those found in traditional non-eclipsing Be stars. The extensive spectroscopic observations that exist for Be stars do not, in general, exist for eclipsing binary stars. Therefore, little can be said about any other similarities between these two groups other than general appearance. Similarly Be stars generally do not have the extensive photometric coverage that many binary stars have. Certainly, none of the brightest Be stars could undergo an eclipse without being detected, but for fainter Be stars, objects for which observations are infrequent at best, eclipses could possibly be completely missed. The problem of eclipse probability and detection probability will be discussed in detail by Plavec later in this session.

Turning now to the spectroscopic methods for the detection of companions to Be stars one first thinks to look for radial velocity variations caused by orbital motion. Unfortunately, the low expected semi-amplitude (less than 10 km s^{-1} in many cases), the poor, rotationally broadened lines, and the possible contamination of the line by emission or absorption from the envelope make this method unlikely to yield many unchallenged discoveries. It can, however, be very helpful in finding the shorter period, 10 to 20 day, Be binary systems. If the eclipsing binary AU Mon was inclined to our line of sight such that it would not eclipse, it would appear simply as a B5 emission line star that underwent periodic radial velocity variations with a period of approximately 11 days. This would probably be the only way that a system similar to this could be identified as a binary star.

Another means of discerning the binary nature of a Be star is through the detection of short term (days to months) periodic spectral changes possibly associated with the period of the system. Cowley has shown that AX Mon (Cowley, 1964) and 17 Lep (Cowley, 1968) display a mildly periodic activity associated with the orbital period. Peton (1974) in his exhaustive study of AX Mon established a definite periodic activity associated with the orbital period. He also discovered shorter and longer term variations associated with changes in the structure of the envelope. Doazan and Peton (1970) observed periodic velocity variations of the shell lines and profile variations of the $H\beta$ line in the Be binary star HD 218393. Peters (1972) discovered periodic spectral changes in the Be star HR 2142. These systems, in particular HR 2142, will be discussed along with other stars showing periodic or quasi-periodic spectral variations by Peters in the next paper in this session.

One other method for detecting spectroscopically the presence of a compansion is through the discovery of spectral lines incompatible with the photospheric and envelope spectrum of the Be star. Since the companion is expected to be considerably cooler than the B-type star we would, ideally, like to look as far into the red as possible. Unfortunately, a good low noise detector is required to see the weak lines. This limits us to regions shortward of approximately one micron. Because of the distribution of energy in a late-G early-K-type star, regions shorter than roughly $H\alpha$ are unlikely to yield many discoveries. Thus, the current optimum spectral region for searching for companions to Be stars appears to be between λ 7000 Å and λ 10 000 Å.

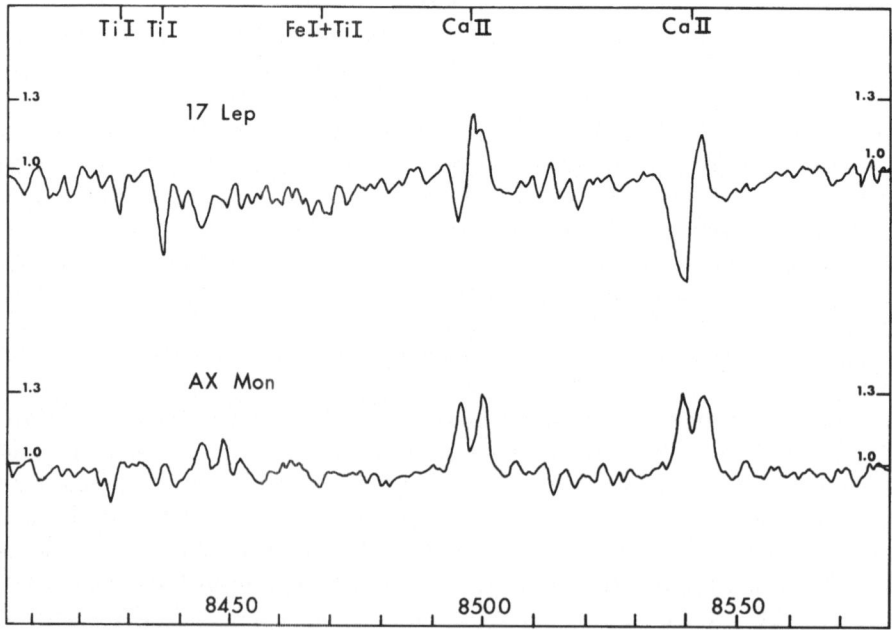

Fig. 1.   A portion of the near infrared spectra of AX Mon and 17 Lep. The most prominent lines of the companions are marked. Ordinate is in units of intensity normalized to the continuum. Original dispersion: 23 Å mm^{-1}.

The most likely candidate lines to search for would be lines of CN. They are strong in late type stars and cannot arise either in the photospheric or envelope spectrum of the Be star. The strongest band of CN has its head at λ 10 900 Å. This is, however, very near the center of the atmospheric Φ band of water. A weaker band with its head at λ 9150 Å is also contaminated by atmospheric water but less severely than the λ 10 900 Å band. A third, even weaker, band, actually a pair of bands, exists near λ 8200 Å. This too is badly contaminated by atmospheric water.

Strong atomic lines are not uncommon in the infrared spectra of late type stars. The infrared calcium triplet at $\lambda\lambda$ 8498, 8542, and 8662 Å is particularly strong. The region surrounding the triplet also contains many weaker lines of Fe I and Ti I, lines that are not likely to be found in the envelope spectrum of the Be star. This region is also relatively free of photospheric features, except for the Paschen lines, and of shell features (Paschen lines, O I λ 8446 Å, N I λ 8680 Å, and in very rare cases the Ca II triplet). This is far from an ideal spectral region, however. Clearly the Ca II lines are the best lines to search for; they can, however, be severely filled with emission. This can be seen in the spectra of 17 Lep and AX Mon (Figure 1). In both cases emission completely dominates the photospheric absorption (in 17 Lep the absorption is due to the expanding shell and not the M-star photosphere). If other Be binary systems are similar to these two objects, then possibly the Ca II emission itself could be helpful in detecting Be binary systems. Before we explore this possibility let us turn our attention to a comparison of the four established Be binary stars.

3. Binary Be Stars

Four Be stars are known to be binary systems containing a cool, G, K, or M-type component: AX Mon, 17 Lep, HD 218393, and HR 894 (HD 18552). The spectroscopic and photometric properties of these four systems are summarized in Table I.

TABLE I

Binary Be stars

	17 Lep	AX Mon	HD 218393	HR 894
Spectrum	B9V + M2III	B0.5 + K2II	B3 + K1III	B8V + gG9:
Secondary Contribution:				
Hα	50%	50%	20%	10%
λ 8500 Å	75%	65%	40%	25%
V Magnitude	4.9	6.8	6.8	6.1
HKL Colors[a]	Late-type star	Free-Free	Free-Free	Free-Free
Emission Spectrum	Variable shell	Variable emission and shell	Variable emission and shell	Constant emission (shell?)
Radial Velocity	Orbit[b]	Orbit[c]	Variable[d]	Variable[e]

[a] Allen (1973).
[b] Cowley (1968).
[c] Cowley (1964).
[d] Doazan and Peton (1970), Kříž and Harmanec (1975).
[e] Plaskett et al. (1920).

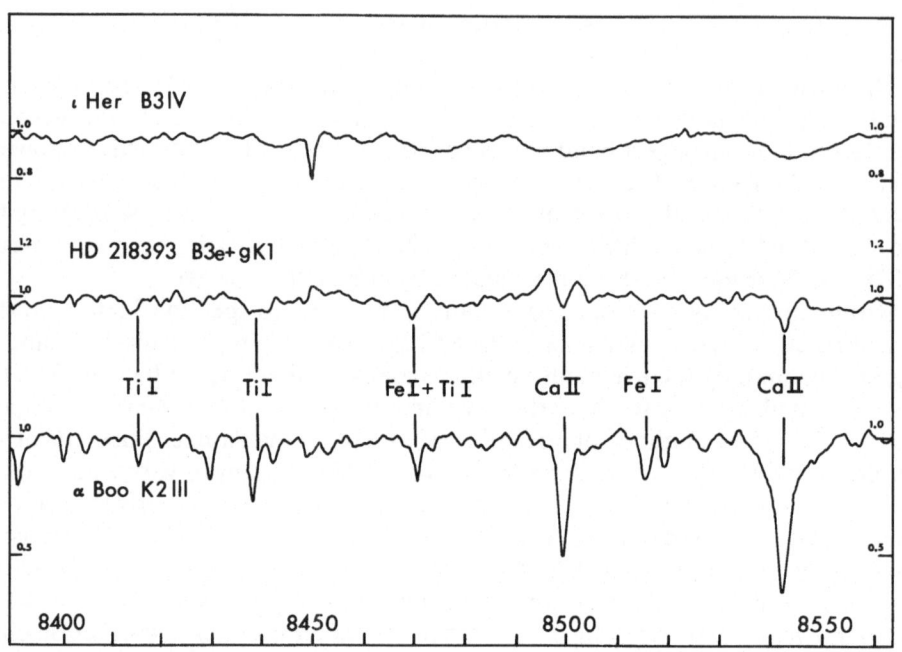

Fig. 2. The infrared spectrum of HD 218393 compared with that of ι Her (B3IV) and α Boo (K2III). The most prominent lines of the companion are marked. Ordinate and original dispersion the same as for Figure 1.

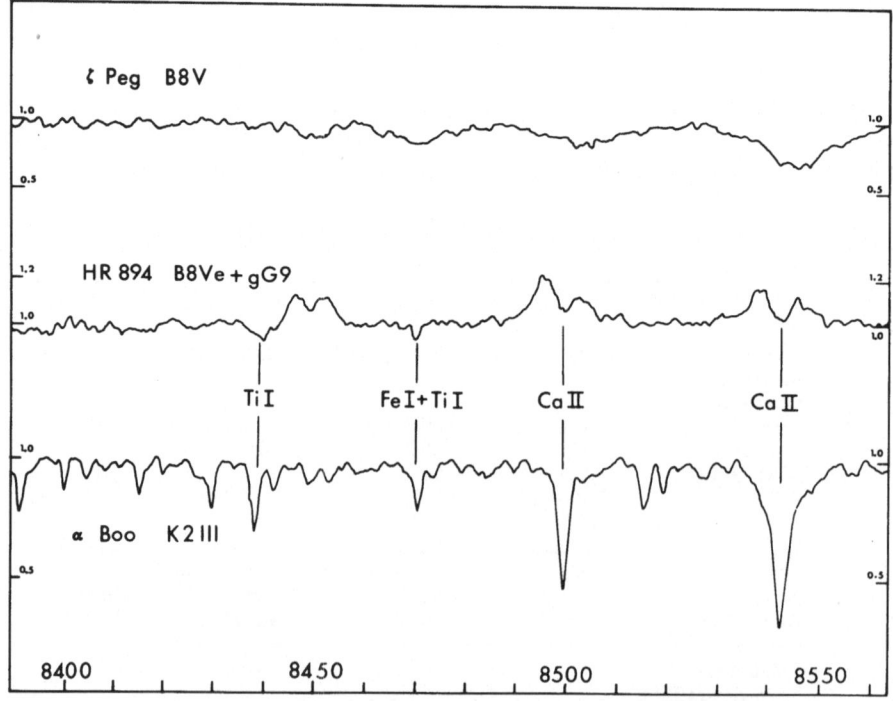

Fig. 3. The infrared spectrum of HR 894 compared with that of ζ Peg (B8V) and α Boo (K2III). The most prominent lines of the companion are marked. Ordinate and original dispersion the same as in Figure 1.

The near infrared spectra of AX Mon and 17 Lep are dominated by the cool component of the system (Figure 1). In both systems strong emission is observed at the calcium triplet. Radial velocity measurements of these lines indicate that the emission is associated with the cool component and not with the emission envelope. In 17 Lep strong shell lines severely cut into the calcium emission. The shell lines do not follow the orbital motion of the M-type star but rather remain at the velocity of the expanding shell that surrounds the whole system. Except for these anomalies both companions appear quite normal. The two newer discoveries, HD 218393 (Figure 2) and HR 894 (Figure 3), have considerably less dominating companions. Here, too, both systems display emission at the calcium triplet. It does not, however, dominate the photospheric absorption of the secondary. Very preliminary radial velocity measurements indicate that in both systems the emission may remain at a fixed velocity, the systemic velocity. HR 894 represents the probable limit of detectability for our equipment. The companion is visible only at the calcium triplet and the strongest iron and titanium lines.

Only one common characteristic is found in the infrared spectra of these Be binary stars: emission at the calcium triplet.

4. Infrared Ca II Emission

As part of a general spectroscopic investigation of Be stars and related objects in the near infrared over 350 spectrograms in the region of the infrared calcium triplet have been obtained. In total over 120 different objects have been observed. The full program and a description of the equipment used was reviewed earlier in this Symposium. A dispersion of 23 Å mm^{-1} was used for most objects. All spectra were widened to at least 0.6 mm.

The results of this investigation show that approximately 20% of all early type emission line stars have calcium triplet emission. If one restricts the observations to only what would be termed 'classical' Be stars only 15% show calcium triplet emission. Table II lists the stars for which we find emission at the calcium triplet and the strength of the emission. Figure 4 shows the region of the first two calcium lines in three representative stars.

The infrared triplet of calcium ($\lambda\lambda$ 8498, 8542, and 8662 Å) is severely blended with Paschen lines P16, P15, and P13. This blending has caused earlier investigators to overestimate the frequency of occurrence of calcium triplet emission in Be stars. On lower dispersion plates, such as our 33 Å mm^{-1} plates and those of earlier investigators, weak Ca II emission appears to be present in stars showing either a strong Paschen shell spectrum or strong Paschen emission. However, higher dispersion plates taken at the same time do not confirm the presence of the calcium emission, they show only Paschen emission.

The presence or strength of infrared calcium triplet emission is found not to correlate with any obvious parameter of the underlying star or the emission envelope. All B-type spectral sub-classes are equally represented. All types of emission envelopes are represented: strong emission (HD 7636), weak emission (HR 894), early-type shell (ϕ Per), and late-type shell (HD 193182). All degrees of

TABLE II
Infrared Ca II emission stars

Object	Spectrum	$\dfrac{\text{Ca II}}{\text{Continuum}}$	Comments
A. Be and Shell Stars			
HD 7636	B2V	1.25	
ϕ Per	B0.5	1.38	Binary $P = 127$ days
HR 894	B8V + gG9:	1.22	Binary
ψ Per	B5	1.20	
28 Tau	B8V	1.37	
48 Per	B3V	1.1:	
HR 2142	B1V	1.38	Binary $P = 81$ days
17 Lep	B9V + M2III	1.17	Binary $P = 260$ days
AX Mon	B0.5 + K2II	1.30	Binary $P = 232$ days
66 Oph	B2V	1.1 :	
HD 173219	B0:	1.2 :	Binary $P = 55$ days
HD 193182	A0V	1.20	
υ Cyg	B1.5V	1.21	
π Aqr	B0:	1.31	
HD 218393	B3 + K1III	1.12	Binary $P = 39$ days
B. Peculiar Emission Line Stars Related to Be Stars			
KS Per	He Star	1.2 :	Binary $P = 360$ days
3 Pup	A2Ib	1.60	Binary $P = 138$ days
XX Oph	B0III + M6III	>2.0	Binary
HD 163296	A0V	1.55	
W Ser	Pec	1.1 :	Eclipsing Binary $P = 14$ days
υ Sgr	He Star	2.00	Binary $P = 138$ days
BD +14°3887	Pec	>2.0	Binary?
HD 190073	Pec	>2.0	
V367 Cyg	B8Ia	1.58	Eclipsing Binary $P = 19$ days

activity are represented, from the very stable (HD 193182) to the very active (π Aqr). A wide variety of rotational velocities are also represented. No correlation was found with any photometric property of the star. Color excesses range from $0^{m}.40$ to essentially zero. The far infrared flux distribution typically shows only the expected free-free emission. Here, again, the range is from little or no excess to a reasonably large excess. Some of the peculiar emission line objects show evidence of extensive dust shells; most, however, do not.

4.1. LINE PROFILES AND STRENGTHS

The calcium triplet line strengths in classical Be stars range in peak intensity from approximately 10% above the local continuum to almost 40% above the local continuum. In the more peculiar Be stars peak intensities in excess of twice the local continuum are not uncommon. The line widths are also quite large: half widths smaller than 350 km s^{-1} are rare. The calcium triplet line widths also do not correlate either with the widths of other emission lines in the spectrum or the $v \sin i$ of the underlying star.

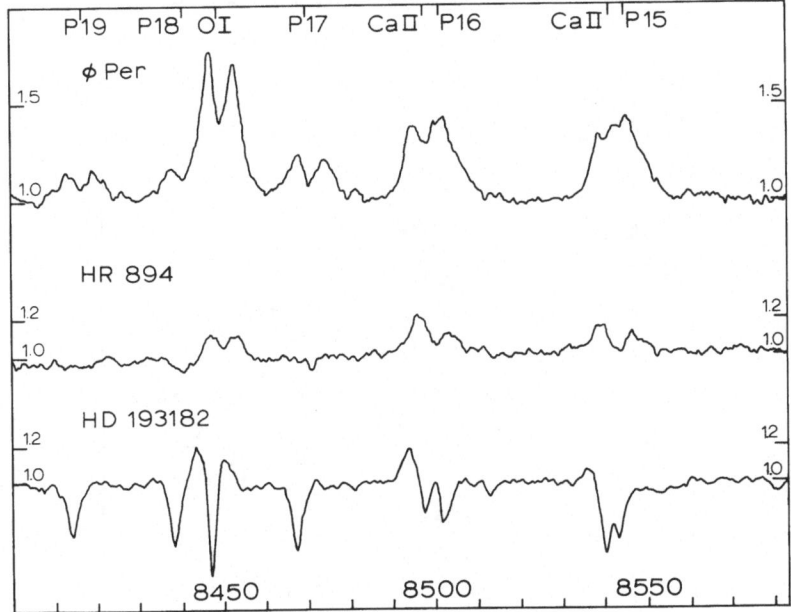

Fig. 4. The region of the first two calcium lines in three calcium triplet emission stars. The shell star HD 193182 also has a strong Ca II shell spectrum. Ordinate and original dispersion the same as for Figure 1.

The expected relative line strengths for the triplet lines are 1 : 9 : 5. This is *not* observed in any case; all show an essentially 1 : 1 : 1 intensity ratio. The atomic parameters for these transitions are sufficiently well known that the observed difference cannot be attributed to uncertain physics. The explanation for this anomaly appears to be that the calcium triplet emission in Be stars is very optically thick. This is further supported by the observation that after the removal of Paschen contamination most stars, in particular the classical Be stars, show strong reversals at the line centers; in many the reversal drops very nearly to the continuum. In no case has the reversal been observed to extend below the continuum. Line widths are found to correlate with the strength of the central reversal as expected from optical depth effects. Thus, the inevitable conclusion is that in Be stars the infrared calcium triplet emission arises from a region of large optical thickness.

4.2. Ca II TRIPLET AND THE H AND K LINES

The large optical thickness of the Ca II infrared triplet is difficult to explain in the context of the classical model of a Be star. This problem, however, is surpassed by an even larger problem. Figure 5 shows a partial Grotrian diagram for Ca II. The lines of the infrared triplet arise from the same upper levels as do the H and K lines of calcium. The ratios of the transition probabilities are well known and strongly in favor of the H and K lines. However, since we have optically thick lines the relative intensities will go as the relative B_ν's and not the relative transition probabilities.

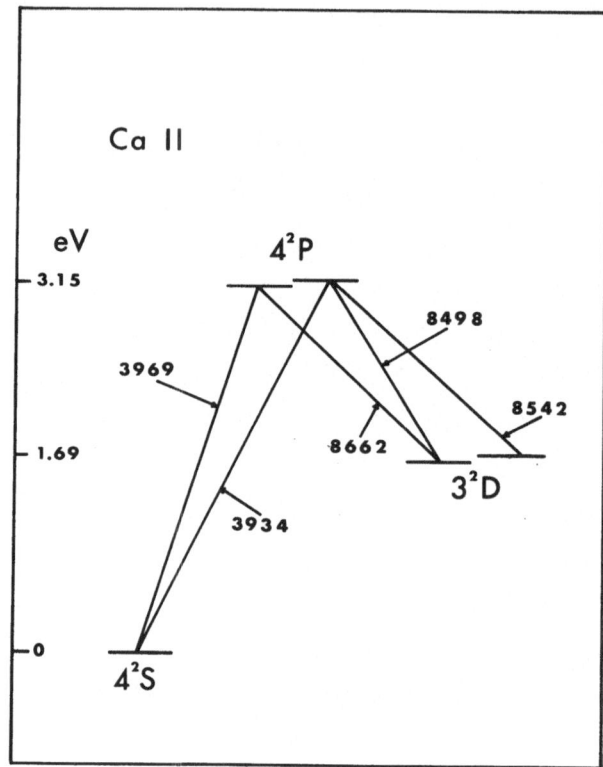

Fig. 5.   Partial Grotrian diagram for Ca II.

The line strengths are compared to the local continuum so allowance must be made for the stronger continuum in the region of the H and K lines. If this is done then the predicted emission line strengths for the H and K lines in the classical Be stars range from 5 to 20% above the local (H and K) continuum, assuming typical Be envelope temperatures (10 000–15 000 K). However, in all cases *NO* significant H and K emission is observed. No anomaly of any kind, emission or absorption, is observed at the H and K lines of the Be stars showing calcium triplet emission. Photospheric absorption, interstellar gas, and in some cases (shell stars) circumstellar gas can account completely for the violet lines of calcium. Figure 6 illustrates the K line and the λ 8498 Å infrared line (both arise from the $4p\ ^2P_{3/2}$ level) in three classical Be stars and one peculiar emission star (HD 163296).

4.3. Variability

The question of the permanence of calcium triplet emission must now be discussed. This is not easily accomplished, as investigations of Be stars in this region are quite rare. A comparison of our data with the published spectral prints of the previous investigators shows only that major changes have not occurred, except in the case of κ Dra, which will be discussed later. Houziaux (1962) has published an intensity

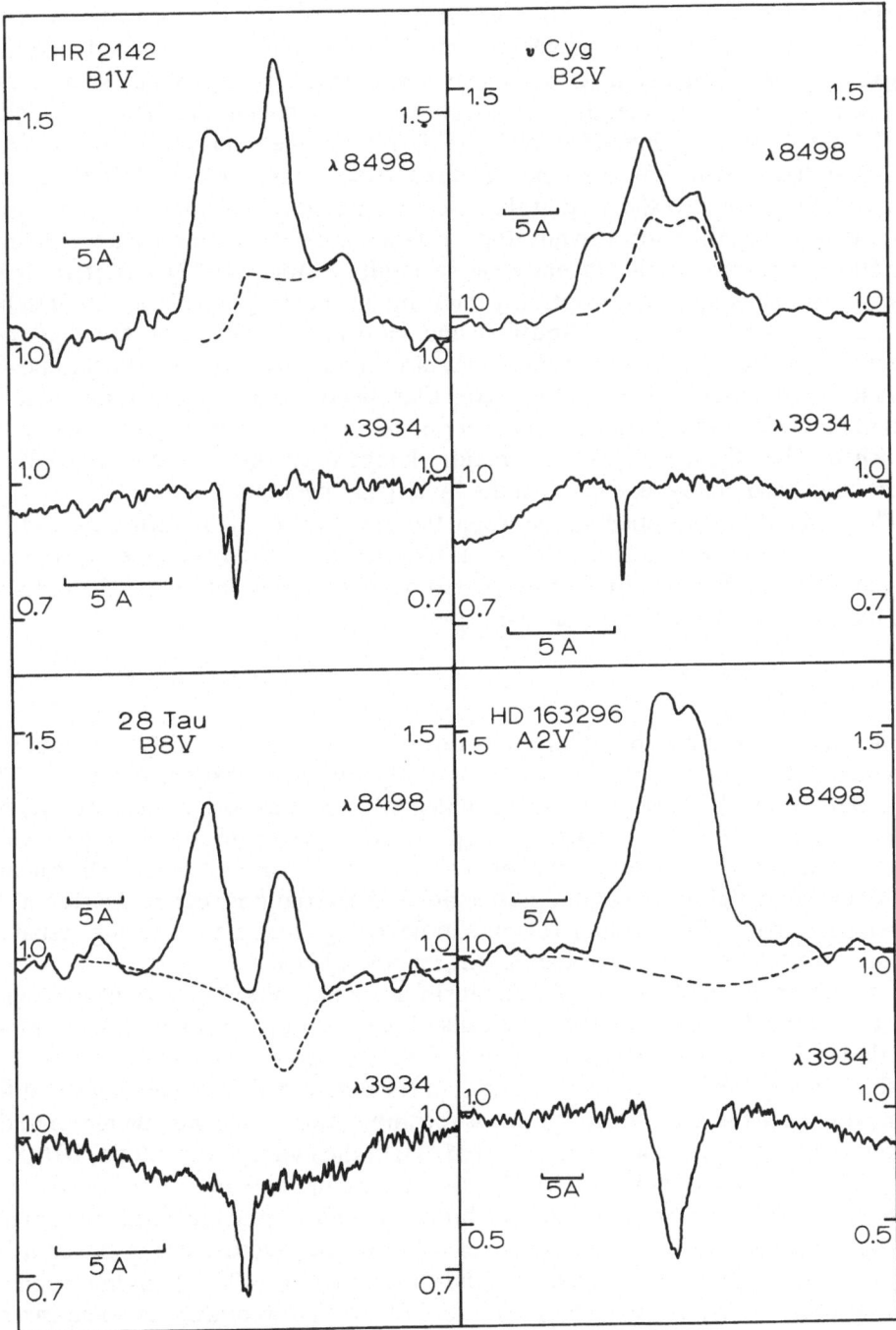

Fig. 6.   Comparison of the λ 8498 Å infrared line to the λ 3934 Å (K) line in four calcium emission stars. Ordinate is in units of intensity normalized to the continuum. The profile of P16 is indicated by the dashed line.

tracing of 28 Tau during one of the star's strong shell phases. We have observed this star during the past three years while its shell has strengthened, declined, and then began to strengthen again. If a comparison is made between our spectra and Houziaux's 1962 spectrum, after removal of the very strong Paschen and calcium shell lines, then one finds that *NO* significant change occurred in the calcium emission lines. Our observations of three other stars, ϕ Per, HR 2142, and HD 163296, have also shown that the calcium emission does not participate in the variations seen in the hydrogen envelope. κ Dra is the only Be star to show a definite variation of infrared calcium emission strength. Hiltner (1947) reported strong calcium emission, apparently optically thick, on his spectrogram along with Paschen and O I λ 8446 Å emission. Andrillat and Houziaux (1967) reported that their spectrum shows calcium triplet emission, also apparently optically thick, and O I λ 8446 Å, but Paschen lines in absorption. Our spectra for κ Dra are quite similar to that of Andrillat and Houziaux except for one thing: the calcium triplet emission has vanished. This, then, is a case of a major change occurring in the calcium triplet without an accompanying change in the hydrogen envelope.

This lack of correspondence between the emission lines of hydrogen and the emission lines of calcium strongly suggests that they arise in two completely separate regions. This suspicion is further supported when one looks at the mechanisms for the formation of the infrared calcium triplet.

4.4. MODELS

A model for the calcium triplet emission in Be stars must satisfy a number of conditions. The emitting region must be optically thick to all three calcium lines. The predicted H and K emission must be sufficiently small as to be undetectable. No enhancement of H and K or triplet absorption in the spectrum should be predicted. Reversals in the triplet lines should not drop below the local continuum. The calcium triplet emitting region should not participate in the variations seen in the rest of the envelope. Finally, the calcium region should not give rise to a strong visual or infrared continuum or to any other strong emission lines.

Using simple calculations, an estimate of the temperature of the emitting region can be obtained. Wellman (1951) (see also Pagel, 1960) developed a method for obtaining the volume emission measure, $VN_e N_{ion}$, for an optically thin gas. This method utilizes the total equivalent width of the emission line and assumes that the dominant mechanism of line formation is recombination. If the gas giving rise to the calcium triplet has a temperature typical of that of the hydrogen envelope, 10 000 to 15 000 K, then recombination will be the dominant line formation mechanism. Remember, however, that the triplet lines are quite optically thick. Wellman's method then can only give an estimate of the lower limit to the size of the emitting region. When applied to the weakest of the triplet lines (λ 8498 Å) this lower limit is in itself quite revealing. Very large emitting regions are required: in some cases a total mass of the envelope in excess of 10^{-3} solar masses is needed to explain the calcium line strengths. This method also predicts H and K emission much stronger than that allowed by the observations. Recombination processes in a hot gas appear unable to even approximately explain the observations.

A reduction of the gas temperature below 10 000 K, specifically to the 5000–6000 K range, will alleviate both the problem of the line strengths and the absence of strong H and K emission. The calcium triplet has its largest emission per unit volume in this temperature regime. Also, since the line intensities will go as the relative B_ν's, this temperature regime will strongly favor the triplet lines over the H and K lines. This reduction of H and K emission, however, is still insufficient to explain the complete absence of emission at H and K. A further mechanism is required to reduce the H and K strength to below the limit of detectability.

The first attempt to identify the cause of this reversal was made by Wyse (1941). Merrill (1934) had observed infrared calcium emission in long-period variables when the H and K lines showed only absorption. Wyse suggested that selective photoionizations by $L\alpha$ photons, or in the case of the Be stars the far ultraviolet continuum shortward of $L\alpha$, depopulated the lower ^2D level thus causing the intensity reversal. Hiltner (1947) found that this explanation could not be reconciled with his observations. With this we completely agree. Our observations are incompatible with Wyse's model. The complete lack of correlation of either the presence or the strength of calcium emission with any attribute of the underlying star or the ionized hydrogen envelope strongly suggests that selective photoionization cannot explain the observed results. It no doubt plays an important role in the line formation process, but it cannot be the sole cause of the intensity reversal.

Herbig (1952) and Kraft (1957) suggested an alternative mechanism to explain the reversal in long-period variables that may have applicability to Be stars. They pointed out that self-absorption by cooler layers of gas outside the emitting region can explain the reversal of intensities. The H and K lines will be strongly affected by the cooler gas, whereas the triplet lines will, because of their much lower opacity ($f_{3934}/f_{8498} \simeq 75$), pass through the gas essentially unaffected. If indeed in Be stars the calcium emission arises in a region of gas temperature 5000–6000 K, it would not be surprising to find a small amount of cooler ($T \sim 4000$ K) gas. Optical depths to the H and K lines in this region would need to be in the range $\tau \simeq 1$–2 to reduce the H and K emission below the limit of detectability.

If this model is correct for those Be stars displaying calcium triplet emission, then it requires the existence of a two-component envelope: a hot, $T \sim 10\,000$ K, ionized hydrogen disk plus a cool ($T \sim 5000$ K) region that shows little variation and does not lie in our line of sight to the Be star. How this not insignificant amount of unionized gas can be maintained in the presence of a hot B-star is as yet not understood.

Throughout this discussion it has been stressed that the presence of calcium triplet emission does not correlate with any attribute of the Be star. This is not entirely true: calcium triplet emission appears to correlate with binary nature. Over one-half of the stars listed in Table II are confirmed binary stars or are strongly suspected of being binary stars. Of the Be stars in our survey not displaying calcium triplet emission less than 10% are suspected of being binary stars. How the presence of a cool companion causes calcium triplet emission is not completely understood. In some cases (e.g. AX Mon) the emission appears to come from a 'chromosphere' associated with the late-type companion. In other cases (e.g. HD 218393) it appears that the emission may arise in an extensive disk surrounding the entire system, possibly being fed by the cool companion.

In the case of some of the peculiar stars (e.g. HD 163296) it is possible that the calcium emission is associated with an extensive dust cloud. The stars in Table II for which evidence has been found suggesting that they have large dust shells associated with them (roughly half the peculiar objects and none of the Be stars of Table II) have very different calcium emission lines than the non-dust shell stars. The lines are still optically thick, i.e. the intensity ratio is still $1:1:1$, but no line reversals are observed. All lines appear to have simple, or only slightly distorted, Gaussian profiles. This association of strong calcium emission and dust does not appear to apply to all stars. Swings (1973) has observed the star HD 45677 in the region of the calcium triplet and finds no trace of emission. Another parameter must be invoked to explain why some objects with strong infrared excesses have calcium triplet emission and others do not.

4.5. CONCLUSIONS

Clearly the explanation, or explanations, of the calcium triplet in emission line stars is not yet at hand. The evidence in favor of the source of the emission being relatively cool gas is strong. The origin of the gas and the mechanism by which it is maintained are unknown. The high incidence of interacting binary stars among those stars displaying calcium triplet emission strongly suggests a connection between the two phenomena. The answer to this possible connection lies in further analysis. In particular, a detailed line transfer investigation of the calcium triplet emission lines should supply a set of physical parameters for the emitting region that will allow establishment of unique models for these Be stars.

The one conclusion that is now, I believe, quite obvious is that the 15 Be and shell stars listed in Table II, many of which have been discussed at this Symposium, can no longer be considered 'classical' Be stars in the sense that they cannot be represented by the simple Struve model of a rotationally unstable Be star.

5. Summary

Unfortunately the conclusion reached regarding the detectability of cool companions to Be stars was not the one desired. Photometric detection appears to be ruled out for all but the most extreme cases. Spectroscopic detection, still the most promising way to detect companions, is also unlikely to yield many new systems. The problem is quite analogous to that of the longer period Algol systems, where, if it were not for the eclipses little evidence would exist for their binary nature. The existence of infrared calcium triplet emission in a number of Be stars is quite intriguing. Possibly it is the signature of an interacting binary Be star; certainly it is the indicator of a very unusual Be star.

Acknowledgements

I wish to thank Dr G. J. Peters and Dr M. Plavec for many helpful discussions on all topics covered in this paper. I would also like to acknowledge their assistance and that of Mr E. A. Harlan of Lick Observatory in obtaining many of the observations. I also wish to thank Dr J. L. Linsky for helpful discussions regarding the calcium triplet paradox. This research was supported by NSF grant MPS 74-04194A01 (Popper/Plavec) as part of a general project studying interacting binary stars.

References

Allen, D. A.: 1973, *Monthly Notices Roy. Astron. Soc.* **161**, 145.
Andrillat, Y. and Houziaux, L.: 1967, *J. Observ.* **50**, 107.
Cowley, A. P.: 1964, *Astrophys. J.* **139**, 817.
Cowley, A. P.: 1967, *Astrophys. J.* **147**, 609.
Doazan, V. and Peton, A.: 1970, *Astron. Astophys.* **9**, 245.
Gehrz, R. D., Hackwell, J. A., and Jones, T. W.: 1975, *Astrophys. J.* **191**, 675.
Herbig, G. H.: 1952, *Astrophys. J.* **116**, 369.
Hill, G. and Hutchings, J. B.: 1970, *Astrophys. J.* **162**, 265.
Hiltner, W. A.: 1947, *Astrophys. J.* **105**, 212.
Houziaux, L.: 1962, *Publ. Astron. Soc. Pacific* **74**, 250.
Johnson, H. L., Mitchell, R. I., Iriarte, B., and Wiśniewski, W. Z.: 1966, *Comm. Lunar Planetary Laboratory* **4**, 99.
Kraft, R. P.: 1957, *Astrophys. J.* **125**, 336.
Kríz, S. and Harmanec, P.: 1975, *Bull. Astron. Inst. Czech.* **26**, 65.
Lockwood, G. W., Dyck, H. M., and Ridgway, S. T.: 1975, *Astrophys. J.* **195**, 385.
Merrill, P. W.: 1934, *Astrophys. J.* **79**, 183.
Pagel, B. E. J.: 1960, *Vistas in Astronomy* **3**, 203.
Peters, G. J.: 1972, *Publ. Astron. Soc. Pacific* **84**, 498.
Peton, A.: 1974, *Astrophys. Space Sci.* **30**, 481.
Plaskett, J. S., Harper, W. E., Young, R. K., and Plaskett, H. H.: 1920, *Publ. Dominion Astrophys. Obs.* **1**, 163.
Slettebak, A.: 1950, *Astrophys. J.* **112**, 559.
Swings, J. P.: 1973, *Astron. Astrophys.* **26**, 443.
Wellman, P.: 1951, *Z. Astrophys.* **30**, 71.
Wyse, A. B.: 1941, *Publ. Astron. Soc. Pacific* **53**, 184.

DISCUSSION

Meisel: One wonders how much the outer edge of an optically thick shell can look like a K-type star. K-type components might be expected to show the molecular CN features in the infrared if they are indeed photospheric. I presume that these have not yet been searched for in the infrared spectrum of infrared excess objects.

Polidan: I assume that you are referring to the feature at λ 9150 Å. I had hoped to observe it but it is beyond the reach of our Varo tube. There are a few weak CN lines in the water band around λ 8300 Å. I am now trying to map out this band and to search for the CN lines. In some of the stars I think I do see them.

Swings: The observers at Asiago have observed CN absorption bands in two stars, but I do not know if they are on your list.

Snijders: Two fortunate coincidences occur in the Ca II ion level structure. Firstly the strongest upward transitions from the $4P$ level are the $4P \to 4D$ lines at about λ 3700 Å. Plates obtained for the H and K lines will often cover this region also. Study of the $4P - 4D$ multiplet will be helpful for establishing both the $4P$ populations and the importance of recombination processes. Secondly the forbidden $3D \to 4S$ lines are just like the $4P \to 3D$ triplet in the near infrared. Study of the Ca II lines in Be stars would be much easier if observers systematically would give data for all four multiplets, not just the two you just discussed.

Polidan: We have looked for the forbidden $3D - 4S$ transition in all the triplet emission stars. Only in one case do we find emission, and this is the star υ Sgr. Greenstein and Merrill reported the same observation in 1946.

EVIDENCE FOR THE EXISTENCE
OF MASS-EXCHANGE BINARY Be STARS
FROM PERIODIC SPECTRAL VARIATIONS

GERALDINE J. PETERS

Dept. of Astronomy, University of California, Calif., Los Angeles, U.S.A.

Abstract. In this paper, we discuss the Be stars which show recurrent shell structure and/or cyclic variations in the profiles of their emission lines with periods near 100^d and, thus, are good candidates for interacting binaries undergoing mass-exchange. The periodic spectral variations observed in the B1e star HR 2142 are reviewed and brought up to date in the light of recent observations. HR 2142 is then compared with other periodic binary Be stars (AX Mon, 17 Lep, HD 218393, HD 173219, and ϕ Per) in order to point out the wide variance in observed spectra which interacting binaries can display.

1. Introduction

The simplest way to explain the occurrence of periodic shell structure or cyclic variations in the profiles of emission lines in Be stars is in terms of binary motion of an interacting system. As the binary presents different aspects, the column density of absorbers and the projected envelope geometry vary as a result of an asymmetrical distribution of material caused by mass streaming. In this paper, we will be concerned with Be stars which show variations with periods around 100^d and, thus, can be considered as good candidates for interacting binary systems undergoing Case B mass exchange.

In connection with periodic activity, I would like to discuss HR 2142, a B1e star whose spectrum I have studied extensively for the past six years. HR 2142 (HD 41335, MWC 133, BD $-6°1391$) periodically displays a conspicuous, short-termed, two-component shell phase every $80^d.86$. Outside of shell phase (90% of the cycle), HR 2142 very much resembles what most researchers studying Be stars would call the prototype 'classical Be star'. The photospheric features are extremely broad while $H\alpha-H\delta$ contain conspicuous emission features superposed on their respective photospheric profiles. In addition, HR 2142 has emission features of Fe II, O I λ 7774 Å, and O I λ 8446 Å which are sometimes observed in 'classical' Be stars. However, every 81^d the sequence of events shown in Figure 1 occurs. The Balmer lines develop deep shell features which persist for about 5^d, disappear, then reappear again at the same strength for the short duration of $1^d.5$. The first segment of the shell phase is termed the *primary* shell phase while the second part is called the *secondary* shell phase. The details of the shell phase of HR 2142 are discussed below. In Section 3 we will present a model for HR 2142 based upon the observations. Finally, toward the end of the paper HR 2142 is compared with some well-studied binary Be stars.

The spectrograms used in the analysis of HR 2142 were obtained with the coudé spectrograph of the 120-in. (305 cm) telescope at Lick Observatory and with the cassegrain spectrograph of the UCLA 24-in. (61 cm) telescope at Ojai, California. The Lick Observatory spectrograms were taken with both the 120-in. telescope and the 24-in. Coudé Auxiliary Telescope and include Varo image tube spectra in the

A. Slettebak (ed.), Be and Shell Stars, 417–428. All Rights Reserved
Copyright © 1976 by the IAU.

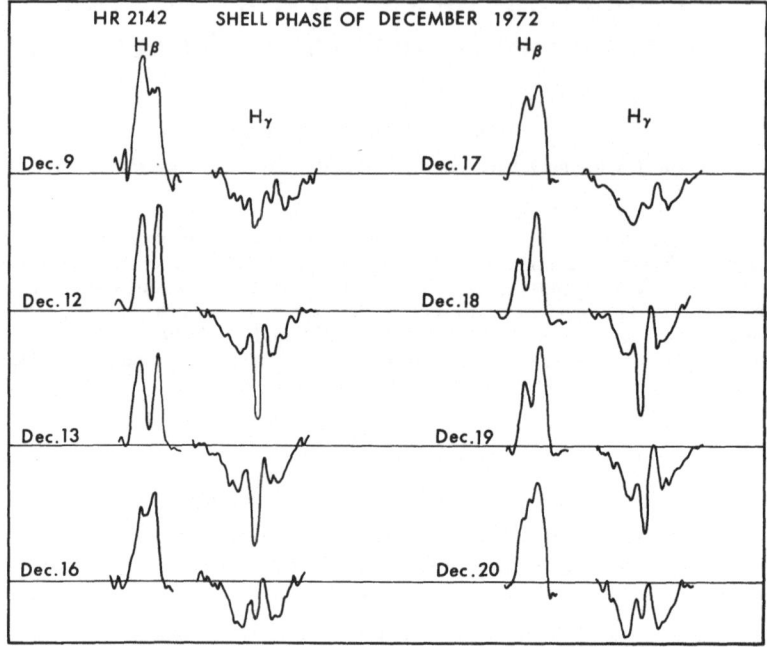

Fig. 1.   The nature of the shell phase of HR 2142 is shown by this sequence of Hβ and Hγ profiles observed in December 1972. The spectrograms (46 Å mm^{-1}) were obtained with the UCLA 24-in. telescope at Ojai, California.

region λλ 4800–8700 Å (11–23 Å mm^{-1}) and direct spectrograms from λλ 3200–4900 Å (11–16 Å mm^{-1}). The dispersion for the Ojai spectrograms is 46 Å mm^{-1}.

2. Summary of Observations to Date for HR 2142

The sequence of events which take place during the *primary* shell phase of HR 2142 is described in detail by Peters (1972). In this section, the basic features of the *primary* shell phase are reviewed, new observations are presented, and the *primary* shell phase is compared with the *secondary* shell phase. In describing the observations, we define $\Phi_s = 0.0$ as the midpoint of the interval of time in which the hydrogen cores are at maximum strength during the *primary* shell phase. The notation Φ_s is used in order to avoid confusion with phases based upon a binary orbit.

The two-component nature of the shell phase can easily be seen from the plot in Figure 2. As a measure of the varying hydrogen shell strength, we consider the change in the residual intensity of the Hγ core. Whereas shell structure in Hβ can easily be seen on high dispersion plates taken five days before $\Phi_s = 0.0$, cores do not appear in Hγ until $\Phi_s = -2^d$. The Hγ core disappears by $\Phi_s = 4^d$.

The *secondary* shell phase comes on abruptly at $\Phi_s = 5^d.5$ and lasts for only $1^d.5$. Although we have been unable to make a complete sequence of observations during the rise to maximum strength, our observations accumulated during the past three

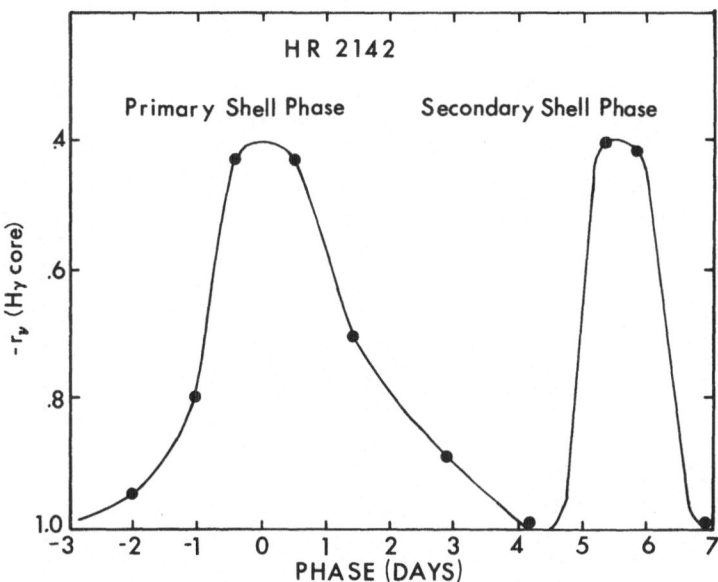

Fig. 2. The strength of the Hγ core versus phase. Note the striking two-component nature of the shell phase. The data are from Lick Observatory spectrograms.

years suggest that the Balmer cores increase from zero to the same strength which they displayed during the *primary* shell phase in about 6 h. λλ 5016 and 3965 Å of He I (which arise from the metastable 2^1S level) are seen as shell features during the *secondary* shell phase but not during the *primary* shell phase. These He I cores are -0^d5 out of phase (they appear earlier) with the H cores. Although the strengths of the Balmer shell lines are comparable during both shell phases, they appear to be sharper during the *secondary* shell phase. The profiles of the H cores are slightly broader than the instrumental ones during the *primary* shell phase. Typically, we observe cores to H14 in the *primary* shell phase and to H20 in the *secondary* shell phase.

The contrast between the radial velocities exhibited by the Balmer cores from *primary* to *secondary* shell phase is as striking as the difference in their duration. The observed effect is seen in Figure 3. The Balmer shell features (Hβ and higher) are red-shifted relative to the photospheric features during the *primary* shell phase and show a steady decrease in velocity toward the photospheric value as the shell phase progresses. The Hα core is also red-shifted but appears to remain more constant in velocity throughout the *primary* shell phase. When the Balmer cores re-appear during the *secondary* shell phase, they are blue-shifted relative to the photosphere and remain constant in velocity for the entire short duration of this segment of the shell phase. A Balmer progression is observed which indicates deceleration of material along our line of sight during the *primary* shell phase and the reverse for the *secondary* shell phase.

The profile of Hα in HR 2142 varies in an interesting fashion during the shell phase. Some representative Hα profiles observed at various phases are presented in

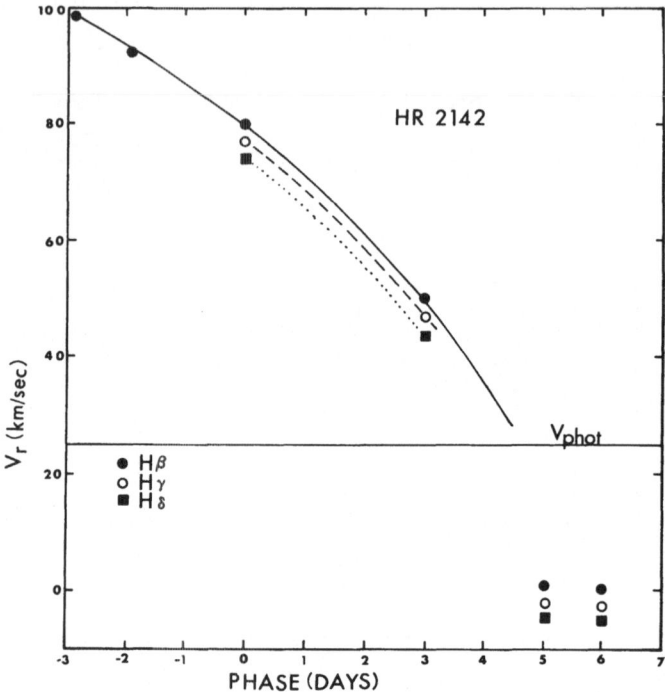

Fig. 3.   The radial velocities of the Balmer shell lines as a function of phase. The velocity of the
photospheric features is indicated. Note the Balmer progression.

Figure 4. Outside of shell phase, Hα is a strong emission feature $(I \approx 4I_c)$ which shows no conspicuous structure. The intensity of the redward side of the line is greater than that of the blueward side, however. During the *primary* shell phase, we observe a strong, red-shifted core in Hα plus a weak blue-shifted (≈ 150 km s^{-1}) satellite feature which appears near $\Phi_s = 0.0$, reaches maximum strength at $\Phi_s = 4^d$, and remains fixed in velocity. However, the most interesting Hα profiles in HR 2142 are observed during the *secondary* shell phase and shortly thereafter. When deep shell lines are present in Hβ and the higher Balmer features, multiple, weak cores are observed in Hα. The material in the outer portion of the 'envelope structure' through which we are viewing during the *secondary* shell phase appears to have a wider range of velocities than the material in the deeper sections. A minimum of three components nearly equally spaced across the emission profile can *always* be seen during the *secondary* shell phase. The fine structure is as repetitive from one cycle to the next as the appearance of the deep shell absorptions in the other Balmer members. For a few days after the *secondary* shell phase, when shell structure has disappeared in the higher Balmer members, Hα continues to show weak structure. One noteworthy component is blue-shifted by nearly 150 km s^{-1} relative to the center of the emission feature.

Recent observations of Hβ during the post shell phase have revealed the presence of multiple weak, blue-shifted absorption cores. We feel that these absorption components are responsible for the low intensity of the V lobe of Hβ (and the other

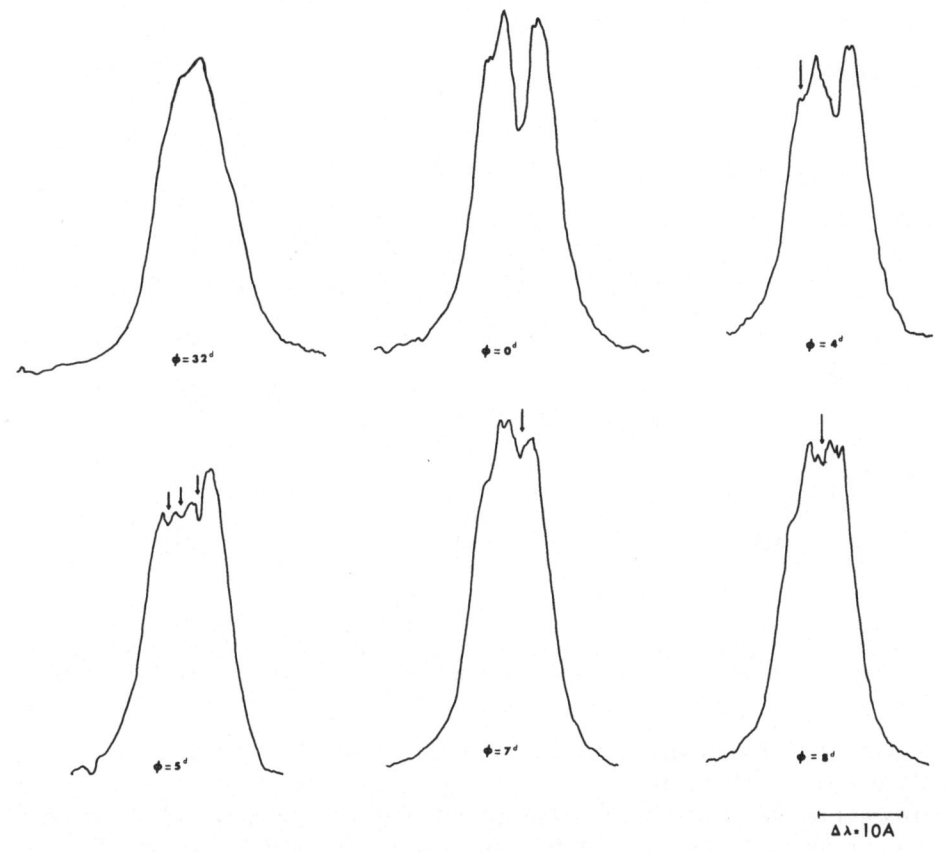

Hα DENSITY PROFILES FOR VARIOUS PHASES

Fig. 4.   Representative profiles of Hα in HR 2142 observed at selected phases. The arrows point to
confirmed weak absorption components in the emission line profiles.

Balmer features) which has already been reported at this phase (Peters, 1972). The
equivalent width of the missing V lobe of Hβ is comparable to that of the Hβ core
during the *secondary* shell phase.

It has already been established that the Hβ profile varies in a regular manner with
phase (Peters, 1972). Typical Hβ profiles for various phases were presented in
Figure 3 of the latter paper. Within the past three years we have gathered additional
data which not only confirm the earlier result but also show that the R lobe remains
essentially constant in intensity throughout the cycle. The cyclic V/R variations are a
result of the changing intensity of the V lobe. The data accumulated on V/R for Hβ
as a function of phase appear in Figure 5. It can be seen that $R > V$ at all phases
except near $\Phi_s = 0^p.9$.

In concluding this section of the paper, I would like to re-emphasize the strict
periodicity of the shell phases and other cyclic spectral variations which are observed
in HR 2142. Even fine structure in Hα repeats from one cycle to the next. The data
which we have accumulated on HR 2142 in the past six years combined with Dr D. B.

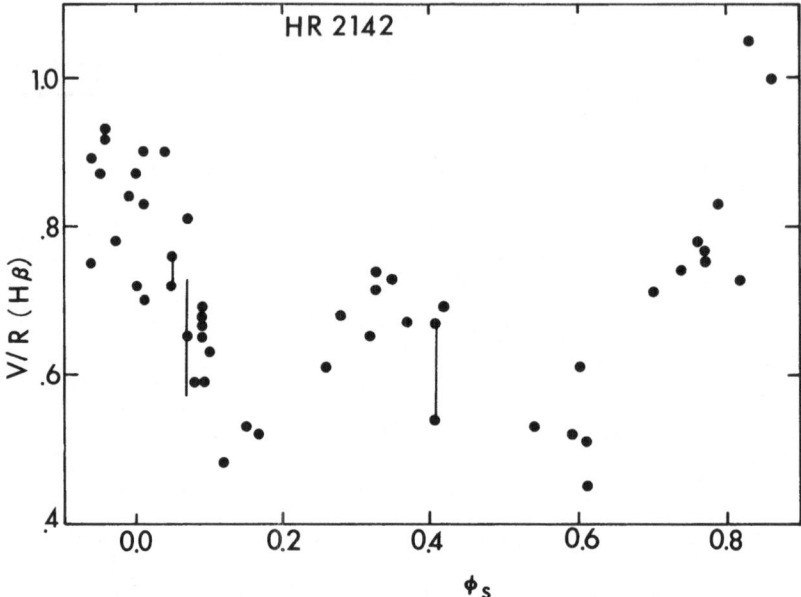

Fig. 5. The change in the ratio V/R for Hβ with phase. Vertical lines joining two points indicate that the observations were made on the same night. The point plus vertical line at $\Phi_s = 0\overset{p}{.}07$ represents the mean of seven observations made during one night; the length of the line indicates twice the standard deviation.

McLaughlin's spectrograms which exist in the plate files of the University of Michigan allow a refinement to the period quoted in Peters (1972). A Michigan spectrogram taken on October 6, 1933 which shows conspicuous shell structure at Hγ was particularly valuable in this endeavor. All available observations of HR 2142 can be fit well with $P = 80\overset{d}{.}860 \pm 0.015$ and $\Phi_s = 0\overset{p}{.}0$ on JD 2440855.5.

3. A Model for HR 2142

The strict periodicity of the two-component shell phase and the complex spectral variations observed not only during the shell phase but throughout the cycle suggest that HR 2142 is an interacting binary undergoing Case B mass exchange. The *primary* and *secondary* shell phases occur when we view the light of the B1 star through, respectively, the main gas stream and the counter streaming material. This model differs slightly from the one presented at IAU Symposium No. 51 (Peters and Polidan, 1973) in that we now feel that the material surrounding the primary extends out to 100 $R_\odot$ (about 15 R_*), the limit of the Roche lobe of the primary, and that there is some streaming of material near $\Phi_s = 0\overset{p}{.}2$ and at L_3. The computations of Lubow and Shu (1975) for mass streaming were considered in constructing the dimensions of the primary stream.

A polar view of the model for HR 2142 is shown in Figure 6. The mass losing secondary has a radius of approximately 40 $R_\odot$ and is positioned 20° outside our line of sight. A single arrow shows the direction of mass flow in the primary gas stream

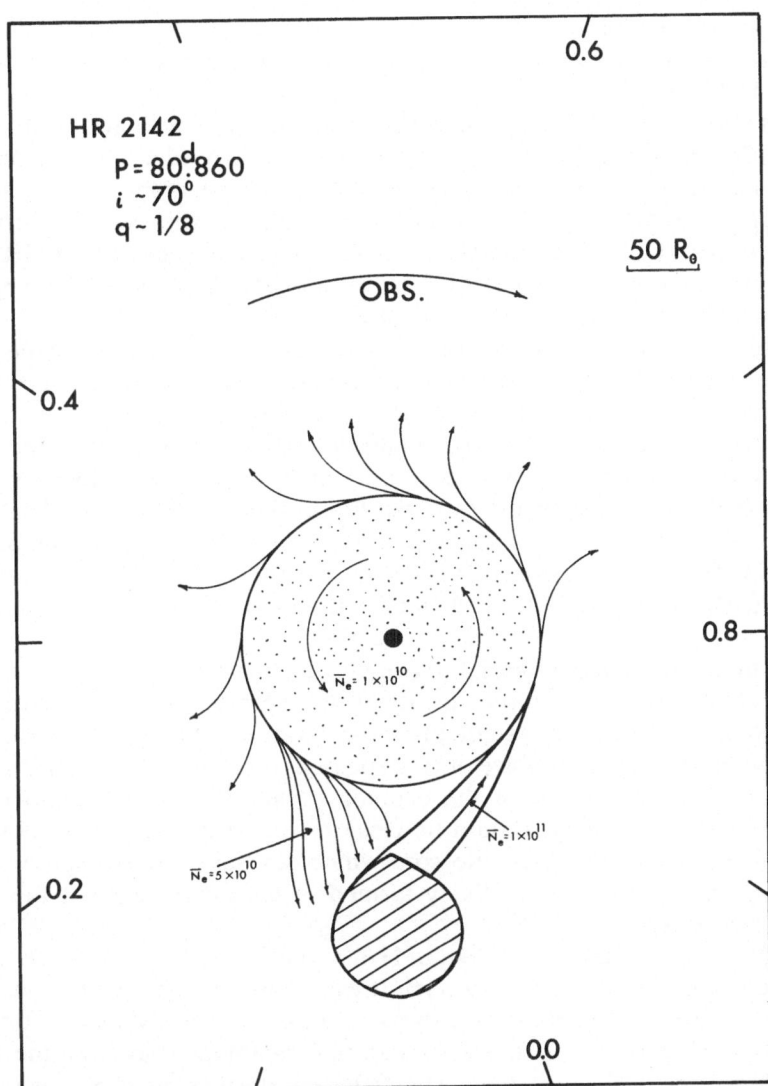

Fig. 6.   A polar view of our model for HR 2142 drawn to scale. The notation and the derivation of the physical parameters are discussed in the text.

while a series of additional arrows, whose spacing are an indication of the extent of mass streaming, show other regions of mass loss. Some physical parameters obtained for the system are also noted in Figure 6. Nearly all of the observed features of HR 2142 can be explained, at least semi-quantitatively, by the model. Some of the more important and fundamental ones are discussed below.

The mass ratio which was used in the construction of the model, $1:8$, is an assumed value. Since the primary is a B1 star, its mass must be near $12\ M_\odot$. A mass of $1.5\ M_\odot$ is then implied for the secondary. Since the spectrum of the secondary cannot be seen in the near infrared, its spectral type is most likely G to early K (see paper by Polidan presented in this session, p. 401). At the end of this section the plausibility of the assumed mass ratio will be discussed in light of some recent radial velocity measurements for HR 2142.

The fact that we do not observe eclipses suggests that the inclination of the system is less than 75°. However, the system must be viewed *nearly* edge-on since, according to our model, we are looking through a well defined stream of material. In addition, the photospheric features in HR 2142 are amongst the broadest seen in early B type stars and suggest that $v \sin i \approx 350$ km s^{-1}. Therefore, we suggest that HR 2142 is viewed nearly equator-on and that it just barely misses being eclipsed. 70° appears to be a reasonable value for the inclination of the system.

The observed velocity sequence displayed by the Balmer cores during the *primary* shell phase indicates that the main gas stream is fairly curved and moderately focused toward the direction of the primary. At the beginning of the shell phase, when the cores are weak, we are viewing through material which has a fairly high component of velocity toward the primary (≈ 100 km s^{-1}); when $\Phi_s = 0^{\mathrm{P}}04$, we are looking through the stream as it is moving tangentially to our line of sight. At $\Phi_s = 0^{\mathrm{P}}0$, when $N_{0,2}h$ is a maximum, we view the stream through some intermediate orientation. Since our line of sight intercepts the stream for one-tenth of the orbit, a lower limit on the length of the stream is 100 $R_\odot$. Recall that the observed Balmer progression suggests a deceleration of material toward the primary.

The counter streaming material through which we are viewing during the *secondary* shell phase must, in the vicinity of the secondary star, have a very small velocity dispersion in our line of sight. From the duration of the *secondary* shell phase we estimate that the projected size of this 'focused' portion of the counter stream is 25 $R_\odot$. The outward velocity of the counter streaming material, relative to the photosphere of the primary, should be close to the observed projected value of 25 km s^{-1}. The blue-shifted satellite core which develops and strengthens during the *primary* shell phase is most likely formed in the outer edges of the counter streaming structure.

The cyclic profile variations evident in Hβ (and other Balmer envelope features) can be understood in terms of the model. It was mentioned in the previous section that the V/R variations result from a cyclic depression of the V lobe. We feel that multiple weak absorptions arising from material streaming away from the disk are responsible for the V/R variations. The V lobe is weakest near phases $0^{\mathrm{P}}15$ and $0^{\mathrm{P}}55$. Near $\Phi_s = 0^{\mathrm{P}}15$ we are viewing the light of the primary through a portion of the counter streaming material which has a large dispersion in velocity. If one looks carefully at the sequence of events in Hβ commencing with the *secondary* shell phase, one observes the sharp blue-shifted core broaden then move blue-ward until it 'disappears' into the V lobe. The fact that the equivalent width of the missing V lobe observed between $0^{\mathrm{P}}15 < \Phi_s < 0^{\mathrm{P}}25$ is comparable to that of the core observed during the *secondary* shell phase suggests that we are viewing through comparable amounts of material. We interpret the depression of the V lobe near $\Phi_s = 0^{\mathrm{P}}55$ as evidence of material streaming off from L_3. We do not expect a well-defined stream at L_3. Between $0^{\mathrm{P}}8 < \Phi_s < 0^{\mathrm{P}}0$, where the primary stream is joining the disk, we expect to see a minimum of high velocity blobs leaving the system. Our observations support this prediction (see Figure 5).

The mean density of the material in the envelope around the primary star was determined from the volume emission measure, $N_{ion}N_e V$, which was computed from Wellman's formula (1951; Pagel, 1960) and the measured equivalent widths of the emission features of Hβ, Hγ, and Hδ. If one assumes that $N_{ion} = N_e$ and that the envelope is a wedge-shaped disk of radius 100 $R_\odot$ with inner and outer heights of 5 $R_\odot$ and 25 $R_\odot$, respectively ($V = 10^{38}$ cm^3), then one obtains a mean electron density of about 10^{10} cm^{-3} for the material in the circumstellar envelope.

Estimates for the mean electron density in the two streams were obtained from the column density of absorbers, $N_{0,2}h$, deduced from the strengths of the hydrogen shell lines observed during the *primary* and *secondary* shell phases. The temperature of the gas was taken to be 12 000 K and the assumed path lengths through the streams were 25 $R_\odot$ and 60 $R_\odot$ for the main gas stream and the counter streaming material, respectively. The mean density for the primary gas stream turned out to be ten times higher than the mean density in the circumstellar envelope; the density for the counter streaming material is somewhat lower than that of the main gas stream but higher than the mean density of the envelope.

Finally, one can estimate the rate of mass loss from the secondary star from the observed velocities and the computed density and dimensions of the primary gas stream. We obtain a mass loss rate between 10^{-7} and 10^{-6} $M_\odot$ yr^{-1}.

The strict periodicity of the shell phases in HR 2142 has inspired this model. However, in principle, if HR 2142 is a binary, then one should be able to measure radial velocity variations in the primary and determine an orbit for the system. The fact that the photospheric features in HR 2142 are quite rotationally broadened coupled with the distortions which appear in the centers of the He I lines at certain phases have made it impossible to derive an adequate radial velocity curve from a Grant-type measuring machine. Recently, Dr J. B. Hutchings used the 'wide-scan' oscilloscopic comparitor ARCTURUS at the Dominion Astrophysical Observatory to measure the positions of the He and H features on five of the Lick plates which were taken at representative phases. His results are shown in Figure 7. The measurements not only produced a radial velocity curve but also indicated that conjunction occurs between the *primary* and *secondary* shell phases. Since K is about 20 km s^{-1}, the mass function is about 0.07 $M_\odot$. If $M_p = 12$ $M_\odot$, then $M_s = 2.5$ $M_\odot$. In

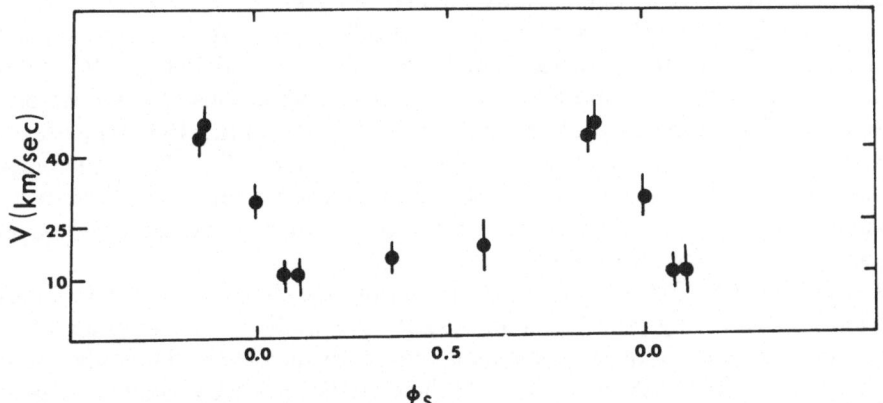

Fig. 7.   The radial velocity curve for HR 2142 suggested by Hutchings' measurements. Double points indicate two measurements of the same plate.

this case the mass ratio of the system would be closer to $1:5$ than $1:8$. Hutchings' measurements suggest an elliptical orbit. I feel that the orbit must be fairly circular due to the regularity of the shell phases. Velocities near $\Phi = 0^{\text{p}}5$ could be affected by gas streaming from L_3. The orbit should certainly be regarded as preliminary at this time. Additional measurements which will allow us to refine the orbit are planned in the immediate future.

4. Comparison between HR 2142 and other Binary Be Stars

In this section we briefly compare the spectral behavior of some confirmed Be binaries with HR 2142. Before we concern ourselves with the details of each object, however, it should be stated that a two-component shell phase of short duration as we observe in HR 2142 has not been observed in any other Be star. Thus, the spectral variations in HR 2142 do remain unique.

AX Mon (B1e + K2II; $P = 232^{\text{d}}5$) has been studied extensively since the early 1920's (Merrill, 1923). More recently the cyclic spectral variations in this star have been investigated by Cowley (1964) and Peton (1974). AX Mon periodically develops a strong hydrogen and metallic shell six weeks before conjunction. Maximum shell strength is reached three weeks before conjunction and the velocities of the shell features are positive relative to the photosphere during the initial segment of the shell phase. The observations strongly indicate the presence of gas streaming. However, AX Mon differs from HR 2142 in that shell features are seen over most of the orbit (they are absent for only 20% of the time). In addition, HR 2142 does not show a metallic line shell in the ground based portion of its spectrum. Strong hydrogen shell lines and metallic shell features are observed 20% of the time in AX Mon; HR 2142 shows a conspicuous hydrogen shell 6% of the time.

17 Lep (B9e + M1; $P = 260^{\text{d}}$) has been studied in detail by Cowley (1967). This star tends to undergo outbursts at 30–40 days past the time of periastron passage. However, shell lines which are blue-shifted relative to the permanent shell features are observed at all phases. The fact that the orbit is eccentric ($e \approx 0.1$) is most likely responsible for the high percentage of unpredictable shell outbursts. Unlike AX Mon and HR 2142, 17 Lep is viewed nearly pole-on.

HD 218393 (B2e + K1II; $P = 39^{\text{d}}$) shows periodic variations in the radial velocities and intensities of the hydrogen shell lines. Doazan and Peton (1970) have extensively studied this star and interpret the velocity variations as evidence of an oscillating envelope. Cyclic variations in the profile of $H\beta$ of the V/R type are also observed. For most of the cycle, $R \gg V$. However, unlike HR 2142, both components vary in intensity. The spectrum of the secondary has recently been discovered (Polidan and Peters, 1975). At this time it is not clear how mass exchange enters into the observed picture for HD 218393.

HD 173219 (B0e; $P = 58^{\text{d}}$) shows large amplitude radial velocity variations (≈ 200 km s^{-1}). Hutchings and Redman (1973) have studied this star in detail and find that the amplitude of the radial velocity curve for the He I and metallic lines is lower than the one for the H cores. Some of the metallic lines are variable with phase and may be partially formed in the circumstellar envelope. Hutchings and Redman suggest that the secondary may be a collapsed object. Incidentally, the spectrum of HD 173219 in the near infrared closely matches that of HR 2142.

ϕ Per (B0e) displays a cyclic strengthening and fading of a H and He shell with a period of $126^{d}.696$. The spectral variations in this star have been studied in detail by Hynek (1940) and Hickok (1972). The spectrum of ϕ Per resembles that of HR 2142 in that both objects show (1) a strong, near featureless $H\alpha$ emission feature, (2) emission of Fe II, O I λ 7774 Å, O I λ 8446 Å, and the infrared Ca II triplet, and (3) extensively broad photospheric features. However, the shell sequence in ϕ Per is quite different from that in HR 2142. Balmer shell features are observed for most of the cycle in ϕ Per. During one segment of the cycle, the velocities of the hydrogen cores and the He I shell lines which arise from metastable levels *increase* noticeably ($\Delta V \approx 40$ km s^{-1}) with time, then the H and He cores abruptly disappear for 15^{d}. Subsequently, the cores remain constant in velocity for 40^{d}, then show multiple components. In addition, He I cores appear when the H shell lines show constant velocity.

All of the Be binaries which have been described above, including HR 2142, show emission at the infrared Ca II triplet. The paradox of the Ca II triplet and its possible connection with binary nature were discussed by R. S. Polidan in the preceding paper.

5. Conclusions and Suggestions for Future Work

Cyclic spectral variations with periods of the order of 100^{d} apparently do characterize most binary Be stars. These variations can be of several types and include periodic shell activity (HR 2142, AX Mon, ϕ Per), cyclic variations of the emission line profiles (HR 2142, HD 218393), and subtle cyclic variations of the profiles of the photospheric features (HD 173219). Of course, one can construct a variety of models based upon the concept of binary mass transfer and, thus, predict a wide range of observed envelope spectra and shell sequences. The stage of mass transfer, the mass ratio of the system, the separation of the components of the system, and the inclination of the system to our line of sight determine what one observes. In fact, we can potentially learn a great deal about the details of binary mass transfer from studying the Be stars which are confirmed interacting binaries.

I suggest that it may prove fruitful to look for periodic shell phases in other Be stars. I refer specifically to searching for variations with periods around 100^{d}. Certainly it is a difficult observational problem to search for periodic short-termed shell structure as one sees in HR 2142. Consider, if short-termed shells were strong and common in Be stars, they most likely would have been discovered. The periodic shell phase in HR 2142 was elusive enough, as D. B. McLaughlin observed the star forty times in thirty years and missed discovering it even though shell structure was present on $\frac{1}{5}$ of his plates. Ideally, one needs 'continuous' coverage at $H\alpha$ or $H\beta$ at dispersions of 16 Å mm^{-1} or higher. It may prove easier to look for cyclic profile variations with periods near 100^{d}. HR 2142 and other Be stars which show periodic spectral variations have shown us that we are in need of more data on the constancy (or variability) of Be stars on the time scale of 100^{d}.

Acknowledgements

I wish to acknowledge many interesting and helpful discussions on periodic shell activity and mass transfer with R. S. Polidan and Dr M. Plavec. The model of HR 2142 presented in this paper was formulated jointly with R. S. Polidan. Dr M. Plavec has allotted generous amounts of his time on the Lick Observatory 120-in. telescope for this project on HR 2142. For this I am very grateful. I would like to thank E. A. Harlan of Lick Observatory and M. V. Wright of UCLA for numerous supplementary plates of HR 2142. In fact, many of the observations presented in Figure 1 were made by M. V. Wright. Finally, I wish to thank Dr J. B. Hutchings for measuring the Lick Observatory spectrograms on the comparitor of the Dominion Astrophysical Observatory. This project is part of a broad investigation of the structure and evolution of close binary stars and was partially supported by NSF MPS 74-04194A01 (Popper/Plavec).

References

Doazan, V. and Peton, A.: 1970, *Astron. Astrophys.* **9**, 245.
Cowley, A. P.: 1964, *Astrophys. J.* **139**, 817.
Cowley, A. P.: 1967, *Astrophys. J.* **147**, 609.
Hickok, F. R.: 1973, preprint.
Hutchings, J. B. and Redman, R. O.: *Monthly Notices Roy. Astron. Soc.* **163**, 219.
Hynek, J. A.: 1940, *Contrib. Perkins Obs.* **2**, 1.
Lubow, S. H. and Shu, F. H.: 1975, *Astrophys. J.* **198**, 383.
Merrill, P. W.: 1923, *Publ. Astron. Soc. Pacific* **35**, 303.
Pagel, B. E. J.: 1960, *Vistas in Astronomy* **3**, 203.
Peters, G. J.: 1972, *Publ. Astron. Soc. Pacific* **84**, 334.
Peters, G. J. and Polidan, R. S.: 1973, in A. H. Batten (ed.), 'Extended Atmospheres and Circumstellar Matter in Spectroscopic Binary Systems', *IAU Symp.* **51**, 174.
Polidan, R. S. and Peters, G. J.: 1975, this volume, 59.
Peton, A.: 1974, *Astrophys. Space Sci.* **30**, 481.
Wellman, P.: 1951, *Z. Astrophys.* **30**, 71.

DISCUSSION

Delplace: How many cycles have you observed for HR 2142?
Peters: Of the order of 20.
Polidan: Based on the old plates, it appears that HR 2142 has shown the same shell activity with the same period for roughly 50 years.

Note added in proof: In December 1975, the higher dispersion spectrograms of HR 2142 were measured with *ARCTURUS* at the Dominion Astrophysical Observatory. An orbit was obtained for HR 2142 which supports our model ($K \simeq 10 \text{ km s}^{-1}$, $e \simeq 0$, and conjunction near $\Phi_s = 0^\text{p}05$). Since $f(m) = 0.008\,M_\odot$, reasonable masses for the primary and secondary are $12\,M_\odot$ and $1.2\,M_\odot$, respectively.

TOWARD A MODEL FOR THE Be BINARY SYSTEM φ PER

E. M. HENDRY

Astronomy Dept., Northwestern University, Evanston, Ill., U.S.A.

Abstract. Arguments are presented against the adoption of an early B-red giant combination for the φ Persei system. Velocity curves of the primary and secondary components based on the measures of the He I lines on nearly two hundred plates are shown and discussed insofar as orbital parameters and system membership are concerned. Evidence from the He I lines, the emission edges, shell absorptions, and underlying wide absorptions of the hydrogen lines indicate that the system is made up of a B1 primary and a B3 secondary, both of which are emission-line stars.

1. Introduction

Interest in the star φ Persei – which lay largely dormant since the papers of Hynek in the 1940's (Hynek, 1940, 1944) – has recently begun to emerge again among investigators of the Be and shell phenomenon. This enigmatic star exemplifies in many ways the problems which beset astronomers studying these types of stars – namely, it is a spectrum and radial velocity variable. Its variations occur periodically in the space of some 127 days and strongly suggest that the system may in fact be a binary.

2. Discussion of the B-Red Giant Hypothesis

Recently, Plavec (1975a, b) and others have advanced the hypothesis that a tentative model for the φ Persei system might consist of a main-sequence B0–B1 star and a hypothetical red giant companion, presumably of about spectral type K, which would transfer matter to a shell or ring about the earlier primary through the inner Lagrangian point. It is based largely on the failure to find evidence for the presence of a B type secondary star in the far ultraviolet *Copernicus* scans and the hypothesis advanced by Kříž and Harmanec (1975) among others that probably all Be and shell stars are interacting binaries.

Although this author has not had access to the *Copernicus* data of Plavec, he states that two ultraviolet observations were made, one of these being the more relevant to test for duplicity, namely the one *Copernicus* scan made at second quadrature phase. At this phase according to my model (Hendry, 1975) which suggests that the system consists of a B1 primary and a B3 secondary, the predicted separation of the lines from the two components is about half an angstrom at λ 1200 Å. Since the resolution of the *Copernicus* scanner is at best about 0.2 Å in the relevant scanning mode, the separation of two even moderately rotationally broadened photospheric lines would clearly at best be extremely difficult. Observations in the higher dispersion *Copernicus* scanning mode were, by circumstance, limited to sharp interstellar lines and can not support or contradict duplicity. Also, if one considers the hypothetical spectral type of the secondary star to be B3 as my model would suggest, and the observed spectral type of the primary to be B1, calculations show that the secondary

A. Slettebak (ed.), Be and Shell Stars, 429–437. *All Rights Reserved*

would be from 1.5 to 2.0 mag. fainter at λ 1200 Å, and consequently, would be unlikely to be easily visible. Therefore, the observations in the far ultraviolet do not contradict a model containing two B-type stars since the model does not predict the secondary spectrum to be easily visible in this region of the spectrum, and even if it did, one might hesitate to destroy a model based on 184 observations which seems satisfactory in the photographic and infrared on the basis of a single observation.

But let us accept for the moment the ad hoc assumption of a late-type companion within the system. On this basis, one would expect, first, to find several predicted features in the photographic and infrared – which are not observed – and secondly, one should also expect that the B – K model would explain certain features which are definitely observed on the strength of nearly two hundred plates. Neither of these expectations appears to be fulfilled. The spectrum of ϕ Persei has been examined in the photographic infrared for any evidence whatsoever of the existence of a late-type companion and no trace of such a star was found. Furthermore, artificial composite spectra were made of a combination of a B-type star and a K-type star with different relative exposures so as to simulate increasing magnitude differences between the two stars. It was found that the K-type star would have to be nearly three to four magnitudes fainter in the infrared than the B-type star for it to be invisible in the spectrum. Therefore, unless greatly underluminous or the primary overluminous by some three magnitudes, the late type star should be visible in the infrared spectrum of ϕ Persei if it exists.

Evidence of the late-type secondary can also be searched for at still longer wavelengths from 2.3 to 19.5 μ. Such observations have been made by Gehrz *et al.* (1974) and they found that indeed ϕ Persei does have an infrared excess. However, this excess is well-fitted by a purely free-free radiation model from proton-electron scattering in a hot ionized circumstellar plasma. Of course, the excess could be explained by a cool infrared companion. Nonetheless, from Kepler's Law, the center of mass separation between the primary in the ϕ Persei system and the hypothetical infrared secondary is only about 1.5 AU, far too small for the short wavelength flux of the B-type star not to seriously affect the secondary's photosphere.

3. Observations

The B-red giant model can not satisfactorily explain the observations upon which my model of two B-type stars is primarily based. The model is based upon measures of 184 plates of various dispersions from Mt. Wilson, Perkins, Michigan, David Dunlap, and Northwestern's observatories. Measures of the spectral features of ϕ Persei are somewhat difficult; that is undoubtably why the star holds second place only to Algol in the number of spectrograms taken of it. The paradoxical out-of-phase behaviour of the dominant (and I stress the word dominant) component of the doubly-structured helium lines has been fully confirmed by my recent measures, as has a double periodicity in the velocity curves of the emission edges of the hydrogen lines.

Some doubts have been raised concerning the visibility of these helium lines from the secondary component. I was not the first to see them; they were first measured by

Whitt, later by Hynek (1944), and most recently oscilloscopically by Hickok (1969). Further, the precaution was taken in the measures of not calculating phases until after the radial velocity reductions were completed. This procedure was scrupulously followed so as not to bias the measurements. Clearly, then, we do not have a 'Martian canal' situation.

It has also been suggested that with spectral types of B1 and B3, the secondary star would be nearly invisible in the spectrum of the dominant primary. Calculations show that in the photographic region the magnitude difference is about 1.1 mag., difficult but not impossible to discern, as is evidenced by the work of Batten (1962) on V380 Cygni, Popper (1943) on V Puppis, Pearce and Plaskett (1932) on 57 Orionis, and Petrie (1950) on BD +58°2546, where similar magnitude differences and spectral types are involved.

The results of the measures of the helium lines are shown in Figure 1. The measures have not been formed into normal points on the graph so as to indicate the scatter of individual measures more clearly. Measures near the V_0 axis have been purposefully omitted as subject to a high degree of blending and hence, unreliable. An orbital solution was performed on the primary and secondary star velocities separately using a version of the program of Wolfe *et al.* (1974). The derived orbital elements appear in Table I. The individual masses of the stars are found to be 15.0 and 8.2 solar masses, respectively, if an orbital inclination of 80° is assumed, a not unreasonable assumption from the appearance of the spectrum and the absence of

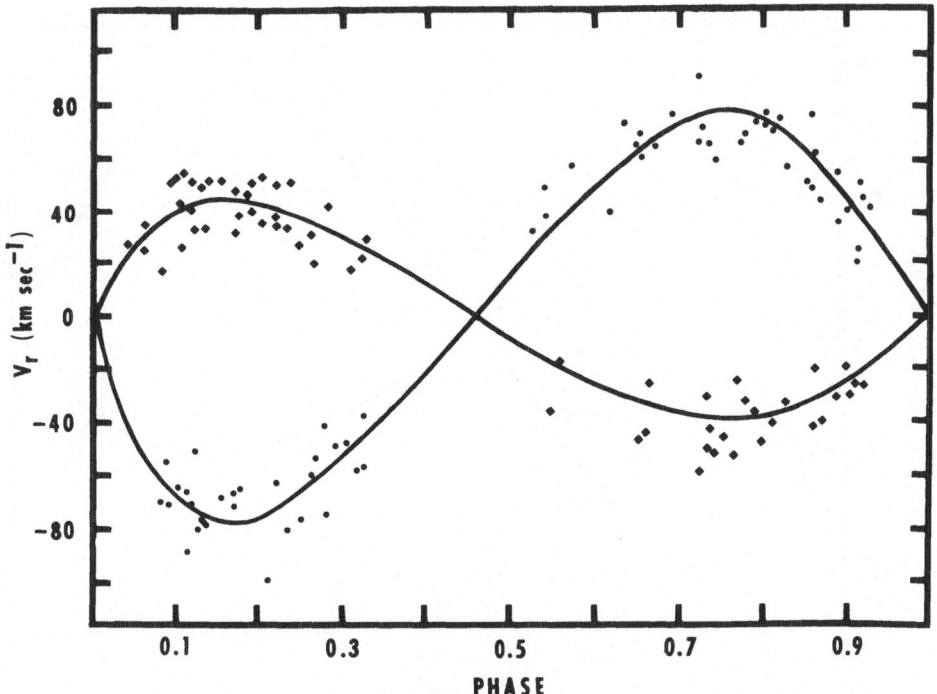

Fig. 1.   Velocity curves derived from measures of the lines of He I. The solid lines denote the curves of the orbital solution.

TABLE I

Orbital elements of the ϕ Persei system

$P = 126.696$ days
$K_1 = 42.5 \pm 4.1$ km s^{-1}
$K_2 = 77.5 \pm 3.5$ km s^{-1}
$V_0 = 0.0 \pm 3.0$ km s^{-1}
$e = 0.15 \pm 0.02$
$\omega_1 = 285° \pm 15°$
$a_1 \sin i = 7.3 \times 10^7 \pm 0.2 \times 10^7$ km
$a_2 \sin i = 1.3 \times 10^8 \pm 0.2 \times 10^7$ km
$M_1 \sin^3 i = 14.28 \pm 0.3$ solar masses
$M_2 \sin^3 i = 7.82 \pm 0.3$ solar masses

eclipses. Curiously, the helium absorption lines of the primary are the more difficult to measure since they are more rotationally broadened and the lines of the secondary are never completely free of the blending effects of the primary. Hence, the values of K_1 and K_2 found are mimimum estimates and that of the mass ratio, a maximum estimate.

However, an immediate question arises. Measures of the $v \sin i$ of ϕ Persei have indicated very high values. Consequently, measures of duplicity in such a system where at least one of the stars is rapidly rotating and has wide profiles are suspect. Yet, tracings of $\lambda\lambda$ 4026 and 4471 Å of He I show duplicity clearly with the component from the secondary star being sharper and more intense. This is an

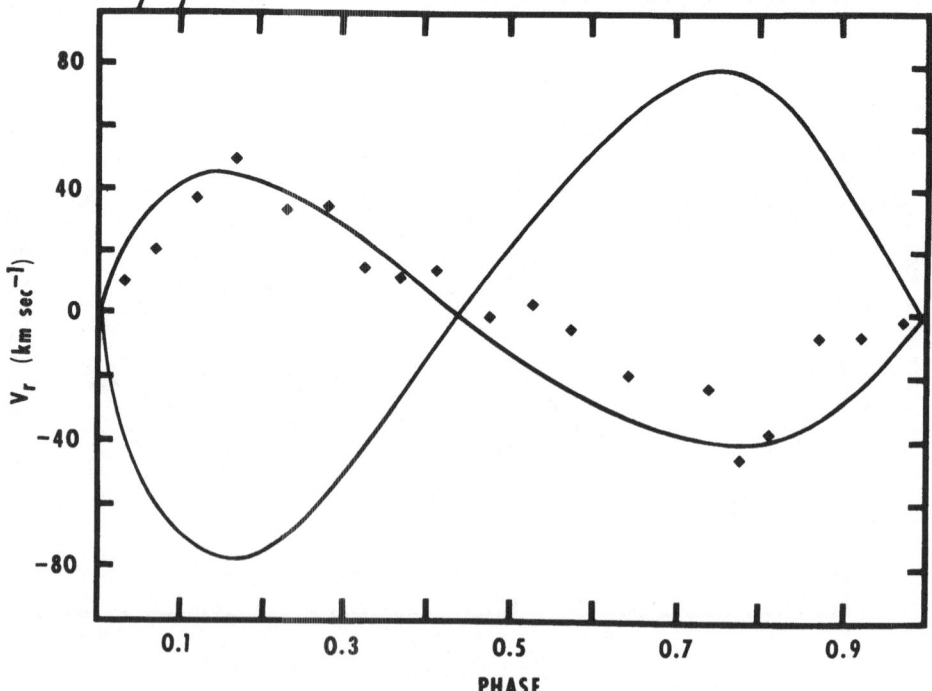

Fig. 2.   Velocity curve derived from the measures of the sharp Balmer curves of the primary. The solid lines denote the curves derived from the orbital solution.

important point: as is evidenced by both the recent measures and those of Hynek and Whitt, the primary contribution to the He I triplet lines is from the secondary star. This is nearly impossible to explain by a hypothesis that emission in the lines or some such other feature could mimic duplicity. Also, corroboration of the existence of a B-type secondary comes from the measures of other features in the spectrum. Very infrequently on only the best plates, lines of other ions such as N II, O II, Si II, and others are seen and measured. These lines follow the secondary's velocity curve exclusively which is consistent with the secondary being a more slowly rotating B star.

The velocity curve derived from the measures of the sharp Balmer cores appears in Figure 2, superimposed on the theoretical curves of the orbital solution. It can be immediately seen that it is probably non-Keplerian and shows the influence of gaseous motion within the system. The points plotted are the means of individual observational points and in this process of forming the normal points, the quite considerable scatter for such sharp features is masked. The scatter is in part intrinsic, as might be expected since the lines are clearly formed in the shell of the primary, and is also a function of the asymmetry of the lines which becomes quite marked at certain phases, thereby making visual measures subject to uncertainty. The form of the hydrogen curve is not entirely definitive, however, since it is derived from measures of plates taken from 1925 to 1975 and there is some evidence that the form of the curve undergoes slight secular changes. This point is still under investigation. However, it may be mentioned that the helium lines and the lines of the other elements used for the orbital solution show no evidence of any secular change from cycle to cycle or of non-Keplerian motion.

The behaviour of the hydrogen lines also suggests that the secondary star is an early B-type star. The structure of these lines is quite complex and is made up of the following contributions: (1) Wide underlying absorption from the primary (2) Wide absorption from the secondary (3) Intense shell absorption from the primary (4) Weaker shell absorption from the secondary (5) Strong double emission from the primary (6) Narrower double emission from the secondary.

Evidence for the contribution of the wide underlying hydrogen absorption of the secondary star to the line profiles of the primary is found both in examination of the profiles themselves and in measurements of the velocities of the absorption edges. The wing of Hδ is definitely displaced to the violet of the sharp core when the core is at maximum positive velocity and vice versa when it is at negative velocity. The widths of Hδ are found to vary such that the line is widest at both velocity maxima and narrowest when the curves are near to the systemic velocity. This is consistent with the blending hypothesis.

The presence of the central shell absorption of the secondary can be directly seen. On the best plates, there appears to be a second sharp and faint absorption core between the wing and the central sharp Balmer core of the primary which is nearly coincident with it, producing observed asymmetries except at the velocity maxima when it is separate. Measures of this secondary core agree in direction and approximate magnitude with the theoretical curve of the secondary star, and consequently, can be expected to be formed from the projection of shell material onto its photosphere.

That the secondary star possesses also double hydrogen emission lines can be seen in the behaviour of the primary's emission lines. At about phase 0.2, the violet component of the emission line is observed to be wider than the red component, while the central absorption becomes diffuse and 'filled-in'. This can be accounted for by the presence of the secondary's double emission line. Its violet component is velocity-shifted so as to artificially broaden the primary's emission profile by superposition and the secondary's red component acts to fill in the core, and hence causes it to appear more diffuse. At about phase 0.7, the opposite effect occurs and it is the primary's red emission which is widened and the core is once more filled in. This effect is corroborated by the velocity measures of the emission edges of hydrogen. The curves obtained from the measures of the emission edges of hydrogen are illustrated in Figure 3. The absolute values of the measures are not important but the forms of the curves are. Presumably, approximately the same point on the profile is measured each time. A double-peaked curve of unequal amplitude is found and

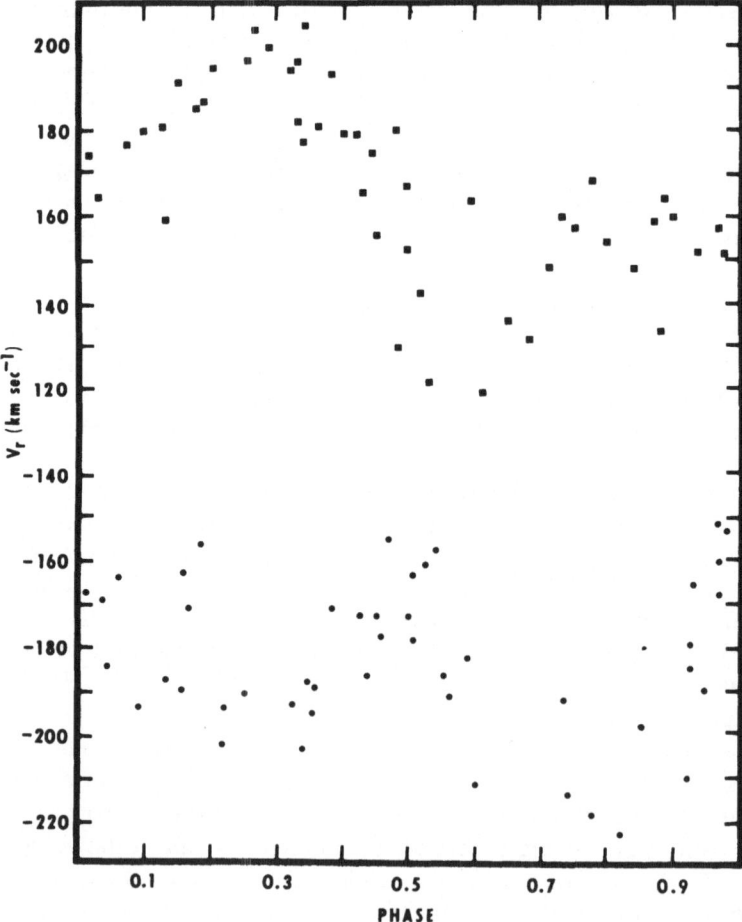

Fig. 3.   Velocity curves derived from measures of the emission edges of the hydrogen lines Hβ through Hδ.

seems consistent with the superposition hypothesis. The primary's emission may be surmised to be wider; at the first quadrature phase it extends further to the red. At this phase, however, the violet edge is not extended as much by the secondary's contribution. The situation is reversed at the second quadrature phase. Consequently, it may be seen from the evidence of the hydrogen emission that the secondary star is also a Be star with its own concomitant shell.

A brief note on the behaviour of the other features in the spectrum in view of the proposed model is in order. The H and K lines of Ca II follow a velocity curve which is essentially that of the primary but with far smaller amplitude. This may result from a blending of the components of both stars and their shells and this is corroborated by the occasional doubling seen on good plates, measures of which confirm that one component essentially follows the primary and the other the secondary. The central absorptions of the Fe II lines in the photographic and the infrared follow much the same behaviour and may also be composite. Their emission edges mimic the double periodicity of the hydrogen emissions and presumably the same explanation of a contribution from each shell applies. Measures of the central absorptions of the infrared Ca II triplet lines, whose emission edges are blended to different degrees with the Paschen emissions and which were made from considerably lower dispersion plates than the rest of my data seem to indicate that these lines follow the primary and their emission edges do not reveal the influence of any emission from the secondary. Clearly, the conditions under which they are formed seem to be unique to the primary and its shell.

4. Conclusions

The observations then seem to indicate that the φ Persei system is made up of two Be stars, each with a shell whose properties or interactions causes the distortions in the velocity curve of the Balmer lines. To state more than this at the present time would be premature and a more quantitative description of the proposed model is under study at the present. However, it would seem that the adoption of a B-red giant combination, although tempting from a theoretical point of view, is somewhat arbitrary and not in agreement with certain of the observations.

References

Batten, A.: 1962, *Publ. Dominion Astrophys. Obs.* **12**, 91.
Gehrz, R. D., Hackwell, J. A., and Jones, T. W.: 1974, *Astrophys. J.* **191**, 675.
Hendry, E. M.: 1975, *Bull. Am. Astron. Soc.* **7**, 268.
Hickok, F.: *The Spectroscopic Characteristics of Phi Persei*, University of Toronto (master's thesis).
Hynek, J. A.: 1938, *Contr. Perkins Obs.*, No. 10, 1.
Hynek, J. A.: 1940, *Contr. Perkins Obs.*, No. 14, 1.
Hynek, J. A.: 1944, *Astrophys. J.* **100**, 151.
Hynek, J. A.: 1951, in J. A. Hynek (ed.), *Astrophysics*, McGraw-Hill Book Co., New York, p. 472.
Kříž, S. and Harmanec, P.: 1975, *Astron. Astrophys.* **26**, 65.
Pearce, J. A. and Plaskett, J.: 1932, *Publ. Astron. Soc. Pacific* **44**, 259.
Petrie, R. M.: 1950, *Publ. Dominion Astrophys. Obs.* **8**, 328.
Plavec, M.: 1975a, *Bull. Am. Astron. Soc.* **7**, 405.

Plavec, M.: 1975b, private communication.
Popper, D. M.: 1943, *Astrophys. J.* **97**, 400.
Wolfe, Jr., R. H., Horak, H. G., and Storer, N. W.: 1967, in M. Hack (ed.), *Modern Astrophysics*, Gauthier-Villars, Paris, p. 251.

DISCUSSION

Cowley: Can you make some estimate as to what the individual masses are and what the sizes of the stars are, with respect to the Roche lobes?

Hendry: From my preliminary orbital elements I find a mass ratio of about 0.55 and if I assume an angle of inclination of about 80° for the system it gives me masses of about 14.3 and 7.8; something like that.

Cowley: What is the period of the system?

Hendry: 126.696 days.

Cowley: OK. That means that neither star comes close to filling the Roche lobe. They are two separate stars. So that if we are talking about binary stars, we are not talking about mass transfer.

Hendry: That's correct.

Polidan: (1) Regarding the detection of the K-type star, we also have formed composite spectra and we find great difficulty in seeing the late type star's lines if the I magnitude difference is greater than roughly $2^m.5$. If indeed ϕ Per does have a K-type companion we expect that it would have a greater magnitude difference than this. It could possibly be as large as three to four magnitudes in I.

(2) As for the Ca II core, from our higher dispersion data we find that because of the severe blending of the nearly equally strong Ca II and Paschen lines (Ca II peak = 1.38, Paschen peak = 1.25; see Figure in my talk) that it is impossible to measure the true core of the line. One must remove completely the Paschen emission line to see the true reversal in the calcium line.

(3) Regarding the far infrared flux of ϕ Per, as I pointed out in my talk, it is impossible to distinguish between free-free emission and a K-type continuum in the region beyond $2\,\mu$. Swings and Allen (*Publ. Astron. Soc. Pacific* **84**, 523, 1972) pointed this unfortunate fact out. Combining this with the expected small flux contribution of the K-type star, it is not surprising that no obvious evidence of a cool secondary in ϕ Per exists.

Hendry: In response to your comments (1) In my artificial composite spectra (B plus K), the spectrum of the B-type star with a K-type spectrum added with a 4-mag. difference in I didn't look like a B-type star; it looked anomalous. On that basis, I am saying that one could possibly see evidence of a K-type component almost four magnitudes down, as was predicted by Hynek in his study of composite spectra.

(2) My Ca II emission peaks are very high and the Paschen peaks are reasonably low above the continuum. The difference in wavelengths between the Paschen lines and the relevant calcium triplet line is different for each line. Nonetheless, I find that my measures of the cores of the triplets do not show very much scatter. Consequently, I am a bit uncertain of that effect, as you described it.

(3) Concerning the infrared data of Gehrz, Hackwell, and Jones, I believe that their comment concerning ζ Tau and the objections to an infrared companion there would also apply to ϕ Per. The stars are too close; there is only about $1\frac{1}{2}$ AU between the two.

Cowley: The quoted rotational velocities for ϕ Per are widely discrepant. If you have two components that are both B-type stars, depending upon the phase of the observations and did not realize that there were separate lines, you would infer a broader profile and therefore a higher rotational velocity. It would be very interesting on very low noise spectrum scans to get the rotational velocities of both of the components, which would probably be lower than any of the quoted values.

Peters: We also have a series of spectrograms ($5.5\,\text{Å mm}^{-1}$) of ϕ Per on IIIa-J plates and we see no evidence of duplicity. ϕ Per has strong Fe II emission and significant He I emission, which is also variable. I am wondering if what you are measuring is the variable helium emission and not the line shifts. The equivalent width of λ 4026 is about 0.8 Å, which is 30% lower than one observes in non-Be stars of similar spectral type. In other words, everything is consistent with significant emission filling.

Hendry: One of the reasons I cannot be measuring the emission effects in the line is because the secondary component is the stronger component. What you should be objecting to is how do I measure the primary component, because it is broader and less intense.

Meisel: Berg and I have recently studied (*Astrophys. J.*, June 1975) the λ 10 830 structure in Spica with 1.2 Å resolution, where the components are of spectral type B1 and B2^{+}. It is relatively easy to see lines of both components, and measure radial velocities as well as rotational velocities. Hence high resolution observations of λ 10 830 in ϕ Per may help to resolve the question of the helium lines and the nature of the components.

Massa: The equivalent width of the helium lines is small, even for a star of early spectral type. We have scanner observations of φ Per, and the Balmer jump appears to be in emission, so you probably have continuum which is veiling the lines.

Peters: We have obtained several scans of φ Per, none of which show a negative (emission) Balmer discontinuity. We used the Hayes-Latham system of absolute calibration to reduce the scans. We do see some emission due to the envelope shortward of the Balmer discontinuity, however. We fit the energy distribution of φ Per to a 28 000 K, log $g = 4$ Princeton model atmosphere.

FINAL REMARKS ON THE BINARY HYPOTHESIS
FOR THE Be STARS

MIROSLAV PLAVEC

Dept. of Astronomy, University of California, Los Angeles, Calif., U.S.A.

Abstract. The hypothesis that a significant fraction of Be stars are interacting binaries is discussed in general terms. The author believes that the observed Be stars are a mixture of three different types of objects. The weakest point of the single-star hypothesis is probably the difficulty in explaining the quasi-periodic V/R variations; the weakest point of the binary hypothesis appears to be the statistical expectancy of a fairly high number of eclipsing systems among them.

From various comments made during this Symposium, I gather that the binary model for Be stars is beginning to be taken seriously. This enables me to change somewhat this closing talk of the session. I feel that it is no longer necessary to call attention repeatedly to the binary model; instead, I may try to be more impartial. In other words, I will not follow the radicalism of Harmanec and Kříž who propose the working hypothesis that *all* Be stars are binaries. Rather, I will adhere to the suggestion made several times in the past (Plavec and Horn, 1969; Kříž, 1969; Plavec, 1970a; Plavec *et al.*, 1973) that a significant number of Be stars are probably interacting binaries. This formulation admits the existence of several possible models, so that we would have objects similar in appearance but different in origin. This is not too satisfactory at first sight, but after all, a Be star is simply a B star surrounded by an extended envelope. Why should we insist that the envelope can be formed in only one way? In fact, I can think of three possible ways: a rapidly rotating single star; an interacting binary system; and a young star still surrounded by the remnants of its original cocoon (i.e., Herbig's Ae and Be stars).

The idea of the binary model was conceived in 1968–69 not because the classical model proposed by Struve was deemed unacceptable. Rather, theoretical results of modeling of interacting binaries indicated that here we have a process which could *also* lead to something similar to a Be star. The results of evolutionary model calculations for interacting binaries have been reviewed several times (see, e.g., Plavec, 1970b, or Paczyński, 1971), so that I can be quite brief. The more massive star of a binary system is the first component to evolve away from the near vicinity of the zero-age main sequence; as it evolves, it expands until it fills its critical Roche lobe. If the system is very close, the orbital period being less than one day, the more massive component is trapped while it is still a main-sequence star. It begins to lose mass on a thermal time scale. The other component cannot be much smaller, being a main-sequence star in a system where the separation of the components is only of the order of 10 solar radii or less. The gas streaming out of the more massive star near its first Lagrangian point is deflected owing to the rotation of the system, and generally impinges on the less massive star under an oblique angle. It will probably accelerate that star's axial rotation, but an extensive circumstellar envelope cannot be expected in this case. Accretion causes the accreting star to swell and very soon it will fill its critical lobe, too (Benson, 1970), and we will have a contact system.

A. Slettebak (ed.), Be and Shell Stars, 439–444. All Rights Reserved
Copyright © 1976 by the IAU.

Much more interesting for us is the case when the binary system is wider to begin with, having an orbital period of at least several days. In this case the more massive component will not reach the Roche limit until it is burning hydrogen in a shell surrounding a helium-rich core. The system is generally wider, the separation of the components being between 10 and 100 solar radii, and this separation grows in the later stages of mass transfer, when the mass-losing star already becomes the less massive component. It is these stages which interest us in connection with the Be phenomenon. For now the material leaving it comes into the vicinity of the mass-accreting star with excess angular momentum, and therefore tends to form a disk around that star. The disk may just about fill the critical Roche lobe of the accreting star, but is relatively flat, the material being concentrated near the orbital plane. Our idea is that this disk will produce emission and/or shell absorption lines typical for Be stars and shell stars.

That this mechanism works is indicated by the Hα profiles shown in the introductory lecture (p. 1). There is every reason to believe that RZ Sct, V 367 Cyg, W Ser, AU Mon, and RX Gem are interacting binaries of roughly the type described above. Observed profiles like these can now be fairly easily obtained thanks to modern image tubes; one generation ago, Hα was much more difficult to observe in fainter stars. Nevertheless, Struve, Merrill, McLaughlin and others did occasionally comment on the similarity between the spectral features in certain Be stars or shell stars and in certain eclipsing binaries. Why, then, did no one suggest a binary model for Be stars already thirty years ago? I think the reason is the observed rapid rotation of the Be stars. Rapid rotation is an inherent property of many B and A stars. We do not have to postulate a binary model to explain rapid rotation. In fact, close binary stars are often associated with slow rotation, since the tidal force tends to synchronize the axial rotation with orbital revolution. I think that the rapid rotation still represents the strongest argument in favor of Struve's classical model for many Be stars. It is only necessary to find an additional weak force which would push the material into the envelope. Various possibilities were investigated by Limber and Marlborough (1968). In recent years, stellar winds became very popular. However, in agreement with Dr Conti's comments, I must ask how the stellar wind can explain the cyclic changes observed in some of the envelopes, particularly in shell stars. We observe the material to drift slowly outwards, but then the motion stops and may even be reversed, as in 48 Librae. Also, while the parental star appears to be the same, a new shell may begin to form after years of inactivity, as in Pleione or very recently in o And. These phenomena are hardly consistent with a uniformly blowing stellar wind.

One may object that model calculations for mass-transferring binaries do not predict this type of mass transfer rate variations either. However, one should bear in mind that the computing codes used cannot describe changes occurring on a time scale substantially shorter than the thermal time scale of the stars. In an approach where effects on the dynamical time scale were included, Bath (1969, 1972) repeatedly obtained very significant instabilities, and the observed period fluctuations in many Algol binaries indicate that short-term variations in the transfer rate are likely (Biermann and Hall, 1973).

Without going into details, one can, I think, easily see that the binary model for the Be stars is more promising than the single-star model for objects displaying periodic or quasi-periodic changes such as the V/R variations. In a rapidly rotating single star, one has just one fundamental period, namely the period of rotation of the star; and it is difficult to understand why the envelope should not be symmetric with respect to the equatorial plane and to the axis of rotation. This is why it was necessary to introduce some asymmetry by postulating an elliptical ring. A simple geometrical model of such a ring does indeed explain the V/R variation at least qualitatively or semi-quantitatively, but nobody has explained yet how such a ring can exist as a three-dimensional feature, how it can be formed and how it can persist for a sufficiently long time. In the solar system, we have analogical structures in the meteor streams. However, there it is easy to understand why the ring is relatively narrow, for it has been formed by gradual or at most mildly violent disintegration of a comet moving in an elliptical orbit. Even so, these meteor streams show fairly large scatter of orbits at the aphelion, where the spatial density of the particles is much smaller than near the perihelion where the stream moves faster but is much more concentrated.

Harmanec and Kříž have shown that a structure somewhat similar to a ring may be more easily formed by a certain type of mass influx from a mass-losing component in an interacting binary. I hope to see soon a more detailed study showing that indeed this model, in a three-dimensional picture, does give not only the required profile shape and variation, but the required emission intensity as well. However, I believe that even the quasi-stationary state of a fully developed disk offers possibilities for periodic or cyclic spectral changes. We have now at least two different fundamental periods in the system, rotation and revolution. As the components revolve about the center of gravity, the disk is observed from different directions. No such symmetry as in the single-star model can exist, for the gravitational field is not symmetrical with respect to the central star of the disk. Moreover, we can anticipate streams in the system; the main stream transferring material between the components, a possible stream which may return part of the material back to the parental star, and another possible stream (or streams?) which may carry part of the gas away from the system (see Figure 8 on page 9).

Naturally, very much must be done to convert this crude picture into something that can be compared with observations. This field is far from inactive, but hydrodynamics of gas streaming in close binaries is a very difficult problem. Moreover, theorists working in this field are much more eager to investigate these processes in X-ray binaries, and ordinary binary systems have been rather neglected. An excellent discussion of the problem, together with references to earlier work, can be found in a recent article by Lubow and Shu (1975). It is not yet clear how fast the disk dissipates if the influx of material ceases. In the picture offered by Lubow and Shu, the viscosity in the disk is small, and the disk dissipates slowly. Some students of X-ray binaries, however, postulate very high viscosity which rapidly transfers angular momentum through the disk outwards, and dissipates the disk fast.

Even less well understood is the actual process of accretion of the mass from the disk by the star inside it. At the advanced stages of mass transfer, when the accreting star is small compared to the dimensions of the system, the accretion occurs from essentially circular Keplerian orbits at small relative speeds, and will not affect the axial spinning of the accreting star seriously. However, at the earlier stages the stream impinges on the accreting star directly and under an oblique angle, and may accelerate the star's rotation significantly (Plavec, 1970a; Van den Heuvel, 1970). We have a good example of this process in the Algol-type eclipsing binary star U Cephei. This system consists of a mass-losing subgiant, spectral class about G8 III–IV, and of a B7V component which accretes mass, and was studied in great detail by Batten (1974). The synchronized equatorial velocity of axial rotation of the B star would be about 60 km s^{-1}; however, the great width of its photospheric lines indicates a velocity of rotation of about 310 km s^{-1}. An almost identical velocity is indicated by the emission lines of the disk which surrounds the star. We have no model of the accretion process or of the structure of the star. We cannot say if this high speed of rotation is the property of only a thin surface layer – but the same uncertainty exists in the single-star model of Be stars.

But can U Cephei be called a bona-fide Be star? I think it can just now, because it recently developed strong emission lines (Batten *et al.*, 1975; Plavec and Polidan, 1975) which are at times visible even outside of eclipse. If the system were observed at an angle substantially different from 90°, we would speak of a Be star in a spectroscopic binary system. The other star would probably be invisible except in the infrared. For a truly conspicuous Be star, the emitting region would have to be larger compared to the central star. This can be achieved in a binary system of larger size, therefore of longer period, at least 30 days or more. Two systems of this type are known for sure, namely AX Mon and 17 Lep. Their spectra are rather complex, however, and they are not typical Be stars. Part of this anomaly may be due to the large mass and very large size of the mass-losing component: in both cases this is a large red giant, fairly easily detectable in the red region of the combined spectrum.

If AX Mon or 17 Lep were to represent typical examples of binary Be stars, we would have a serious problem with the absence of eclipses in Be stars. Suppose we have a large number of identical binary systems with a random orientation of their orbital planes. Let R_1 be the radius of the mass-accreting star, R_2 the radius of the mass-losing star, and A their separation. Typically in the interacting binaries we are considering here, the mass-losing star will be a giant of a later spectral type, and therefore considerably larger in size than the accreting B star. The fraction of systems that will display at least partial eclipses is

$$f_p = (R_1 + R_2)/A, \tag{1}$$

while for the probability of total eclipses we have similarly

$$f_t = (R_2 - R_1)/A. \tag{2}$$

Grazing partial eclipses will be invisible and $R_1 \ll R_2$, so that we can estimate the fraction of eclipsing systems to be roughly

$$f_e = R_2/A. \tag{3}$$

But the cool component is assumed to fill its critical Roche lobe, so that its fractional radius $R_2/A = r_2$ is fixed by the mass ratio of the components; approximately

$$r_2 = 0.38 - 0.3 \log (m_1/m_2) . \tag{4}$$

For AX Mon, we have probably a mass ratio of about 3, so that $r_2 = 0.28$. If Be stars were built on a model similar to AX Mon, 28 out of each 100 Be stars would display visible eclipses. Not much photometric work has been done systematically on Be stars, in particular on the faint ones, thus we have no good statistical surveys: nevertheless, the above figure is unacceptable and would simply mean that the binary model represents only a small fraction of all Be stars.

More plausible are models with mass ratios $m_1/m_2 = 10$ or even more. The fraction of eclipsing systems will then drop below 20%. Moreover, the size of the emitting disk will then be conveniently large, for this dimension is probably given roughly by the radius of the critical Roche lobe about the accreting star, which is

$$R_{\text{disk}} = A (0.38 + 0.2 \log (m_1/m_2)) . \tag{5}$$

Nevertheless, I must admit that the simple calculation of probability of eclipses appears to me as a rather serious objection to the idea that most Be stars can be explained as interacting binaries. However, we have not really tried yet to develop and evolve binary models specifically meant to represent Be stars.

I think it is fair to conclude that at the present time, it is impossible to decide what fraction of Be stars are interacting binaries. Both competing models should be further studied, developed and tested.

Acknowledgements

My thanks are due to Ms Marietta L. Eaker for typing the manuscript. The work reported in this session of the Symposium by R. S. Polidan, G. J. Peters, and myself, has been to a large degree supported by NSF grant MPS 74-04194A01 (Popper/Plavec).

References

Bath, G. T.: 1969, *Astrophys. J.* **158**, 571.
Bath, G. T.: 1972, *Astrophys. J.* **173**, 121.
Batten, A. H.: 1974, *Publ. Dom. Astrophys. Obs.* **14**, 191.
Batten, A. H., Fisher, W. A., Baldwin, B. W., and Scarfe, C. D.: 1975, *Nature* **253**, 174.
Benson, R. S.: 1970, *Bull. Am. Astron. Soc.* **2**, 295.
Biermann, P. and Hall, D. S.: 1973, *Astron. Astrophys.* **27**, 249.
Kříž, S.: 1969, in M. Hack (ed.), *Mass Loss from Stars*, D. Reidel, Publ. Co., Dordrecht-Holland, p. 257.
Limber, D. N. and Marlborough, M. J.: 1968, *Astrophys. J.* **152**, 181.
Lubow, S. H. and Shu, F. H.: 1975, *Astrophys. J.* **198**, 383.
Paczyński, B.: 1971, *Ann. Rev. Astron. Astrophys.* **9**, 183.
Plavec, M.: 1970a, in A. Slettebak (ed.), *Stellar Rotation*, D. Reidel, Publ. Co., Dordrecht-Holland, p. 133.
Plavec, M.: 1970b, *Publ. Astron. Soc. Pacific* **82**, 957.
Plavec, M. and Horn, J.: 1969, in M. Hack (ed.), *Mass Loss from Stars*, Reidel, Publ. Co., Dordrecht-Holland, p. 242.
Plavec, M., Polidan, R. S., and Peters, G. J.: 1973, *Bull. Am. Astron. Soc.* **5**, 413.
Plavec, M. and Polidan, R. S.: 1975, *Nature* **253**, 173.
van den Heuvel, E. P. J.: in A. Slettebak (ed.), *Stellar Rotation*, Reidel, Publ. Co., Dordrecht-Holland, p. 178.

DISCUSSION

Conti: Your argument about V/R variations being of necessity due to a binary seem pretty conclusive. It could be that V/R variations can come from non-mass exchange systems in some cases (e.g. ϕ Per), but I think it is also true that not all Be stars show V/R variations. Perhaps the V/R variation stars are the binaries and those without this variation are single. A decision on duplicity would then be relatively straightforward.

Plavec: It would be excellent if your idea were correct. At this moment, I do not think we are able to decide.

Young: If the secondary star is in fact supplying most of the angular momentum to the Be star that must come out of the orbital angular momentum and so presumably the orbital period will change. Have you calculated or estimated if the changes might be observable over, say, a 10-year period?

Plavec: Yes. Calculations have been made. The big problem is whether some part of the material is leaving the system. This is a great unknown. If you play with that, unfortunately, you can get almost any numbers you like. But at least in the case of β Lyr and U Cep, period changes have been detected and they are in agreement with what one can expect.

Slettebak: You mentioned mechanisms for getting the material out into the shell in the case of single stars, and there may be some difficulties with the stellar wind model. There was a suggestion made at the Stellar Rotation Colloquium in 1969 by Roxburgh (*Stellar Rotation*, ed. A. Slettebak, D. Reidel Publ. Co., Dordrecht-Holland, 1970, p. 19) in which he pointed out that the temperature and pressure in the equatorial regions of a rapidly rotating star will at some time resemble that in the atmospheres of pulsating stars. You will get an A-type giant atmosphere, as you have in RR Lyrae stars, and so a pulsational instability may set in which may eject material. I do not believe that this suggestion has ever been discussed in a quantitative way.

Plavec: I agree that this is one possibility. Incidentally I am not against the stellar wind explanation; I only think that it is incomplete. If there is a strong stellar wind blowing all the time, how do you explain that in spite of that there is an envelope around the star with a great outflow which stops and then falls back, in the case of 48 Lib, or as in the case of Pleione?

Slettebak: I think a second problem with the stellar wind is one that I referred to before. If you believe that gravity darkening takes place, then you have trouble explaining how the dark equatorial regions, as you would expect in the later Be stars, can produce a strong stellar wind. So I was just proposing this pulsation idea in order to get it out on the floor for discussion.

Cowley: If we accept Plavec's idea of the binaries being of two types, that is, one where there is still mass transfer going on and another type where the stars have evolved beyond that so that they do have helium cores and have gone to the stage where they can no longer transfer material, we can propose an observational test. For example, for the ones in which you now have a contracted core you no longer have a contact system and therefore you would expect that the emitting shell would always be constant. Therefore we should look in systems in which there are no known variations, where the helium and hydrogen emission does not come and go, for evidence of this kind of a companion; whereas in the ones where it comes and goes we should expect then only to find late-type subgiant companions.

Plavec: I agree completely. If there exist binary systems in which the mass transfer has terminated, the only change we should observe in the emitting disk would be its secular fading.

PART VII

GENERAL DISCUSSION

GENERAL DISCUSSION

Heap: What is the range in excitation among Be shells; and within a given shell, what is the range in excitation?

Cowley and others: The excitation in Be shells varies enormously from star to star. The typical lines seen are due to Fe II, Ti II, etc. and also the hydrogen lines, but in some of the hotter stars you see helium lines (λ 3889, etc.), which are also due to the shell. One can place limits: at the hot end, one does not see He II shell lines and at the cool end, neutral metallic lines are not usually seen. Within a given shell, the range of excitation is much less.

Conti: I am concerned about the classification of Be stars, because in some cases things are somewhat confusing. There are really several kinds of stars which we are collectively referring to as Be stars. I think we are in a position now where by looking at the spectra we can distinguish between certain types of B-type stars which show emission. I would like to propose that we call a star a Be star if it is a more-or-less main sequence B-type star with emission in the hydrogen lines and sometimes in the helium lines (I do not believe we are at a stage yet where we can distinguish between a single and a binary star so I don't think we can put that into a classification system). A second class of objects would be those B-type stars which show forbidden emission lines and I would suggest that we classify these as B with a small e in brackets B[e], following the notation for forbidden lines. The so-called Herbig Ae and Be stars might all collectively be called Ae stars because most are A types anyway. A distinguishing nomenclature would help us recognize their rather different evolutionary history. The supergiant stars should not be collectively referred to as Be stars at all, as they have emission because of stellar winds. I would suggest that they be called only by their spectral type and luminosity class, which can be followed by a small e, as for example B5Ie, and referred to as supergiants and not as Be stars. The tiny group of O type stars which show central hydrogen emission can be called Oe stars, as I have suggested earlier. I would also propose the small letter p be used only for those objects which do not otherwise fit into these classes. It would be desirable to have some notation for stars which exhibit shells, but we have not yet been able to define a shell star and so I think that we should not include this into our classification scheme at this time.

Garrison: We must consider what to put into classification schemes. It seems to me that it is premature to attempt a detailed classification of Be stars. In lieu of that, I would like to suggest the following: that something like the scheme of Conti be included in the spectral type column of a table and that a fairly complete description of peculiarities be put in the notes to the table, with the following guidelines:

Table

(1) Spectral type, luminosity class, rotation (n or nn), and emission (e).

Notes

(2) difference in line width between hydrogen and helium, if any.

A. Slettebak (ed.), Be and Shell Stars, 447–451. All Rights Reserved
Copyright © 1976 by the IAU.

(3) whether hydrogen emission is narrow or broad.

(4) absorption reversal in cores ('shell' formerly).

(5) Fe II emission or 'shell' absorption.

(6) veiling or apparent filling in of lines.

(7) infrared: O I, Ca II.

(8) ultraviolet: Si IV, Mg II, Fe III.

(9) variation and time scale (if known).

(10) line asymmetries (P Cyg, inverse P Cyg, V/R, etc.)

Since Be stars are probably all variable, the presence of any of these characteristics at any epoch shows that the particular star is *capable* of showing them (e.g. only some stars become 'shell' stars).

I would like to emphasize that the detailed description should be relegated to the notes and that, *at the present time*, we *not* try to incorporate all of this information into the classification scheme. It may be that some day when we understand Be stars, we will be able to couple several of these characteristics, but attempts to do that now, as with Lesh's system, are bound to fall into disuse (because Fe II is not always coupled with strong hydrogen emission, for example). Also we must avoid a scheme with too many parameters because only the proponent of the scheme will ever be able to remember it (e.g., a well-known galaxy classification scheme).

I am willing to start this with the extensive table of southern OB star classifications that I will be preparing in a few months; i.e., indication of emission in the table with extensive description in the notes at the end of the table.

Snow: My suggestion of a useful note to include among all the other spectroscopic descriptions would be one saying whether or not line asymmetries indicative of mass loss are seen in the ultraviolet, and if so, which lines show mass loss effects.

Schild: I should like to support the suggestion of Dr Garrison. It would be opening Pandora's box to try to expand the present classifications as listed in the spectral type column of a table. At the same time, it is extremely useful to have remarks to such a table, organized in a systematic way, so that a person desiring to make further observations of a class of stars (such as all stars showing Fe II emission, or sharp He I absorption, or possibly veiling of the He I spectrum) could easily select a list of objects. It is particularly useful if the information is organized in a systematic way.

Cowley: What observational quantities do the theorists need?

Hummer: It is important to try to make as much use of continuum information as possible. Well selected, high accuracy profiles are also important. I think that low dispersion material is almost useless for model making. It is important to get a wide variety of lines and to get some photospheric lines as well as shell lines, so that we can get some idea of what the inside of the star looks like as well as the outside.

Marlborough: I agree with everything that Hummer says. It is important to give data which are quantitative instead of just qualitative. I also support Hummer's suggestion that continuum data are very useful. Also, it is important to have accurate line profiles as Hummer suggested – that is, lines of different excitation potentials, shell lines, emission lines, and stellar lines, all with the highest dispersion possible. Simultaneous photometry would also be useful, especially if the stars vary. Narrow band interference filter photometry would help to determine whether the lines vary, whether the continuum varies, or whether both vary.

Hummer: Speaking as an ignorant theoretician it would be useful for those of us involved in model building if you could give examples of a well observed, well behaved, non-varying, classical Be star.

Cowley: 1 Del.

Schild: χ Oph.

Someone: But the polarization varies.

Heard: If you insist on a star which does not vary, you might end up picking the most unusual kind of Be star.

Hummer: How about a well behaved shell star?

Schild: Pleione is typical. It loses the shell, the shell dissipates, it becomes a Be star and then it becomes a normal garden-variety, main-sequence B8 star. It is in a cluster. Even though it is variable, Pleione is a universal Be star.

I seem to recall that many of McLaughlin's observations of V/R variations do not show strict periodicity, but rather show quasi-periodicity. Does not the binary star hypothesis for Be stars require exact periodicity, even though the amplitude of the effect may vary? Can the binary star proponents comment?

Harmanec: It is necessary to consider two types of V/R variation. The first type may last days or up to several hundred days and is strictly periodic. But there is also a 5–50 year variation which may be related to the slow rotation of an elliptical ring, formed during a non-continuous process of mass transfer around the mass-gaining component. This type of variation need not be periodic but may be simply cyclic instead, reflecting thus a gradual circularization of the elliptical ring.

Hutchings: If large amounts of mass are being transferred would you not expect to see peculiar abundances because you are coming right down to nuclear processed material? (as e.g., in the supergiant binary HD 163181).

Plavec: I do not think you would recognize too many changes. You may remember that Bolton in Cambridge claims that the anomalous abundances in certain O and B stars might be due to this, but I do not think you can expect too much.

Poeckert: For ζ Tau or ϕ Per, which have periods of 100 days or more, would V/R variations which are less than the orbital period be a problem in the binary hypothesis?

Plavec: Yes. If we are certain that the orbital period in ϕ Per is 130 days, say, and a V/R variation of 80 days were observed, it would be a problem for the binary hypothesis.

Peters: What are V/R variations? For a number of years, I have been attempting to determine the cause of this type of profile variation. In the case of HR 2142, V/R variations can be understood in terms of variable absorption components on the blue sides of the emission line profiles.

How many high dispersion observations of Hα and Hβ exist for Be stars and what is the extent of the time coverage? Many older observations did not show structure because the plates were of too low a dispersion. Usually, only the instrumental profile (a Gaussian) was observed.

Hutchings: In Victoria, we have high quality photographic, photoelectric, and TV profiles of some bright Be stars. However, coverage is sporadic and irregular. I imagine the same is true at, say, Haute Provence or Lick or Perkins.

Plavec: We heard it said at this Symposium that some of the stars observed with the *Copernicus* satellite show a different velocity of rotation as derived from the lines in the ultraviolet as compared with rotations derived in the normal visual region. I think that this is a very exciting fact; I would like to hear more about this and if possible some explanation.

Snow: I have noticed one or two cases of this, though these are not Be stars. One example is 42 Ori, a B1V star, which shows lines sharper in the ultraviolet than you would expect from $v \sin i$'s based on visual spectra.

Peters: I wonder if this effect is not confined to the shell stars, because I do not find this effect in the 'classical' Be stars which I observed, v Cyg or μ Cen.

Heap: Most of the ultraviolet stellar lines that are most useful have broad wings (damping wings) and so one does not want to just measure the full widths of the lines and say, "Ah, I have a rotational velocity". I think one must compare the observed profiles against theoretical profiles to estimate rotational velocity. I have computed a grid of profiles for Si III, Si IV, C III, and C IV lines, for temperatures from 17 000 to 30 000°. In the case of ζ Tau, which is the only star I have looked at, I found that the rotational velocity is the same regardless of which line I looked at in the ultraviolet. And this bothers me. I would have thought that Si III λ 1206, being a lower excitation line than Si IV λ 1393, λ 1403, would be formed nearer to the equator and would therefore have a higher projected rotational velocity, but that was not the case.

Plavec: And the rotational velocities you derived are significantly lower in the ultraviolet than in the ordinary blue region?

Heap: Yes.

Hutchings: I might suggest that if you are looking at a star with a temperature gradient across its surface, in the far ultraviolet the underlying continuum radiation may come from the polar regions, where one would expect $v \sin i$ to be small. You may then expect to see a dependence of $v \sin i$ on the line wavelength.

Poeckert: Perhaps the theoreticians should get to work and calculate some ultraviolet line profiles to compare with the visual ones.

Heap: I have computed profiles for a spherically symmetric model. Gravity-darkened profiles for ultraviolet lines are badly needed.

Cowley: One thing that is very important to do now is to have very careful photometry. It seems to me that what we know is that the Be stars are variable but that is about all we know. Especially for the systems where there is some evidence that they are binary, we need good photometry, perhaps in one or two cycles. It is not good enough to have an observation now and one three months from now and one three years from now and to try to put them together because there may be erratic variations from cycle to cycle.

Feinstein: It would be most desirable to have simultaneous spectroscopy and photoelectric photometry.

Slettebak: I would like to raise two questions which I brought up earlier in this Symposium. The first goes back to the single star hypothesis. Bidelman raised the question explicitly in his summarizing remarks as to how the ring is formed. I would like to ask the theoreticians present whether they have any ideas with regard to the pulsational hypothesis which I mentioned earlier as a trigger to get material into the ring, and whether such a process might be feasible. My second question has to do with the fact that very broad Hα emission is observed in some Be stars, much broader than the absorption lines. How does one understand this? Electron scattering and turbulence have been suggested – do you have any ideas about this?

Harmanec: There was a paper by Morgan in the *Astrophysical Journal* (**195**, 391, 1975) in which he computed some hydrodynamics in rotating single stars. I would like to hear more about this work.

Marlborough: What he did was to perform a linear stability analysis of Limber's hydrostatic solutions and looked for unstable temporal and angular modes. He found that he could obtain unstable modes with a variety of periods and, in particular, he could get variations with time scales of the order of a few minutes. Now there was no consideration of what could initiate such variations, so he was only showing that instabilities could grow and you could get variations with this time scale. So it is possible that circumstellar envelopes around single stars have certain instabilities in them which might be able to account for some of the variations people have seen, but in all these linear stability analyses you do not know whether the perturbations will grow or not unless you go to the non-linear theory. But at least certain modes are unstable. With regard to the large Hα emission widths, I suggested long ago that these might be due to electron scattering in the inner regions of a very dense disk, but no one has made any calculations along these lines. But the densities are high enough and the electron velocities are large enough at, say 10 000°, to qualitatively explain this effect.

Hummer: Auer and Mihalas (*Astrophys. J.* **153**, 245, 923, 1968) have done calculations with electron scattering for O-type stars to see how well electrons do really broaden lines. They found that because the electron scattering cross section is so very small, under many conditions the continuous absorption gets most of the photons before they can be scattered by electrons.

Coyne: Could someone give a list of five stars which would be the most interesting for someone to get a light curve for?

Peters: HR 2142.

Plavec: 4 Her.

Herman: I would like to say that our photographic catalogue of 250 Be stars arranged according to their classification by the intensity and depth of the hydrogen lines, will be published next year.

CONCLUDING REMARKS

WILLIAM P. BIDELMAN

Warner and Swasey Observatory, Case Western Reserve University, Cleveland, Ohio, U.S.A.

It is now time to ring down the curtain on the drama entitled 'IAU Symposium No. 70'. In this presentation in which the actors have doubled as the audience, some of us have acted as principals, some have played supporting roles, some have been bit players, some have walked on, and others have been stage hands. Lest you think I am being unduly histrionic, I would remind you of that immortal line from Hamlet, namely, the one that describes the predicament of a rapidly rotating early type star, which goes "to Be or not to Be: that is the question".* Our play may be characterized as partly historical, partly tragical, as we have seen some of our revered old ideas called seriously into question, and partly comical, though not always intentionally so. There has been some resemblance to 'The Tempest', and, finally, some of us who are not so deeply immersed in the subject as other might claim that there has been 'Much Ado About Nothing'. For certain, however, it could not possibly be called a farce.

Almost everybody's concluding remarks at such a meeting as this include the statement that "there is no point in attempting to summarize the discussion". I agree; it would be quite impossible for me to elucidate the theoretical papers or to add anything significant to the observational ones. What I would like to do is to introduce a bit of systematization into the subject, and to make a very few comments concerning some of the things that we have talked about.

It is obvious that among the stars that would have been considered by Merrill and McLaughlin as B-emission stars we have a great diversity of objects. I would divide them into at least five classes:

B-Emission Objects

(1) Supergiants
(2) Rapidly rotating single stars
(3) Interacting binaries
(4) Early type nebular variables
(5) Quasi (or young) planetary nebulae

All of these types of objects are loosely occasionally termed 'Be' stars, but in view of their diversity it is essential that we attempt to make the theories fit the observational data in each case. In other words, if an object is a binary no theory based on the hypothesis that it is a single star can prove satisfactory in the long run, and vice versa.

Now a few comments: (1) The supergiants comprise perhaps 15% of all of the stars in the Be catalogues. We have had almost no discussion of these objects except for the *Copernicus* data demonstrating their appreciable mass loss and Dr Hummer's

* The editor assumes no responsibility for any comments made in this or other papers. (A.S.)

A. Slettebak (ed.), Be and Shell Stars, 453–455. All Rights Reserved
Copyright © 1976 by the IAU.

theoretical paper on line formation in spherically symmetric expanding atmospheres which should apply in these cases.

(2) Despite today's eloquence, I am inclined to believe that the majority of the Be stars that we know today are really single rather than double stars. At least I think the presumption is in this direction. However, from the discussions it appears that there is a crucial question involved in the single-star hypothesis: why is the Be and/or shell phenomenon so markedly irregular? Why should Pleione suddenly decide to eject a shell and then stop? Why not continuous ejection of matter? The ultraviolet observations, again, have made a real contribution in pointing out that these stars have appreciable mass loss. Polarization and infrared and radio observations are capable of telling us a lot about the envelopes, though they may not contribute much to answering the important question just posed. In connection with the general problem of stellar shells, there is no doubt that we either need some agreement on the definition of a shell star, or at least we should always make clear exactly what type of shell we are talking about. Is there shell absorption in the hydrogen lines, and if so, which? Or is there also strong metallic-line absorption? I hope that we may be able to provide a useful list of shell stars of various types, though the often discontinuous nature of the shell phenomenon necessitates continued spectroscopic observations.

(3) I am amazed that we have not heard a single paper on β Lyrae. I find it hard to believe that we could have had a Be-star symposium without at least four or five such papers. My only conclusion is that we must have given up on this enigmatic binary, but I cannot help but feel strongly that we should spend more time on objects that we already know something about rather than frittering it all away on 10th to 15th magnitude stars about which we know practically nothing! Surely, understanding β Lyrae should still be one of astrophysics' highest priority goals. In connection with the general problem of binaries it is perhaps worth emphasizing that the detection of secondary components that are decidedly cooler than their primaries may be a comparatively simple task with the aid of infrared or millimeter-wave observations, but the situation is not at all as favorable if the secondaries are of comparable spectral type to their primaries. We should never forget that binaries with early type primaries can have secondaries that are, as far as we know, main sequence, giant or sub-giant stars, horizontal-branch stars, white dwarfs, neutron stars, or even, of course, black holes. We must be prepared for the worst.

(4) I only mention the nebular variables in order that they are not forgotten. There are some in Merrill's catalogues, and, curiously enough, the spectra of many are not so very different from those of ordinary Be or shell stars. For example, a spectrum of BF Orionis that I once obtained resembles that of a conventional A-type shell star to a surprising extent. We have heard almost nothing about these interesting objects at this meeting, although it is likely that a number of the curious pathological objects picked up in the far infrared and radio range belong to this class.

(5) Again, little has been said of the fascinating objects of the 5th class aside from their probable radio detection. They are certainly important objects for study. At the moment there is much confusion between these and other varieties of emission stars such as the symbiotics, and much clarification remains to be done.

So far I have said nothing profound; and also, I have as yet failed to say anything nice about the various papers. The review papers all appeared to me to have been excellent, and I am looking forward to seeing them in print where I can begin to absorb them. The contributed papers were, as usual, a somewhat mixed bag; but none seemed to be sheer nonsense, as occasionally happens, and we should all be thankful for that.

Finally, I will suggest that we are all working on a giant jigsaw puzzle, or perhaps better, on a number of them. We all know what happens when we do this: sometimes progress is painfully slow until one person fits a piece in and then the way is open for others to make their contributions. Once in a while it turns out that a few of the pieces have fallen on the floor and been overlooked; someone happens to find them, and they make all the difference. Sooner or later the puzzle is done; we admire it for a while and then tear it up and turn enthusiastically to another apparently hopelessly scrambled picture.

APPENDIX

WILLIAM P. BIDELMAN

Warner and Swasey Observatory, Case Western Reserve University, Cleveland, Ohio, U.S.A.

A List of Early-Type Shell Stars

The following table lists many of the hotter stars whose spectra have occasionally or continuously exhibited moderately conspicuous shell phenomena. It includes such well-known objects as P Cygni, γ Cassiopeiae, β Lyrae and R Monocerotis as well as a host of other less-studied and less-well-understood objects. Novae, classical T Tauri, VV Cephei and symbiotic stars, and stars situated outside the Galaxy have been excluded. The list is undoubtedly very incomplete: many weak or uncertain cases have been left out. It should be remembered that shell phenomena are notoriously variable in time and, in the binaries, often periodically so; thus inclusion in the list does not guarantee that an entry will exhibit shell structure at any specific time.

An attempt has been made to distinguish between the objects whenever possible: category (1) includes Of stars and O, B, and A supergiants showing P Cygni profiles; (2) includes presumably single rapidly rotating stars of classes O, B, A and F; (3) contains stars deemed to be binary, though the shell phenomena may not always be related to this fact; and (4) is devoted to the early-type nebular objects. In many cases assignment to a particular group is very uncertain or impossible, and in the absence of fairly definite evidence to the contrary stars have generally been assigned to category (2). Many of the brighter Be stars have been found to have variable radial velocity, but they have not been considered to be binaries unless a period has been determined.

A single reference, not always the most recent, has been given for each entry. The reader is strongly urged to consult the various bibliographies of emission-line objects for additional important data. Magnitudes are visual unless underlined; some useful information has been omitted for lack of space. 'K' numbers refer to the suspected variable catalogues, 'Carlson' to his 1968 Northwestern U. Dissertation. A few references are given to unpublished data from several other investigators.

The writer is happy to acknowledge that this work, while largely completed in Cleveland, was put into final form while he was serving as Visiting Professor in the Department of Geophysics and Astronomy of the University of British Columbia in January 1976.

A. Slettebak (ed.), Be and Shell Stars, 457–465. All Rights Reserved
Copyright © 1976 by the IAU.

EARLY-TYPE SHELL STARS

Cate-gory	Name	Catalogue designation	α(1900)	δ	Magnitude	Reference[a]	Notes
			h m	° ′			
1	K 102301	HD 108	0 0.9	+63 7	7.4	*A + A* **29**, 171	Of star
2	HR 10	HD 256	0 2.2	−17 57	6.2	*ApJ* **182**, 809	A star
3	SX Cas	+54 7	0 5.5	+54 20	9.7–11.2	*ApJ* **99**, 89	*P* = 36.6 d
3	K 100007	HD 698	0 6.3	+57 39	7.1	*ApJ* **108**, 537	*P* = 55.9 d
2	MWC 5	+61 39	0 14.8	+61 54	8.5	*ApJ* **97**, 217	
?	VX Cas		0 25.7	+61 27	10.7–13.3	*Ast* **9**, 104	
2	MWC 419	+61 154	0 37.5	+61 22	10.6	*ApJ* **173**, 353	
2	o Cas	HD 4180	0.39.2	+47 44	4.5	*A + A* **18**, 106	
2	γ Cas	HD 5394	0 50.7	+60 11	1.6–3.0	*ApJ* **98**, 153	
3	U Cep	HD 5679	0 53.4	+81 20	6.6–9.8	*DAO* **14**, 191	*P* = 2.5 d
2	MWC 10	HD 6343	0 59.4	+65 26	7.1	*DDO* **2**, 315	
2	LS + 58° 26		1 8.9	+57 46	11.2	*ApJ* **137**, 547	
2		+57 240	1 12.3	+57 51	9.5	*ApJS* **2**, 389	HD 236689
3	AQ Cas	+61 242	1 12.6	+61 51	10.0–11.0	*ApJ* **104**, 253	*P* = 11.7 d
2	MWC 426	HD 9709	1 30.0	+46 36	7.0	*ApJ* **145**, 121	
2	α Eri	HD 10144	1 34.0	−57 45	0.5	*ApJ* **157**, 313	
3	φ Per	HD 10516	1 37.4	+50 11	4.1	*ApJ* **98**, 153	*P* = 126.6 d
2	ε Cas	HD 11415	1 47.2	+63 11	3.4	*ApJS* **17**, 371	K 100141
2	MWC 23	HD 12302	1 55.6	+59 12	8.0	*ApJ* **119**, 496	
2	MWC 26	HD 12882	2 1.1	+64 34	7.5	*RicA* **8**, 353	
1	HR 618	HD 12953	2 1.7	+57 57	5.6	*ApJ* **182**, 523	MWC 436
2	MWC 702	+55 521	2 2.1	+55 43	9.8	*ApJ* **154**, 933	
2	V351 Per	HD 13051	2 2.6	+56 31	8.7v	*ApJ* **154**, 933	MWC 27
2	MWC 706	HD 13590	2 7.6	+63 33	7.9	*RicA* **8**, 353	
2	MWC 29	HD 13661	2 8.1	+54 4	8.6	*ApJ* **154**, 933	
2	MWC 707	HD 13669	2 8.2	+55 20	7.9	*ApJ* **154**, 933	
2	V358 Per	HD 13890	2 10.2	+56 19	8.5–8.6	*ApJ* **154**, 933	MWC 443
?	VZ Cet	o Cet(B)	2 14.3	− 3 26	9.5–12	*ApJS* **1**, 39	MWC 35
2	MWC 36	+56 563	2 14.7	+56 40	9.4	*ApJ* **154**, 933	
2	MWC 710	+56 579	2 15.2	+57 11	9.5	*ApJ* **154**, 933	
1	9 Per	HD 14489	2 15.3	+55 23	5.2	*ApJ* **182**, 523	
1		HD 14535	2 15.8	+56 47	7.5	Bidelman	
2	MWC 46	+56 624	2 19.6	+56 39	9.6	*ApJ* **154**, 933	
2	V529 Cas	HD 15238	2 22.2	+60 13	8.4	*ApJ* **119**, 496	MWC 47
2	V528 Cas	HD 15239	2 22.2	+60 12	8.2	*ApJ* **119**, 496	
2	HR 716	HD 15253	2 22.4	+55 5	6.6	*PASP* **80**, 685	A star
2	MWC 49	HD 15472	2 24.4	+70 31	7.9	*DDO* **2**, 315	
2	MWC 451	+55 643(B)	2 26.2	+55 52	9.8	*ApJ* **101**, 224	
2	MWC 51	+60 510	2 26.3	+60 34	9.0	*ApJ* **98**, 91	
2	MWC 717	HD 15963	2 28.9	+57 38	8.0	*ApJ* **154**, 933	
3	RY Per	HD 17034	2 39.0	+47 43	8.5–10.7	*ApJ* **104**, 396	*P* = 6.9 d
2	MWC 59	+60 606	2 51.5	+60 12	9.1	*PASP* **86**, 558	
2	HR 894	HD 18552	2 53.8	+37 45	5.9	*ApJ* **145**, 121	MWC 455
3	RX Cas	+67 244	2 58.8	+67 11	8.6–9.5	*ApJ* **99**, 295	*P* = 32.3 d
2	MWC 61	HD 19243	3 0.7	+62 0	6.7	*DDO* **2**, 315	
2	MWC 63	HD 20017	3 7.9	+48 19	7.9	*DDO* **2**, 315	
2	HR 985	HD 20336	3 11.2	+65 17	4.8	*ApJ* **137**, 1085	K 100264
2		+49 916	3 17.1	+49 17	9.5	*ApJS* **2**, 389	
2	HR 1051	HD 21551	3 23.5	+47 46	5.8	*ApJ* **136**, 381	
2	MWC 727	HD 21641	3 24.5	+47 31	6.8	*ApJ* **145**, 121	
2	MWC 68	HD 21650	3 24.6	+41 23	7.3	*DDO* **2**, 315	
2	ψ Per	HD 22192	3 29.4	+47 52	4.2	*Mich* **4**, 175	MWC 69
2	13 Tau	HD 23016	3 36.5	+19 23	5.5	*AJ* **65**, 535	
2		HD 23478	3 40.4	+32 0	6.6	*A + A* **33**, 473	

Early-type Shell Stars (*continued*)

Category	Name	Catalogue designation	α(1900) h m	δ ° ′	Magnitude	Reference[a]	Notes
2	HR 1160	HD 23552	3 41.0	+50 26	6.1	*ApJ* **145**, 121	MWC 464
2	η Tau	HD 23630	3 41.5	+23 48	2.9	*MN* **113**, 477	MWC 74
2	BU Tau	HD 23862	3 43.3	+23 50	4.8–5.5	*ApJ* **115**, 145	MWC 75
2	MWC 76	HD 23982	3 44.4	+63 11	8.0	*DDO* **2**, 315	
3	RW Tau	HD 25487	3 57.8	+27 51	8.0–12.5	*ApJ* **110**, 438	P = 2.8 d
3	K 377	HD 25799	4 0.3	+32 6	7.0	*ApJ* **137**, 791	P = 10.7 d
2	HR 1289	HD 26356	4 5.0	+83 34	5.4	*ApJ* **196**, 773	
2	MWC 82	HD 26420	4 5.7	+41 52	7.9	*DDO* **2**, 315	
2	MWC 83	HD 26906	4 10.1	+45 59	7.9	*DDO* **2**, 315	
3	RW Per	+41 851	4 13.3	+42 4	9.7–11.4	*ApJ* **102**, 74	P = 13.2 d
2	HR 1423	HD 28497	4 24.5	−13 16	5.6	*DDO* **2**, 315	MWC 86
2	Hen 4	HD 29557	4 34.1	−24 52	8.6	*AJ* **78**, 687	
2	HR 1500	HD 29866	4 37.3	+40 36	6.1	*DDO* **2**, 315	MWC 88
3	KS Per	HD 30353	4 41.8	+43 6	7.6–7.8	*PASJ* **24**, 495	P = 363.5 d
3	RS Cep		4 48.6	+80 6	10.2–11.9	*ApJ* **104**, 253	P = 12.4 d
4	AB Aur	HD 31293	4 49.4	+30 23	7.2–8.4	*ApJ* **173**, 353	MWC 93
?	MWC 480	HD 31648	4 52.4	+29 41	7.5	*ApJ* **119**, 501	
1 + 3	ε Aur	HD 31964	4 54.8	+43 40	2.9–3.8	*JRASC* **43**, 15	P = 27.1 yr
?	UX Ori	−4 1029	4 59.5	− 3 56	8.7–12.8	*Ast* **9**, 104	HD 293782
3	103 Tau	HD 32990	5 2.0	+24 8	5.5	*A + A* **40**, 203	P = 58.3 d
2	105 Tau	HD 32991	5 2.0	+21 34	5.8	*ApJ* **143**, 285	MWC 98
2	K 100452	HD 33232	5 3.8	+40 53	8.2	*ApJ* **128**, 61	MWC 100
2	λ Eri	HD 33328	5 4.4	− 8 53	4.3	*ApJ* **196**, 773	K 100453
2	MWC 101	HD 33461	5 5.2	+41 6	7.8	*DAO* **12**, 1	
2	Hen 6	HD 33599	5 6.2	−61 56	8.3	*MemRAS* **75**, 1	
2	12 Aur	HD 33988	5 9.0	+46 18	6.9	*AN* **279**, 19	MWC 104
3	MWC 490	+33 997	5 11.2	+33 56	10.3	*ApJ* **115**, 154	HD 242257
2	MZ Aur	HD 34626	5 13.8	+36 32	8.0–8.1	*A + A* **33**, 473	
2	HR 1761	HD 34959	5 16.0	+ 3 55	6.6	*ApJS* **17**, 371	
2	HR 1772	HD 35165	5 17.7	−34 27	6.1	*Obs* **73**, 86	Hen 7
?	HR 1786	HD 35407	5 19.4	+ 2 16	6.3	*ApJ* **196**, 773	
2	25 Ori	HD 35439	5 19.6	+ 1 45	4.9	*Mich* **4**, 169	K 6170
2		HD 35502	5 20.0	− 2 54	7.4	*ApJ* **136**, 381	
2	120 Tau	HD 36576	5 27.6	+18 28	5.5	*AN* **277**, 167	MWC 111
4	T Ori	−5 1329	5 30.9	− 5 32	9.5–12.6	*ApJS* **4**, 337	MWC 763
2	K 100617	HD 37115	5 31.0	− 5 41	7.1	*DDO* **2**, 315	MWC 114
?	BN Ori	+6 971	5 31.1	+ 6 46	9.0–13.7	*Ast* **9**, 104	HD 245465
3	ζ Tau	HD 37202	5 31.7	+21 5	2.9–3.0	*A + A* **4**, 341	P = 132.9 d
4	BF Ori	−6 1259	5 32.3	− 6 39	9.8–13.4	*ApJ* **174**, 401	
4	RR Tau	+26 887a	5 33.3	+26 19	10.2–14.2	*ApJ* **173**, 353	HD 245906
2	MWC 120	HD 37806	5 36.0	− 2 46	7.9	*ApJ* **98**, 91	
2	HR 1962	HD 37971	5 37.2	−16 46	6.2	*ApJS* **17**, 371	
2	MWC 776	+26 954	5 41.2	+26 22	10.8	*ApJ* **110**, 387	HD 247525
2		HD 38708	5 42.6	+29 6	8.2	*ApJS* **2**, 389	
2	MWC 779	HD 39018	5 44.7	+18 0	7.5	Bidelman	
2	β Pic	HD 39060	5 44.9	−51 6	3.8	*ApJ* **197**, 137	A star
2	MWC 781	+28 920	5 45.6	+28 14	9.2	*ApJ* **110**, 387	HD 248411
2		+32 1146	5 53.0	+32 53	8.8	*ApJS* **2**, 389	HD 249845
2	MWC 786	+25 1065	5 53.8	+25 5	9.1	*ApJS* **2**, 389	HD 250028
3	DN Ori	HD 40632	5 55.0	+10 13	9.8–10.9	*PASP* **76**, 210	P = 13.0 d
4	MWC 789	+16 974	5 56.2	+16 31	9.7	*ApJ* **173**, 353	HD 250550
3	HR 2142	HD 41335	5 59.4	− 6 42	5.2	*PASP* **84**, 334	P = 81 d
3	17 Lep	HD 41511	6 0.5	−16 29	4.9	*ApJ* **143**, 121	P = 276 d
4	Lk 208		6 1.9	+18 40	12.7	*ApJS* **4**, 337	

Early-type Shell Stars (*continued*)

Cate-gory	Name	Catalogue designation	α(1900) h m	δ ° '	Magnitude	Reference[a]	Notes
2	HR 2174	HD 42111	6 3.7	+ 2 31	5.7	*PASP* **80**, 685	A star
2	MWC 523	HD 42908	6 8.2	+ 8 44	8.2	*ApJ* **98**, 91	
2	MWC 803	HD 44351	6 16.3	+14 21	8.5	*ApJ* **116**, 501	
2	MWC 527	+11 1179	6 21.3	+10 59	9.0	Bidelman	HD 257366
2	MWC 140	HD 45314	6 21.6	+14 57	6.6	Bidelman	
2	ν Gem	HD 45542	6 23.0	+20 17	4.2	*AN* **277**, 167	
2		HD 45626	6 23.4	− 4 24	9.4	*ApJS* **2**, 41	
?	FS CMa	HD 45677	6 23.7	−12 59	7.6–8.5	*A + A* **26**, 443	MWC 142
2	β Mon(A)	HD 45725	6 24.0	− 6 58	4.7	*A + A* **22**, 203	MWC 143
2	HR 2364	HD 45871	6 24.9	−32 18	5.8	*ApJ* **157**, 313	Hen 13
3	AX Mon	HD 45910	6 25.2	+ 5 56	6.6–6.9	*A + SS* **30**, 481	P = 232.5 d
2		HD 46131	6 26.4	−22 15	7.2	*AJ* **78**, 687	
2	Lk 215		6 27.2	+10 14	10.7	*ApJ* **173**, 353	
2	MWC 149	+8 1388	6 28.1	+ 8 25	8.6	*ApJS* **2**, 389	HD 259597
2	HR 2418	HD 47054	6 31.6	− 5 8	5.5	*ApJ* **145**, 121	MWC 150
3	HR 2422	HD 47129	6 32.0	+ 6 13	6.1	*ApJS* **4**, 157	P = 14.4 d
4	R Mon		6 33.7	+ 8 50	11.3–13.8	*ApJ* **152**, 439	MWC 151
4	V590 Mon		6 35.2	+ 9 53	12.7–14.0	*ApJS* **4**, 337	
2		HD 48914	6 40.7	+ 2 37	7.2	*DAO* **12**, 1	
2	MWC 821	HD 49330	6 42.8	+ 0 53	8.9	*ApJ* **110**, 387	
3	RX Gem	HD 49521	6 43.6	+33 21	9.2–11.2	*A + A* **38**, 225	P = 12.2 d
?	MWC 822	HD 49699	6 44.6	−12 33	7.5	Bidelman	
2	κ CMa	HD 50013	6 46.1	−32 24	4.0	*PASP* **87**, 137	K 6509
2		HD 50091	6 46.5	−13 7	8.6	*ApJS* **2**, 41	
?	MWC 158	HD 50138	6 46.7	− 6 51	6.6	*ApJ* **119**, 501	
2	MWC 159	HD 50209	6 47.1	− 0 11	8.4	*DDO* **2**, 315	
2	ψ⁹ Aur	HD 50658	6 49.1	+46 24	5.8	*A + AS* **9**, 133	MWC 537
3	AU Mon	HD 50846	6 49.8	− 1 15	8.2–9.5	*ApJ* **101**, 235	P = 11.1 d
2	MWC 539	HD 50850	6 49.8	−18 10	9.1	*ApJ* **98**, 153	
2	MWC 100	HD 51354	6 51.9	+18 2	7.1	*DAO* **5**, 1	
?	K 924	HD 51480	6 52.4	−10 42	7.0	*PASP* **80**, 197	MWC 161
4	Z CMa	HD 53179	6 59.0	−11 24	8.8–11.2	*ApJ* **173**, 353	MWC 165
2	MWC 838	HD 54858	7 5.4	− 9 10	8.4	*ApJ* **116**, 501	
2	MWC 839	HD 55806	7 9.4	+ 3 4	8.8	*AJ* **78**, 687	
?	EW CMa	HD 56014	7 10.2	−26 11	4.3–4.6	*ApJ* **119**, 496	MWC 170
2	HR 2787	HD 57150	7 14.8	−36 33	4.7	*PASP* **77**, 376	MWC 173
4	K 1025	−44 3318	7 16.4	−44 24	10.0	*PASP* **87**, 87	Hen 32
2		−17 1952	7 19.0	−18 1	11.6	*W + S* **1**, 1	
2	HR 2819	HD 58155	7 19.2	−31 44	5.4	*ApJ* **157**, 313	
?	MWC 560		7 21.0	− 7 31	12.5	*ApJ* **98**, 153	
	RY Gem	HD 58713	7 21.7	+15 52	8.5–11.3	*DAO* **8**, 235	P = 9.3 d
2	β CMi	HD 58715	7 21.7	+ 8 29	2.9	*ApJ* **119**, 146	K 6586
2	FY CMa	HD 58978	7 22.8	−22 53	5.6–5.7	*ApJ* **119**, 496	MWC 179
3		HD 59771	7 26.4	−18 2	9.1	*ApJ* **115**, 154	
2	HR 2911	HD 60606	7 30.2	−36 7	5.6	*PASP* **77**, 376	MWC 183
2	HR 2932	HD 61224	7 33.1	−14 13	6.5	*ApJ* **145**, 121	MWC 849
?		−30 5135	7 45.2	−30 53	9.2v	Bidelman	
3	UX Mon	HD 65607	7 54.4	− 7 14	8.0–8.9	*ApJ* **106**, 255	P = 5.9 d
2	HR 3135	HD 65875	7 55.7	− 2 36	6.5	*A + A* **3**, 485	MWC 190
2	Hen 109	−60 968	7 56.1	−60 33	9.7	*AJ* **74**, 813	
3	XY Pup	HD 67862	8 4.8	−11 42	9.2–11.4	*PASP* **74**, 129	P = 13.8 d
2	MWC 577	HD 68468	8 7.4	−13 52	8.5	*ApJ* **119**, 496	
2	MX Pup	HD 68980	8 9.7	−35 36	4.6–4.9	*ApJ* **197**, 137	MWC 192
2	Hen 150	HD 70461	8 17.0	−26 1	9.6	*AJ* **78**, 687	

Early-type Shell Stars (*continued*)

Cate-gory	Name	Catalogue designation	α(1900)	δ	Magnitude	Reference	Notes
			h m	° ′			
2		HD 75485	8 45.1	−76 46	8.0	*AJ* **78**, 687	
?	Hen 209		8 45.4	−45 43	12.2	Carlson	
2		HD 75740	8 46.6	−25 59	10.1	*AJ* **78**, 687	
2		−46 4657	8 47.0	−46 12	10.7	*Ast Lett* **2**, 153	
2	Hen 222	−48 1997	8 48.7	−48 36	10.8	*MN* **122**, 239	
2	67 Cnc	HD 77190	8 55.8	+28 18	6.2	*ApJ* **182**, 809	A star
2	V345 Car	HD 78764	9 4.8	−70 8	4.7	*PASP* **77**, 376	MWC 196
3	S Vel	HD 82829	9 29.5	−44 46	7.7–9.5	*ApJ* **116**, 35	$P = 5.9$ d
2	Hen 327	−50 2767	9 45.5	−50 41	10.5	*MN* **122**, 239	HD 297625
2	HR 3971	HD 87543	10 0.5	−61 24	6.1	*MemRAS* **72**, 233	Hen 362
?	MWC 198	HD 87643	10 1.1	−58 11	9.1	*Ast Lett* **2**, 153	
2	ω Car	HD 89080	10 11.4	−69 32	3.3	Bahng	
?	Hen 394	−57 2874	10 11.8	−57 22	8.6	Carlson	
?	MWC 200	HD 89249	10 12.6	−55 6	9.1	Carlson	
?	HR Car	HD 90177	10 19.4	−59 8	8.2–9.6	*ApJ* **115**, 133	MWC 202
2	HR 4123	HD 91120	10 26.1	−13 5	5.6	*A + A* **3**, 485	MWC 205
2	HR 4128	HD 91269	10 27.1	−60 51	6.4	*MemRAS* **72**, 233	Hen 437
2	PP Car	HD 91465	10 28.5	−61 10	3.3–3.4	*PASP* **77**, 376	MWC 208
1	HR 4169	HD 92207	10 33.6	−58 13	5.5	*A + SS* **23**, 431	
2	Hen 462	HD 92406	10 35.0	−58 12	9.1	*AJ* **78**, 687	
2	MWC 216	−60 2160	10 37.6	−60 15	10.1	*AJ* **75**, 703	HD 305483
?	η Car	HD 93308	10 41.2	−59 10	0.8–7.9	*MN* **113**, 211	MWC 214
2		HD 93383	10 41.8	−32 47	8.5	*AJ* **78**, 687	
2	Hen 515	HD 94509	10 49.4	−57 54	9.1	*AJ* **78**, 687	
?	Hen 519	−59 3400	10 50.0	−59 55	10	*ApJ* **115**, 133	
3	GG Car	HD 94878	10 52.0	−59 52	9.1–9.5	*A + A* **34**, 333	$P = 62.1$ d
?	AG Car	HD 94910	10 52.2	−59 55	7.1–9.0	*Vist* **2**, 1380	MWC 216
3	TT Hya	HD 97528	11 8.3	−25 55	7.5–9.5	*ApJ* **103**, 71	$P = 7.0$ d
2	ϕ Leo	HD 98058	11 11.6	− 3 6	4.5	*ApJ* **182**, 809	A star
2	V644 Cen	−60 3278	11 38.3	−60 11	9.5–10.2	*ApJ* **115**, 578	Hen 700
2		HD 104015	11 53.5	−70 11	7.0	*MemRAS* **77**, 199	
2	ε Cha(B)	HD 104237	11 55.1	−77 38	6.6	*AJ* **78**, 687	Hen 741
2	δ Cen	HD 105435	12 3.2	−50 10	2.6	*ZfA* **59**, 108	K 6892
2	14 Com	HD 108283	12 21.4	+27 49	4.9	*DAO* **9**, 237	A star
?	κ Dra	HD 109387	12 29.2	+70 20	3.9	*A + A* **11**, 100	K 101294
2	HR 4804	HD 109857	12 32.8	−74 49	6.5	*MemRAS* **72**, 233	Hen 802
2	HR 4893	HD 112028	12 48.4	+83 57	5.3	*ApJ* **138**, 118	A star
2	24 CVn	HD 118232	13 30.4	+49 32	4.7	*ApJ* **182**, 809	A star
2	μ Cen	HD 120324	13 43.6	−41 59	2.9–3.4	*PASP* **52**, 198	MWC 229
2	47 Hya	HD 121847	13 52.9	−24 29	5.2	*ApJ* **175**, 453	
2	η Cen	HD 127972	14 29.1	−41 43	2.3	*ZfA* **59**, 108	K 7142
2	Hen 1034	HD 130903	14 45.6	−40 24	7.6	*AJ* **78**, 687	
2	Hen 1042	HD 131891	14 51.1	−72 19	8.3	*AJ* **78**, 687	
?	V748 Cen	−32 10517	14 53.5	−33 1	11–<13	*Obs* **91**, 112	Hen 1045
2	K 7170	HD 133738	15 0.9	−61 30	7.0	*TrIAU* **12**A, 442	Hen 1050
2	Hen 1051	HD 133901	15 1.8	−50 47	9.2	*AJ* **78**, 687	
2		HD 134783	15 6.6	−53 46	9.2	*MN* **122**, 239	
3	U CrB	HD 136175	15 14.1	+32 1	7.0–8.4	*ApJ* **102**, 480	$P = 3.5$ d
2	κ^1 Aps	HD 137387	15 20.6	−73 3	5.5	*ApJ* **157**, 313	MWC 235
3	γ UMi	HD 137422	15 20.9	+72 11	3.0	*ApJ* **116**, 541	K 101502
2	HR 5736	HD 137432	15 20.9	−36 25	5.4	*ApJ* **157**, 313	
?	MWC 236	HD 138403	15 26.7	−71 35	10.0	*Vist* **2**, 1380	
2	ν^2 Boo	HD 138629	15 28.2	+41 14	5.0	*ApJ* **182**, 809	A star
2	HR 5781	HD 138769	15 29.0	−44 37	4.5	*ApJ* **157**, 313	

Early-type Shell Stars (*continued*)

Cate-gory	Name	Catalogue designation	α(1900)	δ	Magnitude	Reference[a]	Notes
			h m	° ′			
2		HD 140605	15 39.4	−51 50	7.1	*MemRAS* **67**, 51	
3	4 Her	HD 142926	15 52.2	+42 51	5.8	*A + A* **22**, 337	$P = 46.0$ d
2	FX Lib	HD 142983	15 52.6	−13 59	4.8–5.0	*Ast Lett* **8**, 45	MWC 239
4	RY Lup		15 52.7	−40 5	9.9–13.0	*ApJ* **174**, 401	
2	Hen 1138	−52 9243	15 59.3	−52 47	9.3	Carlson	
2	V856 Sco	HD 144667	16 1.9	−38 50	7.5–7.8	*ApJ* **177**, 209	
2		−44 10777	16 13.1	−44 53	9.6	*AJ* **78**, 687	
2	25 Her	HD 148283	16 21.8	+37 37	5.5	*ApJ* **182**, 809	A star
3	R Ara	HD 149730	16 31.4	−56 48	6.0–6.9	*ApJ* **116**, 27	$P = 4.4$ d
2	ζ Oph	HD 149757	16 31.7	−10 22	2.6	*ApJ* **188**, L19	
1	HR 6245	HD 151804	16 44.6	−41 4	5.3	*ApJ* **100**, 189	Of-star
2	Hen 1264	HD 151873	16 45.0	−56 52	8.8	*AJ* **78**, 687	
1	ζ^1 Sco	HD 152236	16 47.0	−42 12	4.7	*ApJ* **186**, 909	K 7518
1	Hen 1278	HD 152386	16 47.9	−44 50	8.1	MacConnell	Of star
4	AK Sco	HD 152404	16 48.0	−36 43	8.8–10.3	*ApJ* **174**, 401	
1	HR 6272	HD 152408	16 48.0	−41 0	5.8	*ApJ* **100**, 189	Of star
2	HR 6274	HD 152478	16 48.4	−50 31	6.3	*ApJ* **157**, 313	Hen 1282
?	MWC 873	−40 11253	17 8.3	−40 13	9.6	*ApJ* **115**, 133	HD 327083
2	MWC 253	HD 155851	17 9.0	−32 34	8.2	*ApJ* **97**, 194	
2	HR 6507	HD 158352	17 23.7	+ 0 25	5.4	*ApJ* **182**, 809	A star
2	α Ara	HD 158427	17 24.1	−49 48	2.9	*ZfA* **59**, 109	MWC 261
3	Hen 1419	HD 158503	17 24.5	−58 29	8.3	*IAUC* **2130**	
4	AS 232		17 27.1	−39 19	11.0	*AJ* **80**, 212	HD 323771
2	MWC 263	HD 160095	17 32.9	−33 30	8.7	*ApJ* **72**, 98	
1	MWC 266	HD 160529	17 35.3	−33 27	6.7	*ApJ* **91**, 592	
?	XX Oph	HD 161114	17 38.6	− 6 13	9.1–11.1	*ApJ* **133**, 503	MWC 269
2		HD 161261	17 39.4	+ 5 46	8.3	*ApJ* **139**, 1139	
?	MWC 272	−27 11944	17 41.9	−27 59	9.6	*ApJ* **93**, 349	HD 316285
2	MWC 594	HD 162428	17 45.9	+24 30	7.0	*ApJ* **145**, 121	
2	V771 Sgr	HD 162718	17 47.3	−25 45	8.9–9.4	*ApJ* **94**, 353	MWC 273
3	88 Her	HD 162732	17 47.4	+48 25	6.4	*A + A* **33**, 117	$P = 86.6$ d
?	MWC 275	HD 163296	17 50.3	−21 56	6.6	*ApJ* **72**, 98	
2	V2048 Oph	HD 164284	17 55.3	+ 4 22	4.6–4.8	*A + A* **3**, 485	MWC 278
2	HR 6720	HD 164447	17 56.1	+19 31	6.4	*ApJ* **110**, 387	MWC 279
1+3	μ Sgr	HD 166937	18 7.8	−21 5	3.8–3.9	*ApJ* **186**, 909	$P = 180.6$ d
2	HR 6819	HD 167128	18 8.7	−56 3	5.3	*ApJ* **157**, 313	
1	MWC 291	HD 168607	18 15.5	−16 25	8.3	*PASP* **52**, 401	
2	HR 6873	HD 168797	18 16.5	+ 5 24	6.1	*A + AS* **9**, 133	MWC 601
2		HD 168936	18 17.2	−17 43	8.1	*AJ* **78**, 687	
2	MWC 292	HD 168957	18 17.3	+25 1	7.0	*DDO* **2**, 315	
1	MWC 293	HD 169226	18 18.6	−12 15	9.1	*ApJ* **99**, 205	
1	MWC 294	HD 169454	18 19.6	−14 2	6.6	*ApJ* **72**, 98	
3	RY Sct	HD 169515	18 19.9	−12 45	9.7–10.3	*ApJ* **97**, 394	$P = 11.1$ d
3	RZ Sct	HD 169753	18 21.1	− 9 15	7.3–8.8	*ApJ* **130**, 791	$P = 15.1$ d
?	HR 6929	HD 170235	18 23.2	−25 19	6.6	*ApJ* **157**, 313	MWC 299
?	MWC 300		18 24.0	− 6 9	10.0	*ApJS* **4**, 337	
?		HD 170682	18 25.6	−19 14	7.9	*Obs* **87**, 286	in M 25
2	MWC 938	HD 171219	18 28.4	+ 5 22	8.0	*ApJ* **145**, 121	
2	HR 6984	HD 171780	18 31.6	+34 22	6.1	Bidelman	MWC 604
2	MWC 303	HD 172694	18 36.5	−15 57	8.1	*ApJ* **115**, 47	
3	MWC 304	HD 173219	18 39.2	− 7 13	7.8	*MN* **163**, 219	$P = 58.4$ d
2	4 Aql	HD 173370	18 39.8	+ 1 57	5.0	*ApJ* **196**, 773	K 101758
2	MWC 956	HD 173371	18 39.8	− 0 28	6.8	*ApJ* **145**, 121	
2	MWC 305	HD 174105	18 43.8	+15 17	6.9	*ApJ* **145**, 121	

Early-type Shell Stars (*continued*)

Cate-gory	Name	Catalogue designation	α(1900)	δ	Magnitude	Reference[a]	Notes
			h m	° ′			
2	HR 7081	HD 174179	18 44.2	+31 39	5.8	*ApJS* **17**, 371	
2	HR 7084	HD 174237	18 44.5	+52 53	5.9	*AN* **277**, 179	MWC 608
3	β Lyr	HD 174638/9	18 46.4	+33 15	3.3–4.3	*BAC* **25**, 6	P = 12.9 d
2	MWC 308	HD 175863	18 52.3	+59 53	7.1	*A + AS* **9**, 133	
4	TY CrA	−37 13024	18 54.9	−37 1	8.7–12.4	*ApJ* **174**, 401	
4	R CrA	−37 13027	18 55.1	−37 6	10.0–14.0	*T + T* **5**, 107	
2	MWC 973	HD 177291	18 58.9	−18 51	8.7	*ApJ* **119**, 496	
2	MWC 310	HD 177648	19 0.5	+23 11	7.2	*A + AS* **9**, 133	
2	MWC 978	HD 179343	19 7.1	+ 2 27	7.0	*ApJ* **118**, 18	
2	MWC 312	HD 180398	19 11.3	+12 56	7.9	*DDO* **2**, 315	
2		HD 180587	19 12.0	+10 49	8.6	*MemRAS* **68**, 173	
3	U Sge	HD 181182	19 14.4	+19 26	6.6–9.2	*ApJ* **114**, 513	P = 3.4 d
4	AS 353		19 15.6	+10 51	11.5	*ApJ* **174**, 401	
3	υ Sgr	HD 181615/6	19 16.0	−16 9	4.3–4.4	*PASJ* **19**, 564	P = 137.9 d
?	MWC 314	+14 3887	19 17.0	+14 42	10.0	*ApJ* **97**, 226	
2	MWC 985	+30 3526	19 17.8	+30 58	9.2	*ApJ* **116**, 501	
?	WW Vul	+20 4136	19 21.7	+21 1	10.2–11.8	*Ast* **9**, 104	HD 344361
2	HR 7403	HD 183362	19 24.1	+37 44	6.4	*AN* **279**, 19	MWC 318
2	V923 Aql	HD 183656	19 25.5	+ 3 14	6.3–6.4	*ApJ* **116**, 501	MWC 988
2	V1294 Aql	HD 184279	19 28.6	+ 3 33	6.8–7.2	*ApJ* **115**, 47	MWC 319
2	11 Cyg	HD 185037	19 32.2	+36 43	5.9	*ApJ* **145**, 121	MWC 619
2	LS + 22° 8		19 35.9	+22 18	11.8	*ApJ* **137**, 547	
3	MWC 321	HD 187399	19 44.7	+29 9	7.7	*MN* **163**, 209	P = 28.0 d
2	MWC 995	+32 3583	19 45.7	+32 42	9.1	*ApJS* **2**, 389	HD 225985
?	AS 363	+26 3687	19 46.9	+26 53	9.4	*ApJS* **1**, 220	HD 338970
2		HD 187851	19 47.0	+27 27	7.7	*A + A* **40**, 203	
?	9 Sge	HD 188001	19 47.9	+18 25	6.2	*DAO* **11**, 143	K 102965
2	AS 372	+16 4079	19 53.1	+17 6	9.0	*ApJ* **116**, 501	HD 351123
?	V1295 Aql	HD 190073	19 58.1	+ 5 28	8.6v	*ApJ* **113**, 55	MWC 325
2	MWC 998	HD 190150	19 58.5	+21 34	8.0	*ApJ* **145**, 121	
1	HR 7678	HD 190603	20 0.7	+31 56	5.6	*DAO* **9**, 1	MWC 326
?	V425 Cyg	+35 3981	20 4.3	+35 50	9.8–10.4	*ApJ* **131**, 632	MWC 628
2	20 Vul	HD 192044	20 7.8	+26 11	5.9	*DDO* **2**, 315	MWC 331
3	KU Cyg		20 9.6	+47 5	11.5–13.8	*ApJ* **139**, 143	P = 38.4 d
2	21 Vul	HD 192518	20 10.1	+28 23	5.2	*ApJ* **182**, 809	A star
3	VW Cyg	+34 3938	20 11.4	+34 12	10.4–13.6	*ApJ* **103**, 76	P = 8.4 d
2	MWC 335	HD 192954	20 12.6	+15 34	7.3	*ApJ* **118**, 18	
2	MWC 336	HD 193009	20 12.9	+32 4	7.2	*AN* **279**, 19	
2	MWC 632	HD 193182	20 13.8	+39 17	6.5	*ApJ* **119**, 496	
1	P Cyg	HD 193237	20 14.1	+37 43	3–6	*DAO* **9**, 1	MWC 338
2	MWC 1011	+29 3982	20 14.1	+30 7	9.1	*ApJ* **110**, 387	HD 334060
?	MWC 340	+40 4124	20 17.0	+41 3	10.6	*ApJ* **173**, 353	
2	25 Vul	HD 193911	20 17.7	+24 8	5.4	*ApJ* **145**, 121	MWC 341
2	MWC 342		20 19.4	+39 10	10.6	*ApJ* **97**, 217	
2	HR 7807	HD 194335	20 20.0	+37 9	5.9	*AN* **277**, 167	MWC 343
2	1 Del	HD 195325	20 25.5	+10 34	5.9	*ApJ* **128**, 61	MWC 1019
2	MWC 1020	HD 195358	20 25.7	+19 5	6.6	Bidelman	
2	MWC 346	HD 195407	20 26.0	+36 39	7.2	*ApJ* **128**, 61	
2	MWC 1027	HD 197434	20 38.6	+53 51	7.9	*ApJ* **110**, 387	
2	λ Cyg	HD 198183	20 43.5	+36 7	4.5	*AN* **277**, 32	MWC 352
3	V367 Cyg	HD 198287/8	20 44.2	+38 55	7.4–8.0	*ApJ* **134**, 568	P = 18.6 d
?	DV Aqr	HD 199603	20 53.2	−14 52	6.0–6.2	*BAAS* **7**, 543	A star
2	60 Cyg	HD 200310	20 57.6	+45 46	5.4	*DDO* **2**, 315	MWC 360
4	V1331 Cyg		20 57.9	+49 58	11.8–12.2	*ApJ* **140**, 1409	LkHα 120

Early-type Shell Stars (*continued*)

Cate-gory	Name	Catalogue designation	α(1900)	δ	Magnitude	Reference[a]	Notes
			h m	° ′			
4	K 102052	HD 200775	21 0.4	+67 46	7.3	*ApJ* **173**, 353	MWC 361
2	HR 8103	HD 201733	21 6.4	+45 6	6.6	*ApJS* **17**, 371	MWC 363
2	K 103041	+41 4064	21 12.4	+42 7	9.0	*ApJS* **2**, 41	
2	υ Cyg	HD 202904	21 13.8	+34 29	4.4	*PASJ* **20**, 178	MWC 364
3	K 8645	HD 203025	21 14.6	+58 10	6.4	*ApJ* **76**, 173	*P* = 5.4 d
2	MWC 1043	HD 203356	21 16.6	+53 32	7.7	*ApJ* **145**, 121	
2	6 Cep	HD 203467	21 17.3	+64 27	5.2	*Mich* **4**, 175	MWC 367
2	MWC 370	HD 204722	21 25.5	+43 54	7.7	Bidelman	
2	MWC 371	HD 205060	21 27.7	+42 16	7.2	*DDO* **2**, 315	
?	ε Cap	HD 205637	21 31.5	−19 55	4.7	*MN* **148**, 79	K 8668
2	V1427 Cyg	+47 3487	21 32.2	+47 28	9.1–9.2	*ApJ* **116**, 501	MWC 374
3	AQ Peg	+12 4653	21 32.5	+13 2	10.3–13.0	*ApJ* **103**, 76	*P* = 5.5 d
?	MWC 376	HD 206773	21 39.3	+57 17	6.9	*DDO* **2**, 315	
4	LkHα 234		21 40.8	+65 39	11.9	*ApJ* **173**, 353	
2	MWC 377	HD 207232	21 42.4	+50 13	7.0	*ApJ* **145**, 121	
4	K 8703	+46 3471	21 48.7	+46 46	10.1	*ApJ* **173**, 353	AS 477
2	EM Cep	HD 208392	21 50.9	+62 8	7.0–7.1	Bidelman	MWC 380
2	HR 8375	HD 208682	21 52.9	+64 52	5.9	*DDO* **2**, 315	MWC 381
2	MWC 646	+50 3496	21 53.1	+51 6	8.1	*ApJ* **98**, 153	HD 235668
2	Hen 1927	HD 208886	21 54.1	−32 0	7.1	*AJ* **78**, 687	
3	MR Cyg	+47 3639	21 55.1	+47 30	8.8–9.7	*A* + *A* **23**, 357	*P* = 1.7 d
2	η PsA	HD 209014	21 55.1	−28 56	5.4	Bidelman	Hen 1928
2	MWC 1052	+44 4014	21 56.2	+45 7	8.8	*ApJ* **110**, 387	
2	MWC 383	HD 209296	21 57.2	+56 14	8.3	*DDO* **2**, 315	
2	o Aqr	HD 209409	21 58.1	− 2 38	4.7	*ApJ* **110**, 387	K 8738

Early-type Shell Stars (*continued*)

Cate-gory	Name	Catalogue designation	α(1900)		δ		Magnitude	Reference[a]	Notes
			h	m	°	′			
?	MWC 1055		22	4.7	+53	44	12.8	*ApJ* **110**, 387	
2	π Peg	HD 210459	22	5.5	+32	41	4.3	*ApJ* **117**, 269	F star
2	MWC 1056	+52 3147	22	9.8	+53	7	9.4	*ApJ* **110**, 387	
2	31 Peg	HD 212076	22	16.6	+11	42	5.0	*A + AS* **9**, 133	MWC 387
?	SV Cep	+72 1031	22	19.6	+73	10	10.1–12.6	*Ast* **9**, 104	
2	π Aqr	HD 212571	22	20.2	+ 0	52	4.7	*ApJS* **7**, 65	MWC 388
2	8 Lac	HD 214168	22	31.4	+39	7	5.7	*A + AS* **9**, 133	K 102192
2		HD 215024	22	37.2	+64	44	8.6	*A + A* **40**, 203	
2	HR 8682	HD 216057	22	44.6	+53	53	6.1	*ApJ* **145**, 121	MWC 393
2	EW Lac	HD 217050	22	52.7	+48	9	5.0–5.3	*A + A* **19**, 224	MWC 394
2	HR 8758	HD 217543	22	56.3	+38	10	6.6	*A + AS* **9**, 133	K 103109
2	o And	HD 217675	22	57.3	+41	47	3.5–3.6	*ApJ* **119**, 460	
2	MWC 1076	+53 3066	22	59.6	+53	40	8.7	*ApJ* **119**, 501	HD 236031
2		HD 218325	23	2.0	+46	23	7.7	*ApJS* **23**, 257	
3	MWC 397	HD 218393	23	2.6	+49	39	6.8	*A + A* **11**, 100	P = 38 ± d
3		HD 218674	23	4.8	+49	6	6.7	*AJ* **65**, 535	
4	MWC 1080		23	12.9	+60	18	13.0	*ApJS* **4**, 337	
?	MWC 399	HD 220116	23	16.1	+57	44	8.7	*MN* **155**, 357	
2	MWC 1081	HD 220300	23	17.6	+55	48	7.8	*ApJ* **116**, 501	
?	R Aqr(B)		23	38.6	−15	50	9 ±	*DAO* **9**, 1	MWC 400
1	6 Cas	HD 223385	23	44.0	+61	40	5.4	*DAO* **14**, 107	K 8884
2	MWC 402	HD 223501	23	45.0	+61	40	7.8	*DDO* **2**, 315	
2	HR 9068	HD 224544	23	53.7	+31	48	6.5	*ApJ* **110**, 387	MWC 406
2	HR 9070	HD 224559	23	53.8	+45	52	6.5	*DDO* **2**, 315	K 8901
2	HR 9094(B)	HD 225010	23	57.5	+65	33	7.3	*PASP* **80**, 685	A star
2	MWC 409	HD 225095	23	58.3	+55	0	7.9	*DDO* **2**, 315	

[a] *A + A = Astron. Astrophys.*
   A + AS = Astron. Astrophys. Suppl. Ser.
   A + SS = Astrophys. & Space Science
     AJ = Astron. J.
    AN = Astron. Nachrichten
   ApJ = Astrophys. J.
  ApJS = Astrophys. J. Suppl. Ser.
    Ast = Astrophysics (translation)
Ast Lett = Astrophys. Letters
 BAAS = Bull. Amer. Astron. Soc.
  BAC = Bull. Astron. Soc. Czechoslovakia
  DAO = Publ. Dominion Astrophys. Obs.
  DDO = Publ. David Dunlap Obs.
  IAUC = IAU Circular
 JRASC = J. Roy. Astron. Soc. Canada
MemRAS = Mem. Roy. Astron. Soc.
   Mich = Publ. U. of Michigan Obs.
    MN = Monthly Notices Roy. Astron. Soc.
   Obs = Observatory
  PASJ = Publ. Astron. Soc. Japan
 PASP = Publ. Astron. Soc. Pacific
 Ric A = Ricerche Astronomiche, Specola Vaticana
 TrIAU = Transactions IAU
  T + T = Bol. Tonantzintla y Tacubaya Obs.
  Vist = Vistas in Astronomy
 W + S = Publ. Warner & Swasey Obs.
  ZfA = Z. Astrophys.